Growing Gourmet and Medicinal Mushrooms

食用及び薬用きのこの栽培

This book should help advance the cause of mycology and mushroom biology worldwide. It will be an important reference for those who are interested in research as well as in the cultivation of mushrooms. *Growing Gourmet and Medicinal Mushrooms* is unique not only in its treatment of the technical aspects of growing gourmet and medicinal mushrooms, but also in its emphasis on the environmental importance of mushrooms in terms of world biological diversity.

—S. T. Chang, Ph.D., The Chinese University of Hong Kong

Growing Gourmet and Medicinal Mushrooms is a visionary quest—and Paul Stamets is your best possible guide—not just for informing you about growing mushrooms, but for transforming you into a myco-warrior, an active participant in a heroic, Gaian process of planetary health through mushroom cultivation.

—Gary Lincoff, author of The Audubon Field Guide to Mushrooms

Stamets draws on the collective experience of centuries of mushroom cultivation, creating a revolutionary model for the use of higher fungi. Not only does he cover every aspect of cultivation, he also addresses the issues of environmentalism, health, and business.

—Alan Bessette, Ph.D., Utica College of Syracuse University

Growing Gourmet and Medicinal Mushrooms is the most comprehensive treatment of the subject I have seen in my thirty years as a mycologist and mushroom specialist.

—S. C. Jong, Ph.D., The American Type Culture Collection

Pick up this book and prepare to be swept away into the world of mushroom cultivation on the tide of Paul's contagious enthusiasm. Doers and dreamers, students and teachers will all find something to enjoy in this book.

—Nancy Smith Weber, Ph.D., Forest Sciences Department, Oregon State University

This book, a true labor of love, makes a major contribution to our knowledge of the practical production of gourmet and medicinal mushrooms.

—Dan Royse, Ph.D., Penn State College of Agricultural Sciences

Growing Gourmet and Medicinal Mushrooms

食用及び薬用きのこの栽培

THIRD EDITION

PAUL STAMETS

TEN SPEED PRESS
Berkeley   Toronto

Ten Speed Press
Box 7123
Berkeley, California 94707
www.tenspeed.com

Distributed in Australia by Simon and Schuster Australia, in Canada by Ten Speed Press Canada, in New Zealand by Southern Publishers Group, in South Africa by Real Books, and in the United Kingdom and Europe by Airlift Book Company.

Photograph of Chinese rank badge (Wild Goose, 4th Rank) on page 350 (bottom) courtesy of Beverley Jackson.

Cover design by Andrew Lenzer and Jeff Brandenburg
Interior design by Jeff Brandenburg

Library of Congress Cataloging-in-Publication Data
Stamets, Paul.
    Growing gourmet and medicinal mushrooms = [Shokuyo oyobi yakuyo kinoko no saibai] : a companion guide to The Mushroom Cultivator / by Paul Stamets. — 3rd ed.
        p.   cm.
    Includes bibliographical references (p.).
    ISBN-10: 1-58008-175-4
    ISBN-13: 978-1-58008-175-7
1. Mushroom culture. I. Title: Growing gourmet and medicinal mushrooms. II. Title: Shokuyo oyobi yakuyo kinoko no saibai.

SB353 .S73 2000
635'.8-dc21
                                                                                    00-042584

Printed in China
First printing, 2000

4 5 6 7 8 9 10 — 08 07 06 05

Mycotopia:
An environment wherein ecological equilibrium is enhanced through the judicious use of fungi for the betterment of all lifeforms.

To my family
and
the Warriors of Hwa Rang Do

Contents

Foreword

Mushrooms—fleshy fungi—are the premier recyclers on the planet. Fungi are essential to recycling organic wastes and the efficient return of nutrients back into the ecosystem. Not only are they recognized for their importance within the environment, but also for their effect on human evolution and health. Yet, to date, the inherent biological power embodied within the mycelial network of mushrooms largely remains a vast, untapped resource. As we begin the twenty-first century, ecologists, foresters, bioremediators, pharmacologists, and mushroom growers are converging at a new frontier of knowledge, wherein enormous biodynamic forces are at play.

Only recently have we learned enough about the cultivation of mushrooms to tap into their inherent biological power. Working with mushroom mycelium *en masse* will empower every country, farm, recycling center, and individual with direct economic, ecological, and medical benefits. Through the genius of evolution, the Earth has selected fungal networks as a governing force managing ecosystems. This sentient network responds quickly to catastrophia. I believe the mycelium is Earth's natural Internet, a neural network of communicating cells. All landmasses are criss-crossed with interspersing mosaics of mycelial colonies. With more than a mile of cells in a cubic inch of soil, the fungi are moving steadily, although silently all around us. This vast mass of cells, in the hundreds of billions of tons, represents a collective intelligence, like a computer honed to improve itself. Only now are scientists discovering that it is the microbial community upon which all higher life forms are dependent. And only now do we know how to join in alliance with them to improve life. As we begin a new century, myco-technology is a perfect example of the equation of good environmentalism, good health, and good business.

This book strives to create new models for the future use of higher fungi in the environment. As woodland habitats, especially old growth forests, are lost to development, mushroom diversity also declines. Wilderness habitats still offer vast genetic resources for new strains. The temperate forests of North America, particularly the mycologically rich Pacific Northwest, may well be viewed in the twenty-first century as pharmaceutical companies viewed the Amazon Basin earlier in the twentieth century. Hence, mushroom cultivators should preserve this gene pool now for its incalculable, future value. The importance of many mushroom species may not be recognized for decades to come.

In many ways, this book is an offspring of the marriage of many cultures arising from the worldwide use of mushrooms as food, as religious sacraments in Mesoamerica, and as medicine in Asia. We now benefit from the collective experience of lifetimes of mushroom cultivation. As cultivators we must continue to share, explore, and expand the horizons of the human/fungal relationship. In the future, humans and mushrooms must bond in an evolutionary partnership. By empowering legions of individuals with the skills of tissue culture and mycelial management, future generations will be able to better manage our resources and improve life on this planet.

Now that the medical community widely recognizes the health-stimulating properties of mushrooms, a combined market for gourmet *and* medicinal foods is rapidly emerging. People with compromised immune systems would be wise to create their own medicinal mushroom gardens. I envision the establishment of a community-based, resource-driven industry, utilizing recyclable materials in a fashion that strengthens ecological equilibrium and human health. As recycling centers flourish, their by-products include streams of organic waste, which cultivators can divert into mushroom production.

I foresee a network of environmentally sensitive and imaginative individuals presiding over this new industry, which has previously been controlled by a few mega-businesses. The decentralization began with *The Mushroom Cultivator* in 1983, and continues with *Growing Gourmet and Medicinal Mushrooms*. Join me in the next phase of this continuing revolution.

Acknowledgments

I first acknowledge the Mushrooms who have been my greatest teachers. They are the Body Intellect, the Neural Network of this book.

My family has been extremely patient and forgiving during this multiyear project. Azureus and LaDena have tolerated my insistent need for their modeling talents and have helped on many mushroom projects. Dusty Wu Yao is credited for her research skills, support, humor, and love.

My parents have taught me many things. My father championed education and science and impressed upon me that a laboratory is a natural asset to every home. My mother taught me patience and kindness, and that precognition is a natural part of the human experience. My brother John first piqued my interest for mushrooms upon his return from adventures in Colombia and Mexico. Additionally, his knowledge on the scientific method of photography has greatly helped improve my own techniques. In some mysterious way, their combined influences set the stage for my unfolding love of fungi.

Other people warrant acknowledgement in their assistance in the completion of this book. Andrew Weil played a critical role in helping build the creative milieu, the wellspring of spiritual chi from which this manuscript flowed. Gary Lincoff was extremely helpful in uncovering some of the more obscure references and waged intellectual combat with admirable skill. Brother Bill Stamets is thanked for his critical editorial remarks. Satit Thaithatgoon, my friend from Thailand, is appreciated for his insights about mushroom culture and life. I must thank Kit and Harley Barnhart for their advice on photographic technique. Michael Beug deserves acknowledgment for his unwavering support through all these years. Erik Remmen kept me healthy and strong through the many years of rigorous training in the ancient and noble martial art of Hwarang Do.

Joseph Ammirati, David Arora, Julie Bennett, Alan and Arleen Bessette, David Brigham, Janet Butz, Jonathan Caldwell, Alice Chen,

Jeff Chilton, Ken Cochran, Don Coombs, Kim and Troy Donahue, Robert Ellingham, Gaston Guzman, Paxton Hoag, John Holliday, Rick Hoss, Lou Hsu, Eric Iseman, Loren Israelson, Omon Isikhuemhen, Barbara King, Mike Knoke, Alexander Krenov, Gary Leatham, Andrew Lenzer, Mike Maki, Andrew H. Miller, Orson and Hope Miller, Scott Moore, Tomiro Motohashi, Peter Mohs, Yoshikazu Murai, Takashi Mizuno, Takeshi Nakazawa, Louise North, George Osgood, Christiane Pischl, David Price, Paul Przybylowicz, Warren Rekow, Scott Redhead, Rusty Rodriguez, Maggie Rogers, Luiz Amaro Pachoa de Silva, Bulmaro Solano, Lillian Stamets, David Sumerlin, Ralph Tew, Harry Thiers, Tom O'Dell, James Trappe, Solomon Wasser, Dusty Yao, and Rytas Vilgalys all helped in their own special ways.

The late Jim Roberts, of Lambert Spawn, gained my respect and admiration for his devotion to helping the gourmet mushroom industry. And, I will never forget the generosity shown to me by the late Alexander Smith and Daniel Stuntz who were instrumental in encouraging me to continue in the field of mycology—in spite of those who fervently opposed me.

Companies that unselfishly contributed photographic material to this work, and to whom I am grateful, are The BOTS Group, The Minnesota Forest Resource Center, The Growing Company, DXN Company, Morel Mountain, Organotech, and Ostrom's Mushroom Farms. I would also like to thank The Mushroom Council and the American Mushroom Institute. The Evergreen State College generously supported my studies with *Psilocybe* mushrooms and in scanning electron microscopy.

Finally, I wish to acknowledge all those bewildered and bemushroomed researchers who have paved the way into the future. For your help on this odyssey through life, I will forever be in your debt.

Introduction

Mushrooms have never ceased to amaze me. The more I study them, the more I realize how little I have known, and how much more there is to learn. For thousands of years, fungi have evoked a host of responses from people—from fear and loathing to reverent adulation. And I am no exception.

When I was a little boy, wild mushrooms were looked upon with foreboding. It was not as if my parents were afraid of them, but our Irish heritage lacked a tradition of teaching children anything nice about mushrooms. In this peculiar climate of ignorance, rains fell and mushrooms magically sprang forth, wilted in the sun, rotted, and vanished without a trace. Given the scare stories told about "experts" dying after eating wild mushrooms, my family gave me the best advice they could: Stay away from all mushrooms, except those bought in the store. Naturally rebellious, I took this admonition as a challenge, a call to arms, firing up an already overactive imagination in a boy hungry for excitement.

When we were seven, my twin brother and I made a startling mycological discovery—*Puffballs*! We were told that they were not poisonous but if the spores got into our eyes, we would be instantly blinded! This information was quickly put to good use. We would viciously assault each other with mature puffballs, which would burst upon impact and emit a cloud of brown spores. The battle would continue until all the puffballs in sight had been hurled. They provided us with hours of delight over the years. Neither one of us ever went blind—although we both suffer from very poor eyesight. You must realize that to a seven-year-old these free, ready-made missiles satisfied instincts for warfare on the most primal of levels. This is my earliest memory of mushrooms, and to this day I consider it to be a positive emotional experience. (Although I admit a psychiatrist might like to explore these feelings in greater detail.)

Not until I became a teenager did my hunter-gatherer instincts resurface, when a relative returned from extensive travels in South America.

With a twinkle in his eyes, he spoke of his experiences with the sacred *Psilocybe* mushrooms. I immediately set out to find these species, not in the jungles of Colombia, but in the fields and forests of Washington State where they were rumored to grow. For the first several years, my searches provided me with an abundance of excellent edible species, but no Psilocybes. Nevertheless, I was hooked.

When hiking through the mountains, I encountered so many mushrooms. Each was a mystery until I could match them with descriptions in a field guide. I soon came to learn that a mushroom was described as "edible," "poisonous," or my favorite, "unknown," based on the experiences of others like me, who boldly ingested them. People are rarely neutral in their opinion about mushrooms—either they love them or they hate them. I took delight in striking fear into the hearts of the latter group, whose illogical distrust of fungi provoked my overactive imagination.

When I enrolled in the Evergreen State College in 1975, my skills at mushroom identification earned the support of a professor with similar interests. My initial interest was taxonomy, and I soon focused on fungal microscopy. The scanning electron microscope revealed new worlds, dimensional landscapes I never dreamed possible. As my interest grew, the need for fresh material year-round became essential. Naturally, these needs were aptly met by learning cultivation techniques, first in petri dishes, then on grain, and eventually on a wide variety of materials. In the quest for fresh specimens, I had embarked upon an irrevocable path that would steer my life on its current odyssey.

Paul Stamets in the virgin rainforest of Washington State, in route to collect new strains of wild mushrooms.

CHAPTER 1

Mushrooms, Civilization, and History

Humanity's use of mushrooms extends back to Paleolithic times. Few people—even anthropologists—comprehend how influential mushrooms have been in affecting the course of human evolution. They have played pivotal roles in ancient Greece, India, and Mesoamerica. True to their beguiling nature, fungi have always elicited deep emotional responses: from adulation by those who understand them to outright fear by those who do not.

The historical record reveals that mushrooms have been used for less than benign purposes. Claudius II and Pope Clement VII were both killed by enemies who poisoned them with deadly *Amanitas*. Buddha died, according to legend, from a mushroom that grew underground. Buddha was given the mushroom by a peasant who believed it to be a delicacy. In ancient verse, that mushroom was linked to the phrase "pig's foot" but has never been identified. (Although Truffles grow underground, and pigs are used to find them, no deadly poisonous species are known.)

The oldest archaeological record of probable mushroom use is a Tassili image from a cave dating back 5000 years B.C. (Lhote, 1987). The artist's intent is clear. Mushrooms with electrified auras are depicted outlining a bee-masked dancing shaman. The spiritual interpretation of this image transcends time and is obvious. No wonder the word "bemushroomed" has evolved to reflect the devout mushroom lover's state of mind.

In the fall of 1991, hikers in the Italian Alps came across the well-preserved remains of a man who died over 5,300 years ago, approximately 1,700 years later than the Tassili cave artist. Dubbed the "Iceman" or "Oetzi" by the news media, he was well-equipped with a knapsack, flint axe, a string of dried Birch Polypores *(Piptoporus*

2000+ year-old Mesoamerican mushroom stone.

alpine wilderness, this intrepid adventurer had discovered the value of the noble polypores. Even today, this knowledge can be life-saving for anyone astray in the wilderness.

Fear of mushroom poisoning pervades every culture, sometimes reaching phobic extremes. The term *mycophobic* describes those individuals and cultures who look upon fungi with fear and loathing. The English and Irish epitomize mycophobic cultures. In contrast, *mycophilic* societies can be found throughout Asia and Eastern Europe, especially among Polish, Russian, and Italian peoples. These societies have enjoyed a long history of mushroom use, with as many as a hundred common names to describe the mushroom varieties they love.

An investment banker named R. Gordon Wasson intensively studied the use of mushrooms by diverse cultures. His studies concentrated on the use of mushrooms by Mesoamerican, Russian, English, and Indian cultures. With the French mycologist Dr. Roger Heim, Wasson published research on *Psilocybe* mushrooms in Mesoamerica, and on *Amanita* mushrooms in Eurasia/Siberia. Wasson's studies spanned a lifetime marked by a passionate love for fungi. His publications include *Mushrooms, Russia, and History; The Wondrous Mushroom: Mycolatry in Mesoamerica; Maria Sabina and Her Mazatec Mushroom Velada;* and *Persephone's Quest: Entheogens and the Origins of Religion.* More than any individual of the twentieth century, Wasson kindled interest in ethnomycology to its present state of intense study. Wasson died on Christmas Day in 1986.

One of Wasson's most provocative findings can be found in *Soma: Divine Mushroom of Immortality* (1976) where he postulated that the mysterious Soma in Vedic literature, a red fruit leading to spontaneous enlightenment for those who ingested it, was actually a mushroom. The Vedic symbolism carefully disguised its true identity: *Amanita muscaria*, the hallucinogenic Fly Agaric. Many cultures portray *Amanita muscaria* as the archetypal mushroom, invoking both fear and admiration.

betulinus), a tinder fungus *(Fomes fomentarius),* and another as-yet-unidentified mushroom that may have had magico-spiritual significance (Peintner et al. 1998). Polypores can be used as spunk for starting fires and medicine for treating wounds. Further, a rich tea with immuno-enhancing and antibacterial properties can be prepared by boiling these mushrooms. Equipped for traversing the high

Although some Vedic scholars disagree with his interpretation, Wasson's exhaustive research still stands (Brough, 1971 and Wasson, 1972).

Aristotle, Plato, Homer, and Sophocles all participated in religious ceremonies at Eleusis where an unusual temple honored Demeter, the Goddess of Earth. For over two millennia, thousands of pilgrims journeyed fourteen miles from Athens to Eleusis, paying the equivalent of a month's wage for the privilege of attending the annual ceremony. The pilgrims were ritually harassed on their journey to the temple, apparently in good humor.

Upon arriving at the temple, they gathered in the initiation hall, a great telestrion. Inside the temple, pilgrims sat in rows that descended step-wise to a hidden, central chamber from which a fungal concoction was served. An odd feature was an array of columns, beyond any apparent structural need, whose designed purpose escapes archaeologists. The pilgrims spent the night together and reportedly came away forever changed. In this pavilion crowded with pillars, ceremonies occurred, known by historians as the Eleusinian Mysteries. No revelation of the ceremony's secrets could be mentioned under the punishment of imprisonment or death. These ceremonies continued until repressed in the early centuries of the Christian era.

In 1977, at a mushroom conference on the Olympic Peninsula, R. Gordon Wasson, Albert Hofmann, and Carl Ruck first postulated that the

Meso-American mushroom stones, circa 300 B.C., from the Pacific slope of Guatemala.

Eleusinian Mysteries centered on the use of psychoactive fungi. Their papers were later published in a book entitled *The Road to Eleusis: Unveiling the Secret of the Mysteries* (1978). That Aristotle and other founders of Western philosophy undertook such intellectual adventures, and that this secret ceremony persisted for nearly 2,000 years, underscores the profound impact that fungal rites have had on the evolution of Western consciousness.

Pre-classic Mayan mushroom stone from Kaminaljuyu Highlands of Guatemala, circa 500 B.C.

CHAPTER 2

The Role of Mushrooms in Nature

M ushrooms can be classified into three basic ecological groups: *mycorrhizal, parasitic*, and *saprophytic*. Although this book centers on the cultivation of the gourmet and medicinal saprophytic species, other mushrooms are also discussed.

The Mycorrhizal Gourmet Mushrooms: Matsutake, Boletus, Chanterelles, and Truffles

Mycorrhizal mushrooms form a mutually dependent, beneficial relationship with the roots of host plants, ranging from trees to grasses. "Myco" means mushrooms, while "rhizal" means roots. The collection of filament of cells that grow into the mushroom body is called the *mycelium*. The mycelia of these mycorrhizal mushrooms can form an exterior sheath covering the roots of plants and are called *ecto*mycorrhizal. When they invade the interior root cells of host plants they are called *endo*mycorrhizal. In either case, both organisms benefit from this association. Plant growth is accelerated. The resident mushroom mycelium increases the plant's absorption of nutrients, nitrogenous compounds, and essential elements (phosphorus, copper, and zinc). By growing beyond the immediate root zone, the mycelium channels and concentrates nutrients from afar. Plants with mycorrhizal fungal partners can also resist diseases far better than those without.

Most ecologists now recognize that a forest's health is directly related to the presence, abundance, and variety of mycorrhizal associations. The mycelial component of topsoil within a typical Douglas fir forest in the Pacific Northwest approaches 10% of the total biomass. Even this estimate may be low, not taking into account the mass of the endomycorrhizae and the many yeast-like fungi that thrive in the topsoil.

A Truffle market in France.

The nuances of climate, soil chemistry, and predominant microflora play determinate roles in the cultivation of mycorrhizal mushrooms in natural settings. Species native to a region are likely to adapt much more readily to designed habitats than exotic species. I am much more inclined to spend time attempting the cultivation of native mycorrhizal species than to import exotic candidates from afar. Here is a relevant example.

Truffle orchards are well established in France, Spain, and Italy, with the renowned Perigold Black Truffle, *Tuber melanosporum*, fetching up to $500 per pound. Only in the past thirty years has tissue culture of Truffle mycelia become widely practiced, allowing the development of planted Truffle orchards. Landowners seeking an economic return without resorting to cutting trees are naturally attracted to this prospective investment. The idea is enticing. Think of having an orchard of oaks or filberts, yielding pounds of Truffles per year for

decades at several hundred dollars a pound! Several companies in this country have, in the past twenty years, marketed Truffle-inoculated trees for commercial use. Calcareous soils (i.e., high in calcium) in Texas, Washington, and Oregon have been suggested as ideal sites. Tens of thousands of dollars have been exhausted in this endeavor. Only two would-be Truffle orchards have had any success thus far, with only a small percentage of trees producing. This discouraging state of affairs should be fair warning to investors seeking profitable enterprises in the arena of Truffle cultivation. Suffice it to say that the only ones to have made money in the Truffle tree industry are those who have resold "inoculated" seedlings to other would-be trufflateurs.

A group of Oregon trufflateurs has been growing the Oregon White Truffle, *Tuber gibbosum*. Douglas fir seedlings are inoculated with mycelium from this native species and planted in plots similar to Christmas tree farms. Several years pass before the first harvests begin. However, since Oregon White Truffles were naturally occurring nearby, whether or not the inoculation process caused the truffles to form is unclear.

In Sweden, Eric Danell (1994; 1997), who is the first to grow Chantarelles (*Cantharellus cibarius*) with a potted pine tree in a greenhouse, is continuing an ambitious project of cultivating mycorrhizal mushrooms using a community of microorganisms as allies. (See photo page 8.) At the Invermay Agricultural Center in New Zealand, scientists have succeeded in inoculating pines with Matsutake (*Tricholoma magnivelare*) mycelia in the hope that mushrooms will appear a decade later. In New Zealand, mycorrhizal inoculations are more successful because of the extremely limited number of natural mycorrhizal candidates, in contrast to the hundreds seen in the forestlands of North America. These pilot projects hold great promise for replenishing the fungal genome of threatened mycorrhizal mushrooms in endangered ecosystems.

Mycorrhizal mushrooms in Europe have suffered a radical decline in years of late. The combined effects of acid rain and other industrial pollutants, even the disaster at Chernobyl, have been suggested to explain the sudden decline of both the quantity and diversity of wild mycorrhizal mushrooms. Most mycologists believe the sudden availability of deadwood is responsible for the comparative increase in the numbers of saprophytic mushrooms. The decline in Europe portends, in a worst case scenario, a total ecological collapse of the mycorrhizal community, followed by a widespread die-back of the forests. In the past ten years, the diversity of mycorrhizal mushrooms in Europe has fallen by more than 50%! Some species, such as the Chanterelle, have all but disappeared from regions in the Netherlands where it was abundant only twenty years ago (see Arnolds, 1992; Leck, 1991; Lizon 1993, 1995). Many biologists view these mushrooms as indicator species, the first domino to fall in a series leading to the failure of the forest's life-support systems.

One method for inoculating mycorrhizae calls for the planting of young seedlings near the root zones of proven Truffle trees. The new seedlings acclimate and become "infected" with the mycorrhizae of a neighboring, parent tree. In this fashion, a second generation of trees carrying the mycorrhizal fungus is generated. After a few years, the new trees are dug up and replanted into new environments. This method has had the longest tradition of success in Europe.

Another approach, modestly successful, is to dip exposed roots of seedlings into water enriched with the spore-mass of a mycorrhizal candidate. First, mushrooms are gathered from the wild and soaked in water. Thousands of spores are washed off the gills, resulting in an enriched broth of inoculum. A spore-mass slurry coming from several mature mushrooms and diluted into a 5-gallon bucket can inoculate a hundred or more seedlings. The concept

Scanning electron micrograph of an emerging root tip being mycorrhized by mushroom mycelia.

Scanning electron micrograph of mycelia encasing the root of a tree, known as ectomycorrhizae.

The first authenticated success in the cultivation of the Chantarelle: *Pinus sylvestris* in companionship with *Cantharellus cibarius.*

On sterilized media, most mycorrhizal mushrooms grow slowly, compared to the saprophytic mushrooms. Their long evolved dependence on root by-products and complex soils makes media preparation inherently more complicated. Some mycorrhizal species, like *Pisolithus tinctorius*, a puffball favoring pines, grow quite readily on sterilized media. A major industry has evolved providing foresters with seedlings inoculated with this fungus. Mycorrhized seedlings are healthier and grow faster than non-mycorrhized ones. Unfortunately, the gourmet mycorrhizal mushroom species do not fall into the readily cultured species category. The famous Matsutake may take weeks before its mycelium fully colonizes the media on a single petri dish! Unfortunately, this rate of growth is the rule rather than the exception with the majority of gourmet mycorrhizal species.

Chanterelles are one of the most popularly collected wild mushrooms. In the Pacific Northwest of North America the harvesting of Chanterelles has become a controversial, multi-million dollar business. Like Matsutake, Chanterelles also form mycorrhizal associations with trees. Additionally, they demonstrate a unique interdependence on soil yeasts and pseudomonads. This type of mycorrhizal relationship makes tissue culture most difficult. At least three organisms must be cultured simultaneously: the host tree, the mushroom, and soil microflora. A red soil yeast, *Rhodotorula glutinis*, is crucial in stimulating spore germination. The Chanterelle life cycle may have more dimensions of biological complexity. Cultivators have yet to grow Chanterelles to the fruitbody stage under laboratory conditions. Not only do other microorganisms play essential roles, the timing of their introduction appears critical to success in the mycorrhizal theater.

Senescence occurs with both saprophytic and mycorrhizal mushroom species. Often the first sign of senescence is not the inability of mycelia to grow vegetatively, but the loss of the formation of the sexually reproducing organ: the mushroom. Furthermore, the slowness from sowing the mycelium to the final stages of harvest confounds

is wonderfully simple. Unfortunately, success is not guaranteed.

Broadcasting spore-mass onto the root zones of likely candidates is another venue that costs little in time and effort. Habitats should be selected on the basis of their parallels in nature. For instance, Chanterelles can be found in oak forests of the Midwest and in Douglas fir forests of the West. Casting spore-mass of Chanterelles into forests similar to those where Chanterelles proliferate is obviously the best choice. Although the success rate is not high, the rewards are well worth the minimum effort involved. Bear in mind that tree roots confirmed to be mycorrhized with a gourmet mushroom will not necessarily result in harvestable mushrooms. Fungi and their host trees may have long associations without the appearance of edible fruitbodies. (For more information, consult Fox, 1983.)

Oyster and Honey mushrooms sharing a stump.

the quick feedback all cultivators need to refine their techniques. Thus, experiments trying to model how Matsutakes grow may take twenty to forty years each, the age the trees must be to support healthy, fruiting colonies of these prized fungi. Faster methods are clearly desirable, but presently only the natural model has shown any clue to success.

Given the huge hurdle of time for honing laboratory techniques, I favor the "low-tech" approach of planting trees adjacent to known producers of Chanterelles, Matsutake, Truffles, and Boletus. After several years, the trees can be uprooted, inspected for mycorrhizae, and replanted in new environments. The value of the contributing forest can then be viewed, not in terms of board feet of lumber, but in terms of its ability for creating satellite, mushroom/tree colonies. When industrial or suburban development threatens entire forests, and is unavoidable, future-oriented foresters may consider the removal of the mycorrhizae as a last-ditch effort to salvage as many mycological communities as possible by simple transplantation techniques, although on a much grander scale.

Until laboratory techniques evolve to establish a proven track record of successful marriages that result in harvestable crops, I hesitate to recommend mycorrhizal mushroom cultivation as an economic endeavor. Mycorrhizal cultivation pales in comparison to the predictability of growing saprophytic mushrooms like Oyster and Shiitake mushrooms. The industry simply needs the benefit of many more years of mycological research to better decipher the complex models of mycorrhizal mushroom cultivation.

Parasitic Mushrooms: Blights of the Forest?

Parasitic fungi are the bane of foresters. They do immeasurable damage to the health of resident tree species, but in the process create new habitats for many other organisms. Although the ecological damage caused by parasitic fungi is well understood, we are only just learning of their importance in the forest ecosystem. Comparatively few mushrooms are true parasites.

Parasites live off a host plant, endangering the host's health as it grows. Of all the parasitic mushrooms that are edible, the Honey mushroom, *Armillaria mellea*, is the best known. One of these Honey mushrooms, known as *Armillaria gallica*, made national headlines when scientists reported finding in Michigan a single colony covering 37 acres, weighing at least 220,000 pounds, with an estimated age of 1,500 years! Washington State soon responded with reports of a colony of *Armillaria ostoyae* covering 2,200 acres and at least 2,400 years old.* With the exception of the trembling Aspen forests of Colorado, this fungus is the largest known living organism on the planet. And, it is a marauding parasite!

On a well-traveled trail in the Snoqualmie Forest of Washington State, hikers have been stepping upon the largest and perhaps oldest polypore: *Bridgeoporus (Oxyporus) nobilissimus*, a conk that grows up to several feet in diameter and can weigh hundreds of pounds!** This "parasitic"

---

* http://abcnews.go.com:80/sections/science/DailyNews/fungus000806.html
** *Oxyporus nobilissimus* has been placed in its own genus, Bridgeoporus (Burdsall et al. 1996).

Intrepid amateur mycologist Richard Gaines points to parasitic fungus attacking a yew tree.

Scotland, another ancient polypore, *Rigidioporus ulmarius*, might also be medically significant, holding the Guinness Book of Records for the largest mushroom in the world—with an estimated weight of more than 625 pounds (284 kilograms). These examples from the fungal kingdom attract my attention in the search for candidates having potentially new medicines. With the loss of old-growth forests, cultivator–mycologists can play an all-important role in saving the fungal genome, especially in old-growth forests, a potential treasure trove of new medicines.

In the past, a parasitic fungus has been looked upon as biologically evil. This view is rapidly changing as science progresses. Montana State University researchers have discovered a new parasitic fungus attacking the yew tree. This new species is called *Taxomyces andreanae* and is medically significant for one notable feature: it produces minute quantities of the potent anticarcinogen Taxol, a proven treatment for breast cancer (Stone, 1993). This new fungus was studied and now a synthetic form of this potent drug is available for cancer patients. Recently, a leaf fungus isolated in the Congo has been discovered that duplicates the effect of insulin, but is orally active. Even well known medicines from fungi harbor surprises. A mycologist at Cornell University (Hodge et al. 1996) recently discovered that the fungus responsible for the multibillion dollar drug, cyclosporin, has a sexual stage in *Cordyceps subsessilis*, a parasitic mushroom attacking scarab beetle larvae. Of the estimated 1,500,000 species of fungi, approximately 70,000 have been identified (Hawksworth et al. 1995), and about 10,000 are mushrooms. We are just beginning to discover the importance of species hidden within this barely explored genome.

species grows primarily on old-growth *Abies procera* (California red fir) or on their stumps. Less than a dozen specimens have ever been collected. This mushroom is the first ever to be listed on any list, private or public, as an endangered species. Known only from the old-growth forests of the Pacific Northwest, the Noble Polypore's ability to produce a conk that lives for hundreds of years distinguishes it from any other mushroom known to North America. This fact—that it produces a fruiting body that survives for centuries—suggests that the Noble Polypore has unique anti-rotting properties, antibiotics, or other compounds that could be useful medicinally. Located at the Kew Gardens in

Many saprophytic fungi can be weakly parasitic in their behavior, especially if a host tree is dying from other causes. These can be called *facultative* parasites: saprophytic fungi activated by favorable conditions to behave parasitically. Some parasitic

The cultivation of the Button mushroom in caves near Paris in 1868. Note candle used for illumination. (Robinson, 1885)

fungi continue to grow long after their host has died. Oyster mushrooms (*Pleurotus ostreatus*) are classic saprophytes, although they are frequently found on dying cottonwood, oak, poplar, birch, maple, and alder trees. These appear to be operating parasitically when they are only exploiting a rapidly evolving ecological niche.

Most of the parasitic fungi are microfungi and are barely visible to the naked eye. In mass, they cause the formation of cankers and shoot blights. Often their preeminence in a middle-aged forest is symptomatic of other imbalances within the ecosystem. Acid rain, groundwater pollution, insect damage, and loss of protective habitat all are contributing factors unleashing parasitic fungi. After a tree dies, from parasitic fungi or other causes, saprophytic fungi come into play.

Saprophytic Mushrooms: The Decomposers

Most of the gourmet mushrooms are saprophytic, wood-decomposing fungi. Saprophytic fungi are the premier recyclers on the planet. The filamentous mycelial network is designed to weave between and through the cell walls of plants. The enzymes and acids they secrete degrade large molecular complexes into simpler compounds. All ecosystems depend upon fungi's ability to decompose organic plant matter soon after it is rendered available. The end result of their activity is the return of carbon, hydrogen, nitrogen, and minerals back into the ecosystem in forms usable to plants, insects, and other organisms. As decomposers, they can be separated into three key groups. Some mushroom species cross over from one category to another depending upon prevailing conditions.

Primary Decomposers: These are the fungi first to capture a twig, a blade of grass, a chip of wood, a log or stump. Primary decomposers are typically fast-growing, sending out ropy strands of mycelium that quickly attach to and decompose plant tissue. Most of the decomposers degrade wood. Hence, the majority of these saprophytes are woodland species, such as Oyster mushrooms (*Pleurotus* species), Shiitake (*Lentinula edodes*), and King Stropharia (*Stropharia rugosoannulata*). However, each species has developed specific sets of enzymes to break down lignin-cellulose, the structural components of most plant cells. Once the enzymes of one mushroom species have broken down the lignin-cellulose to its fullest potential, other saprophytes utilizing their own repertoire of enzymes can reduce this material even further.

Secondary Decomposers: These mushrooms rely on the previous activity of other fungi to partially break down a substrate to a state wherein they can thrive. Secondary decomposers typically grow from composted material. The actions of other fungi, actinomycetes, bacteria, and yeasts all operate within compost. As plant residue is degraded by these microorganisms, the mass, structure, and composition of the compost is reduced, and proportionately available nitrogen is increased. Heat, carbon dioxide, ammonia, and other gases are emitted as by-products of the composting process. Once these microorganisms (especially actinomycetes) have completed their life cycles, the compost is susceptible to invasion by a select secondary decomposer. A classic example of a secondary decomposer is the Button Mushroom, *Agaricus brunnescens*, the most commonly cultivated mushroom. Another example is *Stropharia ambigua*, which invades outdoor mushroom beds after wood chips have been first decomposed by a primary saprophyte.

Tertiary Decomposers: An amorphous group, the fungi represented by this group are typically soil dwellers. They survive in habitats that are years in the making from the activity of the primary and secondary decomposers. Fungi existing in these reduced substrates are remarkable in that the habitat appears inhospitable for most other mushrooms. A classic example of a tertiary decomposer is *Aleuria aurantia*, the Orange Peel Mushroom. This complex group of fungi often poses unique problems to would-be cultivators. *Panaeolus subbalteatus* is yet another example. Although one can grow it on composted substrates, this mushroom has the reputation of growing prolifically in the discarded compost from Button mushroom farms. Other tertiary decomposers include species of *Conocybe*, *Agrocybe*, *Pluteus*, and some *Agaricus* species.

The floor of a forest is constantly being replenished by new organic matter. Primary, secondary, and tertiary decomposers can all occupy the same location. In the complex environment of the forest floor, a "habitat" can actually be described as the overlaying of several, mixed into one. And, over time, as each habitat is being transformed, successions of mushrooms occur. This model becomes infinitely complex when taking into account the interrelationships of not only the fungi to one another, but also the fungi to other microorganisms (yeasts, bacteria, protozoa), plants, insects, and mammals.

Primary and secondary decomposers afford the most opportunities for cultivation. To select the best species for cultivation, several variables must be carefully matched.

Climate, available raw materials, and the mushroom strains all must interplay for cultivation to result in success. Native species are the best choices when you are designing outdoor mushroom landscapes.

Temperature-tolerant varieties of mushrooms are more forgiving and easier to grow than those that thrive within finite temperature limits. In warmer climates, moisture is typically more rapidly lost, narrowing the opportunity for mushroom growth. Obviously, growing mushrooms outdoors in a desert climate is more difficult than growing mushrooms in moist environments where they naturally abound. Clearly, the site selection of the mushroom habitat is crucial. The more exposed a habitat is to direct midday sun, the more difficult it is for mushrooms to flourish.

Many mushrooms actually benefit from indirect sunlight, especially in the northern latitudes. Pacific Northwest mushroom hunters have long noted that mushrooms grow most prolifically, not in the darkest depths of a woodlands, but in environments where shade and "dappled" sunlight are combined. Sensitivity-to-light studies have established that various species differ in their optimal response to wavebands of sunlight. Nevertheless, few mushrooms enjoy prolonged exposure to direct sunlight.

The Global Environmental Shift and the Loss of Species Diversity

Studies in Europe show a frightening loss of species diversity in forestlands, most evident with the mycorrhizal species. Many mycologists fear many mushroom varieties, and even species, will soon become extinct. As the mycorrhizal species decline in both numbers and variety, the populations of saprophytic and parasitic fungi initially rise as a direct result of the increased availability of deadwood debris. However, as woodlots are burned and replanted, the complex mosaic of the natural forest is replaced by a highly uniform, mono-species landscape. Because the replanted trees are nearly identical in age, the cycle of debris replenishing the forest floor is interrupted. This new "ecosystem" cannot support the myriad fungi, insects, small mammals, birds, mosses, and flora so characteristic of ancestral forests. In pursuit of commercial forests, the native ecology has been supplanted by a biologically anemic woodlot. This woodlot landscape is barren in terms of species diversity.

With the loss of every ecological niche, the sphere of biodiversity shrinks. At some presently unknown level, the diversity will fall below the critical mass needed for sustaining a healthy forestland. Once passed, the forest may not ever recover without direct and drastic counteraction: the insertion of multiage trees of different species, with varying canopies and undergrowth. Even with such extraordinary action, the complexity of a replanted forest cannot match that which has evolved for thousands of years. Little is understood about prerequisite microflora—yeasts, bacteria, and micro-fungi—upon which the ancient forests are dependent. As the number of species declines, whole communities of organisms disappear. New associations are likewise limited. If this trend continues, I believe the future of new forests, indeed the planet, is threatened.

Apart from the impact of wood harvest, the health of biologically diverse forests is in increasing jeopardy due to acid rain and other airborne toxins. Eventually, the populations of all fungi—saprophytic and mycorrhizal—suffer as the critical mass of dead trees declines more rapidly than it is replenished. North Americans have already experienced the results of habitat loss from the European forests. Importation of wild picked mushrooms from Mexico, the United States, and Canada to Europe has escalated radically in the past twenty years. This increase in demand is not due just to the growing popularity of eating wild mushrooms. It is a direct reflection of the decreased availability of wild mushrooms from regions of the world suffering from ecological shock. The woodlands of North America are only a few decades behind the forests of Europe and Asia.

With the loss of habitat of the mycorrhizal gourmet mushrooms, market demands for gourmet mushrooms should shift to those that can be cultivated. Thus, the pressure on this not-yet-renewable resource would be alleviated. I believe the judicious use of saprophytic fungi by homeowners as well as foresters may well prevent widespread parasitic disease vectors. Selecting and controlling the types of saprophytic fungi occupying these ecological niches can benefit both forester and forestland.

Catastrophia: Nature as a Substrate Supplier

Many saprophytic fungi benefit from catastrophic events in the forests. When hurricane-force winds rage across woodlands, enormous masses of dead debris are generated. The older trees are especially prone to fall. Once the higher canopy is gone, the growth of a younger, lower canopy of trees is triggered by the suddenly available sunlight. The continued survival of young trees is dependent upon the quick recycling of nutrients by the saprophytic fungi in decomposing deadwood.

Every time catastrophes occur—hurricanes, tornadoes, volcanoes, floods, and even earthquakes—the resulting deadwood becomes a stream of inexpensive substrate materials. In a sense, the cost of mushroom production is underwritten by natural disasters. Unfortunately, to date, few individuals and communities take advantage of catastrophia as a fortuitous event for enhancing mycelial growth. However, once the economic value of recycling with gourmet and medicinal mushrooms is clearly understood, and with the increasing popularity of backyard cultivation, catastrophia can be viewed as a positive event, at least in terms of providing new economic opportunities and positive environmental consequences for those who are mycologically astute.

Mushrooms and Toxic Wastes

In heavily industrialized areas, the soils are typically contaminated with a wide variety of pollutants, particularly petroleum-based compounds, polychlorinated biphenols (PCBs), heavy metals, pesticide-related compounds, and even radioactive wastes. Mushrooms grown in polluted environments can absorb toxins directly into their tissues, especially heavy metals (Bressa, 1988; Stijve 1974, 1976, 1992). As a result, mushrooms grown in these environments should not be eaten. Recently, a visitor to Ternobyl, a city about 60 miles from Chernobyl, the site of the world's worst nuclear power plant accident, returned to the United States with a jar of pickled mushrooms. The mushrooms were radioactive enough to set off Geiger counter alarms as the baggage was being processed. Customs officials promptly confiscated the mushrooms. Unfortunately, most toxins are not so readily detected.

A number of fungi can, however, be used to detoxify contaminated environments, in a process called "bioremediation." The white rot fungi (particularly *Phanerochaete chrysosporium*) and brown rot fungi (notably *Gloephyllum* species) are the most widely used. Most of these wood-rotters produce

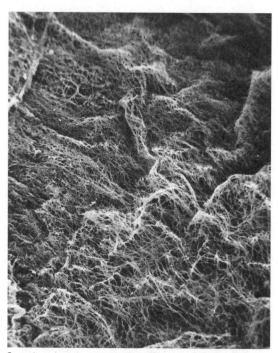

Scanning electron micrograph of the mycelial network.

lignin peroxidases and cellulases, which have unusually powerful degradative properties. These extracellular enzymes have evolved to break down plant fiber, primarily lignin-cellulose, the structural component in woody plants, into simpler forms. By happenstance, these same enzymes also reduce recalcitrant hydrocarbons and other manufactured toxins. Given the number of industrial pollutants that are hydrocarbon-based, fungi are excellent candidates for toxic waste cleanup and are viewed by scientists and government agencies with increasing interest. Current and prospective future uses include the detoxification of PCB (polychloralbiphenols), PCP (pentachlorophenol), oil, and pesticide/herbicide residues. They are even being explored for ameliorating the impact of radioactive wastes by sequestering heavy metals.

A far-reaching patent has been applied for using mycelial mats to break down toxic wastes, particularly those that are hydrocarbon based, including most petroleum products, pesticides, PCBs (polychlorobiphenols), and PCPs (pentachlorophenols), and for eliminating the flow of pathogenic bacteria into sensitive watersheds. This revolutionary patent also describes methods for effectively destroying nerve gas surrogates, including Sarin and VX, as well as chemical and biological warfare components by training the mushroom mycelium (Venter, A. J., 1999; Word et al. 1997; Thomas et al. 1998).

Bioremediation of toxic waste sites is especially attractive because the environment is treated *in situ*. The contaminated soils do not have to be hauled away, eliminating the extraordinary expense of handling, transportation, and storage. Since these fungi have the ability to reduce complex hydrocarbons into elemental compounds, these compounds pose no threat to the environment. Indeed, these former pollutants could even be considered "fertilizer," helping rather than harming the nutritional base of soils.

Dozens of bioremediation companies have formed to solve the problem of toxic waste. Most of these companies look to the imperfect fungi. The higher fungi should not be disqualified for bioremediation just because they produce an edible fruitbody. Indeed, this group may hold answers to many of the toxic waste problems. The most vigorous rotters described in this book are the *Ganoderma* and *Pleurotus* mushrooms. Mushrooms grown from toxic wastes are best not eaten, as residual heavy metal toxins may be concentrated within the mushrooms. However, one experiment using Oyster mushrooms to degrade petroleum residues on an oil-saturated Department of Transportation lot near Bellingham, Washington, not only largely decomposed the oil, but the mushrooms were free of petroleum residues when analyzed (Stamets, 1999).

Mushroom Mycelium and Mycofiltration

The mycelium is a fabric of interconnected, interwoven strands of cells. One colony can range in size from a half-dollar to many acres. A cubic inch of soil can host up to a mile of mycelium. This organism can be physically separated, and yet behave as one.

The exquisite lattice-like structure of the mushroom mycelium, often referred to as the *mycelial network*, is perfectly designed as a filtration membrane. Each colony extends long, complex chains of cells that fork repeatedly in matrix-like fashion, spreading to geographically defined borders. The mushroom mycelium, being a voracious forager for carbon and nitrogen, secretes extracellular enzymes that unlock organic complexes. The newly freed nutrients are then selectively absorbed directly through the cell walls into the mycelial network.

In the rainy season, water carries nutritional particles through this filtration membrane, including bacteria, which often become a food source for the mushroom mycelium. The resulting downstream effluent is cleansed of not only carbon/nitrogen-rich compounds but also bacteria, in some cases nematodes, and legions of other microorganisms. The voracious Oyster mushrooms been found to be parasitic against nematodes (Thorn and Barron, 1984; Hibbett and Thorn, 1994). Extracellular enzymes act like an anesthetic and stun the nematodes, thus allowing the invasion of the mycelium directly into their immobilized bodies.

The use of mycelium as a mycofilter is currently being studied by this author in the removal of biological contaminants from surface water passing directly into sensitive watersheds. By placing sawdust implanted with mushroom mycelium in drainage basins downstream from farms raising livestock, the mycelium acts as a sieve, which traps fecal bacteria and ameliorates the impact of a farm's nitrogen-rich outflow into aquatic ecosystems. This concept is incorporated into an integrated farm model and explored in greater detail in Chapter 5: Permaculture with a Mycological Twist.

Oyster mushrooms fruiting on diesel-contaminated soil at a test bioremediation site near Bellingham, Washington, effectively "de-contaminating" the soil to a level where it could be used for highway landscaping.

CHAPTER 3

Selecting a Candidate for Cultivation

Many mushroom hunters would love to have their favorite edible mushroom growing in their backyard. Who would not want a patch of Matsutake, Shaggy Manes, Giant Puffballs, or the stately Prince gracing their property? As the different seasons roll along, gourmet mushrooms would arise in concert. Practically speaking, however, our knowledge of mushroom cultivation is currently limited to 100 species of the 10,000 thought to exist throughout the world. Through this book and the works of others, the number of cultivatible species will enlarge, especially if amateurs are encouraged to boldly experiment. Techniques for cultivating one species may be applied for cultivating another, often by substituting an ingredient, changing a formula, or altering the fruiting environment. Ironically, with species never before grown, the strategy of "benign neglect" more often leads to success than active interference with the natural progression of events. I have been particularly adept at this nonstrategy. Many of my early mushroom projects only produced when I left them alone.

A list of candidates, which can be grown using current methods, follows. Currently we do not know how to grow those species marked by an asterisk (*). However, I believe techniques for their cultivation will soon be perfected, given a little experimentation. This list is by no means exhaustive, and will be much amended in the future. Many of these mushrooms are described as good edibles in the field guides, as listed in the Resource Directory in this book. (See Appendix 4.)

Woodland Mushrooms

The Wood Ears
Auricularia auricula
Auricularia polytricha

The Prince
Agaricus augustus

The Almond Agaricus
Agaricus subrufescens

The Sylvan Agaricus
Agaricus sylvicola
Agaricus lilaceps*

Black Poplar Agrocybe
Agrocybe aegerita

The Clustered Woodlovers
Hypholoma capnoides
Hypholoma sublateritium
Psilocybe cyanescens and allies

Oyster-like Mushrooms
Hypsizygus ulmarius
Hypsizygus tessulatus (= H. marmoreus)
*Pleurotus citrinopileatus (= P. cornucopiae
    var. citrinopileatus)*
Pleurotus cornucopiae
*Pleurotus cystidiosus (= P. abalonus,
    P. smithii (?))*
*Pleurotus djamor (=P. flabellatus,
    P. salmoneo-stramineus)*
Pleurotus dryinus*
Pleurotus eryngii
Pleurotus euosmus
Pleurotus ostreatus
Pleurotus pulmonarius (= "sajor-caju")
Tricholoma giganteum

The Deer Mushroom
Pluteus cervinus

Shiitake Mushroom
Lentinula edodes
Lentinula spp.

Garden Giant or King Stropharia
Stropharia rugosoannulata
Most polypore mushrooms

Grassland Mushrooms

Meadow Mushrooms
Agaricus arvensis
Agaricus blazei
Agaricus campestris
Lepiota procera

Horse Mushroom
Agaricus arvensis

The Giant Puffball
Calvatia gigantea and allies*

Smooth Lepiota
Lepiota naucina*

The Parasol Mushroom
Lepiota procera

Fairy Ring Mushroom
Marasmius oreades

Dung Inhabiting Mushrooms

The Button Mushrooms
Agaricus brunnescens
Agaricus bitorquis (= rodmanii)

The Magic Mushrooms
Psilocybe cubensis
Panaeolus cyanescens (= Copelandia
cyanescens)
Panaeolus subbalteatus
Panaeolus tropicalis (Copelandia tropicalis)

Compost/Litter/ Disturbed Habitat Mushrooms

Shaggy Manes
Coprinus comatus

Scaly Lepiota
Lepiota rachodes*

The Termite Mushrooms
Termitomyces spp.*

The Blewit
Lepista nuda

Termitomyces robustus is one of the best of the edible mushrooms but defies human attempts at cultivation. So far only ants know the secret to growing this delicacy.

Gardening with gourmet and medicinal mushrooms.

Natural Culture: Creating Mycological Landscapes

Natural culture is the cultivation of mushrooms outdoors. After mycological landscapes are constructed and inoculated, the forces of nature take control. For these mycological landscapes to be sustainable, a continual flow of organic debris is essential. Although the cultivator may choose to install desired species, respect towards nature's selection of preferred mushrooms is the only path to successful cultivation. Wild species in the landscape are natural allies. The responsibility of the cultivator is to design a habitat incorporating both wild and cultivated mushrooms, and seeking the right fits. Yet, the complex nature of creating species mosaics is still being understood. Only through the cumulative experiences of mycological landscapers can the knowledge base of this new model expand.

I also call this *laissez-faire* cultivation. After the mushroom patch has been inoculated, it is left alone, subject to the whims of nature, except for some timely watering. The mushroom habitat is specifically designed, paying particular attention to site location, topography, sun exposure, and the use of native woods and/or garden by-products. Once prepared, the cultivator launches the selected mushroom species into a constructed habitat by spawning. In general, native mushroom species do better than exotic ones. However, even those obstacles to growing exotic species are easily overcome with some forethought to design, and the helpful suggestions of an experienced cultivator.

Every day, gardeners, landscapers, rhododendron growers, arborists, and nurseries utilize the very components needed for growing mushrooms. Every pile of debris, whether it is tree trimmings, sawdust, wood chips, or a mixture of these materials, will support mushrooms. Unless selectively inoculated, debris piles become habitats of miscellaneous "weed" mushrooms, making the likelihood of growing a desirable mushroom remote.

When inoculating an outdoor environment with mushroom spawn, the cultivator relinquishes much control to natural forces. There are obvious advantages and disadvantages to natural culture. First, the mushroom patch is controlled by volatile weather patterns. This also means that outdoor beds have the advantage of needing minimum maintenance. The ratio of hours spent per pound of mushrooms grown becomes quite efficient. The key to success is creating an environment wherein the planted mycelium naturally and vigorously expands. A major advantage of growing outdoors compared to growing indoors is that competitors are not concentrated in a tight space. When cultivating mushrooms outdoors entropy is your ally.

The rate of growth, time to fruiting, and quality of the crop depends upon the quality of the spawn, substrate materials, and weather conditions. Generally, when mushrooms are fruiting in the wild, the inoculated patches also produce. Mushrooms that fruit primarily in the summer, such as the King Stropharia (*Stropharia rugosoannulata*) require frequent watering. Shaggy Manes (*Coprinus comatus*) prefer the cool fall rains, thus requiring little attention. In comparison to indoor cultivation, the outdoor crops are not as frequent. However, the crops can be just as intense, sometimes more so, especially when paying modest attention to the needs of the mushroom mycelium at critical junctions in its life cycle.

While the cultivator is competing with molds indoors, wild mushrooms are the major competitors outdoors. You may plant one species in an environment where another species is already firmly established. This is especially likely if you use old sawdust, chips, or base materials. Starting with fresh materials is the simplest way to avoid this problem. Piles of aged wood chips commonly support four or five species of mushrooms within just a few square feet. *Unless, the cultivator uses a high rate of inoculation (25% spawn/substrate) and uniformly clean wood chips, the concurrence of diverse mushroom species should be expected.* If, for instance, the backyard cultivator gets mixed wood chips in the early spring from a county road maintenance crew, and uses a dilute 5–10% inoculation rate of sawdust spawn into the chips, the mushroom patch is likely to have more wild species emerging along with the desired mushrooms.

In the Pacific Northwest of North America, I find a 5–10% inoculation rate usually results in some mushrooms showing late in the first year, the most substantial crops occurring in the second and third years, and a dramatic drop-off in the fourth year. As the patch ages, it is normal to see more diverse mushroom varieties co-occurring with the planted mushroom species.

I am constantly fascinated by the way nature reestablishes a polyculture environment at the earliest opportunity. Some mycologists believe a predetermined sequence of mycorrhizal and saprophytic species prevails, for instance, around a Douglas fir tree, as it matures. In complex natural habitats, the interlacing of mycelial networks is common. Underneath a single tree, twenty or more species may thrive. I look forward to the time when mycotopian foresters will design whole species mosaics upon whose foundation vast ecosystems can flourish. This book will describe simpler precursor models for mixing and sequencing species. I hope imaginative and skilled cultivators will further develop these concepts.

In one of my outdoor wood-chip beds, I created a "polyculture" mushroom patch about 50 by 100 feet in size. In the spring I acquired mixed wood chips from the county utility company—mostly alder and Douglas fir—and inoculated three species into it. One year after inoculation, in late April through May, Morels showed. From June to early September, King Stropharia erupted with force, providing our family with several hundred pounds. In late September through much of November, an assortment of Clustered Woodlovers (*Hypholoma*-like) species popped up. With noncoincident fruiting cycles, this Zen-like polyculture approach is limited only by your imagination.

Species succession can be accomplished indoors. Here is one example. After Shiitake stops producing on logs or sawdust, the substrate can be broken apart, remoistened, resterilized, and reinoculated with another gourmet mushroom; in this case, I recommend Oyster mushrooms. Once the Oyster mushroom life cycle is completed, the substrate can be again sterilized, and inoculated with the next species. Shiitake, Oyster, King Stropharia, and finally Shaggy Manes can all be grown on the same substrate, increasingly reducing the substrate mass, without the addition of new materials. The majority of the substrate mass that does not evolve into gases, is regenerated into mushrooms. The conversion of substrate mass-to-mushroom mass is mind boggling. These concepts are further developed in Chapter 22.

The following is a list of decomposer mushrooms most frequently occurring in wood chips in the northern temperate regions of North America. In general, these natural competitors are easy to distinguish from the gourmet mushroom species described in this book. Those that are mildly poisonous are labeled with *; those that are deadly have two **. This list is by no means comprehensive. Many other species, especially the poisonous mycorrhizal *Amanita, Hebeloma, Inocybe,* and *Cortinarius* species are not listed here. Mushrooms from these genera can inhabit the same plot of ground where a cultivator may lay down wood chips, even if the host tree is far removed.

The mushrooms in the *Galerina autumnalis* and *Pholiotina filaris* groups are deadly poisonous. Some species in the genus *Psilocybe* contain psilocybin and psilocin, compounds that often cause uncontrolled laughter, hallucinations, and sometimes spiritual experiences. Outdoor cultivators must hone their skills at mushroom identification to avert the accidental ingestion of a poisonous mushroom. Recommended mushroom field guides and mushroom identification courses are listed in the Resource Directory in this book.

Some Wild Mushrooms Naturally Found in Beds of Wood Chips

Ground Lovers
Agrocybe spp. and Pholiota spp.

The Sweaters
Clitocybe spp. *

The Inky Caps
Coprinus atramentarius *
C. comatus
C. disseminatus
C. lagopus
C. micaceus and allies

The Vomited Scrambled Egg Fungus
Fuligo cristata

The Deadly Galerinas
Galerina autumnalis and allies **

Red-Staining Lepiotas
Lepiota spp. *

The Clustered Woodlover
Hypholoma capnoides

The Green-Gilled Clustered Woodlover
Hypholoma fasciculare *

The Chestnut Mushroom
Hypholoma sublateritium

The Deadly Ringed Cone Head
Pholiotina filaris and allies **
Pholiota terrestris and allies

The Deer Mushroom
Pluteus atrocapillus (= cervinus)

Black Spored Silky Stems
Psathyrella spp.

The Caramel Capped Psilocybes
Psilocybe cyanescens and allies

Methods of Mushroom Culture

Mushrooms can be cultivated through a variety of methods. Some techniques are exquisitely simple, and demand little or no technical expertise. Others—involving sterile tissue culture—are much more technically demanding. The simpler methods take little time, but also require more patience and forgiveness on the part of the cultivator, lest the mushrooms do not appear on your timetable. As one progresses to the more technically demanding methods, the probability of success is substantially increased, with mushrooms appearing exactly on the day scheduled.

The simpler methods for mushroom cultivation, demanding little or no technical expertise, are outlined in this chapter. They are *spore-mass inoculation, transplantation,* and *inoculation with pure cultured spawn.*

Spore-Mass Inoculation

By far the simplest way to grow mushrooms is to broadcast spores onto prepared substrates outdoors. First, spores of the desired species must be collected.

Spore collection techniques vary, according to the shape, size, and type of the mushroom candidate.

For gilled mushrooms, the cap can be severed from the stem, and laid, gills down, on top of clean typing paper, glass, or similar surface. A glass jar or bowl is placed over the mushroom to lessen the loss of water. After 12 hours, most mushrooms will have released thousands of spores, falling according to the radiating symmetry of the gills, in a symmetrically attractive outline called a *sporeprint.* This method is ideal for mushroom hunters "on the go" who might not be able to make use of the spores immediately. After the spores have fallen, the spore print can be sealed, stored, and saved for future use. It can even be mailed without harm.

By collecting spores of many mushrooms, one creates a species library. Spore collections can resemble stamp or coin collections, but are potentially more valuable. A mushroom hunter may find a species only once in a lifetime. Under these circumstances, the existence of a spore print may be the only resource a cultivator has for future propagation. I prefer taking spore prints on panes of glass, using duct tape as binding along one edge. The glass panes are folded together, and masking tape is used

Collecting the spores of the delicious *Lepiota rachodes,* a Parasol Mushroom, on two panes of glass, which are then folded together, creating a spore booklet.

to seal the three remaining edges. This spore booklet is then registered with written notes affixed to its face as to the name of mushroom, the date of collection, the county and locality of the find. Spores collected in this fashion remain viable for years, although viability decreases over time. They should be stored in a dark, cool location, low in humidity and free from temperature fluctuation. Techniques for creating cultures from spores are explained further on.

For those wishing to begin a mushroom patch using fresh specimens, a more efficient method of spore collection is recommended. This method calls for the immersion of the mushroom in water to create a *spore-mass slurry*. Choose fairly mature mushrooms and submerge them in a 5-gallon bucket of water. A gram or two of table salt inhibits bacteria from growing while not substantially affecting the viability of the spores. With the addition of 50 milliliters of molasses, spores are stimulated into frenzied germination. After 4 hours of soaking, remove the mushrooms from the bucket. Most mushrooms will have released tens of thousands of spores. Allow the broth to sit for 24 to 48 hours at a temperature above 50°F (10°C) but under 80°F (26.7°C). In most cases, spores begin to germinate in minutes to hours, aggressively in search of new mates and nutrients. This slurry can be expanded by a factor of 10 in 48 hours. I have often dreamed, being the mad scientist that I am, of using spore-mass slurries of Morels and other species to aerially "bomb" large expanses of forestlands. This idea, as crazy as it may initially sound, warrants serious investigation.

During this stage of frenzied spore germination, the mushroom patch habitat should be designed and constructed. Each species has unique requirements for substrate components for fruiting. However, mycelia of most species will run through a variety of lignin-cellulosic wastes. Only at the stage when fruitbody production is sought does the precise formulation of the substrate become crucial.

Oyster (*Pleurotus ostreatus*; *P. eryngii*, and allies), King Stropharia (*Stropharia rugosoannulata*), and Shaggy Mane (*Coprinus comatus*) mushrooms thrive in a broad range of substrate formulations.

Other mushrooms such as Morels (*Morchella angusticeps* and *esculenta*) are more restrictive in their requirements. Since there are several tracks that one can pursue to create suitable habitats, refer to Chapter 21 for more information.

Transplantation: Mining Mycelium from Wild Patches

Transplantation is the moving of mycelium from natural patches to new habitats. Most wild mushroom patches have a vast mycelial network emanating from beneath each mushroom. Not only can one harvest the mushroom, but portions of the mycelial network can be gathered and transferred to a new location. This method ensures the quick establishment of a new colony without having to germinate spores or buy commercial spawn.

When transplanting mycelium, I use a paper sack or a cardboard box. Once mycelium is disturbed, it quickly dries out unless measures are taken to prevent dehydration. After it is removed from its original habitat, the mycelium will remain viable for days or weeks, as long as it is kept moist in a cool, dark place.

Gathering the wild mycelium of mycorrhizal mushrooms could endanger the parent colony. Be sure you cover the divot with wood debris and press tightly back into place. In my opinion, mycorrhizal species should not be transplanted unless the parent colony is imminently threatened with loss of habitat—such as logging, construction, etc. Digging up mycelium from the root zone of a healthy forest can jeopardize the symbiotic relationship between the mushroom and its host tree. Exposed mycelium and roots become vulnerable to disease, insect invasion, and dehydration. Furthermore, transplantation of mycorrhizal species has a lower success rate than the transplantation of saprophytic mushrooms.

If done properly, transplanting the mycelium of saprophytic mushrooms is not threatening to naturally occurring mushroom colonies. Some of the best sites for finding mycelium for transplantation are sawdust piles. Mycelial networks running through sawdust piles tend to be vast and relatively clean of competing fungi. Fans of mycelium are more often found along the periphery of sawdust piles than within their depths. When sawdust piles are a foot deep or more, the microclimate is better suited for molds and thermophilic fungi. These mold fungi benefit from the high carbon dioxide and heat generated from natural composting. At depths of 2–6 inches, mushroom mycelia run vigorously. It is from these areas that mushroom mycelium should be collected for transplantation to new locations. One, in effect, engages in a form of *mycelial mining* by encouraging the growth and the harvesting of mycelium from such environments. Ideal locations for finding such colonies are sawmills, nurseries, composting sites, recycling centers, rose and rhododendron gardens, and soil mixing companies.

Inoculating Outdoor Substrates with Pure Cultured Spawn

In the early history of mushroom cultivation, mycelium was collected from the wild and transplanted into new substrates with varying results. Soon compost spawn (for the Button Mushroom, *Agaricus brunnescens*) evolved with greater success. In 1933, spawn technology was revolutionized by Sinden's discovery of grain as a spawn carrier medium. Likewise, Stoller (1962) significantly contributed to the technology of mushroom cultivation through a series of practical advances in using plastic bags, collars, and filters.* *The Mushroom Cultivator* (Stamets and Chilton, 1983) explained the process of producing tissue culture for spawn generation, empowering far more cultivators than ever before. Legions of creative individuals embarked on the path of exotic mushroom production. Today, thousands of cultivators are contributing to an ever-expanding body of knowledge, and setting the stage for the cultivation of many gourmet and medicinal fungi of the future.

The advantage of using commercial spawn is in acquiring mycelium of higher purity than can be harvested from nature. Commercial spawn can be bought in two forms: grain or wood (sawdust or plugs). For the inoculation of outdoor, unpasteurized substrates, wood-based spawn is far better than grain spawn. When grain spawn is introduced to an outdoor bed, insects, birds, and slugs quickly seek out the nutritious kernels for food. Sawdust spawn has the added advantage of having more particles or *inoculation points* per pound than does grain. With more points of inoculation, colonization is accelerated. The distances between mycelial fragments are lessened, making the time to contact less than that with grain spawn. Thus the window of vulnerability is closed to many of the diseases that eagerly await intrusion.

Before spawn is used, the receiving habitat is moistened to near saturation. The spawn is then mixed thoroughly through the new habitat with your fingers or a rake. Once inoculated, the new bed is again watered. The bed can be covered with cardboard, shade cloth, scrap wood, or a similar material to protect the mycelium from sun exposure and dehydration. After inoculation, the bed is ignored, save for an occasional inspection and watering once a week, and then only when deemed necessary.

Certain limitations prevail in the expansion of mycelium and its ability for colonizing new substrates. The intensity or rate of inoculation is extremely important. If the spawn is too dispersed into the substrate, the points of inoculation will not be close enough to result in the rapid reestablishment of one large contiguous mycelial mat. My own experiences show that success is seen with an inoculation rate of 5–50%, with an ideal of 20%. In other words, if you gather a 5-gallon bucket of naturally occurring mycelium, 20 gallons of prepared substrate can be inoculated with a high probability of success. Although this inoculation rate may seem high, rapid colonization is assured. More skilled cultivators, whose methods have been refined through experience, often use a less intensive inoculation rate of 10%. Inoculation rates of 5% or less often result in "island" colonies of the implanted species interspersed among naturally occurring wild colonies.

At a 20% inoculation rate, colonization can be complete in as short as 1 week and as long as 8 weeks. After a new mycelial mat has been fully

---

* In 1977, B. Stoller and J. Azzolini were awarded U.S. patent #4027427 for this innovation.

Establishing an outdoor mushroom bed in a garden.

Sprinkling spawn on top of mulch layer.

Adding more moist mulch over the spawn layer.

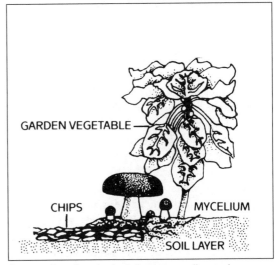

Cross section of garden bed showing mycelium and mushroom growth.

established, the cultivator has the option of further expanding the colony by a factor of 5, or triggering the patch into fruiting. This usually means providing shade and frequent watering. Should prevailing weather conditions not be conducive to fruiting and yet are above freezing, then the patch can be further expanded. Should the cultivator not expect that further expansion would result in full colonization by the onset of winter, then no new raw material should be added, and mushrooms should be encouraged to form. The widely cast mycelial mat, more often than not, acts as a single organism. At

Healthy *Stropharia rugosoannulata* mycelium tenaciously gripping alder chips and sawdust. Note rhizomorphs.

From the same patch, a year later, the wood chips have decomposed into a rich soil-loam.

the time when mushrooms are forming, colonization of new organic debris declines or abates entirely. The energy of the mycelium is now channeled to fruit-body formation and development.

The mycelium of saprophytic mushrooms must move to remain healthy. When the mycelium reaches the borders of a geographically or nutritionally defined habitat, a resting period ensues. If not soon triggered into fruiting, over-incubation is likely, with the danger of "dieback." Only very cold temperatures will keep the patch viable for a prolonged period. Typically, dieback is seen as the drastic decline in vigor of the mycelium. Once the window of opportunity has passed for fruiting, the mushroom patch might be salvaged by the reintroduction of more undecomposed organic matter, or by violent disturbance. The mycelium soon becomes a site for contamination with secondary decomposers (weed fungi) and predators (insects) coming into play. It is far better is *to keep the mycelium running* until fruitings can be triggered at

the most opportune time. Mushroom patches are, by definition, temporary communities.

King Stropharia lasts 3 to 4 years on a hardwood chip base. After the second year more material should be added. However, if the health of the patch has declined and new material is mixed in, then the mushroom patch may not recover to its original state of vigor. Mycelium that is healthy tends to be tenacious, holding the substrate particles together. This is especially true with *Stropharia* spp. and Oyster mushrooms. (*Hericium erinaceus* and *Morchella* spp. are exceptions.) Over-incubation results in a weakened mycelial network whereby the mycelium is incapable of holding various substrate particles together. As mycelial integrity declines, other decomposers are activated. Often, when mixing in new material at this stage, weed fungi proliferate to the decided disadvantage of the selected gourmet species. To the eye, the colony no longer looks like a continuous sheet of mycelium, but becomes spotty in its growth pattern. Soon islands of mycelia become smaller and smaller as they retreat, eventually disappearing altogether. The only recourse is to begin anew, scraping away the now-darkened wood/soil, and replacing it with a new layer of wood chips and/or other organic debris.

When to Inoculate an Outdoor Mushroom Patch

Outdoor beds can be inoculated in early spring to early fall. The key to a mushroom bed is that the mycelium has sufficient time to establish a substantial mycelial mat before the onset of inclement weather conditions. Springtime is generally the best time to inoculate, especially for creating large mushroom patches. As fall approaches, more modestly sized beds should be established, with a correspondingly higher rate of inoculation for faster growth. For most saprophytic species, at least 4 weeks are required to form the mycelial network with the critical mass necessary to survive the winter.

Most woodland species survive wintering temperatures. Temperate woodland mushrooms have evolved protective mechanisms within their cellular network that allow them to tolerate cold temperature extremes. Surface frosts usually do not harm the terrestrially bound mushroom mycelium. As the mycelium decomposes organic matter, heat is released, which benefits subsurface mycelium. Mycelial colonization essentially stops when outdoor temperatures fall below freezing.

Site Location of a Mushroom Patch

A suitable site for a mushroom patch is easy to choose. The best clue is to simply take note of where you have seen mushrooms growing during the rainy season. Or just observe where water traverses after a heavy rain. A gentle slope, bordered by shrubs and other shade-giving plants, is usually ideal. Since saprophytic mushrooms are noncompetitive to neighboring plants, they pose no danger to them. In fact, plants near a mushroom bed often thrive—the result of the increased moisture retention and the release of nutrients into the root zone.

An ideal location for growing mushrooms is in a vegetable, flower, and/or rhododendron garden. Gardens are favored by plentiful watering, and the shade provided by potato, zucchini, and similar broadleaf vegetable plants tend to keep humidity high near the ground. Many gardeners bring in sawdust and wood chips to make pathways between the rows of vegetables. By increasing the breadth of these pathways, or by creating small cul-de-sacs in the midst of the garden, a mushroom bed can be ideally located and maintained.

Other suitable locations are exposed north sides of buildings, and against rock, brick, or cement walls. Walls are usually heat sinks, causing condensation, which provides moisture to the mushroom site as temperature fluctuates from day to night. Protected from winds, these locations have limited loss of water due to evaporation.

Mushrooms love moisture. By locating a mushroom bed where moisture naturally collects, colonization is rapid and more complete, and the need for additional water for fruiting is minimized. The message here: Choose your locations with moisture foremost in mind. Choose shady locations over sunny ones. Choose north-facing slopes rather than south facing. Choose companion plants with broad leaves or canopies that shade the midday sun but allow rain to pass. The difference in results is the difference between a bountiful success or dismal failure.

Stumps as Platforms for Growing Mushrooms

Stumps are especially suitable for growing gourmet mushrooms. There are few better, or more massive platforms, than the stump. Millions of stumps are all that remain of many forests of the world. In most cases, stumps are seen as having little or no economic potential. These lone tombstones of biodegradable wood fiber offer a unique, new opportunity for the mycologically astute. With selective logging being increasingly practiced, cultivating gourmet and medicinal mushrooms on stumps will be the wave of the future.

Giant Oyster mushrooms fruiting from a stump.

The advantage of the stump is not only its sheer mass, but with roots intact, water is continuously being drawn via capillary action. Once mycelium has permeated through wood fiber, the stump's water-carrying capacity is increased, thus further supporting mycelial growth. Candidates for stump culture must be carefully selected and matched with the appropriate species. A stump partially or fully shaded is obviously better than one in full sunlight. Stumps in ravines are better candidates than those located in the center of a clear-cut area. An uprooted stump is not as good a candidate as a well-rooted one. The presence of mosses, lichens, and/or ferns is a good indicator that the microclimate is conducive to mushroom growth. However, the presence of competitor fungi generally disqualifies a stump as a good candidate. These are some of the many factors that determine the suitability of stumpage.

Cultivating mushrooms on stumps requires forethought. Stumps should be inoculated before the first season of wild mushrooms. With each mushroom season, the air becomes laden with spores, seeking new habitats. The open face of a stump, essentially a wound, is highly susceptible to colonization by wild mushrooms. With the spore cast from wild competitors, the likelihood of introducing your species of choice is greatly reduced. If stumps are not inoculated within several months of being cut, the probability of success decreases. Therefore, old stumps are poor candidates. Even so, years may pass after inoculation before mushrooms

form on a stump. But once a colony begins, the same species may predominate for many years.

Large-diameter stumps can harbor many communities of mushrooms. On old-growth or second-growth Douglas fir stumps common to the forests of Washington State, finding several species of mushrooms is not unusual. This natural example of "polyculture"—the simultaneous concurrence of more than one species in a single habitat—should encourage experimentally inclined cultivators. Mushroom landscapes of great complexity could be designed. However, the occurrence of poisonous mushrooms should be expected. Two notable toxic mushrooms frequent stumps: *Hypholoma fasciculare (=Naematoloma fasciculare)* which causes gastrointestinal upset but usually not death, and *Galerina autumnalis*, a mushroom that does kill. Because of the similarity in appearance between *Flammulina velutipes* (Enoki) and *Galerina autumnalis*, I hesitate to recommend the cultivation of Enoki mushrooms on stumps unless the cultivator is adept at identification. (To learn how to identify mushrooms, please refer to the recommended mushroom field guides listed in Appendix 4.)

Several polypores are especially good candidates for stump cultivation, particularly *Grifola frondosa*, Maitake, *Ganoderma lucidum*, Reishi, and its close relatives. As the antiviral and anticancer properties of these mushrooms become better understood, new strategies for the cultivation of medicinal mushrooms will be developed. I envision the establishment of Maitake and Reishi mushroom tree farms wherein stumps are purposely created and selectively inoculated for maximum mushroom growth, interspersed among shade trees. Once these models are perfected, other species can be incorporated in creating a multicanopy medicinal forest.

Small-diameter stumps rot faster and produce crops of mushrooms sooner than bigger stumps. However, the smaller stump has a shorter mushroom-producing life span than the older stump. Often with large-diameter stumps, mushroom formation is triggered when competitors are encountered and/or coupled with wet weather conditions. The fastest I know of a stump producing mushrooms from time of inoculation is 8 weeks. In this case, an

Drilling and inoculating a stump with plug spawn.

asites. Should clear-cuts become colonized with these deadly, root-rotting species, satellite colonies can be spread to adjacent living trees. Now that burning is increasingly restricted because of air pollution concerns, disease vectors coming from stumpage could present a new as-yet-unmeasured threat to the forest ecosystem.

The advantages of growing on stumps can be summarized as

- Developing a new, environmentally friendly wood products industry.

- Recycling wood debris of little or no economic value.

- Prevention of disease vectors from parasitic fungi.

- The rapid return of wood debris back into a nutritious food for the benefit of other citizens of the forest. Since the food chain is accelerated and enriched, ecosystems are invigorated.

oak stump was inoculated with plug spawn of Chicken-of-the-Woods, *Laetiporus (Polyporus) sulphureus*. Notably, the stump face was checkered, with multiple fissures running vertically through the innermost regions of the wood. These fissures trapped water from rainfall and promoted fast mycelial growth. Another technique that improves success rates is to girdle the stump with a chain saw, interrupting direct contact between the root zone and the above-air portion. A local mycological society found that all their inoculated stumps produced which were girdled compared to a small fraction for those that were not. As with the growing of any mushrooms, *the speed of colonization* is a determining factor in the eventual success or failure of any cultivation project.

For foresters and ecologists, actively inoculating and rotting stumps has several obvious advantages. Rather than allowing a stump to be randomly decomposed, species of economic or *ecological* significance can be introduced. For instance, a number of Honey mushrooms, belonging to the genus *Armillaria*, can operate as both saprophytes or par-

The Honey mushroom, *Armillariella mellea*, growing on a fir stump in a rhododendron garden, an ideal locus for creating a mycological landscape.

Few studies have been published on recycling stumps with mushrooms. One notable work from Eastern Europe, published by Pagony (1973), describes the cultivation of Oyster mushrooms (*Pleurotus ostreatus*) on large-diameter poplars with a 100% success rate. Inoculations in the spring resulted in fruitings appearing the ensuing fall, and continued for several years hence. An average of 4 pounds of Oyster mushrooms were harvested over 4 years (i.e., 1 pound/year/stump). Hilber (1982) also reported on the utility of using natural wood (logs and stumps) for growing Oyster mushrooms, and that per cubic meter of elm wood, the yield from one season averaged 17–22 kilograms. A study in France by Anselmi and Deandrea (1979), where poplar and willow stumps were inoculated with spawn of the Oyster mushroom, showed that this mushroom favored wood from newly felled trees, in zones that received speckled sunlight. This study confirmed that *Pleurotus ostreatus* only attacked deadwood and never became parasitic. Their study supports my opinion (Stamets, 1990) that the pur-

poseful inoculation of stumps can forestall the invasion by parasites like Honey mushrooms of the *Armillaria mellea* complex. Mushrooms of this group first kill their host and then continue to live saprophytically. A stump with Honey mushrooms can later destroy neighboring living trees. In Washington State, one colony of Honey mushrooms is blamed for destroying thousands of acres of conifers.

Stumps can be inoculated by one of several simple procedures. Plug spawn can be inserted into the open face of each stump. If the stumps are checkered through with cracks, the plugs are best inserted directly into the fissures. Another method is known as the wedge or disk inoculation technique. With a chain saw, a wedge is cut or a shallow disk is sliced from the open face of the stump. The newly cut faces are packed with sawdust spawn. The cut disk is then replaced. By hammering a few nails into the stump, you can assure firm contact between the cut faces. Another method I have developed is to add spores

Inoculating a stump with the wedge and the spawn disk technique.

to biodegradable chainsaw oil so that as trees are being cut, the stumps are inoculated. (Patent pending).

The broadleaf hardwoods are easier to saprophytize with the gourmet and medicinal mushroom species described in this book than the softwood pines. And within the hardwood group, the rapidly growing species such as the alders and poplars decompose more rapidly—and hence give an earlier crop—than the denser hardwoods such as the oaks. However, the denser and more massive stumps sustain colonies of mushrooms for many more years than the quick-to-rot, smaller-diameter tree species. In a colonial graveyard in New York state, a four-foot-diameter oak has consistently produced clusters of Maitake mushrooms, sometimes weighing up to 100 pounds apiece, for more than 20 years! See page 369.

Inoculating stumps with strains cloned from native mushrooms is favored over the use of exotic fungi. Spring inoculations give the mycelium the longest possible growing season.

Stump cultivation has tremendous potential. These unexploited resources can become production sites of gourmet and medicinal mushrooms. Although more studies are needed to ascertain the proper matching of species to the wood types, I encourage you to experiment. Only a few minutes are required to inoculate a stump or dead tree. The potential rewards could span a lifetime.

Log Culture

Log culture was developed in Japan and China more than a millennium ago. Even today, thousands of small-scale Shiitake growers use log culture to provide the majority of mushrooms sold to markets. In their backyards and along hillsides, inoculated logs are stacked like cordwood or in fence-like rows. These growers supply local markets, generating a secondary income for their families. Attempts to reproduce this model of Shiitake cultivation in North America and Europe has met with modest success.

The advantage of log culture is that it is a simple and natural method. The disadvantage is that the

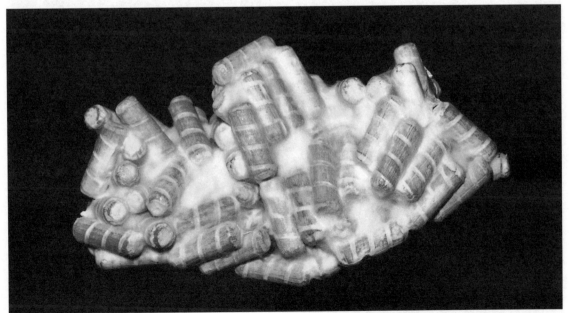

Plug spawn of Shiitake. Spirally grooved wooden dowels help the mycelium survive from the concussion of inoculation.

The "soak and strike" method for initiating Shiitake.

Figure 14. Natural culture of Shiitake in the mountains of Japan. (Photographs, both from Mimura, Japan, circa 1915.)

Drilling and inoculating plugs into logs.

Once inoculated, the plugs and ends are sealed with cheese wax.

Inoculated logs are stacked and covered with a tarp.

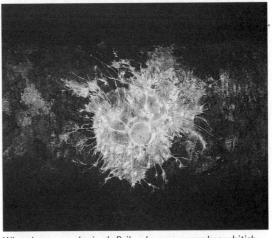

When logs are colonized, *Psilocybe azurescens* has whitish rhizomorphic mycelium.

process is labor-intensive, and slow in comparison to growing mushrooms on sterilized sawdust. Besides Shiitake, many other mushrooms can be grown on logs, including Nameko (*Pholiota nameko*), all the true Oyster mushrooms (*Pleurotus* and *Hypsizygus* spp.), Lion's Mane (*Hericium erinaceus*), Wood Ears (*Auricularia auricula*), Clustered Woodlovers (*Hypholoma capnoides* and *H. sublateritium*) and Reishi (*Ganoderma lucidum*). Since

log culture is not technically demanding, anyone can do it. In contrast, growing on sterilized substrates requires specialized skills and involves training in laboratory technique.

Logs are usually cut in the winter or early spring before leafing, when the sapwood is rich in sugars, to a yard in length and 4–10 inches in diameter. The logs, once felled, should be kept off the ground. Ideally inoculations should occur within 2 months

Gymnopilus spectabilis, the Big Laughing Mushroom, Owaritake, fruiting from a log inoculated with sawdust spawn via the wedge technique.

Oyster mushroom (*Pleurotus ostreatus*) fruiting on alder logs inoculated via the wedge technique.

of felling, preferably in early spring. (In temperate North America, February and March are ideal.) Logs can be inoculated by a variety of methods. Cultivators generally favor logs that have a higher ratio of sapwood to heartwood. (These logs come from fast-growing tree species like alder, poplar, or cottonwood.) Once inoculated with sawdust or plug spawn, the logs (or *billets*) are stacked in ricks and after 6 to 12 months, are initiated by heavy watering or soaking. After soaking, the logs are lined up in fence-like rows. Japanese growers have long favored the "soak and strike" method for initiating mushroom formation. Before the advent of plug and sawdust spawn, newly cut logs would be placed near logs already producing Shiitake so that the airborne spores would be broadcast onto them. This method, although not scientific, succeeded for centuries and still is a pretty good method. After a year, logs showing no growth or the growth of competitor fungi, are removed from the production rows.

A wide variety of broadleaf hardwoods are suitable for log culture. Some provide for short-term

Oyster mushroom fruiting from alder logs inoculated via the spawn disk technique.

cultivation while others are more resistant to quick decay. Oaks, and similar dense hardwoods with thick outer barks, are preferred over the rapidly decomposing hardwoods with their paper-thin bark layers. The rapidly decomposing hardwoods like alder and birch are easily damaged by weather fluctuations, especially humidity. Should the bark layer fall from the log, the mycelium has difficulty supporting good mushroom crops or *flushes*.

Logs are usually pegged, i.e., drilled with holes and inoculated with plug or sawdust spawn. Most logs receive 30 to 50 plugs, which are inserted into evenly spaced holes (4–6 inches apart) arranged longitudinally down the axis of the logs in a diamond pattern. By off-centering the rows of holes in a diamond pattern, the mycelium grows out to become one interconnected macroorganism, after which synchronous fruitings can occur. Once inserted by hand, the plugs are pounded in with a rubber mallet or hammer. The plugged hole is covered with cheese wax or beeswax, usually painted on, to protect the mycelium from insect or weather damage.

Another method calls for the inoculation of logs by packing sawdust spawn into cuts made with a chain saw. A common technique is to cut a V-shaped wedge from a log, pack the wound full of sawdust spawn, and press the wedge of wood back into place. Nailing the wedge back into position assures direct and firm contact. Another variation is to cut logs into 16- to 24-inch sections and sandwich spawn in between.

Sawdust spawn can be used in other ways. The newly cut ends of the logs can be packed with sawdust spawn and capped with aluminum foil or a "sock" to hold the mycelium in place. [San Antonio (1983) named this technique the *spawn disk* method.] Others prefer cutting wedges from the logs and then repacking the cut wedge back into the log with mycelium sandwiched in between. Others cut the logs in sections, 2 feet in length, and pack the sawdust spawn in between the two sections, which are reattached by any means possible. Most hammer long nails to secure the mini-logs back together.

For anyone growing outdoors in climates with severe dry spells, or where watering is a problem, logs should be buried with a third to a fourth of

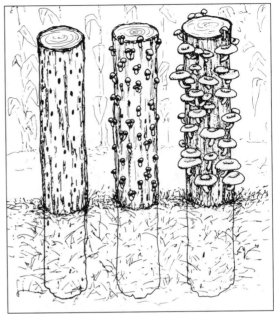

When logs are buried in sand, subsurface moisture is drawn up into the log, encouraging mushroom formation and discouraging competitor fungi.

their length into sandy ground. The ground moisture will constantly replenish water lost through evaporation, lessening the effect of humidity fluctuation. This method is especially useful for the cultivation of Lion's Mane, Nameko, Oyster, and Reishi. It is widely used by growers in China. Many cultivators protect their logs from the sun by either locating them under a forest canopy or by rigging up a shade cloth.

Most log cultivators develop their own unique techniques, dictated by successes and failures. Many books on log cultivation have been written, too numerous to list here. Two books on Shiitake log cultivation that I highly recommend for those who wish to study these techniques further are *Growing Shiitake Mushrooms in a Continental Climate* (second edition) by Mary Kozak and Joe Krawczyk and *Shiitake Grower's Handbook* by Paul Przybylowicz and John Donoghue. The methods described in these books can be extrapolated for the cultivation of other gourmet and medicinal mushrooms on logs.

The Stametsian Model for Permaculture with a Mycological Twist.

CHAPTER 5

Permaculture with a Mycological Twist

The Stametsian Model for a Synergistic Mycosphere

Permaculture is a concept pioneered by Australian Bill Mollison and literally is a fusion of "permanent agriculture" and "permanent culture." His model of biological diversity and complementary agricultural practices promotes a sustainable environment via the interplay of natural ecosystems. Permaculture has gained a huge international following since the publication of his book *Permaculture: A Practical Guide for a Sustainable Future*. Permaculture has become the mainstay philosophy of the organic movement. Mollison's vision, which borrows from Masanobu Fukuoka's "One Straw Revolution," intelligently combines the factors of site location, recycling of by-products from farming and forest activities, species diversity, and biological succession.

When gourmet and medicinal mushrooms are involved as key organisms in the recycling of agricultural and forest by-products, the biodynamics of permaculture soar to extraordinary levels of productivity. Mushrooms are the missing link in the permaculture model. Not only are mushrooms a protein-rich food source for humans, but the by-products of mushroom cultivation unlock nutrients for other members of the ecological community. The rapid return of nutrients back into the ecosystem boosts the life cycles of plants, animals, insects (bees), and soil microflora. The soil that fungi produce sustains, ultimately, all life. The complex activity of mushroom allies allows habitats to achieve degrees of biological intensity that are absent in fungally impoverished habitats. It is the fungal web that holds habitats together. Environments can best support populations of humans when nutrients are cycled back into the ecosystem using fungi as the link. Scientists are substantiating that mushroom mycelia proliferate when farmers employ the "no-till" method, fertilizing the soil as crop debris is decomposed.

What follows is a short list of the ways mushrooms can participate in permaculture. The numbers are keyed to the numbers in the accompanying illustration: The Stametsian Model for Permaculture with a Mycological Twist.

1. **Oyster Mushrooms:** Oyster mushrooms can be grown indoors on pasteurized corn stalks, wheat, rice, and rye straw, and a wide range of other materials including paper and pulp by-products. Soaking bulk substrates in cold water creates a residual "tea" that is a nutritious fertilizer and potent insecticide. Submerging the bulk substrate in hot water produces a different brew of "tea": a naturally potent herbicide. Oyster mushrooms can also be grown on hardwood stumps and logs. (Some varieties of Oyster mushrooms in the *P. pulmonarius* species complex naturally grow on conifer wood.) *Pleurotus* species thrive in complex compost piles, and are easy to grow outside with minimum care. The waste substrate from Oyster production is useful as fodder for cows, chickens, and pigs. Since half of the mass of dry straw is liberated as gaseous carbon dioxide, pumping this CO_2 from mushroom growing rooms into greenhouses to enhance plant production makes good sense. (Cultivators filter the airstream from the mushroom growing rooms so spores are eliminated.) Furthermore, the waste straw can be mulched into garden soils, not only to provide structure and nutrition, but also to reduce the populations of nematodes that are costly to gardeners and farmers. Oyster mushrooms have also been effective in breaking down hydrocarbon-based toxins (Thomas et al. 1998; Harrington, 1999). In a garden experiment matching plants with mushroom species, we found that an Oyster-like mushroom, *Hypsizygus ulmarius*, had a beneficial effect on neighboring plants while *Pleurotus ostreatus* did not (Pischl, 1999). Finding the right combinations of companion saprophytic fungi and plants is a new field of study with profound implications on

Oyster mushrooms fruiting from a stump inoculated with sawdust spawn.

improving crop yield while reducing the need for fertilizers.

2. **King Stropharia:** This mushroom is an ideal player in the recycling of complex wood debris and garden wastes, and thrives in complex environments. Vigorously attacking wood (sawdust, chips, twigs, and branches), the King Stropharia also grows in wood-free substrates, particularly soils supplemented with chopped straw. I have seen this mushroom flourish in gardens devoid of wood debris, benefiting the growth of neighboring plants. Acclimated to northern latitudes, this mushroom fruits when air temperatures range between 60–90°F (15–32°C) which usually translates to ground temperatures of 55–65°F (13–18°C).

For 6 weeks one summer our bees attacked a King Stropharia bed, exposing the mycelium to the air and suckling the sugar-rich cytoplasm from the wounds. A continuous convoy of bees could be traced,

Honey bees suckling on the mycelium of *Stropharia rugosoannulata*.

from morning to evening, from our beehives to the mushroom patch, until the bed of King Stropharia literally collapsed. When a report of this phenomenon was published in *Harrowsmith Magazine* (Ingle, 1988), bee-keepers across North America wrote me to explain that they had been long mystified by bees' attraction to sawdust piles. Now it is clear the bees were seeking the underlying sweet mushroom mycelium.

King Stropharia is an excellent edible mushroom when young. However, edibility quickly declines as the mushrooms mature. Fly larvae proliferate inside the developing mushrooms. In raising silver salmon, I found that when I threw mature mushrooms into the fish-holding tank, they would float. Fly larvae soon emerged from the mush-rooms, struggling for air. Soon the fish were striking the large mushrooms to dislodge the swollen larvae into the water where they were eagerly consumed. After several days of feeding mushrooms to the fish, the salmon would excitedly strike at the King Stropharia in anticipation of the succulent, squirming larvae as the mushrooms hit the water. Inadvertently, I had discovered that King Stropharia is a good base medium for gen-erating fish food, and that mushrooms may be an important element for enhancing fish-eries in riparian ecosystems.

Growing King Stropharia can have other beneficial applications in permaculture. King Stropharia depends upon bacteria for growth. At our farm we had a small herd of Black Angus cows. I established two King Stropharia beds at the heads of ravines that drained from the pastures onto a saltwater beach where my neighbor commercially cultivates oysters and clams. Prior to installing these mushroom beds, fecal col-iform bacteria seriously threatened the water quality. Once the mycelium fully permeated the sawdust/chip beds, downstream fecal bacteria were largely eliminated. The mycelium, in effect, became a micro-filtration membrane. I had discovered that by properly locating mushroom beds, "gray water" runoff could be cleaned of bacteria and nitrogen-rich effluent. Overall water quality improved. Massive mushrooms formed. (See page 42.) After three to four years, chunks of wood are totally reduced into a rich, peat-like soil, ideal for the garden. For nearly 8 years, I have continued to install King Stropharia beds in depressions leading into sensitive watersheds. Government agencies, typically slow to react to good ideas, have finally recognized the potential benefits of *mycofiltration*. Test plots are currently being implanted and monitored to more precisely determine the effects on water quality. If suc-cessful, I envision the widespread installation of mushroom beds into basins leading into rivers, lakes, and bodies of saltwater.

3. **Shiitake/Nameko/Lion's Manes:** Outdoors, inoculated logs can be partially buried or

LaDena Stamets sitting among 5-pound specimens of the King Stropharia, *Stropharia rugosoannulata*.

Fewer gourmet and medicinal mushrooms grow on coniferous woods. Nevertheless, Enokitake (*Flammulina velutipes*), Reishi (*Ganoderma lucidum*), Turkey Tail (*Trametes versicolor*), Clustered Woodlovers (*Hypholoma capnoides*), Chicken-of-the-Woods (*Laetiporus sulphureus*), Oyster (*Pleurotus pulmonarius*), and Cauliflower (*Sparassis crispa*) are good candidates for both conifer and broadleaf tree-stump decomposition. The above combination will provide fruitings of mushrooms from early spring through late winter. Most of these species will also produce crops on hardwoods.

5. **Shaggy Manes:** A cosmopolitan mushroom, Shaggy Manes (*Coprinus comatus*) grow in rich manured soils, disturbed habitats, in and around compost piles, and in grassy and gravel areas. Shaggy Manes are extremely adaptive and tend to wander. Shaggy Mane patches behave much like King

lined up in fence-like rows. Once the logs have stopped producing, the softened wood can be broken up, sterilized, and re-inoculated. Indoors, these mushrooms can be grown on sterilized substrates or on logs using the methods described in this book. Once the indoor substrates cease production, they can be recycled and re-inoculated with another mushroom, a process I call *species sequencing*. (See Chapter 22.) Later, the expired production blocks can be buried in sawdust or soil to elicit bonus crops outdoors.

4. **Polypores and Clustered Woodlovers:** Several species can be incorporated into the management of a sustainable multistage, complex "medicinal mushroom forest." Logs can be inoculated and buried or stumps can be impregnated. The greatest opportunities for stump culture are regions of the world where hardwoods predominate.

Three-pound specimen of King Stropharia.

Mushrooms, in this case Reishi (*Ganoderma lucidum*), can be grown on logs buried into sawdust or soil.

Stropharia and Morels, travelling great distances from their original site of inoculation in their search for fruiting niches.

6. **Morels:** Morels grow in a variety of habitats, from abandoned apple orchards and diseased elms to gravelly roads and streambeds. However, the habitat that can be reproduced easily is the burn-site. (See page 415 for techniques on Morel cultivation.) Burn sites, although increasingly restricted because of air pollution ordinances, are common among country homesteads. If a burn site is not possible, there are alternatives. The complex habitat of a garden compost pile also supports Morel growth. When planting cottonwood trees, you can introduce spawn around the root zones in hopes of creating a perennial Morel patch. Cultivators should note that Morels are fickle and elusive by nature compared to more predictable species like King Stropharia, Oyster, and Shiitake mushrooms.

7. **Mycorrhizal Species:** Mycorrhizal species can be introduced via several techniques. The age-old, proven method of satellite planting is probably the simplest. By planting young seedlings around the bases of trees naturally producing Chanterelles, King Boletes, Matsutake, Truffles, or other desirable species, you may establish satellite colonies by replanting the young trees after several years of association. For those landowners who inherit a monoculture woodlot of similarly aged trees, the permaculturally inclined steward could plant a succession of young trees so that, over time, a multicanopied forest could be reestablished.

8. **The Sacred Psilocybes:** In the Pacific Northwest of North America, the Psilocybes figure as some of the most frequently found fungi in landscaping bark and wood chips. These mushrooms share a strong affinity towards human activities—chopping wood, planting ornamentals, landscaping around

buildings, creating refuse piles. Many spiritually inclined cultivators view the establishment of *Sacred Psilocybe Mushroom Patches* as another step towards living in harmony within their ecosystem.

These are but a few mushroom species that can be incorporated into the permaculture model. Part of a larger, community-based permaculture strategy should also include *Mushroom Response Teams*

(MRTs) which could react quickly to catastrophic natural disasters—such as hurricanes, tornadoes, floods—in the profitable recycling of the enormous debris fields they generate.

Clearly, the use of mushrooms energizes permaculture to a level otherwise not attainable. I hope readers will develop these concepts further. When fungi are incorporated into these models, the ecological health of the whole planet benefits enormously.

At the end of the process, mushroom cultivators are also in the soil-generation business. Popular with gardeners for years, mushroom compost-soils add value to the role of using mushrooms within the permaculture model. For every ton of substrate used in the beginning of the process, cultivators should expect to create soil representing 10 to 20 percent of the original material. A bag of mushroom compost-soil weighing 30 pounds with a volume of 1.5 cubic feet retails for US $10.00.

Alder sawdust and chips.

Colonized with Shiitake mushroom mycelium.

After producing mushrooms and several months of composting, the wood is transformed into rich topsoil. See the following analysis.

Analysis of Post-Production, Composted Shiitake Blocks

PH	Phosphorus	Potassium	Organic Matter	Nitrogen Sources	
				NO$_3$-N	NH$_4$-N
7.0	224 um/g	547 ug/g	28%	29ug/g	18ug/g

CHAPTER 6

Materials for Formulating a Fruiting Substrate

The potential for recycling organic wastes via fungi seems unlimited. Surprisingly, many mushrooms thrive on base materials alien to their natural habitat. Although Oyster mushrooms are generally found in the wild on deciduous woods, these mushrooms grow well on many other materials besides hardwoods, including cereal straws, corncobs, seed hulls, coffee wastes, sugarcane bagasse, paper and pulp by-products, and numerous other materials.

Success increases if the base material is first reduced to an optimal structure and moisture—and if heat-treated—before inoculation. The fact that many mushrooms can cross over to other nonnative habitat components gives the cultivator tremendous latitude in designing habitats.

Materials for composing a mushroom substrate are diverse and plentiful. Because fungi decompose plant tissue, most homeowners can use by-products generated from gardening, landscaping, tree pruning, and even building projects. Homeowners who collect and pile debris have a perfect opportunity for growing a variety of gourmet and medicinal mushrooms. If a mushroom of choice is not introduced, a wild species from the natural environment will invade. The probability that one of these invading wild mushrooms would be a gourmet species is remote.

I prefer sawdust and wood debris as primary substrate components. Deciduous woods, especially those that decompose quickly, are best. These fast-rotting woods, being less able to resist disease, accelerate the mushroom life cycle. Alder, cottonwood, and poplar are favored over the more resistant, denser woods such as the oaks, maples, or ironwoods. Once the wood sawdust is gathered, additional materials are added to fortify the substrate. Three additional factors affect the suitability of a mushroom habitat: *structural composition*, *pH*, and *moisture*.

The selection of the substrate components is more critical for growing gourmet mushrooms indoors than for growing outdoors. Commercial cultivators prefer the controlled conditions of indoor cultivation whereas most home cultivators are more attracted to outdoor cultivation, *au naturale*. Outdoor mushroom beds can be more complex, composed of crude mixtures of components, whereas for indoor cultivation, the uniformity and consistency of the substrate are essential.

Raw Materials

Most by-products from agriculture and forestry industries can make up a base medium for mushroom culture. This base medium is commonly referred to as the "fruiting substrate." This primary material is often supplemented with a carbohydrate- and protein-rich additive to enhance yields. Here is a short list of the materials that can be recycled into mushroom production.

Wood wastes, paper products
Cereal straws and grain hulls
Coconut fibers
Corncobs
Coffee plants and waste
Tea leaves
Sugarcane bagasse
Banana fronds
Seed hulls (cottonseed, sunflower, and
    oil-rich seeds)
Almond, walnut, pecan, peanut hulls
Soybean meal, roughage (Okara), and soy waste
Artichoke waste
Cactus waste: saguaro and prickly pear,
    yucca, agave*

Suitable Wood Types: Candidate Tree Species

A vast variety of woods can be used for growing gourmet and medicinal mushrooms. Generally speaking, the hardwoods are more useful than the softwoods. Several wood types may not perform by themselves but when combined with other more suitable woods—and boosted with a nutritional supplement—will give rise to commercially viable

A new medicinal food? Oyster mushrooms fruiting from corncobs. Zusman et al. (1997) published evidence that this combination could prevent tumors.

crops. Recommended hardwoods are alders, birches, hornbeams, chestnuts, chinkapins, beeches, ashes, larches, sweetgums, tanoaks, cottonwoods, willows, ironwoods, walnuts, sweetgums, elms, and similar woods. Suggested softwoods are Douglas firs and hemlocks. Most other pines (ponderosa, lodgepole), cedars, and redwood are not easily degraded by mushroom mycelium. Aromatic hardwoods, such as eucalyptus, were once not recommended because some people became ill from eating its otherwise edible mushrooms (Arora, 1990). We now know that eucalyptus species are excellent candidates, and that the reports of ill effects were likely isolated cases. Cedars and redwoods are also not recommended as they decompose slowly due to their anti-rotting compounds, and hence stifle mushroom growth. All woods eventually succumb to fungal decomposition.

Woods other than those listed may prove to be satisfactory. Hence, experimentation is strongly encouraged. I find that the fast-growing, rapidly decomposing hardwoods are generally the best

---

* An Oyster mushroom, *Pleurotus opuntiae*, is native to prickly pear, agave, and yucca. Although I have not cultivated Oyster mushrooms on these cacti, they should serve well as a substrate base.

because they have greater ratios of starch-enriched sapwood to heartwood. These sugars encourage rapid initial growth, resulting in full colonization in a short period of time. *The key to successful cultivation is to match the skills of the cultivator with the right strain on the proper substrate under ideal environmental conditions.*

For outdoor log culture, disease-free logs should be selected from the forest in the winter or early spring. If you use sawdust and chips for indoor or outdoor cultivation, freshness counts—or else competitors may have already taken hold. Lumber mills, pulp mills, furniture manufacturers, and many other wood product companies generate waste usable to the mushroom cultivator. However, those industries that run mixed woods and do not separate their sawdust into identifiable piles, are not recommended as substrate suppliers. Cultivators face enough problems in their struggle to understand the different yields of each crop cycle. Hence, mixed wood sources are best avoided, if possible.

Red alder (*Alnus rubra*) is a "weed tree" in western Washington State of North America. Like poplars and cottonwoods, its penchant for valleys, wetlands, and open habitats encourages a prodigious growth rate. Many of these trees are common along roads where they foul telephone and electrical lines. A whole industry is dedicated to rendering these trees into chips, a fortuitous situation for mushroom cultivators. A matrix of smaller and larger particles can be combined to create an ideal habitat for mycelium. The smaller particles stimulate quick growth ("leap-off"). The larger particles encourage the mycelium to form thick, cord-like strands, called rhizomorphs, which forcibly penetrate through and between the cells. The larger chips become nutritional bases, or *fruiting platforms*, giving rise to super-large mushrooms. The larger chips also create air spaces, increasing respiration (Royse, 2000). This concept has been an overriding influence, steering my methods, and has resulted, for instance, in the large 5-pound specimens of *Stropharia rugosoannulata*, the Garden Giant, that is pictured in this book. A simple 50:50 mixture (by volume) of sawdust and chips, of varying particle sizes, provides

the best structure for the mushroom habitat. The substrate matrix concept will be explored in greater detail later on.

City and county utility companies are ready sources for wood chips. However, cultivators should avoid wood chips originating from trees along busy roadways. Automobile exhaust and leachate from the oil-based asphalt contaminate the surrounding soil with toxins, including lead and aluminum. Metals can be concentrated by the mushroom mycelium and transferred to mushrooms. Wood chips from county roads with little traffic are less prone to this heavy metal contamination. Obtaining sawdust and chips from large-diameter trees largely circumvents this problem. Sawmills and pulp chip companies provide the cleanest source of wood debris for substrate preparation.

Currently, the heavy metal concentrations taken up by most mushrooms found in forests of western North America are well below the standards set by the United States government for fish, for instance. However, air pollution is an increasing threat, and mushrooms collected near industrialized centers will have heavier concentrations of heavy metals. Analyses of mushrooms grown in China, California, and Washington State revealed that the Chinese mushrooms had the greatest aluminum, mercury, and lead concentrations, with Californian mushrooms next, and mushrooms grown in the less industrialized Olympic Peninsula of Washington had the least. With the phasing out of lead-based gasoline and the implementation of tougher environmental restrictions, pollution of wood sources may be ameliorated. (For more information on the concentration of metals and toxins, and their potential significance, consult Stijve, 1974, 1976, and 1992.) Many environmental service companies will analyze your product for a nominal fee, usually between US $50 and $125. If an analysis shows unusually high levels, the same specimens should be sent to an unrelated laboratory for confirmation. Please consult your Department of Agriculture, county extension agent, or comparable agency for any applicable threshold requirements.

List of Suitable Tree Species for the Cultivation of Gourmet and Medicinal Mushrooms*

Scientific Name	Common Name	Scientific Name	Common Name
Abies spp.	Red Fir	*Carpinus betulis*	European Hornbean
*Abies alba***	White fir	*Carpinus caroliniana*	American Hornbean
Acer spp.	Maples	*Carpinus fargesii*	
Acer negundo	Box Elder	*Carpinus japonica*	Japanese Hornbean
Acer rubrum	Red Maple	*Carpinus laxiflora*	
Acer macrophyllum	Big Leaf Maple	*Carpinus tschonoskii*	
Acer saccharum	Sugar Maple	*Carpinus turczaninowii*	
Alniphyllum fortunei		Carya spp.	Hickories
Alnus spp.	Alders	*Carya aquatica*	Water Hickory
Alnus alba	White Alder	*Carya cordiformis*	Bitternut Hickory
Alnus glutinosa	European Alder	*Carya glabra*	Pignut Hickory
Alnus incana	Gray Alder	*Carya texana*	Black Hickory
Alnus japonica	Japanese Alder	*Carya illinoensis*	Pecan
Alnus rubra	Red Alder	*Carya laciniosa*	Shellbark Hickory
Alnus serrulata	Hazel Alder	*Carya tomentosa*	Mockernut Hickory
Alnus tinctoria		*Carya ovata*	Shagbark Hickory
Alnus chinensis		Castanea spp.	Chestnuts
Arbutus spp.	Madrones	*Castanea crenata*	Japanese Chestnut
Arbutus menziesii	Pacific Madrone	*Castanea henryi*	
Betula spp.	Birches	*Castanea mollissima*	
Betula alleghaniensis	Yellow Birch	*Castanea sativa*	Spanish Chestnut
Betula dahurica		*Castanea sequinii*	
Betula lenta	Sweet Birch	Castanopsis spp.	Chinkapins
Betula nigra	River Birch	*Castanopsis*	
Betula papyrifera	Paper Birch	*accuminatissima*	
Betula pendula	European Birch	*Castanopsis argentea*	
Betula pubescens	Hairy Birch	*Castanopsis cerlesii*	
Carpinus	Hornbeans	*Castanopsis chinensis*	

*This list was compiled from trials and reports by the author, Pagony (1973), San Antonio (1981), Farr (1983), Gilbertson and Ryvarden (1986), Chang and Miles (1989), Przybylowicz and Donoghue (1989), and Kruger (1992). Some of the listed tree species are probable candidates due to their close affinities to species proven to be suitable for cultivation. I do not encourage the cutting of trees solely as a source of substrate for mushroom cultivation. The acquisition of wood materials from the forest should follow sustainable forest practices, and ideally be a "waste" product generated from other activities.

**Some races of *Ganoderma* (*G. oregonense* and *G. tsugae*), *Hypholoma* (*H. capnoides*), *Pleurotus* (*P. pulmonarius*), *Psilocybe* (*P. cyanescens* and allies), and *Stropharia* (*S. ruguso-annulata*) grow naturally on firs (i.e., *Abies* species). In general, these conifer-degrading mushroom species can also be cultivated on most hardwoods. However, few mushroom species native to hardwoods will fruit on most conifers.

Scientific Name	Common Name	Scientific Name	Common Name
Castanopsis chrysophylla	Golden Chinkapin	Liquidambar spp.	Sweetgums
Castanopsis cuspidata	Shii Tree	*Liquidambar formosana*	
Castanopsis fabri		*Liquidambar styraciflua*	
Castanopsis fargesii		*Liriodendron tulipfera*	Tulip Poplar
Castanopsis fissa		Lithocarpus spp.	Tanoaks
Castanopsis fordii		*Lithocarpus auriculatus*	
Castanopsis hickelii		*Lithocarpus calophylla*	
Castanopsis hystrix		*Lithocarpus densiflorus*	
Castanopsis indica		*Lithocarpus glaber*	
Castanopsis lamontii		*Lithocarpus lanceafolia*	
Castanopsis sclerophylla		*Lithocarpus lindleyanus*	
Castanopsis tibetana		*Lithocarpus polystachyus*	
Cornus spp.	Dogwoods	*Lithocarpus spicatus*	
Cornus capitata	Flowering Dogwood	*Mallotus lianus*	
Cornus florida	Flowering Dogwood	Ostyra spp.	Ironwoods (Hophornbean)
Cornus nuttallii	Pacific Dogwood		
Corylus spp.	Filberts	*Ostyra carpinfolia*	
Corylus heterophylla		*Ostyra virginiana*	
Distylium myricoides		Pasania	
Distylium racemosum		*Plaltycarya strobilacea*	
Elaeocarpus chinensis		Populus spp.	Cottonwoods and Poplars
Elaeocarpus japonicus		*Populus balsamifera*	Balsam Poplar
Elaeocarpus lancaefolius		*Populus deltoides*	Eastern Cottonwood
Engelhardtia chrysolepis		*Populus fremontii*	Fremont Cottonwood
Eriobotrya deflexa		*Populus grandidentata*	Bigtooth Aspen
Euphorbia royleana		*Populus heterophylla*	Swamp Cottonwood
Eurya loquiana		*Populus nigra*	Black Poplar
Fagus spp.	Beeches	*Populus tremuloides*	Quaking Aspen
Fagus crenata		*Populus trichocarpa*	Black Cottonwood
Fagus grandifolia	American Beech	Prosopis spp.	Mesquite
Fraxinus spp.	Ashes	*Prosopis juliflora*	Honey Mesquite
Fraxinus americana	White Ash	*Prosopis pubescens*	Screw Pod Mesquite
Fraxinus latifolia	Oregon Ash	Quercus spp.	Oaks
Fraxinus nigra	Black Ash	*Quercus acuta*	
Fraxinus pennsylvanica	Green Ash	*Quercus acutissima*	
Fraxinus velutina	Velvet Ash	*Quercus agrifolia*	California Live Oak
Juglans spp.	Walnut	*Quercus alba*	White Oak
Juglans nigra	Black Walnut	*Quercus aliena*	
Larix spp.	Larches	*Quercus bella*	
Larix laricina	Larch	*Quercus brandisiana*	
Larix lyalli	Subalpine Larch	*Quercus chrysolepis*	Canyon Live Oak
Larix occidentalis	Western Larch	*Quercus crispula*	

Scientific Name	Common Name	Scientific Name	Common Name
Quercus dentata		*Quercus variabilis*	
Quercus emoryi	Emory Oak	*Quercus virginiana*	Live Oak
Quercus fabri		Rhus spp.	
Quercus falcata	Southern Red Oak	*Rhus glabra*	Sumac
Quercus gambelii	Gambel Oak	*Rhus succedanea*	
Quercus garryana	Oregon White Oak	Robinia spp.	Black Locust
Quercus glandulifera		*Robinia neomexicana*	New Mexico Black Locust
Quercus glauca			
Quercus grosseserrata		*Robinia pseudoacacia*	Black Locust
Quercus kelloggi	California Black Oak	Salix spp.	Willows
Quercus kerii		*Salix amygdaloides*	Peachleaf Willow
Quercus kingiana		*Salix exigua*	Sandbar or Coyote Willow
Quercus lobata	California White Oak		
Quercus laurifolia	Laurel Oak	*Salix fragilis*	Crack Willow
Quercus lyrata	Overcup Oak	*Salix geyerana*	Geyer Willow
Quercus michauxii	Swamp Chestnut Oak	*Salic lasiandra*	Pacific Willow
Quercus mongolica		*Salix lasiolepis*	Arrow Willow
Quercus muehlenbergii		*Salix nigra*	Black Willow
Quercus myrsinae		*Salix scoulerana*	Scouler Willow
Quercus nigra	Water Oak	*Sapium discolor*	
Quercus nuttalli		*Sloanea sinensis*	
Quercus palustris	Pin Oak	Taxus spp.	Yews
Quercus phellos	Willow Oak	*Taxus brevifolia*	Pacific Yew
Quercus prunis		Ulmus spp.	Elms
Quercus rubra	Northern Red Oak	*Ulmus americana*	American Elm
Quercus semiserrata		*Ulmus campestris*	English Elm
Quercus serrata		*Ulmus laevis*	Fluttering Elm
Quercus spinosa		*Ulmus montana*	Mountain Elm

Cereal Straws

For the cultivation of Oyster mushrooms, cereal straws rank as the most usable base material. Wheat, rye, oat, and rice straw perform the best. Of the straws, I prefer wheat. Inexpensive, readily available, preserving well under dry storage conditions, wheat straw admits few competitors. Furthermore, wheat straw has a nearly ideal shaft diameter that selectively favors the filamentous cells of Oyster mushrooms. Chopped into 1- to 4-inch lengths,

the wheat straw needs only to be pasteurized by any one of several methods. The approach most easily used by home cultivators is to submerge the chopped straw in hot water (160°F, 71°C) for 1 to 2 hours, drain, and inoculate. First, fill a metal barrel with hot tap water and place a propane burner underneath. (Drums should be food grade quality. Do not use those that have stored chemicals.) A second method calls for the laying of straw onto a cement slab or plastic sheeting to a depth of no more than

24 inches. The straw is wetted and turned for 2 to 4 days, and then loaded into a highly insulated box or room. Steam is introduced, heating the mass to 160°F (71°C) for 2 to 4 hours. (See Chapter 18 for these methods.)

The semi-selectivity of wheat straw, especially after pasteurization, gives the cultivator a 2-week window of opportunity to establish the gourmet mushroom mycelium. Wheat straw is one of the most forgiving substrates with which to work. Outdoor inoculations of pasteurized wheat straw with grain spawn, even when the inoculations take place in the open air, have a surprisingly high rate of success for home cultivators.

Rye straw is similar to wheat, but coarser. Oat and rice straw are finer than both wheat and rye. The final structure of the substrate depends upon the diameter and the length of each straw shaft. Coarser straws result in a looser substrate, whereas finer straws provide a denser or "closed" substrate. A cubic foot of wetted straw should weigh around 20–25 pounds. Substrates with lower densities tend to perform poorly. The cultivator must design a substrate that allows air exchange all the way to the core. Substrate dynamics are determined by a combination of all these variables.

Oyster mushrooms growing from my previous book, *The Mushroom Cultivator.*

Paper Products: Newspaper, Cardboard, Books

Using paper products as a substrate base is particularly attractive to those wishing to grow mushrooms where sawdust supplies are limited, such as on islands and in desert communities. Paper products are made of lignin-cellulose fibers from pulped wood, and therefore support most of the wood-decomposing mushrooms described in this book. In recent years, most printing companies have switched to soybean-based inks, reducing or almost eliminating toxic residues. Since many large newspapers are recycling, data on toxin residues is readily available. (If the data cannot be validated, or are outdated, the use of such newsprint is not recom-

mended.) Since the use of processed wood fiber may disqualify a grower for state organic certification, U.S. cultivators should check with the Organic Certification Director at the State Department of Agriculture before venturing into the commercial cultivation of gourmet mushrooms on paper-based waste products. If these preconditions can be satisfied, the would-be cultivator can tap into an enormous stream of materials suitable for substrate composition.

Corncobs and Cornstalks

Corncobs (*sans* kernels) and cornstalks are conveniently structured for the rapid permeation of mycelium. Their cell walls and seed cavities provide a uniquely attractive environment for mycelium. Although whole corncobs can be used directly, grinding corncobs to 1–3 inches using hammer-mill type chipper-shredders creates more uniform-sized particles. After moistening, the corn roughage can be cooked for 2 to 4 hours at 160–180°F (71–82°C) to achieve pasteurization. If the kernels are still on the cob, sterilization may be necessary. Cornstalks, having a lower nutritional content, are less likely to become contaminated.

Coffee and Banana Plants

In the subtropical and tropical regions of Central and South America, the abundance of coffee and banana leaves has spurred mycologists to examine their usefulness in growing gourmet mushrooms. The difficulty in selecting any single plant material from warm, humid regions is the speed of natural decomposition due to the activity of competitors. The combination of high humidity and heat accelerates decomposition of everything biodegradable. Leaves must be dried, shredded, and stored in a manner not to encourage composting. Once weed fungi, especially black and green molds, begin to proliferate the suitability of these base materials is jeopardized. Mushrooms demonstrating commercial yield efficiencies on banana and coffee pulp are warm-weather strains of Oyster mushrooms, particularly *Pleurotus citrinopileatus*, *Pleurotus cystidiosus*, *Pleurotus djamor*, *Pleurotus ostreatus*, and *Pleurotus pulmonarius*. Reishi mushrooms (*Ganoderma lucidum*) grow well, and I suspect many other Polypores would also. For more information on the cultivation of Oyster mushrooms on coffee waste, please refer to Martinez-Carrera (1985, 1987, 1988) and Thielke (1989).

Sugarcane Bagasse

Sugarcane bagasse is the major waste product recovered from sugarcane harvesting and processing. Widely used in Hawaii and the Philippines by Oyster growers, sugarcane bagasse needs only pasteurization for cultivating Oyster mushrooms. Some Shiitake strains will produce on sugarcane residue, but yield efficiencies are low compared to wood-based substrates. Since the residual sugar stimulates mycelial growth and is a known trigger to fruiting, sugarcane residues are good complements to wood-based substrates.

Seed Hulls

Seed hulls, particularly cottonseed hulls, are perfect for their particle size and their ability to retain water. Their readily available form reduces the material handling costs. Buffered with 5–7% calcium sulfate and calcium carbonate, cottonseed hulls simply need wetting, pasteurization, and inoculation. Cottonseed hulls, on a dry weight basis, are richer in nitrogen than most cereal straws. Many Button cultivators consider cottonseed hulls a supplement to their manure-based composts. On unamended seed hulls, Oyster and Paddy Straw are the best mushrooms to grow.

Peanut shells have had little or no value except to mushroom growers. Peanut hulls are rich in oils and starch that stimulate mushroom growth. The shells must be chipped into 1/4- to 3/4-inch pieces, wetted, and pasteurized for several hours. Because peanut shells form subterraneously and are in ground contact, they should be thoroughly washed before pasteurization. Sterilization may be required if pasteurization is insufficient. The addition of 5% gypsum (calcium sulfate) helps keep the substrate loose and aerated. Oyster mushrooms, in particular, thrive on this material.

Soybean Roughage (Okara)

Okara is the main by-product of tofu and tempeh production. Essentially the extracted roughage of boiled soybean mash, Okara is perfectly suited for quick colonization by a wide variety of mushrooms, from the *Pleurotus* sp. to *Ganoderma lucidum*, even Morels. Several companies currently use Okara for generating mycelium for extraction and/or for producing flavorings.

All of the above-mentioned materials can be used to construct a base for mushroom production, outdoors or indoors. A more expansive list could include every primary by-product from agricultural or forestry practices. To the imaginative cultivator, the resources seem almost limitless.

Supplements

Supplementing the substrate can boost yields. A wide variety of protein-rich (nitrogenous) materials can be used to enhance the base substrate. Many of these are grains or their derivatives, like rice, wheat or oat brans, ground corn, etc. Many growers use grape pumice from wineries and spent barley from breweries as supplements. Supplementing a substrate, such as straw or sawdust, changes the number and the type of organisms that can be supported. Most of the raw materials used for growing the mushrooms listed in this book favor mushroom mycelium and are nitrogen-poor. Semi-selectivity is lost after nitrogen supplements are added, but ultimately mushroom yields improve. Therefore, when supplements are used, extra care is required to discourage contamination and insure success. Here good hygiene and good flow patterns to, from, and within the growing rooms are crucial. Supplementation of outdoor beds risks competition from contaminants and insects.

If you are supplementing a substrate, the sterilization cycle should be prolonged. Sterilization must be extended from 2 hours for plain sawdust at 15 psi (pounds per square inch) to 4 hours for the same sawdust supplemented with 20% rice bran.

Supplemented sawdust, straw, and compost substrates undergo *thermogenesis*, a spontaneous temperature increase as the mycelium and other organisms begin growing. If this naturally occurring biological combustion is not held in check, a plethora of molds awaken as the substrate temperature approaches 100°F (38°C). Below this threshold level, these organisms remain dormant,

soon being consumed by the mushroom mycelium. Although true sterilization has not been achieved, full colonization is often successful because the cultivator offsets the upward spiral of temperature. Simply spacing spawn bags or jars apart from one another, and lowering spawn room temperatures as thermogenesis begins, can stop this catalytic climb from starting. For many of the gourmet wood decomposers, a temperature plateau of 75° to 85°F (24° to 29°C) is ideal.

The following supplements can be added at various percentages of total dry mass of the bulk substrate to enhance yields. Please refer to Appendix 5 for their nutritional analyses.

cornmeal
cottonseed meal or flour
oat bran, oat meal
grape pumice
rice bran
rye grain
soybean meal and oil
spent grains from beer fermentation (barley
    and wheat)
vegetable oils
wheat grain, wheat bran
nutritional yeast

The nutritional composition of these supplements and hundreds of others are listed in Appendix 5. If you are using rice bran as a reference standard, the substitution of other supplements should be added according to their relative protein and nitrogen contents. For instance, rice bran is approximately 12.5% protein and 2% nitrogen. If soybean meal is substituted for rice bran, with its 44% protein and 7% nitrogen content, the cultivator should add roughly $\frac{1}{4}$ as much to the same supplemented sawdust formula. Until performance is established, the cultivator is better off erring on the conservative side than risking oversupplementation.

A steady supply of supplements can be cheaply obtained by recycling bakery waste, especially stale breads. A number of companies transform bakery by-products into a pelletized cattle feed, which also works well as inexpensive substitutes for many of the additives listed above.

Structure of the Habitat

Whichever materials are chosen for making up the substrate base, particular attention must be given to *structure*. Sawdust is uniform in particle size but, by itself, is not ideal for growing mushrooms outdoors. Fine sawdust is "closed," which means the particle size is so small that air spaces are soon lost due to compression. Closed substrates tend to become anaerobic and encourage weed fungi to grow.

Wood shavings have the opposite problem of fine sawdust. They are too fluffy. The curls have large spaces between the wood fibers. Mycelium will grow on shavings, but too much cellular energy is needed to generate chains of cells to bridge the gaps between one wood curl and the next. The result is a highly dispersed, cushion-like substrate capable of supporting vegetative mycelium, but incapable of generating mushrooms since substrate mass lacks density.

The ideal substrate structure is a mix of fine and large particles. Fine sawdust particles encourage mycelia to grow quickly. Interspersed throughout the fine sawdust should be larger wood chips (1–4 inches), which are used as concentrated islands of nutrition. Mycelium running through sawdust is often wispy in form until it encounters larger wood chips, whereupon the mycelium changes and becomes highly aggressive and rhizomorphic as it penetrates through the denser woody tissue. *The structure of the substrate affects the design of the mycelium network as it is projected*. From these larger island-like particles, abundant primordia form and can enlarge into mushrooms of great mass.

For a good analogy for this phenomenon, think of a campfire or a wood stove. When you add sawdust to a fire, there is a flare of activity that soon subsides as the fuel burns out. When you add logs or chunks of wood, the fire is sustained over the long term. Mycelium behaves in much the same fashion. Larger chips sustain long-term growth.

Although homogeneity in particle size is important at all stages leading up to and through spawn generation, the fruitbody formation period benefits from having a complex mosaic of substrate components. In many cases, there is a direct relationship between the complexity of habitat structure and health of the resulting mushroom bed. Optimizing the structure of a substrate is essential for good yields. If fine sawdust and wood chips (in the 1- to 4-inch range) are available, mixing 2 units of sawdust to every 1 unit of wood chips (by volume) provides a good platform for mycelial colonization. Garden shredders are useful in reducing piles of debris into the 1- to 4-inch chips. Many times roadside maintenance crews trim and chip brush from along roads and will give them to you for free.

A good substrate can be made from a mix of woody debris, chopped corncobs and cornstalks, stalks of garden vegetables, and vines of berries or grapes. When the base components are disproportionately too large or small, without connective particles, then colonization by the mushroom mycelium is hindered. The goal is create a substrate of complex particle size and structure, with air pockets and dense pockets of nutrition.

CHAPTER 7

Biological Efficiency: An Expression of Yield

ushroom strains vary in their ability to convert substrate materials into mushrooms as measured by a simple formula known as the "Biological Efficiency (B.E.) Formula" (originally developed by the White Button mushroom industry). This formula states that

- 1 pound of fresh mushrooms grown from 1 pound of dry substrate is 100% biological efficiency.

Considering that the host substrate is moistened to approximately 75% water content and that most mushrooms have a 90% water content at harvest, 100% B.E. is also equivalent to

- Growing 1 pound of fresh mushrooms for every 4 pounds of moist substrate, a 25% conversion of wet substrate mass to fresh mushrooms, or

- Achieving a 10% conversion of dry substrate mass into dry mushrooms.

Many of the techniques described in this book will give yields substantially higher than 100% B.E. Up to 50% conversion of wet substrate mass into harvestable mushrooms is possible. (I have succeeded in obtaining such yields with sets of Oyster, Shiitake, and Lion's Mane. Although 250% B.E. is exceptional, a good grower should operate within the 75–125% range.) When you consider the innate power of the mushroom mycelium to transform waste products into highly marketable delicacies, it is understandable why scientists, entrepreneurs, and ecologists are awestruck by the prospects of recycling with mushrooms.

Superior yields can be attained by carefully following the techniques outlined in this book, paying strict attention to detail, and matching these techniques with the right strain. The best way to improve yields is simply to increase the spawn rate. Often the cultivator's best strategy is *not* to seek the highest overall yield. The first, second, and third crops (or *flushes*) are usually the best, with each successive flush decreasing. For

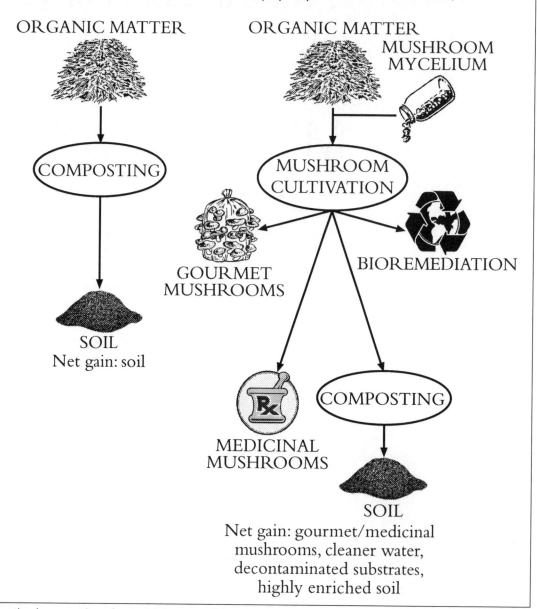

COMPOSTING OF ORGANIC MATTER VIA MUSHROOM CULTIVATION
vs. TRADITIONAL COMPOSTING METHODS
(Agricultural waste, wood debris, paper products, food waste, etc.)

ORGANIC MATTER

ORGANIC MATTER

MUSHROOM MYCELIUM

COMPOSTING

MUSHROOM CULTIVATION

GOURMET MUSHROOMS

BIOREMEDIATION

SOIL
Net gain: soil

Rx

COMPOSTING

MEDICINAL MUSHROOMS

SOIL
Net gain: gourmet/medicinal mushrooms, cleaner water, decontaminated substrates, highly enriched soil

Diagram showing comparison of composting raw materials versus sequencing with mushroom mycelia.

indoor cultivators, who are concerned with optimizing yield and crop rotation from each growing room, maximizing yield indoors may incur unacceptable risks. For instance, as the mycelium declines in vigor after several flushes, contaminants flourish. Future runs are quickly imperiled.

If you are growing on sterilized sawdust, I recommend removing the blocks after the third flush to a specially constructed, four-sided, open-air, netted growing room. This overflow or "yield recapture" environment is simply fitted with an overhead nozzle misting system. Natural air currents provide plenty of circulation. These recapture buildings give bonus crops and require minimum maintenance. Growers in Georgia and Louisiana have perfect climates for this alternative. Subtropical regions of Asia are similarly well suited. (For more information, see Appendix 1.)

Biological efficiency depends upon the stage of mushrooms at harvest. Young mushrooms ("buttons") are generally more delectable and store better. Yet if the entire crop is picked as buttons, a substantial loss in yield potential (B.E.) occurs. Mature mushrooms, on the other hand, may give the cultivator maximum biological efficiency, but also a crop with a very short shelf life and limited marketability.

Each species passes through an ideal stage for harvesting as it matures. Near maturity features are transformed through the reproportionment of cells without any substantial increase in the total weight of the mushroom. This is when the mushroom margins are decurved (pointing downwards) or slightly incurved, and well before spore generation peaks. Over time, cultivators learn the ideal stage for harvest.

As biological efficiency of mushroom production increases, the balance of by-products shifts. After composting the residual substrate, the net nitrogen appears to increase. In fact, overall mass is reduced during composting from loss of carbon-rich gases, especially CO_2, and proportionately affects the C:N ratio.

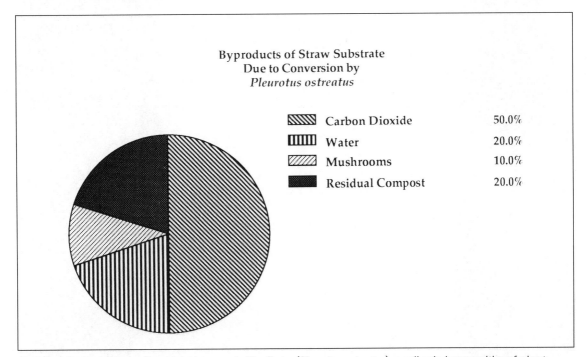

Byproducts of Straw Substrate Due to Conversion by *Pleurotus ostreatus*

▨	Carbon Dioxide	50.0%
▥	Water	20.0%
▨	Mushrooms	10.0%
■	Residual Compost	20.0%

Chart showing comparison of by-products generated by Oyster (*Pleurotus ostreatus*) mycelium's decomposition of wheat straw. (Adapted from Zadrazil, 1976.)

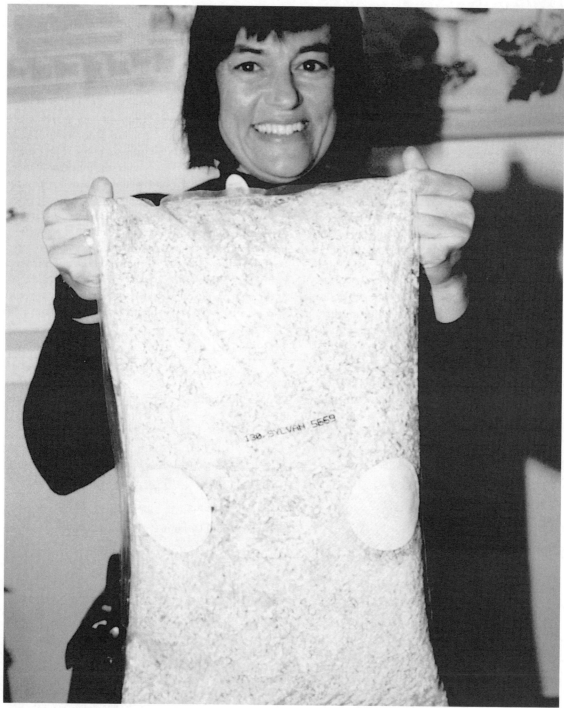

Lillian Stamets holding bag of commercial *Agaricus brunnescens* spawn.

Homemade vs. Commercial Spawn

Spawn is any form of mycelium that can be dispersed and mixed into a substrate. For most would-be cultivators, the easiest way to grow mushrooms is to buy spawn from a company and mix (*inoculate*) it into a substrate. Spawn can be purchased in a variety of forms. The most common forms are grain or sawdust spawn. Grain spawn is typically used by commercial cultivators to inoculate sterilized or pasteurized substrates. The White Button industry traditionally depends on highly specialized companies, often family-owned, which have made and sold spawn for generations.

Most amateurs prefer buying spawn because they believe it is easier than generating their own. This is not necessarily the case. Because spawn is a living organism, it exists precariously. Spawn remains in a healthy state for a very limited period of time, usually for no more than 2 months. Even under refrigeration, a noticeable decline in viability occurs. After this "honeymoon" period, spawn simply over matures for lack of new food to digest. The acids, enzymes, and other waste products secreted by the mushroom mycelium become self-stifling. As the viability of the spawn declines, predator fungi and bacteria exploit the rapidly failing health of the mycelium. Spawn, healthy at 4 weeks, becomes diseased at 8 weeks. A mycelial malaise seizes the spawn, slowing its growth once sown onto new substrates, and lowering yields. The most common diseases of spawn are competitor molds, bacteria, and viruses. Many of these diseases are only noticeable to experienced cultivators.

For the casual grower, buying commercial spawn is probably the best option. Customers of commercial spawn purveyors should demand: the date of inoculation, a guarantee of spawn purity, the success rate of other clients using the spawn, and the attrition rate due to shipping. Spawn shipped long distances often arrives in a state very different from its original condition. The result can be a customer relations nightmare.

I believe the wisest course is for commercial mushroom growers to generate their own spawn. The advantages of making your own spawn are

- *Quality control:* With the variable of shipping removed, spawn quality is better assured. The constant jostling breaks cells and wounds the spawn.

- *Proprietary strain development:* Cultivators can develop their own proprietary strains. The strain is the most important key to success. All other factors pale in comparison.

- *Reduction of an expense:* The cost of generating your own spawn is a mere fraction of the price of purchasing it. Rather than using a spawn rate of only 3–6% of the mass of the to-be-inoculated substrate, the cultivator can afford to use 10–12+% spawning rates.

- *Increase the speed of colonization:* With higher spawning rates, the window of opportunity for contaminants is significantly narrowed and yields are enhanced. Using the spawn as the vehicle of supplementation is far better than trying to boost the nutritional base of the substrate prior to inoculation.

- *Elimination of an excuse for failure:* When a production run goes awry, the favorite excuse is to blame the spawn producer, whether at fault or not. By generating your own spawn, you assume full responsibility. This forces owners to scrutinize the in-house procedures that led to crop failure. Thus cultivators who generate their own spawn tend to climb the learning curve faster than those who do not. They have only themselves to blame for failure.

- *Insight into the mushroom life cycle:* Mycelium has natural limits for growth. If the spawn is "overexpanded," i.e., it has been transferred too many times, vitality declines. Spawn in this condition grows slowly and often shows symptoms of genetic decline. A spawn producer can use his/her spawn at the peak of its vitality. Those who buy spawn from afar cannot.

CHAPTER 9

The Mushroom Life Cycle

When a collector finds mushrooms in the wild, the encounter is a mere coincidence, a "snapshot" in time of a far vaster process. The mushroom life cycle remains largely invisible to most mushroom hunters; not so to cultivators. The mushroom cultivator follows the path of the mushroom life cycle. Only at the completion of the mushroom life cycle, which may span weeks or months, do mushrooms appear, and then they occur but for a few days. The stages leading up to their appearance remain fascinating even to the most sagacious mycologists.

For mushrooms to survive in our highly competitive world, where legions of other fungi and bacteria seek common ecological niches, millions of spores are produced per mushroom. A small mushroom, only 4 inches across can produce more than a million spores! With the larger agarics, the numbers become astronomical. Since mushrooms reproduce through spores, the success of the mushroom life cycle depends upon the production of these spore-producing organs.

Each spore that is released possesses one-half of the genetic material necessary for the propagation of the species. Upon germination, a filamentous cell called a *hypha,* extends. Hyphae continue to reproduce mitotically. Two spores, if compatible, come together, fuse, and combine genetic material. The resulting mycelium is then described as being *binucleate* and *dikaryotic*. After this union of genetic material, the dikaryotic mycelium accelerates in its growth, again reproducing mitotically. Mated mycelium characteristically grows faster than unmated mycelium arising from single spores.

The mating of compatible spores is genetically determined. Most of the gourmet species are governed by two incompatibility factors (A and B). As a result, only subsets of spores are able to combine with one another. When spores germinate, several strains are produced. Incompatible strains grow away from each other, establishing their own

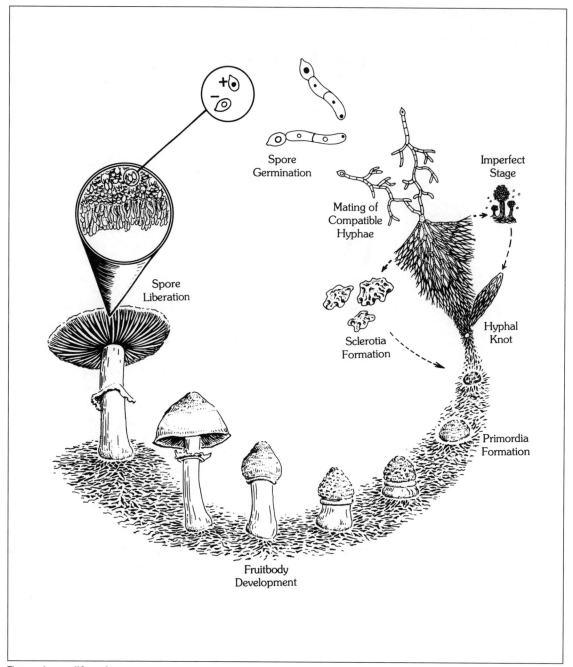

The mushroom life cycle.

Sclerotia of Teonancactl, *Psilocybe mexicana.*

Sclerotia of Zhu Ling, known to North Americans as the Umbrella Polypore, *Polyporus umbellatus.*

territorial domains. In this sense, spores from one mushroom can actually compete with one another for the same ecological niche. When compatible spores come together and mate, a much more vigorous form of mycelium fans out.

Each mushroom is like an island. From each center, populations of spores decrease with distance. When spores germinate, the mycelium grows radially, away from the site of origin. Waves of mycelia emanate outwards, and can cover thousands of acres. Overlaying mosaics of mycelium become a complex biological plateau supporting a wide variety of other organisms. Spores, taken up by the wind, or carried by insects and mammals, are dispersed to habitats well distant from the parent mushroom. By coincidence, different varieties of the same species meet and exchange genetic material. In the ever-changing ecological landscape, new varieties are favorably selected for and survive. The driving force of diversity is critical for successful adaptation to environments in flux.

Enzymes and acids are secreted by the mushroom mycelium into the surrounding environment, breaking down lignin-cellulose complexes into simpler compounds. The mushroom mycelium absorbs these reduced organic molecules as nutrients directly through its cell walls. After one mushroom species has run its course, the partially decomposed substrate becomes available to secondary and tertiary saprophytes who reduce it further. Ultimately, a rich soil is created for the benefit of plants and other organisms. (See page 44.)

As the mycelium expands, a web of cells is formed, collectively called the *mycelial network*. The arrangement of these cells is designed to optimally capture an ecological niche. Species differ in the manner by which the mycelial mat is projected. Initially, Morel mycelium throws a thinly articulated mycelial network. (Morel mycelium is the fastest growing of any mushroom I have seen.) Once a substantial territorial domain has been overrun, side branching of the mycelium occurs, resulting in a thickening of the mycelial mat.

With the approach of winter, the mycelial mat retreats to survive in specific sites. At this time, many mushrooms, both gilled and nongilled, produce *sclerotia*. Sclerotia are a resting phase in the mushroom life cycle. Sclerotia resemble a hardened tuber, wood-like in texture. (See below and page 63.) While in this dormant state, the mushroom species can survive inclement weather conditions like drought, fire, flooding, or other natural catastrophes. In the spring, the sclerotia swell with water and soften. Directly from the sclerotia, mushrooms emerge. Morels are the best-known mushrooms that arise from sclerotia. (See page 419.) By the time you find a mature Morel, the sclerotia from which it came will have disappeared.

Most saprophytic mushrooms produce a thick mycelial mat after spore germination. These types of mycelial mats are characterized by many crossovers between the hyphae. When two spores come together and mate, the downstream mycelium produces bridges between the cells, called *clamp connections*. Clamp connections are especially useful for culti-

vators who want to determine whether or not they have mated mycelium. Mycelium arising from a single spore lacks clamp connections entirely, and is incapable of producing fertile mushrooms.

As the mycelial network extends, several by-products are produced. Besides heat, carbon dioxide is being generated in enormous quantities. One study (Zadrazil, 1976) showed that nearly 50% of the carbon base in wheat straw is liberated as gaseous carbon dioxide in the course of its decomposition by Oyster mushrooms! Ten percent was converted into dried mushrooms; 20% was converted to proteins. Other by-products include a variety of volatile alcohols, ethylenes, and other gases.

While growing through a substrate, the mycelium is growing vegetatively. The vegetative state represents the longest phase in the mushroom life cycle. The substrate will continue to be colonized until physical boundaries prevent further growth or a biological competitor is encountered. When vegetative colonization ceases, the mycelium enters into temporary stasis. Heat and carbon dioxide evolution

Sclerotia of King Tuber Oyster, *Pleurotus tuberregium.*

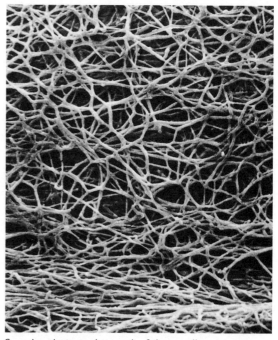

Scanning electron micrograph of the mycelium.

decline, and nutrients are amassed within the storage vestibules of the cells. This resting period is usually short-lived before entering into the next phase.

From the natural decline in temperature within the host substrate, as well as in response to environmental stimuli (water and humidity, light, drop in temperature, reduction in carbon dioxide, etc.), the mushroom mycelium is triggered into mushroom production. The mechanism responsible for this sudden shift from active colonization to mushroom formation is unknown, often being referred to as a "biological switch." (Some believe fruiting is triggered when food sources are consumed; I believe it is more a function of the maturity of the mycelial mat.) The mosaic of mycelium, until now homogeneously arranged, coalesces into increasingly dense clusters. Shortly thereafter—literally minutes with some species—these hyphal aggregates form into young primordia. In quick succession, the first discernible differentiation of the cap can be seen.

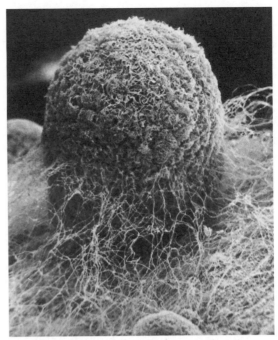

A miniature mushroom (primordium) emerges from the mycelial plateau.

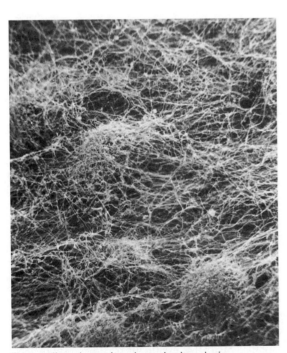

The mycelium clusters into dense circular colonies.

The period of primordia formation is one of the most critical phases in the mushroom cultivation process. Both mycelium and cultivator must operate as a highly coordinated team for maximum efficiency. Bear in mind that it is the mycelium that yields the crop; the cultivator is merely a custodian. The duration for primordia formation can be as short as 2 days as long as 14. If managed properly, the microscopic landscape, the *mycosphere*, will give rise to an even, high-density population of rapidly forming primordia. Visible to the naked eye, the mycelium's surface is punctuated with a latticework of valleys and ridges upon which moisture droplets continually form, rest, and evaporate. In the growing room, this period corresponds to 98–100% rH, or a condensing fog. Even in a fog, air currents have an evaporative effect, drawing moisture to the surface layer. The careful management of this mycosphere, with high oxygen, wicking, evaporation, and moisture replenishment combined with the effect of other environmental stimuli, results in a crescendo of primordia formation. Cultivators call these environmental stimuli collectively the *initiation strategy*.

Primordia, once formed, may rest for weeks, depending upon the species and the prevailing environment. In most cases, the primordia mature rapidly. Rhizomorphs, braided strands of large-diameter hyphae, feed the burgeoning primordia through cytoplasm streaming. The cells become multinucleate, accumulating genetic material. Walls, or *septae*, form, separating the nuclei, and the cells expand, resulting in an explosive generation of mushroom tissue.

As the mushroom enlarges, differentiation of familiar features occurs. The cap, stem, veil, and gills emerge. The cap functions much like an umbrella, safeguarding the spore-producing gills from wind and rain. Many mushrooms grow towards light. A study by Badham (1985) showed that with some

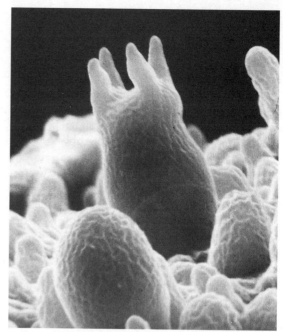

As the basidia mature, sterigmata project from the apex.

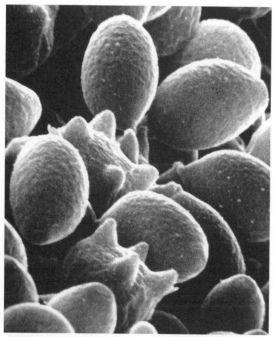

Scanning electron micrograph of young basidium.

mushroom species (i.e., *Psilocybe cubensis*), cap orientation is foremost affected by the direction of air currents, then by light, and finally by gravity. Beneath the cap, the gill plates radiate outwards from a centralized stem like spokes on a wheel.

Over the surface of the gills, an evenly dispersed population of spore-producing cells, called *basidia*, emerges. The basidia arise from a genetically rich, dense surface layer on the gill called the *hymenium*. The gill *trama* is nestled between the two hymenial layers and is composed of larger interwoven cells, which act as channels for feeding the hymenial layers with nutrients. When the mushrooms are young, few basidia have matured to the stage of spore release. As the mushrooms emerge, increasingly more and more basidia mature. The basidia are club-shaped, typically with four "arms" forming at their apices. These arms, the *sterigmata*, project upwards, elongating. In time, each tip swells to form a small globular cavity that eventually becomes a spore. (See page 67.)

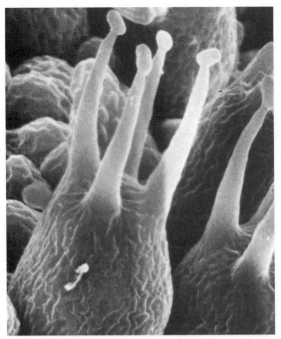

At each tip of the sterigmata, a spore cavity forms.

Initially, the young basidia contain two haploid, sexually paired nuclei. They fuse, in a process known as *karyogamy*, to form one diploid nucleus, containing a full complement of chromosomes. Immediately thereafter, meiosis, or reduction division, occurs resulting in four haploid nuclei. The haploid nuclei are elastic in form, squeezing up the sterigmata to be deposited in their continually swelling tips. Once they are residing in the newly forming spore cavity, the spore casing enlarges. Each spore is attached to the end of each sterigma by a nipple-like protuberance, called the *sterigmal appendage*. With many species, the opposite end of the spore is dimpled with a *germ pore*. (See page 68.)

The four spores of the basidia emerge diametrically opposite one another. This arrangement assures the highly viscous spores do not touch. Should a young spore come into contact with another before their outer shells harden, they fuse and development is arrested. The spores become

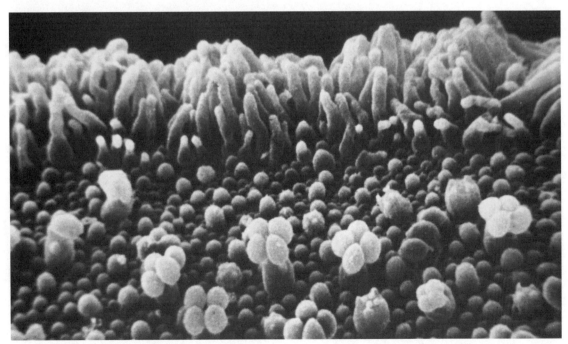

Populations of basidia evenly emerge from the gill plane.

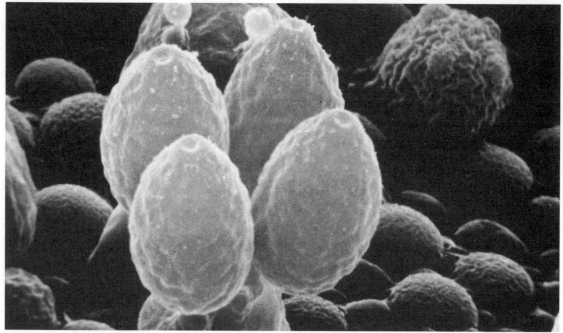

A fully mature basidium. Note germ pores at open ends of spores.

pigmented at maturity and are released simultaneously in sets of paired opposites.

The method of spore ejection has been a subject of much study and yet still remains largely a mystery. At the junction between the spore and the basidium's sterigma, an oily gas bubble forms and inflates. This bubble swells to capacity, explodes, ejecting spores with astonishing force. A recent study (Money, 1998) determined that the basidia shoot spores (ballistospores) with a force equal to 25,000 g's, 10,000 times more than space shuttle astronauts experience when escaping the earth's gravitational pull to achieve orbit! Soon thereafter, the remaining pair of spores is released. After ejaculation, the basidium collapses, making way for neighboring basidia, until now dormant, to enlarge. Successions of basidia mature, in ever-increasing quantities, until peaking at the time of mushroom maturity. The well-organized manner by which populations of basidia emerge from the plane of the gill optimizes the efficiency of spore dispersal. After peak spore

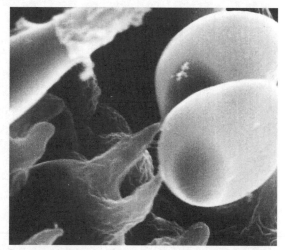

Two-spored basidium. Comparatively few mushrooms have two-spored basidia; most are four spored.

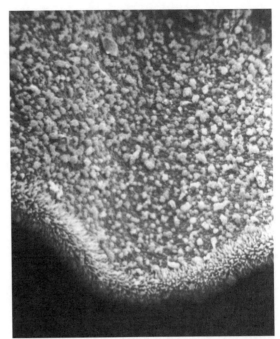

Scanning electron micrograph showing cross-section of gills.

Cheilocystidia as a band on edge of gill.

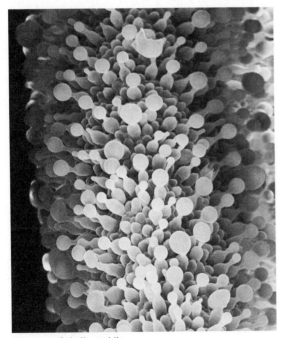

Close-up of cheilocystidia.

Pleurocystidia and basidia on surface of gill.

production, spores cover the gill face several layers deep, hiding the very cells from which they arose. With a Stropharia, this stage would correspond to a mushroom whose gills had become dark purple brown and whose cap had flattened. Spore release at this stage actually declines as the battery of basidia had been largely exhausted and/or because the basidia are rendered dysfunctional by the sea of overlying spores.

Sterile or nonspore-producing cells that adorn the gills are called *cystidia*. Cystidia on the edge of the gills are called *cheilocystidia*, while cystidia on the interior surface are called *pleurocystidia*. (See page 69.) The cystidia appear to help the basidia in their development. The extensive surface areas of the cheilocystidia cause the humidity between the gills to rise, thus preserving the hospitable moist micro-climate necessary for spore maturity. Typically, the pleurocystidia project well beyond the surface plane of basidia, and in doing, keep the gills from contacting one another. Should the gills touch, spore dispersal is greatly hampered. As the mushrooms mature, cystidia swell with metabolic waste prod-ucts. Often an oily droplet forms at their tips. The constant evaporation from these large reservoirs of metabolites is an effective way of purging waste by-products and elevating humidity. Species char-acteristically having pleurocystidia often have a high number of gills per millimeter of radial arc. In other words: more gills, more spores. The survival of the species is better assured.

Once spores have been discharged, the life cycle has come full circle. Mature mushrooms become a feasting site for small mammals (rodents such as squirrels, mice, etc.), large mammals (deer, elk, bear, humans), insects, gastropods (snails), bacteria, and other fungi. From this onslaught, the mushroom quickly decomposes. In due course, spores not transmitted into the air are carried to new ecologi-cal niches via these predators.

Many mushrooms have alternative, asexual life cycles. Asexual spores are produced much in the same manner as mold spores—on microscopic tree-like structures called *conidiophores*. Or, spores can

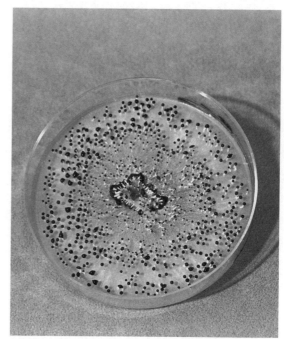

Mycelium characteristic of *Pleurotus cystidiosus* and allies, species possessing both a sexual and asexual life cycle. Black droplet structures contain hundreds of spores.

form embedded within the mycelial network. *Oidia, chlamydospores,* and *coremia* are some examples of asexual reproduction. In culture, these forms appear as "contaminants," confusing many cultivators. An excellent example is the Abalone Mushroom, *Pleurotus cystidiosus* and allies. (See above.) The advantage of asexual reproduction is it is not as biologically taxing as the process of mush-room formation. Asexual reproduction disperses spores under a broader range of conditions required than the rather stringent conditions for mushroom formation. In essence, asexual expressions represent shortcuts in the mushroom life cycle. Asexual spores usually regenerate into vigorously growing mycelium.

A cultivator's role is to assist the mycelium as it progresses through the life cycle by favorably con-trolling a multitude of variables. The cultivator seeks maximum mushroom production; the mycelium's goal is the release of the maximum number of spores through the formation of mushrooms. Both join in a biological partnership. But first, a strain from the wild must be captured. To do so, the cultivator must become skilled at sterile technique. And to be successful at sterile technique requires a basic understanding of the vectors of contamination.

Agaricus blazei, the Almond Portobello or Himematsutake mushroom. As the basidia mature, the gills darken with spores. New gill tissue is generated nearest to the stem.

CHAPTER 10

The Six Vectors of Contamination

Cultivating mushroom mycelium in a laboratory is tantamount to not cultivating contaminants. Diagnosing the *source* of contamination, and the *vector* or pathway through which contaminants travel is the key to a successful laboratory. Over the years, I have identified six distinct and separate vectors of contamination. If a contaminant arises in the laboratory, the cultivator should examine each vector category as being the possible cause of the problem. Through a process of elimination, the distressed cultivator can determine the vector through which the contaminants are spread. Once discovered, the vector can be closed, and the threat eliminated. If one vector is open, then a multitude of contaminants pass through it, often confounding the diagnoses of inexperienced cultivators.

The principal vectors of contamination are

1. The Cultivator

2. The Air

3. The Media

4. The Tools

5. The Inoculum

6. Mobile Contamination Units (MCUs)

The overriding coefficients affecting each vector are *the number of contaminants* and *the exposure time*. The more of each, the worse the infestation. This book does not go into detail as to the identity of the common contaminants. However, my previous book, *The Mushroom Cultivator* (1983), coauthored with Jeff Chilton, has extensive chapters on the identity of the molds, bacteria, and insects. The reader is encouraged to refer to that manual for the identification of contaminants. All contaminants are preventable by eliminating the six vectors of contamination. If you have difficulty determining the vector of contamination, or a solution to a problem, please refer to Chapter 25: Cultivation Problems and Their Solutions.

You, the Cultivator: The human body teems with populations of microorganisms. Diverse species of fungi (including yeast), bacteria, and viruses call your body home. When you are healthy, the populations of these microorganisms achieve an equilibrium. When you are ill, one or more of these groups proliferate out of control. Hence, unhealthy people should not be in a sterile laboratory, lest their disease organisms escape and proliferate to dangerous proportions.

Most frequently, contaminants are spread into the sterile laboratory via touch or breath. Also, the flaking of the skin is a direct cause. Many cultivators wear gloves to minimize the threat of skin-borne contaminants. I find laboratory gloves uncomfortable and prefer to wash my hands frequently, every 20 or 30 minutes with antibacterial soap. Additionally my hands are disinfected with 80% isopropyl alcohol immediately before inoculations, and every few minutes throughout the procedure.

The Air: Air can be described as a sea of microorganisms, hosting invisible contaminants, that directly contaminate sterilized media once exposed. Many particulates remain suspended. When a person walks into the laboratory, they not only bring in contaminants that will become airborne, but their movement disturbs the contaminant-laden floor, re-releasing contaminants into the lab's atmosphere.

Several steps can prevent this vector of contamination. One rule of thumb is to always have at least three doors prior to entry into the sterile laboratory from the outside. Each room or chamber shall, by default, have fewer airborne particulates the nearer they are to the laboratory. Second, by positive-pressurizing the laboratory with an influx of air

through micron filters, the airstream will naturally be directed against the entering personnel. (For the design of the air system for a laboratory, see Appendix 2.)

For those not installing micron filters, several alternative remedies can be employed. Unfortunately, none of these satisfactorily compare with the efficiency of micron filters. These "still-air" laboratories make use of aerosol sprays—either commercial disinfectants like Pine-Sol disinfectant or a dilute solution of isopropanol or bleach. The cultivator enters the work area and sprays a mist high up in the laboratory, walking backwards as they retreat. As the disinfecting mist descends, airborne particulates are trapped, carrying the contaminants to the floor. After a minute or two, the cultivator reenters the lab and begins his routine. (Note that you should not mix disinfectants—especially bleach and ammonia. Furthermore, this method can potentially damage your lungs or exposed mucous membranes. Appropriate precautions are strongly recommended.)

Without the exchange of fresh air, carbon dioxide levels will naturally rise from out-gassing by the mushroom mycelium. As carbon dioxide levels elevate, contaminants are triggered into growth. An additional problem with heavily packed spawn rooms is that with the rise of carbon dioxide, oxygen levels proportionately decrease, eventually asphyxiating the laboratory personnel. Unless the air is exchanged, the lab becomes stifling and contamination-prone. Since the only way to exchange air without introducing contaminants is by filtering, the combination of fans and micron filters is the only recourse.

Other cultivators use ultraviolet lights, which interfere with the DNA replication of all living organisms. UV lamps are effective when the contaminants are directly exposed. However, since shadowed areas are fully protected from UV exposure, contaminants in those regions remain unaffected. I disdain the use of UV in favor of the micron filter alternative. However, many others

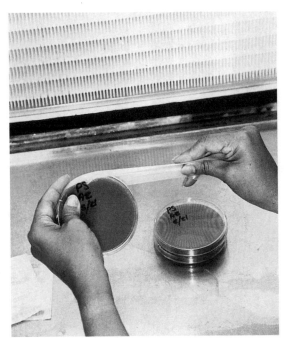

Using an elastic film to seal the top and bottom of petri dishes. This eliminates the chance of airborne contamination entering during incubation.

prefer their use. Note that the lab door should be electrically switched to the UV light so that the lamp turns off at entry. Obviously, exposure to UV light is threatening to human health, potentiating skin cancer and damage to the cornea of the eye.

Frequently, the vector of airborne contamination is easy to detect because of the way it forms on petri dishes. Airborne contaminants enter a petri dish either at the time the lid is opened (during pouring or inoculation) or during incubation. When the dish is opened, airborne contamination can spread evenly across the face of nutrient media. During incubation, contaminants creep in and form along the inside periphery of the petri dish. This latter occurrence is most common with laboratories with marginal cleanliness. A simple solution is to tape together the top and bottom of the petri dish directly after pouring and/or inoculation using elastic wax film.

(Parafilm is one brand.) Plastic, stretchable kitchen wraps available in most grocery stores also can be used. These films prevent entry of contaminant spores that can occur from the fluctuation of barometric pressure due to natural changes in weather patterns.

One helpful practice in eliminating each vector of contamination at the source is to leave containers of media uninoculated. For instance, the cultivator should always leave some culture dishes uninoculated and unopened. These "blanks" as I like to call them, give the cultivator valuable information as to which vector of contamination is operating. *At every step in the cultivation process, "blanks" should be used as controls.*

The air in the growing rooms does not require the degree of filtration needed for the laboratory. For mushroom cultivators, cleaning the air by water misting is practical and effective. (Rain is nature's best method of cleansing the air.) This cultivator's regimen calls for the spraying down of each growing room twice a day. Starting from the ceiling and broadcasting a spray of water back and forth, the floor is eventually washed towards the center gutter. The room feels clean after each session. Each washdown of a 1,000-square-foot growing room takes about 15 minutes. This regimen is a significant factor in maintaining the quality of the growing room environment.

The Media: Often the medium upon which a culture is grown becomes the source of contamination. Insufficient sterilization is usually the cause. Standard sterilization times for most liquid media is only 15 to 20 minutes at 15 psi or 250°F (121°C). However, this exposure time is far too brief for many of the endospore-forming bacteria prevalent in the additives currently employed by many cultivators. I recommend at least 40 minutes at 15 psi for malt extract or potato dextrose agars. If creating soil extracts, you must soak the soil for at least 24 hours, and then subject the extracted water to a minimum of 2 hours of sterilization. Indeed, soil extracts are resplendent with enormous numbers of contaminants. Because of the large initial populations, do not be surprised if some contaminants survive this prolonged sterilization period. Should they persist, then sterilizing the extracted water first, and then resterilizing it with standard malt sugar additives, is recommended. Clearly, sterilization efficiency is best achieved when the medium has a naturally low contamination content. (See "Preparing Nutrified Agar Media" in Chapter 12.)

A good habit for all laboratory managers is to leave a few samples from each sterilization cycle uninoculated. *Not inoculating a few petri dishes, grain jars, and sawdust/bran bags and observing them for a period of two weeks can provide valuable information about the vectors of contamination.* These quality control tests can easily determine whether or not the medium is at fault or there has been a failure in the inoculation process. Under ideal conditions, the uninoculated units should remain contamination-free. If they contaminate within 48 to 72 hours, this is usually an indication that the media or containers were insufficiently sterilized. If the containers are not hermetically sealed, and contaminants occur near the end of three weeks, then the contamination is probably endemic to the laboratory, particularly where these units are being stored. Under ideal conditions, in a perfect universe, no contamination should occur no matter how long the uninoculated media is stored.

Many researchers have reported that sawdust needs only to be sterilized for 2 hours at 15 psi to achieve sterilization. (Royse et al. 1990; Stamets and Chilton, 1983.) However, this treatment schedule works for small batches. When you are loading an autoclave with hundreds of tightly packed bags of supplemented sawdust, sterilization for this short period will certainly lead to failure.

In the heat-treatment of bulk substrates, absolute sterilization is impractical. Here, sterilization is more conceptual than achievable. The best one can hope is that contaminants in the sawdust have been reduced to a level as to not be a problem, i.e., within

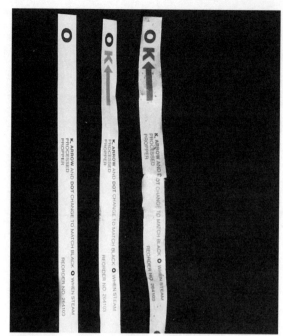

Heat-sensitive sterilization indicator strips showing no sterilization, partial sterilization, and complete sterilization.

the normal time frame needed for the mushroom mycelium to achieve thorough colonization. Again, the time period needed is approximately 2 weeks. *Should colonization not be complete in two weeks, the development of contaminants elsewhere in the substrate is not unusual.* Of course, by increasing the spawn rate, colonization is accelerated, and the window of opportunity favors the mushroom mycelium. The recommended sterilization times for supplemented sawdust are described in Chapter 17. Badham (1988) found that sterilization of supplemented sawdust under pressure for 4 hours at 19 psi was functionally similar (in terms of contamination reduction, growth rate, and yield of Shiitake) to high-temperature pasteurization at 190–194°F (88–90°C) for 14 hours at atmospheric pressure (1 psi). Remote sensing thermometers, placed at various depths, are used to determine temperature. When the coolest probe reads 190°F (88°C), steam is continuously supplied for a minimum of 12 hours, preferably 14 to 16 hours, depending on substrate mass.

Readers are forewarned not to jump to conclusions when comparing the heat-treatment sufficient for small batches and extrapolating this procedure to large batches. Since heat penetration varies with each substrate material's density, and is co-dependent on the moisture content, the use of sterilization indicator strips is recommended to confirm that sterilization has actually occurred.

Yet another limiting factor is that media biochemically change, potentially generating toxins to mycelial growth. Should malt agar be cooked for 2 to 3 hours at 18 psi, the resulting medium changes into a clear amber liquid as sugars have been reduced. Under these conditions, cultivators say the medium has "caramelized" and generally discard the medium and make up a new batch. Contaminants won't grow on this medium; nor does most mushroom mycelia. The cultivator is constantly faced with such dilemmas. What makes a good cultivator is one who seeks the compromises that lead most quickly to colonization and fruitbody production.

The Tools: In this category, all tools of the trade are included—from the scalpel to the pressure cooker to the media vessels. Insufficient sterilization of the tools can be a direct vector since contact with the media is immediate. Flame-sterilizing scalpels is the preferred method over topical disinfection with alcohol or bleach. However, the latter is used widely by the plant tissue culture industry with few problems.

If using a pressure cooker for sterilizing media and other tools, many forget that although the interior of the vessel has been sterilized, for all practical purposes, the outside of the vessel has not been. Contaminants can be easily picked up by the hands of the person handling the pressure cooker and redistributed to the immediate workstation—all

the more reason one should disinfect before beginning transfers.

The Inoculum: The inoculum is the tissue that is being transferred, whether this tissue is a part of a living mushroom, mycelium from another petri dish, or spores. Bacteria and molds can infect the mushroom tissue and be carried with it every time a transfer is made. Isolation of the inoculum from the mushroom mycelium can be frustrating, for many of these contaminant organisms grow faster than the newly emerging mushroom mycelium. Cultivators must constantly "run" or transfer their mycelium away from these rapidly developing competitors. Several techniques can purify contaminated mycelium.

Mobile Contamination Units (MCUs): Mobile Contamination Units are organisms that carry and spread contaminants within the laboratory. These living macro-organisms act as vehicles spreading contaminants from one site to another. They are especially damaging to the laboratory environment, as they are difficult to isolate. Ants, flies, mites, and in this author's case, small bipedal offspring (i.e., children) all qualify as potential MCUs. Typically, an MCU carries not one contaminant, but several.

Mites are the most difficult of these MCUs to control. Their minute size, their preference for fungi (both molds and mushroom mycelium) as food, and their penchant for travel, make them a spawn manager's worst nightmare come true. Once mite contamination levels exceed 10%, the demise of the laboratory is only one generation away. The only solution, after the fact, is to totally shut down the laboratory. All cultures must be removed, including petri dishes, spawn jars, etc. The laboratory should then be thoroughly cleansed several times. I use a 10% household bleach solution. The floor, walls, and ceiling are washed. Two buckets of bleach solution are used—the first being the primary reservoir, the second for rinsing out the debris collected in the first wash-down. The lab is locked tight each day after wash-down. By thoroughly cleansing the lab three times in succession, you can eliminate or subdue the problem of mites to manageable levels. To prevent their recurrence, regenerate mycelia from carefully selected stock cultures.

I have discovered that "decontamination mats," those labs use at door entrances to remove debris from footwear, are ideal for preventing cross-contamination from mites and similarly pernicious MCUs. Stacks of petri dishes are placed on newly exposed sticky mats on a laboratory shelf with several inches of space separating them. These zones of isolation, with culture dishes incubating upon a highly adhesive surface, make the migration of mites and other insects a most difficult endeavor. The upper sheet is removed every few weeks to expose a fresh, clean storage plane for new cultures.

All of these vectors are universally affected by one other variable: *time of exposure*. The longer the exposure of any of the aforementioned vectors of contamination, the more significant their impact. Good laboratory technicians are characterized not only by their speed and care, but also by their rhythm. Transfers are done in a systematically repetitive fashion. Controlling the time of exposure can have a drastic impact on the quality of laboratory technique.

Storing petri dish cultures on "sticky mats" prevents cross-contamination from mites.

CHAPTER 11

Mind and Methods for Mushroom Culture

Sterile tissue culture has revolutionized the biological sciences. For the first time in the history of human evolution, select organisms can be isolated from nature, propagated under sterile conditions in the laboratory, and released back into the environment. Since a competitor-free environment does not exist naturally on this planet*, an artificial setting is created—the laboratory—in which select organisms can be grown in mass.

Louis Pasteur (1822–1895) pioneered sterile technique by recognizing that microorganisms are killed by heat, most effectively by steam or boiling water. Tissue culture of one organism—in absence of competitors—became possible for the first time. By the early 1900s growing organisms in pure culture became commonplace. Concurrently, several researchers discovered that mushroom mycelium could be grown under sterile conditions. However, the methods were not always successful. Without benefit of basic equipment, efforts were confounded by high levels of contamination and only after considerable effort was success seen. Nevertheless, methods slowly evolved through trial and error.

The monumental task of creating a sterile environment has been difficult, until recently. The invention of high-efficiency particulate air filters (called HEPA filters) has made sterile tissue culture practicable to all. *In vitro* ("within glass") propagation of plants, animals, fungi, bacteria, and protozoa became commercially possible. Today, the HEPA (High Efficiency Particulate Air) filter is by far the best filtration system in use, advancing the field of tissue culture more so than any other

---

* Scientists have recently discovered a group of heat-resistant bacteria thriving in the fumaroles of submerged, active volcanoes. These bacteria thrive where no other life forms live.

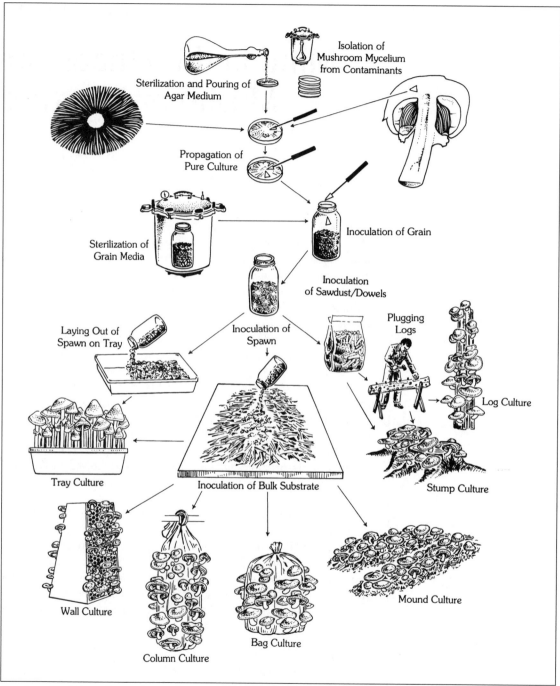

Overview of techniques for growing mushrooms.

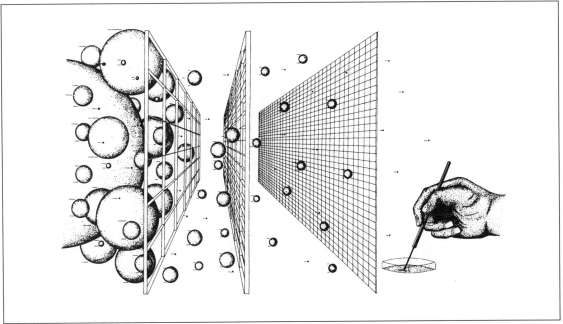

Diagrammatic representation of the effectiveness of filtration media. Dirty air is first filtered through a coarse prefilter (30% @ 10 μ), then an electrostatic filter (99% @ 1 μ), and finally through a High Efficiency Particulate Air (HEPA) filter (99.99% @ .3 μ). Only some free-flying endospores of bacteria and viruses pass through.

invention.** When air is forcibly pressed through these filters, all contaminants down to .3 microns (μ) are eliminated with a 99.99% efficiency. This means only 1 of every 10,000 particulates exceeding .3 μ pass through. For all practical purposes, a sterile wind is generated downstream from the filter. The cultivator works within this airstream. This unique environment calls for unique techniques. A different set of rules now presides, the violation of which invites disaster.

Sterile tissue culture technique fails if it solely relies on mechanical means. Sterile tissue culture

---

** Many types of filtration systems are available. Ionizers, for instance, are insufficient in their air-cleaning capacity. For a comparison of filtration systems, which rates HEPA filtration as the best, please refer to Consumer Reports, October 1992, pg. 657. A new generation of micron filters, the ULPA filters screen out particulates down to .1 micron with a 99.999% efficiency. This means only 1 particle measuring .1 micron in diameter of every 100,000 flows through the filtration medium.

technique is also a philosophy of behavior, ever-adjusting to ever-changing circumstances. Much like a martial artist, the cultivator develops keen senses, a prescience, to constantly evaluate threats to the integrity of the sterile laboratory. These enemies to sterile culture are largely invisible and are embodied within the term "contaminant."

A contaminant is anything you don't want to grow. Classically, *Penicillium* molds are contaminants to mushroom culture. However, if you are growing Shiitake mushrooms, and a nearby fruiting of Oyster mushrooms generates spores that come into the laboratory, then the Oyster spores would be considered the "contaminant." So the definition of a contaminant is a functional one—it being any organism you *don't* want to culture.

The laboratory environment is a sanctuary, a precious space, to be protected from the turmoils of the outside world. Maintaining the cleanliness of a laboratory is less work than having to deal with the aftermath wreaked by contamination. Hence, contaminants, as soon as they appear, should be immediately isolated and carefully removed so neighboring media and cultures are not likewise infected.

Overview of Techniques for Cultivating Mushrooms

The stages for cultivating mushrooms parallel the development of the mushroom life cycle. The mass of mycelium is exponentially expanded millions of times until mushrooms can be harvested. Depending upon the methodology, as few as two petri dishes of mushroom mycelium can result in 500,000–1,000,000 pounds of mushrooms in as short as 12 weeks! If any contaminants exist in the early stages of the spawn production process, they will likewise be expanded in enormous quantities. Hence, the utmost care must be taken, especially at the early stages of spawn production. Several tracks lead to successfully growing mushrooms. For indoor, high intensity cultivation, three basic steps are required for the cultivation of mushrooms on straw (or similar material) and for the cultivation of mushrooms on supplemented sawdust.

Within each step, several generations of transfers occur, resulting in a millionfold increase in mycelial mass. The sheer biological force of cellular activity boggles the mind. Scientists have estimated that this intense myceliation results in so many hyphal cells that more than a mile of mycelium impregnates every cubic inch. For me, the goal of the cultivator is to surf the mycelial wave as it crests. The mycelial momentum will carry the process forward so quickly that contaminants have little chance of catching up. This velocity of colonization nullifies any inherent imperfections at each step that would otherwise confound success.

Culturing Mycelium on Nutrified Agar Media: Mushroom mycelium is first grown on sterilized, nutrified agar media in petri dishes and/or in test tubes. Once pure and grown out, cultures are transferred using the standard cut-wedge technique. Each culture incubating in 100 × 15-mm petri dish can inoculate 10 quarts (liters) of grain spawn. (See page 83.) If the mycelium is chopped in a high-speed stirrer and diluted, one petri dish culture can

effectively inoculate 40–100 quarts (liters) of sterilized grain. These techniques are fully described in the ensuing pages.

Producing Grain Spawn: The cultures in the petri dishes can be expanded by inoculating sterilized grain housed in bottles, jars, or bags. Once grown out, each jar can inoculate 10 (range: 5 to 20) times its original mass for a total of three generations of expansions. Grain spawn can be used to inoculate pasteurized straw (or similar material) or sterilized sawdust. Grain spawn is inoculated into sawdust, straw, etc. at a rate between 3 and 15% (wet mass of spawn to dry mass of substrate).

Producing Sawdust Spawn: Sawdust spawn is inoculated with grain spawn. Sawdust spawn is best used to inoculate a "fruiting substrate," typically logs or supplemented sawdust formulas. One 5-pound bag of sawdust spawn can effectively inoculate 5 to 20 times its mass, with a recommended rate of 10. Sawdust-to-sawdust transfers are common when growing Shiitake, Nameko, Oyster, Maitake, Reishi, or King Stropharia. Once grown out, each of these bags can generate 5 to 10 more sawdust spawn bags (S_1 to S_2). No more than two generations of expansion are recommended in the production of sawdust spawn. Spawn can be added to bulk substrates at rates, varying with the species and inclinations of the grower, equivalent to 5–20% (i.e., wet mass of spawn to dry mass of substrate).

Formulating the Fruiting Substrate: The fruiting substrate is the platform from which mushrooms arise. With many species, this is the final stage where mushrooms are produced for market. The formulas are specifically designed for mushroom production and are often nutrified with a variety of supplements. Some growers on bulk substrates expand the mycelium one more time, although I hesitate to recommend this course of action. Oyster cultivators in Europe commonly mix fully colonized, pasteurized straw into 10 times more pasteurized straw, thus attaining a tremendous amount of mycelial mileage. However, success occurs only if the utmost purity is maintained. Otherwise, the cultivator risks losing everything in the gamble for one more expansion.

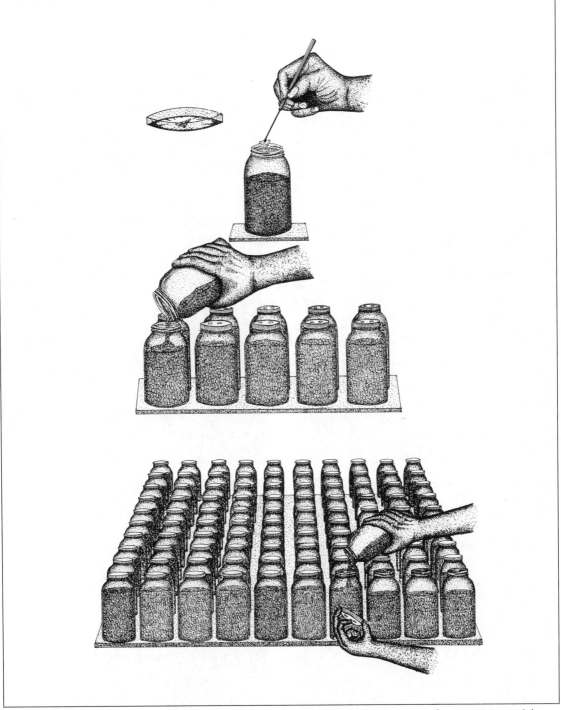

Creating and expanding grain spawn. Sterilized grain is first inoculated from agar-grown cultures. Once grown out, each jar can inoculate 10 times more grain every two weeks.

This final substrate can be amended with a variety of materials to boost yields. With Shiitake, supplementation with rice bran (20%), rye flour (20%), soybean meal (5%), molasses (3–5%), or sugar (1% sucrose) significantly boosts yields by 20% or more. (For more information on the effects of sugar supplementation on Shiitake yields, see Royse et al. 1990.)

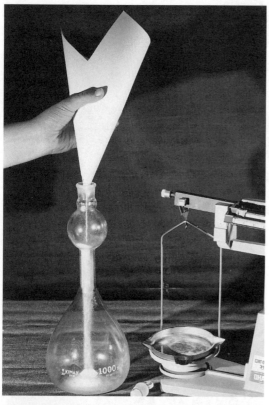

Preparing and pouring nutrified agar media into petri dishes.

CHAPTER 12

Culturing Mushroom Mycelium on Agar Media

Preparing Nutrified Agar Media

Many formulations have been developed for the cultivation of mushrooms on a semisolid agar medium. Agar is a seaweed-derived compound that gelatinizes water. Nutrients are added to the agar/water base that, after sterilization, promote healthy mushroom mycelium. The agar medium most commonly used with the greatest success is a fortified version of Malt Extract Agar (MEA). Other nutrified agar media that I recommend are Potato Dextrose Agar (PDA), Oatmeal Agar (OMA), and Dog Food Agar (DFA).

By supplementing these formulas with yeast and peptone, essential vitamins (such as thiamin) and amino acids (proteins) are provided. These supplements not only greatly stimulate the rate of growth, but the quality of the projected mycelial mat. Most agar media are simple and quick to prepare. What follows is some of my favorite nutrified agar recipes—of the 500 or more published.

Sterilize one of the following media for 45 minutes at 15 psi (= 1 kg/sq. cm.) in a pressure cooker.

Malt Extract, Yeast Agar
1,000 milliliters (1 liter) water
20 grams agar agar
20 grams barley malt sugar
2 grams yeast (nutritional)
1 gram peptone (optional, soybean derived)

(The above medium is abbreviated as MYA. With the peptone, which is not critical for most of the species described in this book, this medium is designated MYPA.)

Potato, Dextrose, Yeast Agar

1,000 milliliters (1 liter) water

300 grams of potato water (i.e., the broth from boiling potatoes in 2–3 liters of water for 1 hour)

20 grams agar

10 grams dextrose

2 grams yeast

1 gram peptone (optional, soybean derived)

(This medium is designated PDYA, or PDYPA if peptone is added. Note that only the broth from boiling the potatoes is used—the potatoes are discarded. The total volume of the media should equal 1 liter.)

Oatmeal, Malt, Yeast Enriched Agar

1,000 milliliters water (1 liter)

80 grams instant oatmeal

20 grams agar agar

10 grams malt sugar

2 grams yeast

(This rich medium is called OMYA. The oatmeal does not have to be filtered out although some prefer to do so.)

Dog Food Agar

1,000 milliliters water (1 liter)

20 grams dry dog food*

20 grams agar agar

(This medium is called DFA.)

Cornmeal, Yeast, Glucose Agar

1,000 milliliters water (1 liter)

20 grams agar agar

10 grams cornmeal

5 grams malt or glucose

1 gram yeast

(This medium is known as CMYA and is widely used by mycological laboratories for storing cultures and is not as nutritious as the other above-described formulas.)

---

*The late Dr. Steven Pollock first recommended the use of dry dog food as a component for agar medium. We have found Nature's Diet dog food to be a good brand.

The pH of the above media formulations, after sterilizing, generally falls between 5.5 and 6.8. This medium can be further fortified with the addition of 3–5 grams of the end-substrate (in most cases hardwood sawdust) upon which mushrooms will be produced. If samples of soil or dung are desired, they first must be preboiled for at least 1 hour before adding to any of the above formulas. One potential advantage of the addition of these end-substrate components is a significant reduction in the "lag period." The lag period is seen when mushroom mycelium encounters unfamiliar components. (Leatham and Griffin, 1984; Raaska, 1990). This simple step can greatly accelerate the mushroom life cycle, decreasing the duration of colonization prior to fruiting.

The dry components are mixed together, placed into a flask, to which 1 liter of water is added. This cultivator finds that well water, spring water, or mineral water works well. Chlorinated water is not recommended. Purchasing distilled water is unnecessary in my opinion. Once the media has been thoroughly mixed, the media flask is placed into a pressure cooker. The top of the media flask is either stopped with nonabsorbent cotton, wrapped in aluminum foil or, if equipped with a screw cap, the cap is loosely tightened. Steam sterilize for 45 minutes @ 15 psi (1 kg/cm²), equivalent to 250°F (120°C).

Pressure cookers or sterilizers that do not release pressure during the sterilization cycle are ideal. The old-fashioned pressure canners, those having weights sitting upon a steam valve, cause the media to boil as steam is vented. A huge mess ensues. The pressure cooker should ideally form a vacuum at cooldown. If, upon returning to atmospheric pressure, the cooker does not form a vacuum, the cultivator must place the pressure cooker in the clean room or open it in front of a laminar flow hood to prevent the introduction of contamination.

As the media cools within the pressure cooker, outside air is sucked in. If this air is ladened with contaminant spores, the media contaminates before the cultivator has handled the flask! One precaution is to saturate a paper towel with alcohol and drape it over the point where outside air is being drawn in. The cloth acts as a filter, lessening the chance of contaminants. Twenty minutes after the pressure has achieved atmospheric, most media vessels can be handled without a hot-glove. Prior to that time, media can be poured, but some form of protection is needed to prevent burns to the hands.

Agar coagulates water when added in excess of 10 grams per 1,000 milliliters H2O. Only high-grade agar should be used. Various agars differ substantially in their ability to gelatinize water, their mineral and salt content, as well as their endemic populations of microorganisms, including bacteria. (Bacteria, if surviving, degelatinize the media.) Increasingly, pollution has affected the refinement of tissue-grade agar, causing the price to spiral. Agar substitutes such as Gelrite are widely used by the plant tissue culture industry. Although only a few grams are needed per liter, it does not result in a media firm enough for most mushroom cultivators. Mushroom cultivators desire a media with a semisolid, firm surface upon which the mycelium will grow. Plant tissue culturists seek a softer, gelatinous media so that plant starts will grow three-dimensionally, deep into media.

Sugars are essential for the healthy growth of mycelium. For media formulation, complex sources of sugars (carbohydrates and polysaccharides) are recommended. Cornsteep fermentative, cooked potatoes, wood, and barley malt extracts provide sugars and an assortment of basic minerals, vitamins, and salts helpful in the growth of the mushroom mycelium. From my experiences, simplified sugars, while they may support growth, are not recommended, as strains cannot be maintained for long without promoting mutation factors, senescence, and dieback. When the mycelium must work to extract sugars from the base medium, the overall vigor of the mycelial colony is enhanced. The reaction is not unlike that of a person exercising versus being lethargic. The effort increases vitality.

A variety of nitrogen- and carbohydrate-based supplements can be added to fortify the media. Strains grown repeatedly on mono-specific media for prolonged periods risk limiting the repertoire of digestive enzymes to just that specific media. In other words, a strain grown on one media adapts to it, and may lose its innate ability to digest larger, more complex and variable substrates. To prevent a strain from becoming media-specific, the following compounds are added to the MEA or PDA at various intervals, often in combinations:

Nitrogen and Carbohydrate Supplements
2 grams yeast or 1–2 grams peptone
2 grams oatmeal or oat bran
2 grams rye or wheat flour
1 gram soybean meal
1 gram spirolina
2 grams high-quality dry dog food

End-Substrate Supplements
3–5 grams sawdust
3–5 grams powdered straw
3–5 grams sugarcane bagasse, etc.

Until some familiarity with media is established, the purchase of media from reputable companies is advised. Be forewarned, however, that the media designed for the growth of imperfect fungi (molds) available from large laboratory supply companies, favors the growth of many mold contaminants over that of mushroom mycelium. The media for most saprophytes should be adjusted to a pH of 5.5 to 6.5. Most, but not all, saprophytes (Thomas et al. 1998), acidify substrates, so near neutral, even basic substrates, become more acidic as the mushroom life cycle progresses.

Pouring Agar Media

One liter of malt extract agar medium will pour 20 to 40 (100 × 15-mm) petri dishes, depending upon the depth of the pour. Before you pour an agar medium, the tabletop is thoroughly wiped clean with an 80% concentration of isopropanol (isopropyl alcohol). Plastic petri dishes usually come pre-sterilized and ready to use. Glass petri dishes should be first washed and sterilized in a petri dish holding rack simultaneous to the sterilization of the agar medium in an autoclavable flask. Prepouring media into glass dishes and then sterilizing is awkward. Media separation occurs, and any movement during the sterilization cycle (or while transferring the pressure cooker to the clean room) causes the liquefied media to spill out of the petri dishes. A huge mess results. However, this problem can be avoided if the pressure cooker cools overnight, for instance, and then is opened the next day. Be forewarned that if you choose this alternative, your pressure cooker must either form a vacuum, safely protecting the media before opening, or be placed into a HEPA filtered airstream to prevent contamination entry during cooldown.

With the micron filters mounted horizontally, and facing the cultivator, every movement is prioritized by degree of cleanliness. The cleanest articles remain upstream, the next cleanest downstream in second position, etc. The cultivator's hands are usually furthest downwind from the media and cultures.

Starting a Mushroom Strain by Cloning

The surest method of starting a mushroom strain is by cloning. Cloning means that a piece of pure, living flesh is excised from the mushroom and placed

into a sterilized, nutrient-enriched medium. If the transfer technique is successful, the cultivator succeeds in capturing a unique strain, one exhibiting the particular characteristics of the contributing mushroom. These features, and the expression thereof, is called the *phenotype*. By cloning, you capture the phenotype. Later, under the proper cultural conditions, and barring mutation, these same features are expressed in the subsequently grown mushrooms.

Several sites on the mushroom are best for taking clones. First, a young mushroom, preferably in "button" form, is a better candidate than an aged specimen. Young mushrooms are in a state of frenzied cell division. The clones from young mushrooms tend to be more vigorous. Older mushrooms can be cloned but have a higher contamination risk, and are slower to recover from the shock of transfer. Three locations resulting in a high number of successful clones are the area directly above the gills, the zone directly below the disk of the cap, and the interior tissue located at the base of the stem. The stem base, being in direct contact with the ground,

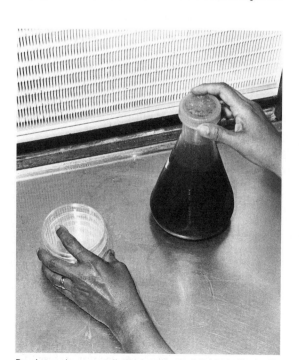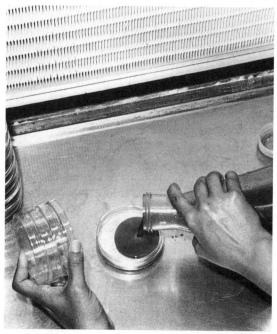

Pouring malt agar media into sterile petri dishes in front of laminar flow hood.

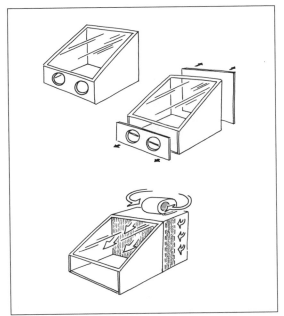

Glove boxes are considered "old tech." To retrofit a glove box into a laminar flow hood, simply cut out the back panel, replace with a similarly sized HEPA filter, and build a 6-inch-deep plenum behind the filter. A squirrel-cage blower is mounted on top, forcing air into the plenum. Air is forced through the filter. Downstream from the filter, a sterile wind flows in which inoculations can be conducted.

the size of a kernel of grain. Quickly transfer the excised tissue to the nutrient-filled petri dish, and submerge the tissue into the same location where the scalpel tip had been cooled. When you insert the tissue part way into the agar medium, in contrast to resting it on the surface, the mushroom tissue has maximum contact with the life-stimulating media. Each time a clone is taken, the scalpel is re-sterilized, cooled, and then the tissue is transferred into a separate petri dish following the aforementioned steps.

One carefully keeps the hot scalpel tip and the freshly poured media plates upstream of the mushroom being cloned or the mycelium being transferred. Next downstream are the cultivator's hands. No matter how many times one has disinfected his or her hands, one should presume they are replete with contaminants. (To test this, wash your hands, disinfect your fingertips with alcohol and fingerprint newly poured media plates. In most cases, the plates will contaminate with a plethora of microorganisms.)

Some use a "cooling dish" into which the hot scalpel tip is inserted before touching the living flesh of a mushroom. Repeatedly cooling the scalpel tip into the same medium-filled petri dish before each inoculation is not recommended. A mistake with any inoculation could cause contamination to be retransmitted with each transfer. If, for instance, a part of the mushroom were being invaded by *Mycogone*, a mushroom-eating fungus, one bad transfer would jeopardize all the subsequent inoculations. Only one cooling dish should be used for each transfer; the same dish that receives the cloned tissue. In this fashion, at least one potential cross-contamination vector is eliminated.

When cloning a mushroom for the first time, I recommend a minimum of ten repetitions. If the mushroom specimen is exceedingly rare, cloning into several dozen dishes is recommended. As the specimen dries out, viable clones become increasingly less likely. With the window of opportunity for cloning being so narrow, the cultivator should clone mushrooms within hours of harvesting. If the mushrooms must be stored, then the specimen should refrigerated at 35–40°F (1–5°C). After 3 to 4 days from harvest, finding viable and clean tissue for cloning is difficult.

is often the entry point through which fly larvae tunnel, carrying with them other microorganisms. I prefer the genetically rich area directly giving rise to the gills and their associated spore-producing cells, the fertile basidia.

The procedure for cloning a mushroom is quite simple. Choose the best specimen possible, and cut away any attached debris. Using a damp paper towel, wipe the mushroom clean. Lay the specimen on a new sheet of paper towel. Flame-sterilize a sharp scalpel until it is red hot. Cool the scalpel tip by touching the nutrient agar medium in a petri dish. This petri dish will be the same dish into which you transfer the mushroom tissue. Carefully tear the mushroom apart from the base, up the stem, and through the cap. With one-half of this split mushroom, cut a small section ("square") of flesh about

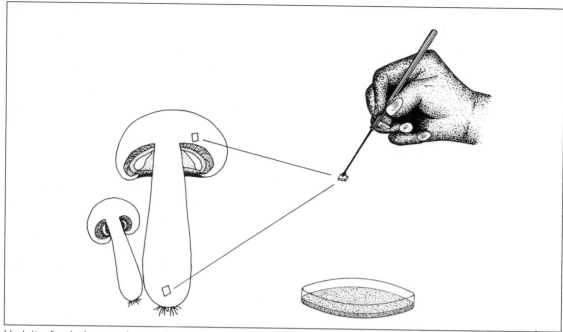

Ideal sites for cloning a mushroom: directly above the gills or at the base of the stem. Young, firm mushrooms are best candidates for cloning.

A few days to 2 weeks after cloning the mushroom, the tissue fragment springs to life, becoming fuzzy in appearance. Contaminants usually become visible at this stage. As a rule, the cultivator always transfers viable mycelium away from contamination, not the other way around. The essential concept here, is that *the cultivator "runs" with the mycelium*, subculturing away from contamination as many times as is necessary until a pure culture is established.

Each transfer from an older petri dish culture to a newer petri dish moves upstream. The scalpel is brought into contact with heat. The tip is cooled into the dish destined to receive the mycelium. The lid of this dish is lifted, the scalpel is cooled, and then the lid is replaced. Next, the lid of the dish hosting the mature mycelium is opened. The mycelium is cut. The wedge is transferred to the newly poured media plate. With the lid replaced, the culture is labeled and moved aside. The process is repeated until a number of plates are inoculated.

As each lid is lifted, care is taken not to extend the fingers beyond the lip of each top. The over-hanging of fingers results in flaking off contaminants into the petri dish. Furthermore, the lids are lifted with their undersides catching the sterile airstream. If the lids must be laid down, they are positioned undersides up, upstream of the operations area, so that contaminants are not picked up off the table. Always presume the air coming off the face of the micron filter is cleaner than the work surface in front of it.

Culture transfers that are fast, evenly repeated, and in quick succession usually are the most successful. The simplest acts dramatically impact sterile technique. Merely breathing over exposed petri dishes significantly affects contamination levels. Singing, for instance, is associated with a high rate of bacterial contamination. One bewildered professor discovered that her soliloquies in the laboratory—she sang as the radio blared—were a direct cause of high contamination rates. An alert student discovered her digression from sterile technique upon passing the door to her lab. This illustrates that the cultivator's unconscious activities profoundly influence the outcome of tissue culture transfers. *Every action in the laboratory has significance.*

Cloning Wild Specimens vs. Cultivated Mushrooms

Many people ask, "What is wrong with just cloning a nice-looking specimen from each crop of cultivated mushrooms to get a new strain?" Although morphological traits can be partially selected for, senescence factors are soon encountered with each subsequent clone, a condition known as *replicate fading.* Generating cultures in this fashion is a fast track to genetic demise, quickly leading to loss of vigor and yield. By not returning to stock cultures, to young cell lines, one goes further downstream one linear chain of cells. Mushrooms, like every sexually reproducing organism on this planet, can generate a limited number of cell divisions before vitality falters. Sectoring, slow growth, anemic mushroom formation, malformation, or no mushroom forma-

tions at all, are all classic symptoms of senescence. Although senescence is a frequently encountered phenomenon with cultivators, the mechanism is poorly understood. (See Kuck et al. 1985.)

In the competitive field of mycology, strains are all-important. With the afore-stated precautions and our present-day technologies, strains can be preserved for decades, probably centuries, all-the-while kept within a few thousand cell divisions from the original culture. Since we still live in an era of relatively rich fungal diversity, the time is now to preserve as many cell lines from the wild as possible. As biodiversity declines, the gene pool contracts. I strongly believe that the future health of the planet may well depend upon the strains we preserve this century.

How to Collect Spores

A culture arising from cloning is fundamentally different from a culture originating from spores. When spores are germinated, many different strains

Paul Stamets's 12-foot-long laminar flow bench designed for commercial cultivation. Simple, but effective.

A laminar flow bench suitable for home or small-scale commercial cultivation.

are then laid on the cleaned, open surface for spore collection. After 12 to 24 hours, the contributing mushroom is removed, dried, and stored for later reference purposes. The remaining edges of the glass are then taped. The result is a glass-enclosed "Spore Booklet," which can be stored at room temperature for years. Spores are easily removed from the smooth glass surface for future use. And, spores can be easily observed without increasing the likelihood of contamination.

Spores have several advantages. Once a spore print has been obtained, it can be sealed and stored, even sent through the mail, with little ill effect. For the traveler, spore prints are an easy way to send back potential new strains to the home laboratory. Spores offer the most diverse source of genetic characteristics, far more than the phenotypic clone. If you want the greatest number of strains, collect the spores. If you want to capture the characteristics of the mushroom you have found, then clone the mushroom by cutting out a piece of living tissue.

are created, some incompatible with one another. A cultivator will not know what features will be expressed until each and every strain is grown out to the final stage, that of mushroom production. This form of genetic roulette results in some very unusual strains, some more desirable than others.

Mushroom spores are collected by taking a spore print. Spore prints are made by simply severing the cap from the stem, and placing the cap, gills down, upon a piece of typing paper, or glass. Covering the mushroom cap with a bowl or plate lessens evaporation and disturbance from air currents. Within 24 hours, often only 12, millions of spores fall in a beautiful pattern according to the radiating symmetry of the gills. A single mushroom can produce from 100,000 to 100,000,000 spores!

I prefer to collect spores on plates of glass, approximately 6 by 8 inches. The glass is washed with soapy water, wiped dry, and then cleaned with rubbing alcohol (isopropanol). The two pieces of glass are then joined together with a length of duct tape, to create, in effect a binding. The mushrooms

Germinating Spores

To germinate spores, an inoculation loop, a sterilized needle, or scalpel is brought into contact with the spore print. (I prefer an inoculation loop.) I recommend flame-sterilizing an inoculation loop until red hot and immediately cooling it in a petri dish filled with a sterilized nutrient medium. The tip immediately sizzles as it cools. The tip can now touch the spore print without harm, picking up hundreds of spores in the process. By touching the tip to the medium-filled petri dish first, not only is it cooled, but also the tip becomes covered with a moist, adhesive layer of media to which spores easily attach. The tip can now be streaked in a "S" pattern across the surface of another media dish.

With heavy spore prints, the "S" streaking technique may not sufficiently disperse the spores. In this case, the scalpel or inoculation loop should be immersed into a sterile vial holding 10 cc (ml) of water. After shaking thoroughly, one drop (or $\approx 1/30$th of 1 cc.) is placed onto the surface of the nutrient medium in each petri dish. Each dish should be tilted

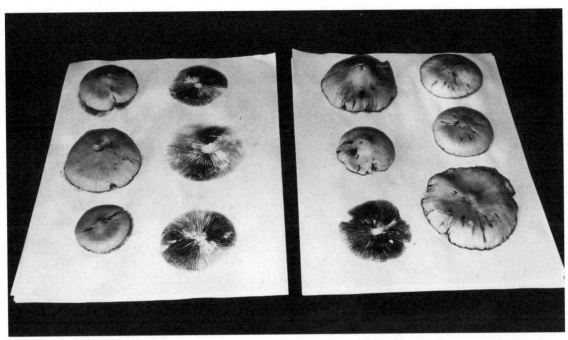

Taking spore prints on typing paper.

back and forth so that the spore-enriched droplet streaks across the surface, leaving a trail of dispersed spores. In either case, spores will be spread apart from one another, so that individual germination can occur.

Five to fifteen days later, spores can be seen germinating according to the streaking pattern. Colonies of germinating spores are subcultured into more petri dishes. (See page 95.) After the mycelium has grown away from the subculture site, a small fragment of pure mycelium is again subcultured into more petri dishes. If these cultures do not sector, then backups are made for storage and future use. This last transfer usually results in individual dikaryotic strains that are labeled. Each labeled strain is then tested for productivity. Mini-culture experiments must be conducted prior to commercial-level production.

When a concentrated mass of spores is germinated, the likelihood of bacteria and weed fungi infesting the site is greatly increased. Bacteria replicate faster than mushroom spores can germinate.

The spore print can be folded and rubbed together so that spores drop onto nutrient agar media. This method is not the best as concentrated populations of spores are grouped together.

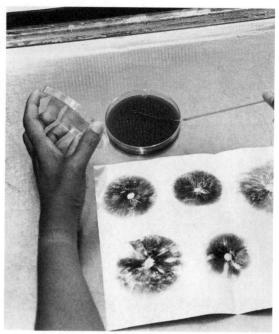

An inoculation loop is sterilized, in this case with a BactiCinerator, and then cooled into the "receiving" dish. The spores are picked up by touching the inoculation loop to the spore print. The spore-laden inoculation loop is streaked across the surface of the sterilized, nutrient-filled medium in a "S" pattern.

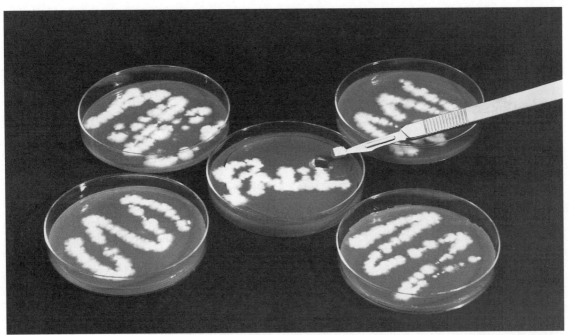

Spores germinate according to the streaking pattern. A small portion is excised and transferred to a new, nutrient-agar-filled petri dish.

As a result, the germinating spores become infected. Mycelium arising from such germination is frequently associated with a high contamination rate, unfortunately, often not experienced until the mycelium is transferred to grain media. However, if the spore prints are made correctly, contamination is usually not a problem.

Once inoculated, the petri dish cultures should be taped with an elastic film (such as Parafilm), which protects the incubating mycelium from intrusive airborne contaminants. (See page 75.)

Purifying a Culture

Many cultures originating from spores or tissue are associated with other microorganisms. Several techniques are at one's disposal for cleaning up a culture. Depending upon the type, and level of contamination, different responses are appropriate.

One way of cleaning a bacterially infested culture is by sandwiching it between two layers of media

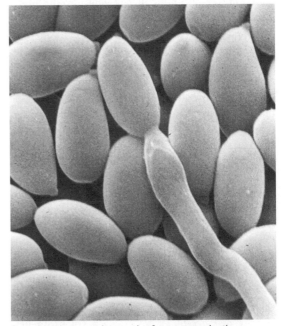

Scanning electron micrograph of spores germinating.

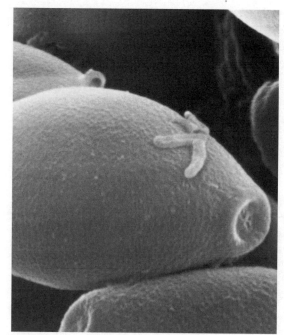

Scanning electron micrograph of spores infected with bacteria.

difficult.* Remember, the advantage that molds have over mushroom mycelia is that their life cycles spin far faster, and thousands of mold spores are generated in only a few days. Once molds produce spores, any disturbance—*including exposure to the clean air coming from the laminar flow hood*—creates satellite colonies.

One rule is to immediately subculture all points of visible growth away from one another as soon as they become visible. This method disperses the colonies, good and bad, so they can be dealt with individually. Repeated subculturing and dispersal usually result in success. If not, then other alternative methods can be implemented.

Mycelia of all fungi grow at different rates and are acclimated to degrading different base materials. One method I have devised for separating mushroom mycelium from mold mycelium is by racing the mycelia through organic barriers. Glass tubes can be filled with finely chopped, moistened straw, wood sawdust, rolled cardboard, and even crushed corncobs (without kernels) and sterilized. The contaminated culture is introduced to one end of the tube. The polyculture of contaminants of mushroom mycelium races through the tube, and with luck, the mushroom mycelium is favorably selected, reaching the opposite end first. At this point, the cultivator simply transfers a sample of emerging mycelium from the end of the tube to newly poured media plates. The cultures are then labeled, sealed, and observed for future verification. This technique relies on the fact that the mycelia of fungi grow at different rates through biodegradable materials. The semi-selectivity of the culture/host substrate controls the success of this method.

Cultivators develop their own strategies for strain purification. Having to isolate a culture from a background of many contaminants is inherently difficult. Far easier is it to implement the necessary precautions when initially making a culture so that running away from contamination is unnecessary.

containing an antibiotic such as gentamycin sulfate. The hyphae, the cells composing the mycelium, are arranged as long filaments. These filamentous cells push through the media while the bacteria are left behind. The mycelium arising on the top layer of media will carry a greatly reduced population of bacteria, if any at all. Should the culture not be purified the first time using this procedure, a second treatment is recommended, again subculturing from the newly emerged mycelium. Repeated attempts increase the chances of success.

If the culture is mixed with other molds, then the pH of the media can be adjusted to favor the mushroom mycelium. Generally speaking, many of the contaminant fungi are strong acidophiles whereas Oyster mushrooms grow well in environments near a neutral pH. If these mold fungi sporulate adjacent to the mycelia of mushrooms, isolation becomes

---

* The spores of most mold fungi become distinctly pigmented at maturity. Some *Penicillium* molds are typically blue green. *Aspergillus* species range in color from black to green to yellow. *Neurospora* can be pink. A few molds, such as *Monilia* or *Verticillium*, produce white colonies. For more information on these competitors, please consult *The Mushroom Cultivator* (1983) by Stamets and Chilton.

CHAPTER 13

The Stock Culture Library:
A Genetic Bank of Mushroom Strains

E very sexually reproducing organism on this planet is limited in the number of its cell replications. Without further recombination of genes, cell lines decline in vigor and eventually die. The same is true with mushrooms. When one considers the exponential expansion of mycelial mass, from two microscopic spores into tons of mycelium in a matter of weeks, mushroom mycelium cell division potential far exceeds that of most organisms. Nevertheless, strains die and, unless precautions have been taken, the cultures may never be retrieved.

Once a mushroom strain is taken into culture, whether from spores or tissue, the resultant strains can be preserved for decades under normal refrigeration, perhaps centuries under liquid nitrogen. In the field of mycology, cultures are typically stored in test tubes. Test tubes are filled with media, sterilized, and laid at a 15- to 20-degree angle on a table to cool. (Refer to Chapter 12 for making sterilized media. I often double the formula so that the test tubes will gel sufficiently.) These are called *test tube slants*. Once inoculated, these are known as *culture slants*.

Culture slants are like "backups" in the computer industry. Since every mushroom strain is certain to die out, one is forced to return to the stock library for genetically younger copies. Good mushroom strains are hard to come by, compared to the number of poor performers isolated from nature. Hence, the Culture Library, aka the Strain Bank, is at the pivotal center of any mushroom cultivation enterprise.

Preserving the Culture Library

One culture in a standard 100 × 15-mm petri dish can inoculate 50 to 100 test tube slants measuring 100 × 20 mm After incubation for 1 to 4 weeks, or until a luxurious mycelium has been established, the test tube cultures are placed into cold storage. I seal the gap between the screw cap and the glass tube with a commercially available elastic, wax-like film. (Those test tube slants not sealed with this film are prone to

Stock cultures, in quadruplicate, sealed in a plastic bag, stored in a cedar box, and refrigerated for years at 35°F (1–2°C) until needed.

contaminate with molds after several months of cold storage.) Culture banks in Asia commonly preserve cultures in straight test tubes whose ends are stuffed with a hydrophobic cotton or gauze. The gauze is sometimes covered with plastic film and secured tightly with a rubber band. Other libraries offer cultures in test tubes fitted with a press-on plastic lid especially designed for gas exchange. The need for gas exchange is minimal—provided the culture's growth is slowed down by timely placement into cold storage. Culture slants stored at room temperature have a maximum life of 6 to 12 months, whereas cultures kept under refrigeration survive for 5 years or more. Multiple backups of each strain are strongly recommended as there is a natural attrition over time.

I prefer to seal test-tube slants in plastic Ziploc bags. Three to four bags, each containing four slants, are then stored in at least two locations remote from the main laboratory. This additional safety precaution prevents events like fires, electrical failure, misguided law enforcement officials, or other

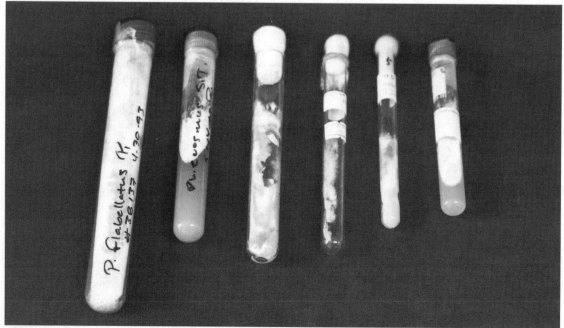

Examples of cultures originating from stock culture libraries from sources in United States, Canada, Thailand, and China. Most culture libraries do not send cultures in duplicate nor indicate how far the cultures have grown since inception.

natural disasters from destroying your most valuable asset—the Culture Library.

Household refrigerators, especially modern ones, suffice. Those refrigerators having the greatest mass, with thermostatic controls limiting variation in temperature, are best for culture storage. With temperature variation, condensation occurs within the culture tubes, spreading a contaminant, should it be present, throughout the culture. Therefore, limiting temperature fluctuation to 2–3°F (1°C) is crucial for long-term culture preservation. Furthermore, when mushroom cultures freeze and thaw repeatedly, they die.

If one has ten or more replicates, stock cultures of a single strain can be safely stored for 5 years by this method. As a precaution, however, one or two representative culture slants should be retrieved every year, brought to room temperature for 48 hours, and subcultured to newly filled media dishes. Once revived, and determined to be free of contamination, the mycelium can once again be subcultured back into test tube slants, and returned to refrigeration. *This circular path of culture rotation ascertains viability and prolongs storage with a minimum number of cell divisions. I cannot overemphasize the importance of maintaining cell lines closest to their genetic origins.*

Cryogenic storage—the preservation of cultures by storage under liquid nitrogen—is the best way to preserve a strain. Liquid nitrogen storage vessels commonly are held at –302°F (–150°C). Test tube slants filled with a specially designed cryoprotectant medium help the mycelium survive the shock of sudden temperature change. (Such cryoprotectants involve the use of a 10% glycerol and dextrose medium.) Wang and Jong (1990) discovered that a slow, controlled cooling rate of –1°C per minute resulted in a higher survival rate than sudden immersion into liquid nitrogen. This slow reduction in temperature allowed the mycelium to discharge water extracellularly, thus protecting the cells from the severe damage ice crystals pose. Further, they found that strains were better preserved on grain media than on agar media. However, for those with limited liquid nitrogen storage space and large numbers of strains, preservation of grain media is not as practical as preserving strains in ampules or test tubes of liquid cryoprotectant media.

Of all the mushrooms discussed in this book, only strains of the Paddy Straw Mushroom, *Volvariella volvacea*, should not be chilled. *V. volvacea* demonstrates poor recovery from cold storage—both from simple refrigeration at 34°F (2°C) or immersion into liquid nitrogen at –300°F (–150°C). When the mycelium of this tropical mushroom is exposed to temperatures below 45°F (7.2°C), drastic dieback occurs. Strains of this mushroom should be stored at no less than 45–50°F (7–10°C) and tested frequently for viability. When cultures are to be preserved for prolonged periods at room temperature, many mycologists cover the mycelium with liquid paraffin. (For more information, consult Jinxia and Chang, 1992.)

When you are retrieving cultures from prolonged storage, the appearance of the cultures can immediately indicate potential viability or clear inviability. If the mycelium is not aerial, but is flat, with a highly reflective sheen over its surface, then the culture has likely died. If the culture caps have not been sealed, contaminants, usually green molds, are often visible, giving the mycelium a speckled appearance. These cultures make re-isolation most difficult. Generally speaking, success is most often seen with cultures having aerial, cottony mycelium. Ultimately, however, cultivators cannot determine viability of stored cultures until they are subcultured into new media and incubated for 1 to 3 weeks.

The Stamets "P" Value System

The Stamets "P" value system is simply an arithmetic scale I have devised for measuring the expansion of mycelium through successive inoculations from one 100 × 15-mm petri dish to the next. The number of cell divisions across a petri dish fall within a range of cell wall lengths. Of the septate strains of fungi, some have cells as short as 20 μ while others have cells 200 μ and longer. The Stamets "P" Value (SPV) benefits cultivators by indicating how close to the origin their culture is at any point in time by simply recording the number

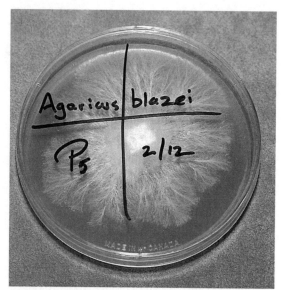

An *Agaricus blazei*, Himematsutake, culture is labeled to indicate genus, species, "P" Value, and date.

of petri dishes the mycelium has grown through. When a culture has been isolated from contaminants, usually in one or two transfers, the first pure culture is designated as P^1. When the mycelium has filled that dish, the next dish to receive the mycelium is called P^2. Each culture is labeled with the date, species, collection number, strain code, "P" Value, and medium (if necessary). Thus, a typical example from one of my culture dishes reads:

Agaricus blazei
P^5
2/12

The date 2/12 refers to the time the medium was inoculated. Spawn created from such young cultures, in contrast to one grown out 100 times as far, produces more mushrooms. The "P" value system is essentially a metric ruler for measuring relative numbers of cell divisions from the culture's birth. (Note that a square centimeter of mycelium is generally transferred from one culture dish to the next.) I have strains in my possession, from which I regularly regenerate cultures, that are 10 years old, and kept at a P^2 or P^3. Having ten to twenty backup culture slants greatly helps in this pursuit.

For purposes of commercial production, I try to maintain cell lines within P^{10}, that is, within ten successive transfers to medium-filled petri dishes. Many strains of Morels, Shiitake, and King Stropharia express mutations when transferred on media for more than ten petri dishes. Morels seem particularly susceptible to degeneration. *Morchella angusticeps* loses its ability to form micro-sclerotia in as few as six or seven plate transfers from the original tissue culture.

The slowing of mycelium may also be partly due to media specificity, i.e., the agar formula selectively influences the type of mycelial growth. To ameliorate degenerative effects, the addition of extracted end-substrates (sawdust, straw, etc.) favors the normal development of mycelium. The introduction of the end-substrate acquaints the mushroom mycelium with its destined fruiting habitat, challenging the mycelium and selectively activating its enzymatic systems. This familiarity with the end-substrate greatly improves performance later on. Parent cells retain a "genetic memory" passed downstream through the mycelial networks. Mycelia grown in this fashion are far better prepared than mycelia not exposed to such cultural conditions. Not only is the speed of colonization accelerated, but also the time to fruiting is shortened. Only 1–3 grams of substrate is recommended per liter of nutrient medium. Substrate additives high in endospores (such as most soils) should be treated by first boiling the sample in an aqueous concoction for at least an hour. After boiling, sugar, agar, and other supplements can be added. The media is then sterilized using standard procedures described in Chapter 12.

By observing the cultures daily, you'll see the changeover of characteristics that defines the health of the mycelial mat. What this book strives to show is the mycelium of each species and its transformations on the path to fruiting. Variations from the norm should alert the cultivator that the strain is in an active state of mutation. Rarely do mutations in the mycelium result in a stronger strain. Paying attention to the structure of the leading edge of the actively growing mycelium often gives the cultivator clues to the health of the culture. Rapidly diverging finger-like growth patterns are usually a positive reflection of vitality, and as a rule, their loss is foreboding of a sickening strain.

Iconic Types of Mushroom Mycelium

Each mushroom species produces a recognizable type of mycelium whose variations fall within a range of expressions. Within a species, multitudes of strains can differ dramatically in their appearance. In culture, mushroom strains reveal much about the portion of the mushroom life cycle that is invisible to the mere forager for wild mushrooms. This range of characteristics—changes in form and color, rate of growth, fragrance, even volunteer fruitings of mushrooms in miniature—reveals a wealth of information to the cultivator, defining the strain's "personality."

Form: Mycelia can be categorized into several different, classic forms. For ease of explanation, these forms are delineated on the basis of their macroscopic appearance on the two-dimensional plane of a nutrient-filled petri dish. As the mycelium

undergoes changes in its appearance over time, this progression of transformations tells the cultivator much about what is normal and what is abnormal. The standard media I use is Malt Yeast Agar (MYA) fortified with peptone (MYPA).

Linear: Linear mycelium is arranged as diverging, longitudinal strands. Typically, the mycelium emanates from the center of the petri dish as a homogeneously forming mat. Shiitake (*Lentinula edodes*) and initially Oyster (*Pleurotus ostreatus*) mycelia fall in this category. Morels produce a rapidly growing, finely linear mycelium, which thickens in time. In fact, Morel mycelium is so fine that during the first few days of growth, the mycelium is nearly invisible, detected only by tilting the petri dish back and forth so that the fine strands can be seen on the reflective sheen of the agar medium's surface.

Rhizomorphic: Often similar to linear mycelium, rhizomorphic mycelium is often called "ropey." In fact, rhizomorphic mycelium is composed of braided, twisted strands of mycelium, often of varying diameters. Rhizomorphic mycelium

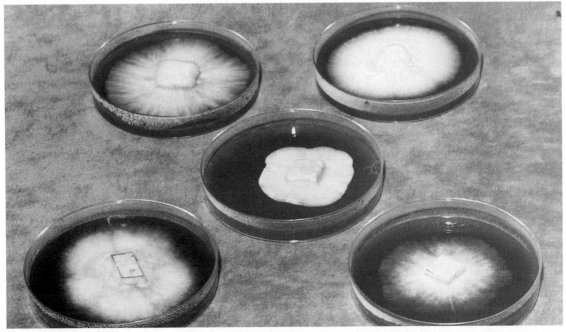

Classic forms of mushroom mycelia.

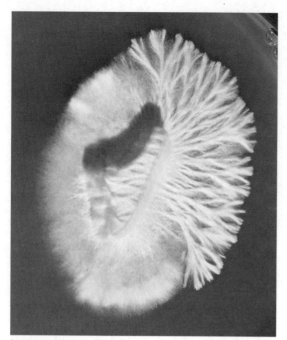

Rhizomorphic mycelium diverging from cottony mycelium soon after spore germination.

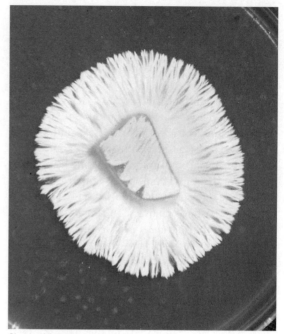

Classic rhizomorphic mycelium.

supports primordia and is encouraged by selecting these sites for further transfer. The disappearance of rhizomorphs is an indication of loss of vigor. Lion's Mane (*Hericium erinaceus*), the King Stropharia (*Stropharia rugosoannulata*), Button mushrooms (*Agaricus brunnescens*, *Agaricus bitorquis*), Himematsutake (*Agaricus blazei*), Magic Mushrooms (*Psilocybe cubensis* and *Psilocybe cyanescens*), and Clustered Woodlovers (*Hypholoma capnoides* and *H. sublateritium*) are examples of mushrooms producing classically rhizomorphic mycelia. Some types of rhizomorphic mycelia take on a reflective quality, resembling the surface of silk.

Cottony: This type of mycelium is common with strains of Oyster mushrooms (*Pleurotus* species), Shaggy Manes (*Coprinus comatus*), and Hen-of-the-Woods (*Grifola frondosa*). Looking like tufts of cotton, the mycelium is nearly as aerial in its growth as it is longitudinal. Cottony mycelium is commonly called "tomentose" by mycologists. When rhizomorphic mycelium degenerates with age, tomentose formations typically take over.

Zonate: Cottony mycelium often shows concentric circles of dense and light growth, or zones. Zonate mycelium is often characteristic of natural changes in the age of the mycelium. The newest mycelium, on the periphery of the culture, is usually light in color. The more aged mycelium, towards the center of the culture, becomes strongly pigmented. Zonations can also be a function of the natural circadian cycles, and even when cultures are incubated in laboratories, and even though temperatures are kept constant and away from the influence of sunlight. Growth occurs in spurts, creating rings of alternating density. This feature is commonly seen in species of *Ganoderma*, *Hypholoma*, and *Hypsizygus*.

Matted or *appressed:* This type of mycelium is typical of Reishi (*Ganoderma lucidum*) after 2 weeks of growth on 2% malt extract agar media. So dense is this mycelial type, that a factory-sharpened surgical blade can't cut through it. The mycelium tears off in ragged sheaths as the scalpel blade is dragged across the surface of the agar medium.

Pseudo-rhizomorphic and cottony mycelia.

Many species develop matted mycelia over time, especially the wood rotters. Cultures that mysteriously die often have mycelium that appears matted but whose surface is flat and highly reflective.

Powdered: This form of mycelium is best exemplified by *Laetiporus sulphureus* (*Polyporus sulphureus*), aka Chicken-of-the-Woods. The mycelium breaks apart with the least disturbance. In front of a laminar flow bench, the sterile wind can cause chains of mycelium (*hyphae*) to be airborne. Free-flying hyphae can cause considerable cross-contamination problems within the laboratory.

Unique Formations: Upon the surface of the mycelial mat, unique formations occur that can be distinguished from the background mycelium. They are various in forms. One common form is *hyphal aggregates*, cottony ball-like or shelf-like structures. I view hyphal aggregates as favorable formations when selecting out rapidly fruiting strains of Shiitake. Hyphal aggregates often evolve into primordia, the youngest visible stages of mushroom formation. *Marasmius oreades*, the Fairy Ring

Zonate mycelium.

Hyphal aggregates of the Fairy Ring Mushroom (*Marasmius oreades*) and Shiitake (*Lentinus edodes*).

Mushroom, produces shelf-like forms that define the character of its mycelium. *Stropharia rugosoannulata*, the King Stropharia, has uniquely flattened, plate-like zones of dense and light growth, upon which hyphal aggregates often form. Morels and many other mushrooms produce dense, spherical formations called *sclerotia*. (See page 105.) Sclerotia can be brightly colored and abundant, as is typical of many strains of *Morchella angusticeps,* or dark colored and sparse like those of *Polyporus umbellatus.*

The mycelia of some mushrooms generate asexual structures called *coremia* (broom-like bundles of spores), which resemble many of the black mold contaminants. Some of these peculiar formations typify *Pleurotus cystidiosus, Pleurotus abalonus,* and *Pleurotus smithii.* I know of one Ph.D. mycologist, not knowing that some Oyster mushrooms go through an asexual stage, promptly discarded the cultures I gave her because they were "contaminated."

Mushroom strains, once characterized by rhizomorphic mycelia, often degenerate after repetitive subculturing. Usually the decline in vigor follows this pattern: A healthy strain is first rhizomorphic in appearance, and then after months of transfers

the culture *sectors*, forming diverging "fans" of linear, cottony, and appressed mycelium. Often an unstable strain develops mycelium with aerial tufts of cotton-like growth. The mycelium at the center of the petri dish, giving birth to these fans of disparate growth, is genetically unstable, and being in an active state of decline, sends forth mutation-ridden chains of cells. Often, the ability to give rise to volunteer primordia on nutrified agar media, once characteristic of a strain, declines or disappears entirely. Speed of growth decelerates. If not entirely dying out, the strain is reduced to an anemic state of slow growth, eventually incapable of fruiting. Prone to disease attack, especially by parasitic bacteria, the mushroom strain usually dies.

Color: Most mushroom species produce mycelia that undergo mesmerizing transformations in pigmentation as they age, from the youngest stages of growth to the oldest. One must learn the natural progression of coloration for each species' mycelium. Since the cultivator is ever watchful for the occurrence of certain colors that can forebode contamination, knowing these changes is critical. Universally, the color green is bad in mushroom culture, usually indicating the presence

This unique mushroom forms concentric rings of sclerotia in culture. These rings form every few days and when separated from the mother mycelium can give rise directly to mushrooms.

A strain of the Abalone Oyster Mushroom, *Pleurotus cystidiosus*. This mushroom species is dimorphic—having an alternative life cycle path. The black droplets are resplendent with spores and are *not* contaminants.

of molds belonging to *Penicillium, Aspergillus,* or *Trichoderma.*

White: The color shared by the largest population of saprophytic mushrooms is white. Oyster (*Pleurotus* spp.), Shiitake (*Lentinula edodes*), Hen-of-the-Woods (*Grifola frondosa*), the King Stropharia (*Stropharia rugosoannulata*), and most Magic Mushrooms (Psilocybes) all have whitish-colored mycelium. Some imperfect fungi, like *Monilia*, however, also produce a whitish mycelium. (See *The Mushroom Cultivator*, Stamets and Chilton, 1983.)

Yellow/Orange/Pink: Nameko (*Pholiota nameko*) produces a white mycelial mat that soon yellows. Oyster mushrooms, particularly *Pleurotus ostreatus*, exude a yellowish to orangish metabolite over time. These metabolites are sometimes seen as droplets on the surface of the mycelium or as excessive liquid collecting at the bottom of the spawn containers. Strains of Reishi, *Ganoderma lucidum*, vary considerably in their appearances, most often projecting a white mycelium that, as it matures, becomes yellow as the agar medium is colonized. A pink

Oyster mushroom, *Pleurotus djamor*, and Lion's Mane, *Hericium erinaceus*, both have mycelium that is initially white, and as the cultures age, develop strong pinkish tones. Chicken-of-the-Woods (*Polyporus* or *Laetiporus sulphureus*) has an overall orangish mycelium. Kuritake (*Hypholoma sublateritium*) has mycelium that is white at first, and in age can become dingy yellow-brown.

Brown: Some mushroom species, especially Shiitake, undergo changes in coloration as they age. Normally, Shiitake becomes brown over time. It would be abnormal for Shiitake mycelium not to brown in age or when damaged. Similarly, the Poplar Mushroom (*Agrocybe aegerita*) produces an initially white mycelium that browns with maturity. Morel mycelium is typically brown after a week of growth.

Blue: Lepista nuda, the Wood Blewit, produces a blue, cottony mycelium. Many species not yet cultivated are likely to produce blue mycelia. Although the number of species generating blue mycelium is few, most of the psilocybian mushrooms are characterized by mycelium that bruises bluish when damaged. Beyond these examples, blue tones are highly unusual and warrant examination through a microscope to ascertain the absence of

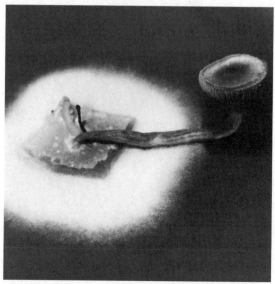

Miniature mushroom (*Gymnopilus luteofolius*) forming on malt agar media. Note proportion of mushroom relative to mycelial mat.

Mushroom primordia on malt agar media. Upper left: *Ganoderma lucidum*, Reishi. Upper right: *Agrocybe aegerita*, the Black Poplar Mushroom. Lower left: *Pleurotus djamor*, the Pink Oyster Complex. Lower right: *Hericium erinaceus*, the Lion's Mane mushroom.

competitor organisms, particularly the blue-green *Penicillium* molds. Although unusual, I have seen cultures of an Oyster mushroom, *P. ostreatus* var. *columbinus*, and a luminescent mushroom, *Mycena chlorophos*, that produce whitish mycelium streaked with bluish tones.

Black: Few mushrooms produce black mycelium. Some Morel strains cause the malt extract medium to blacken, especially when the petri dish culture is viewed from underneath. The parasitic Honey mushroom, *Armillaria mellea*, forms uniquely black rhizomorphs. A pan-tropical Oyster mushroom, called *Pleurotus cystidiosus*, and its close relatives *P. abalonus* and *P. smithii*, have white mycelia that become speckled with black droplets. (See page 106.)

Multicolored: Mycelia can be zonate, with multicolored tones in concentric circles around the zone of transfer. The concentric circles of growth are usually diurnal, reflecting rates of growth dictated by the passage of day to night. All of the species described in the past five categories undergo unique color changes. This sequence of color transforma-

tion defines the unique "personality" of each strain. I have yet to see a mycelium of greater beauty than that of the extraordinary *Psilocybe mexicana*, the sacramental Teonanacatl of Mexico. Its mycelium is initially white, then yellow, golden, brown, and sometimes streaked through with bluish tones. (See Color Plate 2, opposite 176 in *The Mushroom Cultivator* by Stamets and Chilton, 1983.)

Fragrance: The sensation most difficult to describe and yet so indispensable to the experienced spawn producer is that of fragrance. The mycelium of each species out-gasses volatile wastes as it decomposes a substrate, whether that substrate is nutrified agar media, grain, straw, sawdust, or compost. The complexity of these odors can be differentiated by the human olfactory senses. In fact, each species can be known by a *fragrance signature*. As the mass of mycelium is increased, odors become more pronounced. Although odor is generally not detectable at the petri dish culture stage, it is distinctly noticed when a red-hot scalpel blade touches living mycelium. The sudden burst of burned mycelium emits a fragrance specific to each species. More useful to cultivators is the fragrance signature emanating from grain spawn. Odors can constantly be used to check spawn quality and even species identification.

On rye grain, Oyster mycelium emits a sweet, pleasant, and slightly anise odor. Shiitake mycelium has an odor reminiscent of fresh, crushed Shiitake mushrooms. Chicken-of-the-Woods (*Laetiporus* or *Polyporus sulphureus*) is most unusual in its fragrance signature: Grain spawn has the distinct scent of butterscotch combined with a hint of maple syrup! King Stropharia (*Stropharia rugosoannulata*) has a musty, phenolic smell on grain but a rich, appealing woodsy odor on sawdust. Maitake (*Grifola frondosa*) mycelium on grain reminds me of day-old, cold corn tortillas! Worst of all is Enokitake—it smells like week-old dirty socks. Mycologists have long been amazed by the fact that certain mushrooms produce odors that humans can recognize elsewhere in our life experiences. Some mushrooms smell like radishes, some like apricots, maple syrup, and even bubble gum! Is there any significance to these odors? Or is it just a fluke of nature?

The Event of Volunteer Primordia on Nutrified Agar Media

The voluntary and spontaneous formation of miniature mushrooms in a petri dish is a delightful experience for all cultivators. In this chapter, attention and insights are given for many species. By no means is this knowledge static. Every cultivator contributes to the body of knowledge each time a mushroom is cultured.

The cultivator plays an active role in developing strains by physically selecting those that look "good." Integral to the success of the mushroom life cycle is the mycelial path leading to primordia formation. To this end, the mushroom and the cultivator share common interests. The occurrence of primordia not only is a welcome affirmation of the strain's identity but is also indicative of its readiness to fruit. Hence, I tend to favor strains that voluntarily form primordia.

Two approaches lead to primordia formation from cultured mycelium. The first is to devise a standard media, a background against which all strains and species can be compared. After performance standards are ascertained, the second approach is to alter the media, specifically improving and designing its composition for the species selected. As a group, those strains needing bacteria to fruit do not form primordia on sterile media.

Several mushroom species have mycelial networks that, when they are disturbed at primordia formation, result in a quantum leap in the vigor of growth and in the number of subsequently forming primordia. With most strains, however, the damaged primordia revert back to vegetative growth. The following list of species are those that produce volunteer primordia on 2% enriched malt extract agar, supplemented with .2% yeast and .005% gentamicin sulfate.* The formation of primordia on this media is often strain specific. The following species are known by this author to benefit from the timely disturbance of primordia on the nutrified agar medium. Those that do benefit from disturbance are excellent candidates for liquid-inoculation techniques.

Agrocybe aegerita
Ganoderma curtisii
Ganoderma lucidum
Flammulina velutipes
Hericium erinaceus
Hypsizygus tessulatus
Hypsizygus ulmarius
Lentinula edodes
Pholiota nameko
Pleurotus citrinopileatus
Pleurotus djamor complex
Pleurotus euosmus
Pleurotus ostreatus
Pleurotus djamor
Pleurotus pulmonarius

---

*One-twentieth of a gram of gentamycin sulfate per liter of media sufficiently inhibits bacteria to a confinable level.

Evaluating a Mushroom Strain

When a mushroom is brought into culture from the wild, little can be known about its performance until trials are conducted. Each mushroom strain is unique. Most saprophytic fungi, especially primary saprophytes, are easy to isolate from nature. Whether or not they can be grown under "artificial" conditions, however, remains to be seen. Only after the cultivator has worked with a strain, through all the stages of the culturing process, does a recognizable pattern of characteristics evolve. Even within the same species, mushroom strains vary to surprising degrees.

A cultivator develops an intimate, co-dependent relationship with every mushroom strain. The features listed below represent a mosaic of characteristics, helping a cultivator define the unique nature of any culture. By observing a culture's daily transformations, a complex field of features emerges, expressing the idiosyncrasies of each strain. Since tons of mushrooms are generated from a few petri dish cultures in a matter of weeks, these and other factors play essential roles in the success of each production run.

Once familiar with a particular culture, the cultivator is alerted to variations from the norm indicating possible genetic decline or mutation. When differences in expression occur, not attributable to environmental factors such as habitat (substrate) or air quality, the cultivator should become alarmed. One of the first features in the telltale decline of a strain is "mushroom aborts." Aborting mushrooms represent failures in the mushroom colony, as a singular organism, to sustain total yield of all of its members to full maturity. The next, classic symptom to be witnessed with a failing strain is the decline in the population of primordia. Fewer and fewer primordia appear. Those that do form are often dwarfs with deformed caps. These are just some of the features to be wary of should your strain not perform to proven standards.

A good strain is easy to keep, and difficult to impossible to regain once it senesces. *Do not underestimate the importance of stock cultures.* And do not underestimate the mutability of a mushroom strain once it has been developed. I use the following checklist of 28 features for evaluating and developing a mushroom strain. Most of these features can be observed with the naked eye.

28 Features for Evaluating and Selecting a Mushroom Strain

The mushroom strain, its unique personality—mannerisms, sensitivities, yield expressions—is the foundation of any mushroom farm. When a strain goes bad, production precipitously declines typically followed by a proliferation of disease organisms. Therefore, cultivators must continuously scrutinize new strains to find candidates worthy of production. Once a strain has been developed, multiple backups are made in the form of test tube slants. Test tube slants ensure long-term storage for future use. The cold storage of test tube slants limits the rate of cell divisions, protecting the strain from mutation and senescence factors.

Although this list is not all-inclusive, and can be expanded upon by any knowledgeable cultivator, it reveals much about the goals cultivators ultimately seek in bringing a strain into culture. However, the following list arises from a uniquely human, self-serving perspective: creating food and medicines for human consumption. If selecting a strain for bioremediation, or for some ecological function, this list would be considerably altered.

Recovery: The time for a mushroom strain to recover from the concussion of inoculation. This is often referred to as "leap off." Oyster and Morel strains are renowned for their quick "leap off" after transfer, evident in as short as 24 hours. Some strains of mushrooms show poor recovery. These strains are difficult to grow commercially unless they are re-invigorated—through strain development and/or media improvement.

Rate of growth: Strains differ substantially in their rate of growth at all stages of the mushroom growing process. Once the mycelium recovers from the concussion of inoculation, the pace of cell divisions quickens. Actively growing mycelium achieves a *mycelial momentum*, which if properly

Grain spawn 3 days and 8 days after inoculation. Visible recovery of spawn 2 days after inoculation is considered good; 1 day is considered excellent.

managed, can greatly shorten the colonization phase, and ultimately the production cycle. (See chart page 118.)

The fastest of the species described in this book has to be the Morels. Their mycelia typically cover a standard 100 × 15-mm petri dish in 3 to 5 days at 75°F (24°C). Oyster strains, under the same conditions, typically take 5 to 10 days depending on the size of the transfer and other factors. All other conditions being the same (i.e., rate of inoculation, substrate, incubation environment), strains taking more than 3 weeks to colonize nutrified agar media, grain, or bulk substrates are susceptible to contamination. With many strains, such as Oyster and Shiitake, a sufficient body of knowledge and experience has accumulated to allow valid comparisons. With strains relatively new to mushroom science, benchmarks must first be established.

Quality of the mycelial mat: Under ideal conditions, the mycelial mat expands and thickens with numerous hyphal branches. The same mycelium

Healthy mushroom mycelium running through cardboard.

under less than perfect conditions casts a mycelial mat finer and less dense. Its "hold" on the substrate is loose. In this case, the substrate, although fully colonized, falls apart with ease. In contrast, mycelium properly matched with its substrate forms a mat tenacious in character. The substrate and the mycelium unify, requiring considerable strength to rip the two apart. This is especially true of colonies of Oyster, King Stropharia, and *Psilocybe* mushrooms.

Some species of mushrooms, by nature, form weak mycelial mats. This is especially true of the initially fine mycelium of Morels. *Pholiota nameko*, the slimy Nameko mushroom, generates a mycelium considerably less tenacious than *Lentinula edodes*, the Shiitake mushroom, on the same substrate and at the same rate of inoculation. Once a cultivator recognizes each species' capacity for forming a mycelial network, recognizing what is a "strong" or "weak" mycelium soon becomes obvious.

Adaptability to single component, formulated, and complex substrates: Some strains are well known for their adaptability to a variety of substrates. Oyster and King Stropharia are good examples. Oyster mushrooms, native to woodlands, can be grown on cereal straws, corn stalks, sugarcane bagasse, coffee leaves, and paper (including a multitude of paper by-products). These species' ability to utilize such a spectrum of materials and produce mushrooms is nothing short of amazing. Although most strains can grow vegetatively on a wide assortment of substrates, many are narrowly specific in their substrate requirements for mushroom production.

Speed of colonization to fruiting: Here, strains can fall into two subcategories. One group produces mushrooms directly after colonization. This group includes the Oyster mushrooms (*Pleurotus pulmonarius*, some warm-weather *P. ostreatus* strains), Lion's Manes (*Hericium erinaceus*), and the Paddy Straw (*Volvariella volvacea*) mushrooms. Others, like the Woodlovers (*Hypholoma capnoides* and *H. sublateritium*) require a sustained resting period after colonization, sometimes taking up to several weeks or months before the onset of fruiting.

Microflora dependability/sensitivity: Some gourmet and medicinal mushroom species require a living community of microorganisms. The absence of critical microflora prevents the mycelium from producing a fruitbody. Hence, these species will not produce on sterilized substrates unless microflora are introduced. The King Stropharia (*Stropharia rugosoannulata*), Zhu Ling (*Polyporus umbellatus*), and the Button mushroom (*Agaricus brunnescens*) are three examples. Typically these species benefit from the application of a microbially enriched soil or "casing" layer.

The Blewit, *Lepista nuda*, has been suggested by other authors as being a species dependent upon soil microbes. However, I have successfully cultivated this mushroom on sterilized sawdust apart from any contact with soil microorganisms. The Blewit may fall into an intermediate category whose members may not be absolutely dependent on microflora for mushroom production, but are quick to fruit when paired with them.

Photosensitivity: The sensitivity of mushrooms to light is surprising to most who have heard that

Phototropic response of *Psilocybe cubensis* to light.

days. With many temperate mushroom strains, the core temperature of the substrate must be dropped below 60–65°F (15–18°C) before mushroom primordia will set. Once primordia is formed, temperatures can be elevated to the 70–80°F (21–27°C) range. This requirement is particularly critical for strains that have evolved in temperate climates, where distinct seasonal changes from summer to fall precede the wild mushroom season. Because of their cold shock requirement, growing these strains during the summer months or, for instance, in Southern California would not be advisable. Strains isolated from subtropical or tropical climates generally do not require a cold shock. As a rule, warm-weather strains grow more quickly, fruiting in one-half the time than do their cold-weather cousins. Experienced cultivators wisely cycle strains through their facility to best match the prevailing seasons, thus minimizing the expense of heating and cooling.

Requirement for high temperature: Many warm-weather strains will not produce at cooler temperatures. Unless air temperature is elevated above the minimum threshold for triggering fruiting, the mycelium remains in stasis, what cultivators term "overgrowth" or "overvegetation." *Volvariella volvacea*, the Paddy Straw mushroom, will not produce below 75°F (24°C) and in fact, most strains of this species die if temperatures drop below 45°F (7.2°C). *Pleurotus pulmonarius*, a rapidly growing Oyster species, thrives between 75 and 85°F (24–30°C) and is not prevented from fruiting until temperatures drop below 45°F (7.2°C). With most temperature-tolerant strains, higher temperatures cause the mushrooms to develop more quickly. Another example is *Pleurotus citrinopileatus*, the Golden Oyster, which fruits when temperatures exceed 65°F (18°C).

Number and distribution of primordial sites: For every cultivator, the time before and during primordia formation is one of high anxiety, expectation, and hope. The changeover from vegetative colonization to this earliest period of mushroom formation is perhaps the most critical period in the mushroom life cycle. With proper environmental stimulation, the cultivator aids the mushroom

mushrooms like to grow in the dark. In fact, most of the gourmet and medicinal mushrooms require and favorably react to light. The development of mushrooms is affected by light in two ways. Initially, primordia form when exposed to light. Even though thousands of primordia can form in response to brief light exposure, these primordia will not develop into normal-looking mushrooms unless light is sustained. Without secondary exposure to light *post primordia formation*, Oyster mushrooms, in particular, malform. Their stems elongate, and the caps remain undeveloped. This response is similar to that seen in high CO_2 environments. In both cases, long stems are produced. This response makes sense if one considers that mushrooms must be elevated above the CO_2 rich environment within a substrate for the caps and subsequently forming spores to be released. Oyster, Shiitake, and Reishi all demonstrate strong photosensitivity.

Requirement for cold shock: The classic initiation strategy for most mushrooms calls for drastically dropping the temperature for several

organism in its attempt to generate abundant numbers of primordia. Aside from the influences of the environment and the host substrate, a strain's ability to produce primordia is a genetically determined trait. Ideally, a good strain is one that produces a population of numerous, evenly distributed primordia within a short time frame.

Site-specific response to low carbon dioxide levels: As the mycelium digests a substrate, massive amounts of carbon dioxide are produced, stimulating mycelial growth but preventing mushroom formation. The pronounced reaction of mycelium to generate primordia in response to lowering carbon dioxide gives the cultivator a powerful tool in scheduling fruitings. Strains vary in their degree of sensitivity to fluctuations in carbon dioxide. Mushroom cultivators who grow Oyster mushrooms in plastic columns or bags desire strains that produce primordia exactly where holes have been punched. The holes in the plastic become the ports for the exodus of carbon dioxide. At these sites, the mycelium senses the availability of oxygen, and forms primordia. This response is very much analogous to the mushroom mycelium coming to the surface of soil or wood, away from the rich CO_2 environment from within, to the oxygenated atmosphere of the outdoors, where a mushroom can safely propel spores into the wind currents for dispersal to distant ecological niches. With strains supersensitive to carbon dioxide levels, the cultivator can take advantage of this site-specific response for controlled cropping, greatly facilitating the harvest.

Number of primordia forming versus those maturing to an edible size: Some strains form abundant primordia; others seem impotent. Those that do produce numerous primordia can be further evaluated by the percentage of those forming compared to those developing to a harvestable stage. Ideally, 90% of the primordia mature. Poor strains can be described as those that produce primordial populations where 50% or more fail to grow to maturity under ideal conditions. Aborted primordia become sites of contamination by molds, bacteria, and even flies.

Number of viable primordia surviving for second and third flushes: Some strains of Oyster

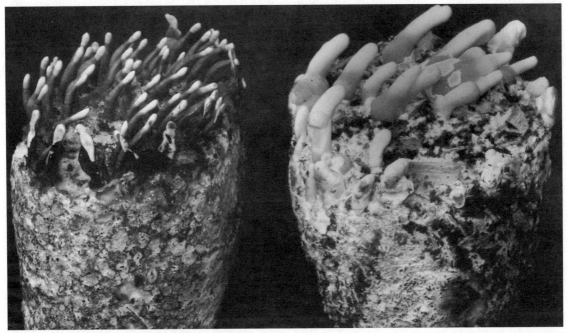

Phototropic response of *Ganoderma lucidum* to light.

and Button mushrooms, especially cold-weather varieties, form the majority of primordia during the first initiation strategy. Many primordia lay dormant, yet viable, for weeks, before development. After the first flush of mushrooms mature and is harvested, the resting primordia develop for the second and subsequent flushes.

Duration between first, second, and third flushes: An important feature of any mushroom strain is the time between "breaks" or flushes. The shorter the period, the better. Strains characterized by long periods of dormancy between breaks are more susceptible to exploitation by insects and molds. By the third flush a cultivator should have harvested 90% of the potential crop. The sooner these crops can be harvested, the sooner the growing room can be rotated into another crop cycle. The rapid cycling of younger batches poses less risk of contamination.

Spore load factors: Over the years, the white Button mushroom, *Agaricus brunnescens*, has been genetically selected for small gills, thick flesh, and a short stem. In doing so, a fat mushroom with a thick veil covering short gills emerged, a form that greatly extended shelf life. As a general rule, once spores have been released in mass, the mushroom soon decomposes. Hence, strains that are not heavy spore producers at the time of harvest are attractive to cultivators. Additionally, the massive release of spores, particularly by Oyster mushrooms, is an environmental hazard to workers within the growing rooms and is taxing on equipment. I have seen, on numerous occasions, the spores from Oyster mushrooms actually clog and stop fans running at several hundred rpms, ruining their motors.

Another mushroom notorious for its spore load is Reishi, *Ganoderma lucidum*. Within the growing rooms, a rust-colored spore cloud forms, causing similar, although less severe, allergic reactions seen to those with Oyster mushrooms. Rather than emitting spores for just a few days, as with most fleshy mushrooms, the woody *Ganoderma* generates spores for weeks as it slowly develops.

Appearance (form, size, and color of the harvestable mushrooms): Every cultivator has a responsibility to present a quality product to the marketplace. Since gourmet mushrooms are relatively new, national standards have yet to be set in the United States for distinguishing grades. As gourmet mushrooms become more common, the public is becoming increasingly more discriminating.

What a cultivator may lose in yield from picking young mushrooms is offset by many benefits. Young mushrooms are more flavorful, tighter fleshed, often more colorful, and able to ship and store longer than older ones. Crop rotation, with much less associated spore load, is likewise accelerated through the harvest of adolescent forms. Diseases are less likely and consistency of production is better assured.

One general feature is common to all mushrooms in determining the best stage for picking: the cap margin. Cap margins reveal much about future growth. At the youngest stages, the cap margin is incurved, soon becoming decurved, and eventually flattening at maturity. The ideal stage for harvest is midway between incurved and decurved. During this period, spore release is well below peak production. Since both the incurving cap and the partial veil (if present) protect the gills, adolescent mushrooms are not nearly as vulnerable to damage. Himematsutake, *Agaricus blazei*, is an excellent example—the "drumstick" form is the prime stage for havesting. (See page 211.) For more information, please consult Chapter 23.

Duration from storage to spoilage—preservation: An important aspect of evaluating any strain is its ability to store well. Spoilage is accelerated by bacteria, which thrive under high moisture and stagnant air conditions. A delicate balance must be struck between temperature, air movement, and moisture to best prolong storage of mushrooms.

Some species and strains are more resistant to spoilage than others. Shiitake mushrooms store and ship far better, on the average, than Oyster mushrooms. Some Oyster mushrooms, especially the slow-forming cold-weather strains, survive under cold storage longer than the warm-weather varieties. In either case, should spores be released, and germinate, bacterial infection quickly sets in.

Abatement of growth subsequent to harvest: Yet another feature determining preservation is whether or not the mushrooms stop growing after picking. Many mushrooms continue to enlarge, flatten out, and produce spores long after they have been harvested. This is especially distressing for a cultivator picking a perfect-looking, young specimen one day only to find it transformed into a mature adult the next. This continued growth often places growers and distant distributors into opposing viewpoints concerning the quality of the product. A strain of *Pleurotus pulmonarius*, especially the widely cultivated "Pleurotus sajor-caju" is one such example. I like to describe this strain of Oyster mushrooms as being "biologically out of control."

Necrosis factors and the protection of dead tissue from competitors: After a mushroom has been picked, tissue remnants become sites for attack by predator insects and parasitic molds. Some species, Shiitake, for instance, have a woodier stem than cap. When Shiitake is harvested by cutting at the base of the stem, the stem butt, still attached to the wood substrate, browns and hardens. As the stem butt dies, a protective skin forms. This ability to form a tough outer coat of cells protects not only the leftover stem remnant from infestation, but also prevents deep penetration by predators. Since most *Pleurotus ostreatus* strains are not graced with this defense, extreme caution must be observed during harvest so no dead tissue remains. The "sajor-caju" variety of *Pleurotus pulmonarius* is surprising in its ability to reabsorb dead tissue, even forming new mushrooms on the dead body remnants of previously harvested mushrooms.

Genetic stability/instability: Since all strains eventually senesce, genetic stability is of paramount concern to every cultivator. Signs of a strain dying are its inability to colonize a substrate, produce primordia, or develop healthy mushrooms. Typical warning signs are a delay in fruiting schedules and an increasing susceptibility to disease. These symptoms are a few of many that suggest strain senescence.

Flavor: Strains of the same species differ substantially in flavor. The cultivator needs to determine each strain's flavor/texture and be sensitive to customer feedback. Americans favor mildly flavored mushrooms whereas the Japanese are accustomed to more strongly tasting varieties. *Pleurotus cit-*

I photographed this unsavory package directly after purchasing it from a major grocery store chain. Mushrooms in this condition, if eaten, cause extreme gastrointestinal discord. This is the "sajor-caju" variety of *Pleurotus pulmonarius*, also known as the Phoenix Oyster mushroom, and has been a favorite of large-scale producers. Subsequent to harvest, hundreds of primordia soon form on the decomposing mushrooms.

rinopileatus, the Golden Oyster mushroom, is extremely astringent until thoroughly cooked, a good example that flavor is affected by the length of cooking. Generally speaking, younger mushrooms are better flavored than older ones. The King Stropharia, *Stropharia rugosoannulata*, is a good example, being exquisitely edible when young, but quickly losing flavor with ·maturity. Shiitake, *Lentinula edodes*, has many flavor dimensions. If the cap surface was dry before picking, or cracked as in the so-called "Donko" forms, a richer flavor is imparted during cooking. Although the cracking of the cap skin is environmentally induced, the cultivator can select strains whose cap cuticle easily breaks in response to fluctuating humidity.

Texture: The stage at harvest, the duration and temperature of cooking, and the condiments with which mushrooms are cooked all markedly affect textural qualities. Judging the best combination of texture and flavor is a highly subjective experience, often influenced by cultural traditions. Most connoisseurs prefer mushrooms that are slightly crispy and chewy but not tough. Steamed mushrooms are usually limp and soft and easily break apart, especially if they have been sliced before cooking. By tearing the mushrooms into pieces, rather than cutting, firmness is preserved. These attributes play an important role in the sensual experience of the mycophagist.

Aroma: Few experiences arouse as much interest in eating gourmet mushrooms as their aroma. When Shiitake and Shimeji are stir-fried, the rich aroma causes the olfactory senses to dance, setting the stage for the taste buds. Once paired with the experience of eating, the aroma signature of each species is a call to arms (or "forks") for mycophagists everywhere. My family begins cooking mushrooms *first* when preparing dinner. The aroma undergoes complex transformations as water is lost and the cells are tenderized. (Please refer to the recipes in Chapter 24.)

Sensitivity to essential elements—minerals and metals: Gray Leatham (1989) was one of the first researchers to note that nanograms of tin and nickel were critical to successful fruitbody formation in Shiitake. Without these minute amounts of tin and nickel, Shiitake mycelium is incapable of fruiting. Manganese also seems to play a determinate role in the mushroom life cycle. Many other minerals and metals are probably essential to the success of the mushroom life cycle. Since these compounds are abundant in nature, cultivators need not be concerned about their addition to wood-based substrates. Only in the designing of "artificial" wood-free media does the cultivator run the risk of creating an environment lacking in these essential compounds.

Ability to surpass competitors: An essential measure of a strain's performance is its ability to resist competitor fungi, bacteria, and insects. Strains can be directly measured by their ability to overwhelm competitor molds, especially *Trichoderma*,

the forest green mold growing on most woods. On thoroughly sterilized substrates, a mushroom strain may run quickly and without hesitation. Once a competitor is encountered, however, strains vary substantially in their defensive/offensive abilities. Oyster mushrooms (*Pleurotus ostreatus*), for instance, are now recognized for nematode-trapping abilities. I have even witnessed Sciarid flies, attracted to aromatic Oyster mycelium, alighting too long, and becoming stuck to the aerial mycelium. The degree by which flies are attracted to a particular Oyster mushroom strain can be considered a genetically determined trait—a feature most cultivators would like to suppress.

Nutritional composition: Mushrooms are a rich source for amino acids (proteins), minerals, and vitamins. The percentages of these compounds can vary between strains. Substrate components contain precursors that can be digested and transformed into tissue to varying degrees by different strains. This may explain why there is such a variation in the protein analysis of, for instance, Oyster mushrooms. The analyses are probably correct. The strains vary in their conversion efficiencies of base substrate components into mushroom flesh.

Production of primary and secondary metabolites: A strain's ability to compete may be directly related to the production of primary and secondary metabolites. All fungi produce extracellular enzymes that break down food sources. Myriad numbers of metabolic by-products are also generated. These extracellular compounds are released through the cell walls of the mycelium, enabling the digestion of potential food sources. Enzymes, such as ligninase that breaks down the structural component in wood, are extremely effective in reducing complex carbon chains, including carbohydrates and hydrocarbons.

Secondary metabolites usually occur well after colonization. A good example is the yellow fluid, the *exudate*, frequently seen collecting at the bottom of aged spawn containers. *Pleurotus* spp., *Stropharia rugosoannulata*, and *Ganoderma lucidum* are abundant producers of secondary metabolites. Complex acids and ethylene-related products are examples of such secondary metabolites. These metabolites forestall competition from other fungi and bacteria.

Production of medicinal compounds: Bound within the cell walls of mushrooms are chains of heavy-molecular-weight sugars, polysaccharides. These sugars compose the structural framework of the cell. Many mushroom polysaccharides are new to science and are named for the genus in which they have been first found, such as lentinan (from Shiitake, *Lentinula edodes*), flammulin or "FVP" (from Enokitake, *Flammulina velutipes*), grifolin or grifolan (from Maitake, *Grifola frondosa*), etc. Research in Asia shows that these cell wall components enhance the human immune system. Cellular polysaccharides are more concentrated, obviously, in the compact form of the mushroom than in the loose network of the mycelium. In traditional Chinese pharmacopeia, the sexually producing organ—in this case, the mushroom—has long been viewed as a more potent source for medicine than its infertile representations.

Cell components other than polysaccharides have been proposed to have medicinal effects. Strain selection could just as well focus on their molecular yields. Precursors in the substrate may play determinant roles in the selective production of these components when matched with various strains.

Optimization of Expansion of Mushroom Mycelium

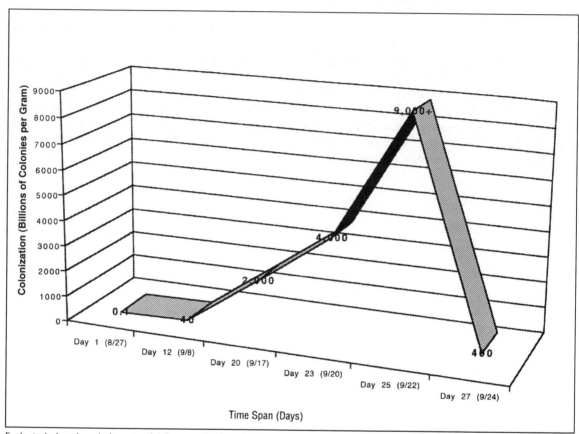

Each strain is unique in its potential for growth. Here, the expansion of mycelial mass peaks on Day 25, with more than 9,000 billion (9 trillion) colonies per gram. Soon thereafter cell viability precipitously declines as substrate nutrition becomes increasingly limited. This trend is a classic biological model that is repeated throughout nature. The goal of the cultivator is to move the mycelium at its peak rate of growth into new media.

CHAPTER 15

Generating Grain Spawn

Making grain spawn is the next step in the exponential expansion of mycelial mass. The intent and purpose of grain spawn is to boost the mycelium to a state of vigor where it can be launched into bulk substrates. The grain is not only a vehicle for evenly distributing the mycelium, but also a nutritional supplement. Whole grain is used because each kernel becomes a *mycelial capsule*, a platform from which mycelia can leap into the surrounding expanse. Smaller kernels of grain provide more points of inoculation per pound of spawn. Many large spawn producers prefer millet, a small kernel grain, because it stores well and end-users like its convenience. Most small-scale, gourmet mushroom growers utilize organically grown rye or wheat grain. Virtually all the cereal grains (including maize) can be used for spawn production. Every spawn maker favors the grain that, from experience, has produced the most satisfactory results.

The preferred rate of inoculation depends upon many factors, not the least of which is cost. If a cultivator buys spawn from a commercial laboratory, the recommended rate is often between 3 and 7% of substrate mass. What this means is that for every 1,000 pounds of substrate (dry weight), 30–70 pounds of spawn (wet weight) is suggested. Since grain spawn is usually around 50% absolute moisture, this rate of inoculation would be equivalent to 1.5–3.0% of dry spawn/dry substrate.

Cultivators who generate their own spawn frequently use a 10–20% rate of moist spawn/dry substrate, or by this example 100–200 pounds of fresh spawn per 1,000 pounds of substrate. This increased rate of spawning accelerates colonization, narrows the window of opportunity for competitor invasion, and significantly boosts yields. Clearly those making their own spawn have an advantage over those buying spawn from afar. One major drawback of high spawning rates is increased thermogenesis, the heating up of the substrate as the mycelium consumes it. Anticipating and controlling thermogenesis is essential for success, and can be used to great advantage. This subject will be explored in detail later on.

Single kernel of rye grain encapsulated with a sheath of mycelium.

Of the many cereal grains used for creating spawn, rye grain is the most popular. Wheat, milo, sorghum, corn, and millet are also utilized. There are two approaches for preparing grain spawn. The first is to submerge grain in a cauldron of boiling water. After an hour of boiling (or steeping), the saturated grain is drained of water (discarded) and scooped into waiting spawn containers. Fitted with a lid having a $3/8$- to $1/2$-inch hole and lined with a microporous filter disk, the grain-filled jar is sterilized in a pressure cooker. This method is widely used

and recommended by many spawn producers because even moisture absorption and consistency are assured.

The second method calls for first placing dry grain into glass spawn jars, adding the recommended amount of water, preferably hot, and allowing the jars to sit overnight. The jars are capped with lids, complete with a $3/8$- to $1/2$-inch hole and fitted underneath with a microporous filter disk. *By allowing the grain to soak for 12 to 24 hours, the heat-resistant endospores of bacteria germinate and become sensitive to heat sterilization.* Before use, the filter disks should be soaked in a weak (5%) bleach solution to dislodge and disinfect any embedded contaminants. The next day, the jars are reshaken by striking them against a rubber tire, or similar surface, to mix together the more moist and drier grain kernels. Once shaken, they are promptly placed into the sterilizer. The advantage of this method is that it is a one-step procedure. A case can be made that starches and other nutrients are preserved with this method since the water is not discarded. Proponents of the first method argue that not only is their

The progressive colonization of sterilized grain by mushroom mycelium.

starting material cleaner, but this second technique causes the grains to have an uneven moisture content. The reader must decide which is most suitable. Neither method, in my opinion, merits endorsement over the other.

With excess water, grain kernels explode, exposing the nutrients within, and making them more susceptible to contamination. Exploded grain kernels also cause clumping and become sites of depressed gas exchange, environments wherein bacteria proliferate. The shape of the intact grain kernel, with its protective outer surface, selectively favors the filamentous mushroom mycelium and produces spawn that separates readily upon shaking.

Both of the above-described preparation techniques have their strong and weak points.

Suitable Containers for Incubating Grain Spawn

16 fl. oz. mineral spring water bottle
quart mason jars, 1-liter bottles
$^1/_2$-gallon jars
1-gallon jars
$2^1/_2$-gallon jars
Polypropylene plastic bags

One method for preparing grain spawn is to simply pour dry grain into glass jars, then add water, allow to sit overnight, and then sterilize. Advantages are a one-step process, less fuel consumption, and less handling. One disadvantage is uneven water absorption.

Formulas for Creating Grain Spawn

Moisture content plays a critical role in the successful colonization by mushroom mycelium of sterilized grain. If the grain is too dry, growth is retarded, with the mycelium forming fine threads and growing slowly. Should too much water be added to the grain, the grain clumps, and dense, slow growth occurs. Higher moisture contents also encourage bacterial blooms. Without proper moisture content, spawn production is hampered, even though all other techniques may be perfect.

The optimum moisture for grain spawn falls within 38–55% moisture, with an ideal around 50%. To determine the moisture content of grain from any supplier, before formulating, weigh 100 grams of the grain, dry out the grain, and reweigh the remaining mass. (This can easily be determined by drying out the moistened grain in an oven for 300°F (150°C) for 8 hours.) The difference in weight is the water lost, or the percentage moisture. Now water is added to achieve a targeted moisture content. Once cooked, a sample of grain is taken and oven dried. To check the proposed formula, just take the mass of the lost water divided by the total mass of dried grain *and* the lost water. This will give you a moisture percentage. Remember that moisture percentage is the mass of water divided by total mass, lost water included. This is *not* a ratio of water to dry mass, but a percentage of water over total mass. (This is a common mistake among certain schools of Shiitake growers and wood lot managers.) Once a targeted moisture content is achieved, spawn growers rely on volumetric scoops customized to the new formula.

Since grain comes to the consumer with an inherent moisture content of 8–15%, less water is added than might be expected to achieve the right moisture content for spawn production. Each cultivator may want to adjust the following proportions of water to grain to best fit his or her needs. Keep in mind that 1 liter (1,000 ml) of water weighs 1 kilogram (1000 g). A quart is almost a liter and for the purposes of the mushroom cultivator can be used interchangeably. (The amount of grain within each vessel is specified in the following formulas. A variation of only 5–7% between the two volumes is not statistically significant.) Gypsum is added to help keep the kernels separated after sterilization and to provide calcium and sulfur, basic elements promoting mushroom metabolism. (See Stoller, 1962; Leatham and Stahlman,1989.)

A delicate balance between the mass of grain and added water must be preserved to promote the highest quality spawn. As the spawn container is increased in its volume, slightly less water is added proportionately. Whereas the percentage of moisture content can be nearly 60% in a small spawn jar, a large container will have a moisture content of only 38%. Anaerobic environments are encouraged with larger masses of grain, a phenomenon that necessitates a drier medium and extended exposure to pressurized steam. Cultivators should adjust these baseline formulas to best meet their specific circumstances. Jars and bags must be fitted with microporous filters for adequate gas exchange.

I sterilize the 16-ounce or quart (liter) jars for only 1 hour at 15 psi (= 1 kg/cm2), the ¹/₂ gallons for 1¹/₂ hours, the gallon jars for 2 hours, and the standard spawn bags for 4 hours. The spawn bags featured in this book have a maximum volume of 12,530 milliliters when filled to the brim, although cultivators usually load the spawn bags to ¹/₃ to ¹/₂ capacity. Using the aforementioned formula, each spawn bag weighs 10 pounds (= 4,826 grams). These bags are best inoculated with 200–300 milliliters of fermented, liquid mushroom mycelium using the techniques described further on. Once inoculated, the bags are laid horizontally for the first week and

Spawn incubating in ¹/₂-gallon (2-liter), 1-gallon (4-liter), and 2.5-gallon (10-liter) containers. Note filter media that prevent contamination but allow respiration.

gentling agitated every 3 days with the filter patch topside, until fully colonized. Spawn generated in bags is far easier to use than from jars.

For a comparison of grains, their moisture contents, and kernels sizes, refer to page 43 in *The Mushroom Cultivator* by Stamets and Chilton (1983). Test batches should be run prior to commercial-scale cycles with sterilization indicator papers. Adjustments in pressure must be made for those more than 3,000 feet above sea level.

Most people create volumetric scoops corresponding to one of the above-mentioned masses. Many cultivators build semiautomatic grain dispenser bins to facilitate the rapid filling of spawn containers. These are similar in design to those seen in many organic food co-ops in North America.

The grain used for spawn production must be free of fungicides, and ideally be organically grown. Grain obtained in the spring was probably harvested 6 or more months earlier. The resident contamination population gradually increases over time. With the proliferation of more contaminants per

pound of grain, cultivators will have to adjust their sterilization schedules to compensate. Experienced cultivators are constantly searching for sources of fresh, high-quality grain with endemically low counts of bacteria and mold spores.

With one brand of commercially available rye grain, a cup of dry grain has a mass of 210 grams. Four cups is approximately a liter. Therefore a single petri dish culture can generate from 1 to 5 liters of spawn, utilizing the traditional wedge transfer technique. These techniques are described next.

First-Generation Grain Spawn Masters

The first time mushroom mycelium is transferred onto grain, that container of spawn is called a grain master, or G^1. The preferred containers for incubating grain masters are traditionally small glass jars or bottles, with narrow mouths to limit exposure to contaminants. Since the grain master is used to

Grain Formulas for Spawn Production

16-Ounce Mineral Spring Bottles	Quart or Liter Jars	1/2-Gallon or 2-Liter Jars
100 grams rye (approx. 125 ml)	200 grams rye	480 grams rye
150 ml water	220 ml water	400 ml water
.5 grams gypsum	1 gram gypsum	2 grams gypsum
(60% moisture*)	(52 % moisture*)	(45 % moisture*)
Gallon or 4-Liter Jars	2 1/2-Gallon 10-Liter Jars	Standard Spawn Bags 7.5 × 8.25 × 4.75 inches)
800 grams rye	2,200 grams rye	3,300 grams rye
600 ml water	1,500 ml water	1,400 ml water
4 grams gypsum	8 grams gypsum	12 grams gypsum
(43% moisture*)	(40% moisture*)	(38% moisture*)

* These moisture contents are not meant to be taken literally. The natural moisture content inherent within "dry" grain can affect absolute moisture by 15% or more. Properly dried grain should have 5–10% ambient moisture. With the slightest increase above this level, the moister microclimate will encourage proliferation of bacteria. Only a few surviving bacterial colonies are needed to soon become millions within a week.

Filling ¹/₂-gallon jars with grain.

mushroom mycelium into wedges or squares using a sterilized scalpel. Prior to this activity, the surface space where the transfers are to take place has been aseptically cleaned. The hopeful spawn maker has showered, washed, and adorned newly laundered clothes. Immediately prior to doing any set of inoculations, the cultivator washes his or her hands and then wipes them with 80% rubbing alcohol (isopropanol). If working in front of a laminar flow hood, the freshest, sterilized material is kept upstream, with the mycelium directly downstream. *The cultivator prioritizes items on the inoculation table by degree and recentness of sterility.* The same attention to movement that was used to inoculate nutrient-filled petri dishes in Chapter 12 is similarly necessary for successful production of grain spawn. Attention to detail, being aware of every minute movement, is again critical to success.

generate 100 to 1,000, even 10,000 times its mass, special attention is focused on its purity. Otherwise, the slightest amount of contamination is exponentially expanded with each step, not by a factor of 10, but ultimately by a factor of thousands! Molds have advantages over mushrooms in that within 2 to 4 days, every spore can send up thousands of microscopic tree-like structures called *conidiophores* on whose branches are dozens of more mold spores. (See *The Mushroom Cultivator* [1983], Chapter 13, pages 233–317.) Mushroom mycelium, on the other hand, typically expands as a linear extension of cells. In a jar holding thousands of kernels of grain, a single kernel of grain contaminated with a mold like *Penicillium*, surrounded by tens of thousands of kernels impregnated with pure mushroom mycelium, makes the entire container of spawn *useless* for mushroom culture.

A single 100 × 15-mm petri dish culture can inoculate 4–20 cups of sterilized grain. The traditional transfer method calls for cutting the

Pressure cooker useful for sterilizing agar and grain media.

Steps for Generating First-Generation Grain Spawn Masters

Step 1. Visually ascertain the purity of a mushroom culture, selecting a petri dish culture showing greatest vigor. Ideally, this culture should be no more than 2 weeks old, and there should be a margin of uncolonized media along the inside peripheral edge. This uncolonized zone, approximately ¹/₂-inch (1.30 cm) in diameter, can tell the cultivator whether or not any viable contaminant spores have recently landed on the media. Once the mycelium has reached the edge of the petri dish, any contaminant spores, should they be present, are dormant and lie invisible upon the mushroom mycelium only to wreak havoc later.

Step 2. Although the contents within may be sterilized, the outer surface of the pressure cooker is likely to be covered with contaminants that can be transferred via hand contact. Therefore, the outside of the pressure cooker should be thoroughly wiped clean prior to the sterilization cycle. Open the pressure

Sterilization indicator test strips are placed into a few grain-filled jars to test sterilization. Note the letter "K" appears when sterilization has been achieved.

Grain-filled ¹/₂-gallon jars ready for loading into a commercial autoclave.)

cooker in the laboratory clean room. Ideally, the pressure cooker has formed a vacuum in cooling. If the pressure cooker in use does not form a vacuum, outside air will be sucked in, potentially contaminating the recently sterilized jars. The pressure cooker should be placed in the clean room directly after the sterilization cycle and allowed to cool therein. (I usually place a paper towel, saturated with isopropanol, over the vent valve as an extra precaution to filter the air entering the pressure cooker.) Another option is to open the pressure cooker in front of a laminar flow bench at the moment atmospheric pressure is re-achieved. Remove the sterilized grain jars from the pressure cooker. Place the grain-filled jars upstream nearest to the laminar flow filter. Sterilize the scalpel by flaming until red hot.

Step 3. Directly cut into the petri dish culture. (The blade cools instantly on contact.) Drag the blade across the mycelium-covered agar, creating eight or more wedges. Replace the petri dish lid.

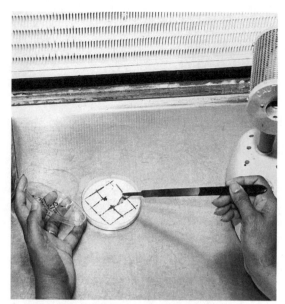

Cutting mycelium from the nutrient agar medium.

Step 4. Loosen the lids of the jars to be inoculated so you can lift them off later with one hand. Re-flame the scalpel. Remove the petri dish cover. Spear two or more wedges simultaneously. Replace the petri dish cover. While moving the wedges of mycelium upstream to the jars, remove the lid of the jar to be inoculated, and thrust the wedges into the sterilized grain. Replace and screw the lid tight. Shake the jar so the wedges move throughout the interior mass of the grain, with the intention that strands of mycelium will tear off onto the contacted grain kernels.

Step 5. Set the inoculated jar of grain onto the shelf and incubate them, undisturbed for several days.

Step 6. Three days from inoculation, inspect each jar to determine two preconditions: first, recovery of the mycelium, or "leaping off" onto contacted grain kernels; and second, the absence of any competitor molds, yeasts, mites, or bacteria. If these preconditions are satisfied, to the best of your knowledge, continue to the next step.

Step 7. Seven days after inoculation, shake each jar again, provided no competitors are detected. Ten to fourteen days after inoculation, incubated at 75°F (24°C), each jar should be fully colonized with mushroom mycelium. If colonization is not complete 3 to 4 weeks after inoculation, something has probably gone awry with the process. Some of the more common causes of slow colonization include unbalanced moisture content, contaminants, weak strain, residual fungicides in the grain, poor-quality grain, etc.

Spawn at the peak of cell development is the best to use, correlating to about 2 weeks after inoculation. The key concept here: *to keep the mycelium running at its maximum potential throughout the spawn generation process.* With over-incubation the grain kernels become difficult to break apart. Along the path of cell expansion, it's important for the myceliated grain kernels to separate so they can be evenly dispersed throughout the next generation of substrates. Over-incubation results in clumping, kernels without structural integrity, and an invitation to disease.

Grain masters are kept at room temperature for a maximum of 4 to 8 weeks from the time of original inoculation, but are best used within a week of full colonization. Some farms refrigerate grain masters until needed. I strongly discourage this practice. If the grain masters are not used within 2 weeks of full colonization, over-incubation may result, slowing of cell divisions, with a corresponding decline in vigor. The rule here: Use it or lose it.

Second- and Third-Generation Grain Spawn

The next generation of spawn jars is denoted as G^2. Each grain master can inoculate 5 to 20 times its mass. Many start with narrow mouth quart mason jars for grain masters, and use $1/2$-gallon or 2-liter jars for second-generation spawn. (For use of bags as spawn containers, see page 130.) Third-

Transferring several squares of mycelium from the nutrified agar medium into sterilized grain.

generation spawn is typically in bag form and is sold to consumer-growers.

A standard inoculation rate would be 1 quart (liter) grain master to 5½ gallons, in other words, a 1:10 expansion. A diluted inoculation, on the verge of being unsuccessful would be 1 quart grain master to 20½ gallons, in other words, a 1:40 expansion. Exceeding a 1:40 expansion of mycelium is likely to be associated with a >20% failure rate, a percentage unacceptable to any spawn laboratory. Not only can the loss be measured in terms of failure to mature, but also each failed spawn jar is likely to be center stage for releasing thousands of contaminants back into the laboratory. Liquid-inoculation techniques allow a much greater exponent of expansion than the traditional method described here. (See Liquid Inoculation Techniques described in this chapter.)

Step-by-step instructions follow for a classic grain-to-grain inoculation. As before, the cleanest items should be prioritized nearest to the micron filter. Adherence to sterile inoculation techniques should be strictly observed.

Steps for Creating Second- and Third-Generation Grain Spawn

The cultivator has sterilized grain in containers larger than the grain master, i.e., ½-gallon or gallon jars, following the standard procedures already outlined. For every quart grain master, 5½-gallon jars are recommended, essentially a 1:10 expansion.

Step 1. Select a grain master showing even, luxuriant growth. Avoid spawn jars having zones of heavy growth, discoloration, or excess liquid.

Step 2. Using a cleaned rubber tire, carefully slam the jar against it, loosening the grain. If the spawn is overgrown, more forcible shaking is required before the spawn kernels will separate. Do not strike the jar against the palm of your hand! Be careful! This author, at the time of this writing, is recovering from a sliced wrist after a brisk visit to the hospital emergency room, caused by a glass jar shattering on his palm during shaking, requiring multiple stitches.

Step 3. Once the grain master has been shaken, loosen the lids of the jars which will receive the spawn. Remove the lid of the grain master and set it aside. With your favored hand (right), move the grain master upstream to the first jar, hovering inches above it. With your other hand, remove the lid, and hold it in the air. By tilting downward and rotating the grain master counterclockwise, kernels of spawn fall into the awaiting jar. Replace the lid of the jar just inoculated and continue to the next. By the time the tenth jar is inoculated, the spawn jar should be empty. Repeated transfers eventually lead to an even dispersal of spawn each time. Precise measurement is desirable but not absolutely critical with this suggested rate of expansion. However, as one becomes more experienced, inoculation rates achieve a high degree of regularity.

Inoculating G² gallon jars of sterilized grain from ¹/₂ gallon (2-liter) G¹ masters.

the bottom recesses of the jar, in effect, rotating and mixing the grain mass.

Step 7. In 7 to 10 days, reinspect each jar to determine even dispersal of growth centers. Should some jars show regions of growth and no growth, another shaking is in order. Those showing good dispersion need not be disturbed. Here the discretion of the cultivator plays an important role. If any unusual pungent odors are noticed, or if the grain appears greasy, contamination may be present although not yet clearly visible.

Step 8. By day 14, all the jars should be thoroughly colonized by mycelium. With Oyster, Shiitake, Enokitake, Reishi, and King Stropharia, the mycelium has a grayish white appearance 48 hours before flushing out with bright white mycelium.

Each second-generation spawn jar can be used for inoculating another set of grain jars, for instance, 5 gallon jars containing twice the amount of grain as the ¹/₂ gallon containers, in effect another 1:10

Step 4. Once inoculated, the lids are tightened securely. Each jar is then shaken to evenly disperse the grain master spawn kernels through the sterilized grain. Thorough shaking encourages fast grow-out. As the jars are shaken, note the rotation of the myceliated grain kernels through the jar.

Step 5. Set the second-generation spawn jars upon a shelf or rack in a room maintained at 75°F (24°C). The jars should be spaced at least ¹/₂ inch apart. Closely packed jars self-heat and encourage contamination. This spawn producer prefers that the jars incubate at an incline, allowing for more transpiration.

Step 6. After 3 to 4 days, each jar is shaken again. As before, striking the jars against a rubber tire or similar surface can loosen the grain. Grasping each jar firmly, accelerate each jar downward in a spiral, pulling back at the end of each movement. This technique sends the top grain kernels deep into

Jars furthest downstream are inoculated first and removed.

Grain inoculated with mycelium but contaminated with bacteria. Note greasy appearance of grain kernels. Bacterially contaminated grain emits a distinct, unpleasant odor.

Throughout every stage in the grain expansion process, any hint of contamination, *especially smell*, the texture of the grain, or unusual colorations, should be considered warning signs. The spawn maker soon develops a sixth sense in choosing which spawn jars should be expanded, and which should be avoided. Most spawn producers only select a portion of the spawn inventory for further propagation. The remainder are designated as "terminal," and are not used for further expansion onto sterilized grain. Unless contaminated, these terminal spawn jars usually are of sufficient quality for inoculating bulk substrates. Of course, the spawn manager can always exercise the option of using first-, second-, or third-generation grain spawn for inoculating sawdust or straw.

In effect, what the spawn maker has accomplished is taking a single petri dish, and in three generations of transfers, created 250 gallons of spawn. A stack of twenty petri dishes can give rise to 5,000 gallons of spawn! This places a whole new perspective on the sheer biological power inherent within a single test tube slant, which can easily inoculate a sleeve of twenty petri dishes. Most laboratories do not fully realize the potential of every culture. In many cases, spawn expansion is terminated at G^2. Many spawn managers choose not to "chase the optimum." Few laboratories are large enough to accommodate the end result of the methods described here.

An alternative method for generating spawn is liquid culture. This method saves time and money, and is less susceptible to contamination. These techniques are described further on.

The next step is for each of these third-generation spawn units to inoculate 10 to 20 times its mass in sawdust or straw. See Chapters 16 and 17.

expansion. You can use $2^1/_2$-gallon (10-liter) jars or bags at a similar rate. These would be denoted as G^3. Third-generation grain spawn is inoculated in exactly the same fashion as second-generation grain spawn. However, contamination is likely to go unobserved. Some large spawn laboratories successfully generate fourth-generation spawn. However, contamination outbreaks discourage most from pushing this expansion any further. As the mass of sterilized grain is increased within each larger container, anaerobic conditions can more easily prevail, encouraging bacteria. These larger containers require more aeration, a feat that is accomplished with frequent shaking (every 48 to 72 hours), greater filter surface area, and near horizontal incubation. When the large containers are laid horizontally, the surface area of the grain-to-air is maximized, providing better respiration for the mushroom mycelium.

Autoclavable Spawn Bags

Autoclavable bags have been used by the mushroom industry for nearly 40 years. Primary uses for autoclavable bags are for the incubation of grain and

Eleven pounds (5 kg) of grain spawn in 2¹/₂-gallon (10- to 11-liter) glass jar and polypropylene bag. Glass jars are reusable, whereas spawn bags are currently not recycled. However, since bags are easier to handle, this form is the most commonly sold to mushroom growers by commercial spawn laboratories.

sawdust. Preferences vary widely between cultivators. Flat, non-gusseted bags are popular for incubating grain spawn. The more grain filled into a bag, the greater the danger of poor gas exchange, a major factor leading to contamination. Three-dimensional gusseted bags are used primarily for holding non-supplemented and supplemented sawdust. The proper handling of these bags is critical to their successful use. Bags contacting hot surfaces become elastic, deform, and fail.

Currently the industry uses polypropylene or polymethylpentene bags with and without microporous filters. Over the years, a number of patents have been awarded, some long since expired. The use of plastic bags has had a drastic impact on the way many cultivators generate spawn. Numerous patents have been awarded for bags specifically designed for mushroom culture. The earliest patent I can find is one from 1958 awarded to a French man by the name of Guiochon (U.S. #2,851,821).

His cylindrical bag resembles the methods still widely in use by Asian cultivators. In 1963 several patents were awarded in London (#985,763; #1,366,777; and #1,512,050). R. Kitamura and H. Masubagashi received a patent (#4,311,477) for a spawn culture-bag in 1982.*

About a dozen bags are currently available to mushroom cultivators, some borrowed from the hospital supply industry. Cellophane deserves reexamination since it is made from wood cellulose and completely biodegradable. If problems with seam integrity, tensile strength, and heat tolerance could be improved, spawn bags made of this environmentally friendly material could eliminate the widespread use of throwaway plastics. An

---

* Other patents, too numerous to list here, were also awarded. Many redesigned the seam, the filtration media, and/or sometimes the wording to qualify for a new variation.

Gallon jars of third-generation grain spawn incubating.

Grain spawn ready for use. Healthy mycelium is usually tenacious, holding the grain together.

advantage of cellophane-like materials is that the mushroom mycelium eventually consumes the very bag in which it is incubated.

Autoclavable bags are inoculated with grain masters and are second or third generation. Agar-to-grain inoculation from petri dish cultures to bags is awkward and impractical unless liquid-inoculation techniques are employed. (These techniques are fully described later on.) Bags are filled with premoistened grain, and the lips folded closed. Some spawn producers use spring-activated clothespins, paper clips, or plastic tape to hold the folds closed. I prefer to simply press the bags together with flaps folded. As the bags are sterilized, the contents exceed the boiling point of water, and gases are released. If the bags are sealed before loading, explosion or "blow-outs"—holes where live steam vented—is likely.

In the standard 18 × 8 × 5-inch gussetted auto-clavable spawn bag featuring a 1-inch filter patch, no more than 3,500 grams of dry grain should be used.* One All-American 941 pressure cooker can process 50 pounds of dry rye grain in one run. However, the pressure cooker—with its tightly packed contents—should be kept at 15+ psi for at least 4 to 5 hours to ensure even and full sterilization.

If the grain is first boiled or simmered in hot water before filling, even moisture absorption is assured. Excess water collecting at the bottom of the bags often leads to disaster. If this water is reabsorbed back into the media by frequent shaking or by turning the bags so that the excessively moist grain is on top, the cultural environment is soon rebalanced in favor of mycelial growth. Standing water, at any stage in the mushroom cultivation process, encourages competitors. Many spawn producers add 20–30 grams of calcium sulfate to the

---

* Please see formulas on page 122. Spawn incubation bags are available from suppliers listed in Appendix IV: Resource Directory.

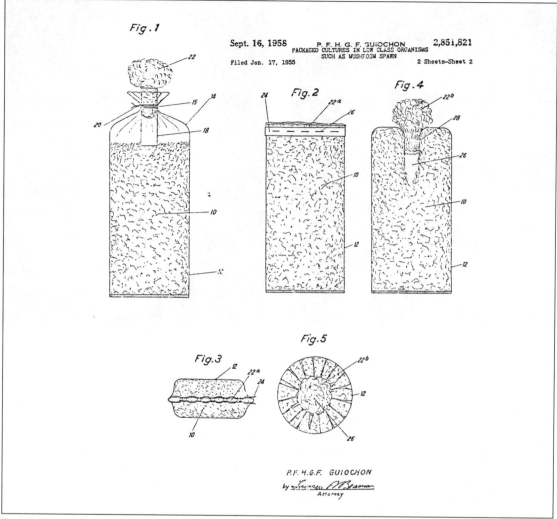

Fig.1

Sept. 16, 1958   P. F. H. G. F. GUIOCHON   2,851,821
PACKAGED CULTURES IN LOW CLASS ORGANISMS
SUCH AS MUSHROOM SPAWN

Filed Jan. 17, 1955   2 Sheets-Sheet 2

Fig.2

Fig.4

Fig.5

Fig.3

P.F. H.G.F. GUIOCHON

by _____
Attorney

The use of heat-tolerant plastic bags greatly advanced the practicality of the bulk processing of grain and sawdust. One of the first patents for this innovation was awarded to Guiochon. (U.S. Patent #2,851,821) in 1958.

grain, when dry, to help keep the kernels separated after autoclaving.

After sterilization for 2 hours at 15 to 18 psi, if the bags are separated, or 4 to 5 hours if the bags are tightly packed, the bags are removed and allowed to cool in the pure windstream coming from the laminar flow bench. An alternative is to allow the grain bags to cool within the pressure vessel, provided it is of the type that holds a vacuum.

The vacuum is then "broken" by allowing clean-room air to be sucked in. If the pressure cooker does not hold a vacuum, then it should cool within the sterile laboratory to preserve sterility. In either case, I place a presterilized cotton towel, soaked in alcohol, over the vent cock to act as a filter. For equalizing the pressure in a larger autoclave, air passes directly through a microporous filter into the vessel's interior.

Once the bags are cooled, unfold them by hand, being careful to only touch the outer surfaces of the plastic. A jar of spawn is selected, shaken, and opened. Using a roll-of-the wrist motion, allow spawn to free-fall into each bag at a recommended rate within a range of 1:10 to 1:20. The bag is then laid down to open into the airstream. The top 2 inches of the bag is positioned over the element of a heat sealer and expanded open, again by only touching the outer surfaces of the plastic. The clean air coming from the laminar flow filter inflates the bag. I gently press on the sides, which further inflates them before sealing. The top arm of the sealer is brought forcibly down, often two or three times in rapid succession, pausing briefly to allow the plastic seam to resolidify. Each bag is squeezed to determine whether the seam is complete and to detect leaks. (Often pinhole leaks can be detected at this stage. Having a roll of plastic packing tape handy, 3–4 inches wide, solves this problem by simply taping over the puncture site.)

If the bags hold their seal with no leaks, shaking each bag should mix through the spawn. This cultivator strives to capture enough air within each bag so that when they are sealed, each bag appears inflated. Inflated bags are much easier to shake and support better mycelial growth than those without a substantial air plenum.

Spawn bags should be set on a shelf, spaced 1/2 inch or more from each other, to counteract heat generation. After 4 days, each bag should be carefully inspected, laid on a table surface, and rotated to disperse the colonies of mycelium. In another week, a second shake may be necessary to ensure full and even colonization.

The advantages of using bags for processing grain spawn are

- In the limited space of a sterilizer, more grain can be treated using bags than jars.

- Bags, if they break, are not dangerous. Being cut by glass jars is one of the occupational hazards of spawn producers.

- Since the bags are pliable, spawn can be more easily broken up into individual kernels and distributed into the next substrate. The process of spawning is simply easier.

Liquid-Inoculation Techniques

A rainstorm is a form of liquid-inoculation. The earliest fungophile, unwittingly or not, engaged in liquid-inoculation techniques. Every time mushrooms are eaten, cooked or washed, spores are disseminated in liquid form. Nature's model can be modified for use within the laboratory. Currently several strategies incorporate liquid-inoculation methods. The advantages of liquid-inoculation are *the speed of colonization, the purity of spawn, and the ease of handling*.

Spore-Mass Inoculation

The ultimate shortcut for culturing mushrooms is via spore-mass/liquid-inoculation directly into bulk substrates. Primarily used in China, this technique works well with Oyster and Shiitake mushrooms but is also applicable to all the mushroom species discussed in this book. In effect, this process parallels the technology of the brewery and pharmaceutical industries in the cultivation of yeasts, *Saccharomyces cerevisiae* and allies. Large fermentation vessels are filled with sugar broth, inoculated with pure spores, incubated, and aerated via air compressors.

Spore-mass inoculation of sterilized substrates is limited to those species that form mushrooms under totally sterile conditions. (Spores collected from wildly picked mushrooms have too many contaminants.) Those mushrooms that require the presence of microflora, such as the White Button mushroom (*Agaricus brunnescens*) and the King Stropharia (*Stropharia rugosoannulata*), are excluded. The key requirement is that the parent mushroom fruits on a sterilized substrate, within a sterile environment, and sporulates abundantly. The following mushrooms are some of those that qualify. All are wood or straw saprophytes.

Agrocybe aegerita
Flammulina velutipes
Ganoderma lucidum and allies
Lentinula edodes
Macrocybe gigantea
Pholiota limonella
Pholiota nameko
Pleurotus citrinopileatus
Pleurotus cystidiosus
Pleurotus djamor
Pleurotus eryngii
Pleurotus euosmus
Pleurotus ostreatus
Pleurotus pulmonarius

A practical approach is to first sterilize a half-filled gallon container of wood chips, which is then inoculated with grain spawn. After several weeks of incubation, depending on the species, mushrooms form within the environment of the gallon jar. (Supplementation, for instance, with rice bran, oatmeal, or rye flour, facilitates mushroom formation.) Once mature, the mushrooms are aseptically removed and immersed in sterilized water. Commonly the water is enriched with sugar-based nutrients and trace minerals to encourage rapid spore germination. Millions of spores are washed into the surrounding broth. After vigorous shaking (a few seconds to a few minutes), the spore-enriched liquid is poured off into another sterile container, creating a *spore-mass master.*

Spores begin to germinate within minutes of contact with water. Immediately upon germination, and as the mycelium grows, respiration cycles engage. Therefore, the liquid broth must be aerated or the mycelium will be stifled. The method most used by the fermentation industry is aeration via oilless compressors pushing air through banks of microporous filters. The air is distributed by a submerged aerating stone, a perforated water propeller, or the turbulence of air bubbles moving upwards, as in a fish aquarium. As the mass of the mycelium increases, and as the filters become clogged with airborne "dust," pressure is correspondingly increased to achieve the

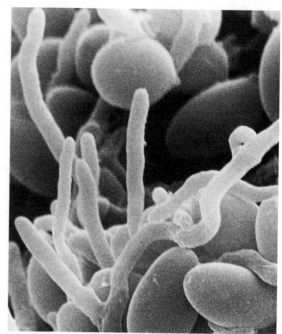

Scanning electron micrograph of spores in a frenzied state of germination. Note clamp connection on right.

same rate of aeration. The vessels must be continuously vented to exhaust volatile metabolites. The goal is to oxygenate the broth.

Each spore-mass master can inoculate 100 times its mass. For instance, if one removes a Shiitake mushroom, 4–5 inches in diameter, from a jar of sterilized sawdust, and then places that mushroom into a gallon of sterilized water, the spore-enriched broth, the spore-mass master can inoculate 100 gallons of nutrified liquid media. The functional range of expansion is 1:25 to 1:200, with a heavier inoculation rate always resulting in faster growth. After 2 to 4 days of fermentation at 75°F (24°C), a second-stage of expansion can occur into enriched sterilized water, resulting in yet another 25- to 200-fold expansion of mycelial mass.

Success of the fermentation process can be checked periodically by streaking .1 milliliter across a sterilized nutrient-filled petri dish and incubating for a few days. (See page 135.) Additionally, contaminants can be immediately detected through odor and/or through examination of the liquid

Pressurized vessels in China designed for spore-mass fermentation.

munity of spore matings behaves quite differently than paired individuals. San Antonio and Hanners (1984) are some of the first Western mycologists to realize that grain spawn of Oyster mushrooms could be effectively created via spore-mass inoculation.

The most aggressive strains out-race the least aggressive strains to capture the intended habitat. Recent studies have shown that these aggressive strains overpower and invade the cellular network of competing strains. Dr. Alan Rayner (1988) in studies at the University of Bath described this form of genetic theft as "non-self fusions" between genetically different mycelial systems within the same species. This ability to adapt has made fungi one of the most successful examples of evolution in the biological arena.

Spore-mass fermentation techniques are not yet widely used by North American or European cultivators. Concern for preserving strain stability, lack of experience, equipment, and intellectual conflicts are contributing factors. In mushroom culture,

sample with a microscope. Any gases produced by bacteria or contaminants are easily recognizable, usually emitting a uniquely sour or musty and sometimes sickeningly sweet scents.

The liquid spore-mass inoculum can be transferred directly onto sterilized substrates such as grain, sawdust, straw, cottonseed hulls, etc. If the liquid-inoculum is sprayed, even colonization occurs. If poured, the liquid-inoculum streams down through the substrate, following the path of least resistance. Unless this substrate is agitated to distribute the mycelium, colonization will be uneven, resulting in failure.

Theoretically, the germination of spores in mass creates multitudes of strains that will compete with one another for nutrients. This has been long accepted as one of the Ten Commandments of Mushroom Culture. Scientists in China, whose knowledge had not been contaminated by such preconceptions first developed spore-mass inoculation techniques to an industrial level. Only recently have Western mycologists recognized that a large com-

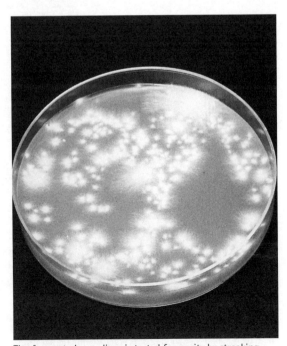

The fermented mycelium is tested for purity by streaking sample droplets across a nutrient-media-filled petri dish. After 48 to 72 hours, pure colonies of mycelium (or contaminants) are easily visible.

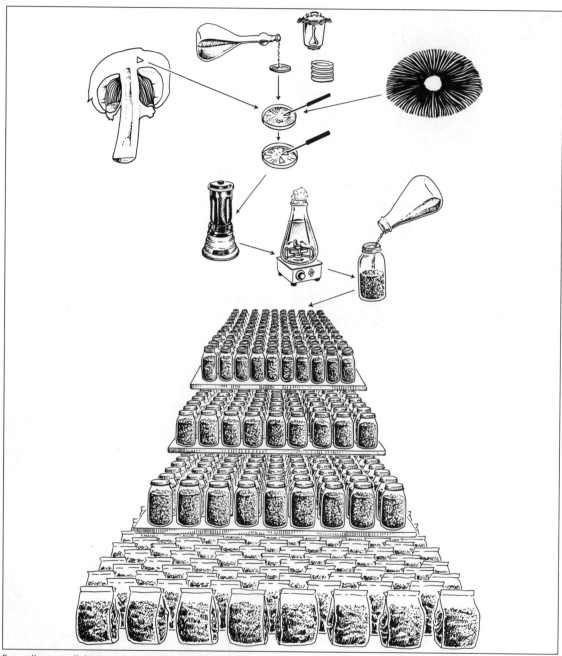

Expanding mycelial mass using a combination of liquid fermentation and traditional grain-transfer techniques. After fermentation for 3 to 4 days, 100 quart (liter) jars of sterilized grain are liquid-inoculated. These are denoted as G^1 masters. In 7 to 10 days, 1,000 $^1/_2$-gallon (2-liter) jars are inoculated from the G^1 masters. These are called G^2. Then, 10,000 G^3 gallon jars are inoculated from the G^2s. Once grown through, 100,000 bags of sawdust spawn can be generated. Each sawdust bag can be expanded by a factor of 10 into supplemented sawdust, creating 1,000,000 fruiting blocks. At 1–2 pounds of mushrooms per block, more than 1,000,000 pounds of mushrooms can be grown from one petri dish in as short as 80 days depending upon the species and strain.

intransigence to new ideas has prevailed, often because the slightest variation from the norm has resulted in expensive failures. Since the health of any economy is based on its diversity, the emergence of organically minded gourmet mushroom growers is creating a fertile intellectual habitat for many innovative technologies.

Liquid-Inoculation Techniques: Mycelial Fragmentation and Fermentation

This method differs from the spore-mass inoculation techniques in that the starting material is *dikaryotic mycelium, not spores*. In short, the cultivator chops up the mycelium into thousands of tiny fragments using a high-speed blender, allows the mycelium to recover, and transfers dilutions of the broth into jars or bags of sterilized grain. I prefer this technique as it quickly generates high-quality spawn, eliminating several costly steps. Once perfected, most spawn producers find grain-to-grain transfers obsolete. The time not spent shaking the spawn jars frees the cultivator to attend to other chores. Most importantly, high-quality spawn is realized in a fraction of the time than traditional methods. Step-by-step methods are described in detail in the ensuing paragraphs. The ambient air temperature recommended throughout this process is 75°F (24°C).

Step 1. A vigorous, nonsectoring culture incubated in a 100 × 15-millimeter petri dish is selected. This parent culture is subcultured by cutting and transferring 1-centimeter squares from the mother culture to ten blank petri dishes. In effect, ten subcultures are generated. The cultures incubate until the mycelia reaches approximately 1 centimeter from the inside peripheral edge of the petri dish, more or less describing an 80-millimeter diameter mycelial mat.

Step 2. When the cultures have achieved the aforementioned diameter of growth, use the following formula to create a liquid culture media: After mixing and subdividing 750 milliliters of the broth into three 1,500-milliliter Erlenmeyer flasks, the flasks are placed within a pressure cooker and sterilized for 1 to 2 hours at 15 psi (252°F =121°C).*

Liquid Culture Media for Wood Decomposers

1,000 ml water
40 grams barley malt sugar
3–5 grams hardwood sawdust
2 grams yeast
1 gram calcium sulfate

Place a floating stir bar into each Erlenmeyer flask. The openings should be stuffed tightly with nonabsorbent cotton and covered with aluminum foil. *The ingredients do not dissolve. The pH falls between 6.0 and 6.5 when using near-neutral water at makeup.*

First, a 1,000-milliliter Eberbach stirrer is filled with 750 milliliters of water and sterilized. Simultaneously, three 1,500-milliliter Erlenmeyer flasks, each containing 750 milliliters of the above concoction, are sterilized. After sterilization, the pressure cooker naturally cools. If your pressure cooker *does not* form a vacuum upon cooling, then the Eberbach stirrer and the Erlenmeyer flasks *must* be removed at 1 to 2 psi, before reaching atmospheric pressure. Otherwise, contaminants are drawn in. The slightest mistake with this process could ruin everything that is inoculated downstream in this process. If the pressure cooker *does* achieve negative pressure, the vacuum must be broken paying careful attention to the path through which air is drawn. The outer surface of the pressure cooker should be wiped clean and placed into the airstream coming from the laminar flow bench. Since the airstream coming from the face of the micron filter

---

* With experience, the cultivator will likely want larger vessels for fermentation. I prefer 5,000- to 7,000-milliliter squat glass flasks, into which 2,250 milliliters of liquid culture media is placed, sterilized, and inoculated with 750 milliliters of liquid inoculum. When the liquid volume exceeds 5,000 milliliters additional measures are required for adequate aeration, such as peristaltic pumps pushing air through media filters. The surface area of the liquid broth should be at least 110 mm/2,000 ml for sufficient transpiration of gases and metabolic by-products.

Inoculating an Eberbach stirrer with agar wedges of mycelium in airstream of laminar flow hood.

is the single step that is most dependent upon the actions of the laboratory technician. Since five cultures are cut and transferred, the slightest mistake at any time will allow contamination to be passed on, thereby jeopardizing the entire run. Should the scalpel touch anything other than the cultured mycelium, it should be resterilized before continuing. Once the transfers are complete, replace the screw-top lid of the Eberbach, carefully adhering to the principles of standard sterile technique.

Step 4. The Eberbach stirrer is placed on the power unit, and stirred in 3-second bursts. (The blender I use rotates at 22,000 rpm.) Pausing for 5 seconds, the surviving chunks of agar fall downward into the blades. Another 3-second burst decimates these pieces. One more 5-second pause is followed by the last 3-second, high-speed stir. In effect, the stirring process has created thousands of chopped strands of mycelium, in short cell chains.

Step 5. The water/mycelium blend is transferred, 250 milliliters at a time in equal proportions, into the three 1,500-milliliter Erlenmeyer flask. A remote syringe, pipette, or liquid pump can be used. Less elaborate is to simply "free-pour" equal volumes of myceliated fluid from the Eberbach into each Erlenmeyer. The nonabsorbent cotton stoppers are, of course, removed and replaced with each pouring, being careful not to allow contact between the cotton stopper and any contaminated surface. Each Erlenmeyer is placed on stir plates or on a shaker table and rotated at 100–200 rpm for 48 to 72 hours. The water broth is continuously stirred to allow transpiration of metabolic gases and for oxygen absorption. The fluid has a milky-brown color and is not translucent. Settling of the heavier components is clearly visible when the stirring process is interrupted.

Upon completion, 3,000 milliliters of mycelium are rendered in liquid form. The hyphae, recovering from the damage of being cut by the spinning blades of the blender, are stimulated into vigorous regrowth. At a point several cells away from the cut ends, nodes form on the cell walls, new buds push out, and begin branching. A vast interconnected fabric of cells, a mycelial network, forms. The

is free of airborne particulates, the media remains sterile. Additionally, I like to saturate a sterilized cotton cloth (cotton baby diapers work well) with isopropanol and place it over the vent valve as an additional precaution. When the stopcock is opened, clean air is drawn through the alcohol-saturated cloth. Once the pressure returns to normal, the pressure cooker is opened into the airstream, with the leading edge nearest to the filter. The contents are removed and allowed to cool. The cultivator should always remain conscious of the cleanliness of the surfaces of the pressure cooker, his or her hands, and the countertops upon which items are placed.

Step 3. Of the ten cultures, the five best are chosen. Any culture showing uneven growth, sectoring, or any abnormality is viewed with suspicion and is excluded. The mycelium from each petri dish is sectioned into quadrants with a heat-sterilized scalpel and aseptically transferred into the now-opened Eberbach stirrer containing the sterilized water. Heat sterilization of the scalpel need only occur once. This

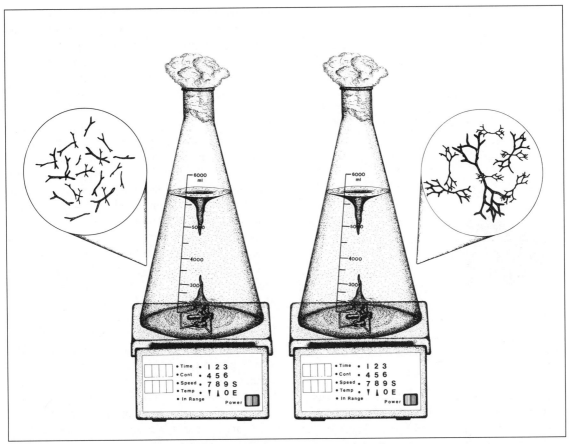

Actively growing mycelium 6 and 12 hours after inoculation.

branches fork continuously. *After 2 to 4 days of regrowth in the nutrient-enriched broth, each Erlenmeyer flask becomes its own universe, hosting thousands of stellar-shaped, three-dimensional colonies of mycelium.* (See above.) *This is the stage ideal for inoculation into sterilized substrates, especially in the generation of grain spawn masters.* Far more bioactive than the same mycelium transferred from the two-dimensional surface plane of a petri dish, each hyphal cluster grows at an accelerated rate.

If, however, the liquid media is not used at its peak rate of growth, and stirs for more than a week, the colonies lose their independence and coalesce into a clearly visible contiguous mycelial mat. Long mycelial colonies adhere to the interface of the fluid surface and the inside of the flask. Chains of mycelium collect downstream from the direction of rotation. Soon after their appearance, often overnight, the medium becomes translucent and takes on a rich amber color. A large glob of mycelium collects on the surface and can be mechanically retrieved with a pair of tweezers, forceps, or scalpel, if desired. The remaining clear amber fluid contains super-fine satellite colonies and hyphal fragments. By passing the fluid through a microporous filter, the mycelium can be recaptured. This technique is especially attractive for those whose goal is running tests on small batches of mycelium. With many species I have grown, the conversion

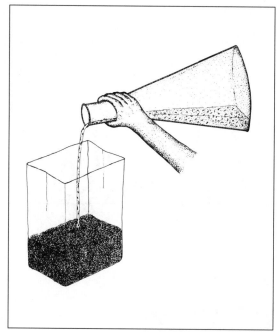

Free-pouring of fermented mushroom mycelium into sterilized grain to create grain spawn masters.

ratio of sugar/wood to mycelium (dry weight) approaches 20%. This percentage of conversion is nearly 80% biological efficiency, considered good in the commercial cultivation of gourmet mushrooms.

Step 6. Each of the three Erlenmeyer flasks now contains 1,000 milliliters of nutrient- and mycelium-rich broth. At 30 milliliters per transfer, 100 ½-gallon (2-liter) grain-filled jars can be inoculated. Here too, a pipette, a back-filled syringe, burette, or pump can be used.* I prefer "free-pouring" 30–50 milliliters of myceliated broth directly out of each Erlenmeyer into each half gallon jar of sterilized grain. In time, an adept cultivator develops a remarkably accurate

---

* Various laboratory pumps can be used for highly accurate injections of liquid media without danger of contamination. The Manostat Jr. Dispenser (#54947-110), equipped with a foot switch, delivers shots of 10–50 milliliters of liquid inocula per second utilizing a ⁵/₁₆-inch silicon tube. If equipped with a multiple dispersion manifold, several spawn containers can be inoculated at once with ease and speed.

ability to dispense liquid spawn in consistently equal proportions. The spawn maker's movements become rapid, repetitious, and highly rhythmic.

One study, using a similar method (Yang and Yong, 1987), showed that the hyphal clusters averaged less than 2 millimeters in diameter and that each milliliter contained 1,000 to 3,500 "hyphal balls." The range of time for the maximum production of hyphal clusters varied between species, from 2 days to 14. The recommended inoculation rate was 15 milliliters for each 250 grams of grain. For ease of handling, distribution, and colonization, I find that the dilution schedule described above efficiently inoculates large volumes of grain in the creation of grain masters. (I use 30–50 milliliters of inoculum to inoculate 500–600 grams of grain in 2-liter or ½-gallon jars.) Most of the wood decomposers described in this book flourish with the aforementioned technique. Kawai et al. (1996) showed that, when comparing the effect on liquid fermentation of Shiitake mycelium versus solid spawn, the time to fruiting was sped up by 25%.

The lids to each container are replaced as soon as they are inoculated. If the lids to each jar are loosened prior to free-pouring, then one hand lifts each lid, while the other hand pours the liquefied mycelium into each jar, moving side to side. If an assistant is present, the jars are removed as soon as they are inoculated. As they are removed, each lid is tightly secured; the jar is quickly shaken to evenly mix the liquid spawn through the grain. Each jar is stored at an angle on a spawn rack. One person inoculating in this fashion can keep two people busy "feeding" the inoculator new jars and removing those just inoculated. Since this system is fast paced, the time vector, the "window of vulnerability," is much less compared to the time-consuming, labor-intensive, traditional methods. The disadvantage of this technique is that the stakes for the clumsy spawn producer are higher. Any mistake will be amplified with force. Should any one of the petri dish cultures harbor contaminants, once those cultures are placed together in the Eberbach stirrer, all resulting spawn jars will be contaminated. This is an all-or-nothing technique. Fortunately, if you

A space-efficient rack for incubating grain spawn in jars. Three hundred forty ¹/₂-gallon jars can be stored on this 8-foot long x 8-foot high x 16-inch wide shelf system. Spawn quality is improved by storing the jars at an angle. Earthquake sensitive.

As with any method described in this book, quality controls must be run parallel with each procedure. A sample of the mycelium-enriched broth is drop-streaked across the surface of a few nutrient-agar-filled petri dishes. (See page 135.) These will later reveal whether the liquid contains one organism—the mycelium—or a polyculture—the mycelium and contaminants. Furthermore, one or more of the sterilized grain-filled jars should be left unopened and uninoculated to determine the success of the sterilization procedure. These "blank" vessels should not spontaneously contaminate. If they do, then either the sterilization time/pressure was insufficient or airborne contamination was introduced, independent of the liquid-fermented spawn. If the jars injected with the fermented mycelium contaminate, and the uninoculated controls do not, then obviously the vector of contamination was related to the act of inoculation, not the cycle of grain sterilization. (See Chapter 10: The Six Vectors of Contamination.)

are following the techniques outlined in this book, success is the norm.

The jars normally grow out in 4 to 10 days, many times faster than the traditional transfer technique *and the jars are only shaken once—at the time of liquid-inoculation.* With the traditional wedge-transfer technique, each individual jar must be shaken two or three times to insure full colonization: first at inoculation; second three days hence; and finally at days 5, 6, or 7. Remember, not only is the cultivator gaining efficiency using the liquid-inoculation method, but 100 grain masters are created in a week from a few petri dish cultures! With less need for shaking, hand contact with the grain masters is minimized. Time is conserved. Probability of contamination is reduced. Growth is accelerated. With each kernel, dotted with stellar clusters of hyphae from the first day of inoculation, spawn quality is greatly improved.

Pelletized (Granular) Spawn

Trends in spawn technology are evolving towards pelletized spawn. Pelletized spawn is specifically designed to accelerate the colonization process subsequent to inoculation. Examples of pelletized spawn range from a form resembling rabbit food to pumice-like particles. In either case, they are nutrient saturated to encourage a burst of growth upon contact with mushroom mycelium. Pelletized spawn varies in size from 1 millimeter to 5 millimeters in diameter.

Adapting food mills designed for the manufacture of animal feeds can make pelletized spawn. With modest reengineering, these machines can be modified to produce spawn pellets. Idealized spawn seeks a balance between surface area, nutritional content, and gas exchange. (See Yang and Jong, 1987; Xiang, 1991; Romaine and Schlagnhaufer, 1992.) Another simple and inexpensive form of pelletized spawn can be made from vermiculite saturated with a soy-protein-based nutrient broth, a "secret" method used by some of the largest spawn producers.

The key to the success of pelletized spawn is that it enables easy dispersal of mycelium throughout the substrate, quick recovery from the concussion of inoculation, and ideally, the sustained growth of mycelium sufficient to fully colonize the substrate. Many grains are, however, pound-for-pound, particle for particle, more nutritious than most forms of pelletized spawn.

I believe the spawn should be used as the vehicle of supplementation into a semiselective substrate. Others subscribe to the school of thought that the substrate's base nutrition should be raised to the ideal prior to spawning. The danger with this approach is that, as the base nutrition of the substrate is raised, so too is its receptivity to contaminants. From my experiences, using a nutrition particle already encapsulated by mushroom mycelium is more successful. The ultimate solution may be a hybrid between liquid-inoculum and grain spawn: a semisolid slurry, millimeters in diameter, that would maximally carry water and nutrients, and support the mycelium as it projects hundreds of cells from each inoculation point.

Matching the Spawn with the Substrate: Critical Choices on the Mycelial Path

Once spawn has been created, the cultivator arrives at a critical crossroad in the mushroom cultivation process. Several paths can be pursued for the growing of mushrooms, depending on the species and base materials. Some of these paths are intrinsically unproblematic; others are not. Success is measured by the following criteria: speed and quality of colonization, crop yield, and resistance to disease.

The first step can be the most critical. When trying to match a mushroom strain with an available substrate, I place a small sample of the substrate into the agar media formula. Upon exposure, the mushroom mycelium generates enzymes and acids to break down the proposed food source. Once

acclimated, the mycelium now carries a genetic memory of the end substrate to which it is destined. With Shiitake, Enokitake, Maitake, and Reishi, I acquaint the mushroom mycelium with the host substrate by introducing to the media 1- to 2-gram samples of the sawdust directly into the liquid fermentation vessels. This liquid-inoculum is then used to generate grain spawn. I am convinced that this method empowers the mushroom mycelium.

Grain spawn can be used for direct inoculation into pasteurized straw, into sterilized sawdust, or into enriched sawdust. If you are growing Oyster mushrooms, the recommended path is to inoculate straw with grain spawn. If one wants to create plug spawn for the inoculation of stumps and logs, the best path is to go from grain spawn to sterilized sawdust, and once grown out, to sterilized wooden dowels. For the rapid, high-yield methods of growing Shiitake, Enokitake, Maitake, Kuritake, and others indoors on sterilized substrates, I recommend the following path: going from grain spawn to sterilized sawdust to enriched sawdust. Each transfer step results in an expansion of mycelial mass, usually by a factor of 5 to 10 and takes 1 week to 2 weeks to fully colonize.

The tracks recommended in the previous paragraph are the result of thousands of hours of experience. More direct methods can be used, but not without their risks. For instance, one can use grain spawn of Shiitake to inoculate enriched sawdust, skipping the above-described intermediate step of sterilized sawdust. However, several events are observed subsequent to inoculation. First, there are noticeably fewer points of inoculation than if sawdust spawn were used. As a result, recovery is slower and colonization is not as even. ("Leap-off" is faster from sawdust spawn than from grain spawn. The mycelium has already acclimated to the sawdust substrate.) Most importantly, a marked increase in temperature occurs soon after inoculation, known by mushroom cultivators as *thermogenesis*. By enriching the substrate with grain spawn, increasing its nitrogen content, you accelerate biochemical reactions, and correspondingly, two main by-products: heat and carbon dioxide. Should internal temperatures exceed 100°F (38°C) in the core of each bag, latent contaminants, especially thermophilic bacteria and black pin molds (*Aspergillus*, *Rhizopus*, and *Mucor*) spring forth, contaminating each and every bag. These same bags incubated at 75°F (24°C) would otherwise be successfully colonized with mushroom mycelium. *In general, the cultivator should assume that a minor population of contaminants will survive "sterilization" especially as the mass of each batch increases.* Thermotolerant contaminants are activated when temperatures within the substrate spiral upwards. To thwart this tragedy, you should space the bags containing nitrogenous supplements well apart, placed on open wire rack shelving. The laboratory manager should carefully monitor air temperature to offset the upwardly spiraling trend in internal temperatures.

This arena of problems is largely avoided by using sawdust spawn for inoculation into supplemented sawdust substrates rather than grain spawn. Thermogenesis is reduced to a more manageable level. Colonization is faster and more even, and one gets more "mycelial mileage" from grain spawn by generating intermediate sawdust spawn. In essence, another exponent of expansion of the mycelial mass has been introduced to the benefit of overall production.

In contrast, grain spawn is preferred over sawdust spawn for the cultivation of Oyster mushroom on cereal straws. Grain spawn boosts the nutritional base of straw, radically improving yields compared to using an equal mass of sawdust spawn. Although sawdust may have more points of inoculation, yields are substantially less than if the straw had been impregnated with grain spawn. Three exceptions are *Hypsizygus ulmarius*, *H. tessulatus*, and *Pleurotus eryngi*, all of which benefit when sawdust spawn is used to inoculate wheat straw. In these cases, yields may be enhanced as the enzymes secreted target the higher lignin components in wood.

In Chapter 21, the growth parameters of each species and the recommended courses for matching spawn and substrate for maximizing yields are described in detail.

Spawn Storage

Spawn can be stored for only a short period of time before a decline in viability occurs. Those who buy spawn from afar are especially at risk. As spawn ages, and with the depletion of food resources, the mycelium's rate of growth declines. Metabolic wastes accumulate. With the loss of vitality, the mycelium's anti-disease defensive mechanisms fail. Opportunistic molds, bacteria, viruses, and other microscopic organisms proliferate. Good-quality spawn on day 30 (from the date of inoculation) is less than half as viable at day 60.

Generally, spawn should be used at peak vitality. If it cannot, only one option remains: slowing growth by refrigeration. Spawn can be refrigerated for several weeks at 35–40°F (1.6–4.4°C), effectively slowing its rate of decline, provided the refrigeration process does not, in itself, cause contamination to flourish. Spawn must not be kept in a refrigerator in the same space as mushrooms are stored. Free-floating mushroom spores can become a vehicle of contamination by bacteria and other fungi directly into the stored spawn. Spoiling mushrooms are often covered with the very contaminants so dreaded in the laboratory environment.

Another problem with refrigeration rooms is that the cooling of spawn causes condensation within the spawn containers. The cultivator should always view excess water, in the form of condensation, with concern. Contaminants proliferate within the water droplets and are efficiently spread by them. Bacteria, in particular, reproduce feverishly in free water environments, even at cool temperatures. Further, refrigeration blowers and cooling elements attract and collect dust particles, which inevitably must be cleaned. The force of the air blasting from the cooling elements covers the outer surfaces of the bags with contaminant particles that are easily transferred by anyone handling them. Most often, the filter media, designed to limit airborne contamination, become the sites of black and green mold growth. In time, they can penetrate from the outside into the interior environment of the spawn containers.

Oyster mushrooms "breaking out" of a jar filled with grain spawn.

- Inspect the stored spawn once a week for visible signs of contamination, especially at the location of the microporous filter patches. (Although spores may not pass through the filtration material, mold mycelia can.)

- Maintain a low relative humidity. The humidity should never exceed 60%, and ideally should be kept at 35–50% range.

- Minimize any material that could become a platform for mold growth, particularly wood, cardboard, and other paper products.

Lastly, some species are more receptive to cold storage than others. Some of the tropical species die upon exposure to cold temperatures. (*Volvariella volvacea* is one notable example.) The cold-weather Oyster strains (*Pleurotus ostreatus* and allies) can be shocked into fruiting upon placement into a cold room. One commonly sees Oyster mushrooms fruiting frantically in containers that were otherwise hermetically sealed. The force of fruiting, the bursting forth of mushrooms within the spawn containers, can actually cause enough stress to split plastic seams, unscrew lids on bottles, and force apart filter membranes.

With the rapid-cycle spawn techniques described in this book, cold storage of spawn is not necessary and is not recommended. Cold storage is an option widely utilized by the Agaricus industry, an industry historically fractured into specialty companies. When inventories exceed demand, spawn is kept for as long as possible under refrigeration. Often the consumer, not knowing better, becomes the victim of a spawn producer's overproduction. If the spawn fails, the excuse heard, more often than not, is that the purchaser mishandled the spawn! This type of business relationship is intrinsically problematic, and is yet another reason why mushroom farms should generate their own spawn.

If refrigeration is the only alternative, then, by definition, you have missed the best opportunity: to use the spawn at its peak of vitality. Nevertheless, every spawn producer faces this dilemma. So, if you have to refrigerate your spawn, the following precautions are suggested:

- Treat the refrigeration room as if it were a clean room. Analyze all potential contamination vectors. Install a HEPA filter if necessary. Make sure floors and walls are kept clean by frequently washing with a 10% bleach solution.

- Rotate your spawn! Only similarly aged spawn should be kept together.

- When refrigerating spawn, use bags, not jars.

CHAPTER 16

Creating Sawdust Spawn

Sawdust spawn is simply created by inoculating grain spawn onto sterilized sawdust. Hardwood sawdust, especially oak, alder, cottonwood, poplar, ash, elm, sweetgum, beech, birch, and similar woods are best. Fresh sawdust is obviously better than aged, and sawdust with dark zones (often a sign of mold infestation) should be avoided. Sawdust from milling lumber is best because of its consistent particle size, measuring, on average, 1–5 millimeters in diameter. Sawdust from furniture manufacturers is much more difficult to formulate. Often that sawdust is either too fine and/or combined with shavings. This combination results in an inconsistently comprised substrate, difficult to use, difficult to replicate, and poor in its overall suitability as sawdust spawn. With shavings, the mycelium must expend excessive cellular energy to span the chasms between each food particle. Per cubic inch, shavings are too loose a form of wood fiber, insufficient to support a dense mycelial mat let alone a substantial mushroom.

Sawdust is moistened 60–70%, scooped into gusseted polypropylene bags to a gross weight of 6 pounds. The open tops of the bags are folded down and stacked tightly into a square pushcart. This helps the bags form into a cube. Should excess water becomes visible, collecting at the bottom of the bags, then less water is added at makeup. Fortunately, mycelium tolerates a fairly broad range of moisture content for the production of sawdust spawn.

Some spawn producers secure the open flaps of the bags with plastic tape, spring-activated clothespins, or even paper clips. To meet this need, a cultivator-mycologist named Dr. Stoller invented a specialized collar, filter, and lid combination that is still in use today. If the bags are carefully handled, however, many cultivators simply press adjacent bags tightly together, negating the need for fasteners. The bags are loaded into an autoclave or pressure cooker and sterilized for 2 to 3 hours at 15 psi. Upon return to atmospheric pressure, and following the same procedures outlined for cycling grain spawn, the bags are removed from the pressure vessel directly into the clean room.

Each gallon (4 liters) of grain spawn can effectively inoculate ten 5-pound bags of moist sawdust spawn. Exceeding twenty bags of sawdust inoculated per gallon of grain spawn is not recommended. Strict adherence to the sterile techniques previously outlined in this book must be followed during the inoculation process. I recommend washing your hands periodically with antibacterial soap (every 30 minutes) and frequently wiping them with isopropanol (every 10 minutes). Even with surgical gloves, periodic cleaning is recommended. Once inoculations are completed, those with delicate skin should apply a moisturizer to prevent damage from disinfectants.

Step-by-Step Instructions for Inoculating Sawdust

Step 1. Choosing the grain spawn. Grain spawn should be selected from the laboratory inventory. Ideally, spawn should be 1 to 3 weeks of age, at most 4 weeks. Carefully scrutinize the filter disk zone, inside and outside, to discern the presence of any molds or unusual signs of growth. Only cottony spawn, void of wet spots or areas of no colonization, should be chosen. Since the spawn generally chosen is second or third generation, bag spawn is preferred for this stage. Slamming the jar against a cleaned rubber tire loosens the grain kernels of each spawn jar.

Step 2. Retrieving the bags from the autoclave. Bags of sawdust, having been removed directly from the autoclave, cool to room temperature by placing them in the windstream of a laminar flow hood. Once the bags are below 100°F (38°C) inoculations can proceed.

Step 3. Opening the bags. The bags are opened by *pulling the outside plastic panels outwards from the outside.* The inoculator's hands never touch the interior surfaces. If they do, contamination is likely. Once ten bags have been fully opened, the inoculator wipes his or her hands with isopropanol, and brings a gallon of grain spawn to the table directly downstream from the newly opened bags. The jar lid is

loosened to a point where it can be lifted off easily with one hand.

Step 4. Inoculating the bags in a specific sequence. If using jar spawn, remove the lid and place it upside down, upstream and away from the bags to be inoculated. Since you may wish to return the lid to the spawn jar should all its contents not be used, pay attention to the manner in which it is handled. Grasping the spawn jar with one hand, palm facing up, position the jar opening above the first bag to be inoculated. If you are right-handed, inoculate each sawdust bag in sequence going from left to right. (Left-handers would logically do this in the reverse.) With a roll of the wrist, angle the jar so that grain spawn free-falls into the opened bag. The spawn must be well separated for this technique to result in a consistent rate of inoculation for each bag. Through trial-and-error and experience, a highly rhythmic and exact amount of spawn is approportioned among the ten sawdust bags.

If there were, for instance, four rows of ten bags in front of the laminar flow bench, then a right-handed person would inoculate bags starting from the far left, rear bag. Each bag to the right would then be inoculated until the back row is finished. In turn, the third row would then be inoculated with the next gallon of grain spawn, again from left to right. In this fashion, the hands of the inoculator cannot jeopardize the sanctity of the upstream bags. To inoculate the first row nearest the face of the micron filter would endanger downstream bags from the debris coming from the inoculator's hands and/or undetected contaminants from a spawn jar.

Step 5. Sealing the sawdust spawn bags. Few steps are as critical as this one. The simple mechanical act of sealing sterile airflow bags can have extraordinarily disparate results for the spawn incubation process. All other steps in this process can be perfectly executed, and yet failure to achieve a continuous seal can be disastrous.

I attempt to create a positively inflated bag at the time of sealing, to create a positive-pressurized bubble-like environment. (See page 148.) Although the filter patch allows the transpiration of gases, it

the plastic liquefies upon contact with the heating element, the bags should not be squeezed during sealing.

Pinholes or small tears cause the bags to collapse. Collapsed bags contaminate with alarming frequency. A simple test determines if the problem is at the seal or not. Roll the sealed region several times into a tight fold and push down. The bag inflates and if there is a leak not at the seal, a distinct hissing sound emanates from the defective site. Should the bag remain tightly inflated with no apparent loss of pressure, then the seal at the top is at fault. Simply reseal and test again for leaks.

Step 6. Shaking the sawdust spawn bags. Once the bags have been properly sealed, they are thoroughly shaken to evenly distribute the spawn kernels. If partially inflated, this process takes only a few seconds. Proper shaking is critical for successful spawn incubation. (See page 148.)

Step 7. Incubating the sawdust spawn bags. Unlike nutrified sawdust, most sawdust bags contacting

Sawdust spawn of Reishi (*Ganoderma lucidum*). Note inflated atmosphere within bag.

is not at a rate that causes the bag to noticeably deflate, even with gentle squeezing. When this bubble environment is created at the time of sealing, two advantages are clearly gained. First, the grain spawn mixes and rotates easily through the sawdust, making shaking easy. Second, each bag now has a voluminous plenum, a mini-biosphere with an atmosphere nearly matching the volume of the sawdust. At least 25% air space should be allotted per spawn bag; otherwise anaerobic activity will be encouraged.

The open bag is laid horizontally, with its opening overhanging the heating element. Grasping both the left and right outside surfaces, the bag is pulled open to fill with the sterile wind. A "Spock-like" finger position keeps the bag maximally inflated while the heat sealer joins the plastic. Two strokes are often necessary for a continuous seal. By gradually increasing the duration of the seal, an ideal temperature/time combination can be found. Since

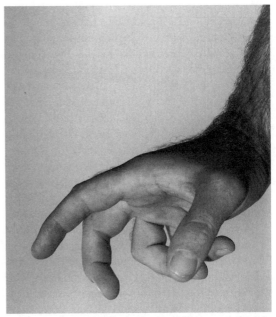

Hand in position used for opening bags, lifting petri dish lids, and removing spawn jar caps.

Inoculating sawdust with grain spawn.

Sawdust spawn is used for one of five purposes:

- To sell to log growers as spawn

- To inoculate outdoor beds by dispersing the spawn or by burying the mass into the sandy soils

- To inoculate sterilized hardwood dowels in the creation of plug spawn for log and stump growers

- To grow mushrooms on (However, most of the species described in this book benefit from having the sawdust enriched with a readily available, nitrogenous supplement such as bran.)

- To inoculate 5 to 20 times more sterilized enriched sawdust, usually sawdust supplemented with nitrogenous bran, soybean flour, etc.

Sealing the bag of sawdust after inoculation with grain spawn.

Dispersing the spawn throughout the sawdust by shaking. The inflated bag not only facilitates shaking, but also provides a sufficient atmospheric plenum within each bag, accelerating the growth of the mushroom mycelium.

each other during incubation grow out without contamination. Even so, I still recommend an air space for heat dispersal. The laboratory space can be maximized with sawdust spawn. By placing a small thermometer between the two faces of the touching bags, the laboratory manager can track temperatures to be sure they do not stray into the danger zone of >95°F (35°C). Above this temperature, thermophilic fungi and bacteria reign. In a burgeoning laboratory where space is limited, cultivators tend to place bags close together, which can start a dangerous cycle of self-heating. Bags should be spaced an inch or two apart to offset this temperature spiral.

In 3 days, recovery from the concussion of inoculation is clearly visible from the grain kernels. The kernels become surrounded by fuzzy mycelium. Looking at a population of bags on a shelf from afar quickly tells the laboratory manager how even the spawn run is. Concentrated pockets of growth, adjacent to vast regions of no growth, result in poor completion. If evenly inoculated, the sawdust spawn is ready to use within 2 weeks.

CHAPTER 17

Growing Gourmet Mushrooms on Enriched Sawdust

When sawdust is supplemented with a nitrogen-rich additive, the yields of most wood-decomposers are enhanced substantially. Rice bran is the preferred additive in Asia. Most brans derived from cereal grains work equally well. Rye, wheat, corn, oat, and soybean brans are commonly used. Flours lack the outer seed coat, and by weight have proportionately more nutrition than brans. Other more concentrated nitrogen sources such as yeast, soy oil, and peptone require precise handling and mixing at rates more dilute than bran supplements. Less is used in formulating the substrate. The nutritional tables in Appendix 5 will help cultivators devise and refine formulas. Mini-trials should be conducted to prove suitability prior to any large-scale endeavor.

For the cultivation of Shiitake (*Lentinula edodes*), Enokitake (*Flammulina velutipes*), Maitake (*Grifola frondosa*), Kuritake (*Hypholoma sublateritium*), Lion's Mane (*Hericium erinaceus*), the Black Poplar (*Agrocybe aegerita*), King Oyster (*Pleurotus eryngii*), Tuber Oyster (*Pleurotus tuberregium*), Nameko (*Pholiota nameko*), and many other wood-loving species, the following formula is recommended. Alterations to the formula will further enhance yields for each strain. I find this formula to be highly productive and recommend it highly.

The base substrate is composed of a fast-decomposing hardwood, such as alder, poplar, and cottonwood, in contrast to the slow-rotting woods like oak and ironwood. If these types of quick-rotting woods are unavailable, deferment should first be made to the tree types upon which the mushroom species natively inhabits. Most of the photographs in this book are from blocks made with this basic formula.

I have devised the following *fruiting formula* utilizing hardwood sawdust, hardwood chips, and a nitrogen-rich supplement, in this case, oat bran. Water is added until 65–75% moisture is achieved, a few percentage points below saturation.

Components for the fruiting formula: sawdust, chips, and bran.

Adding the supplement (rice bran) to hardwood sawdust.

The Supplemented Sawdust "Fruiting" Formula: Creating the Production Block

This formulation is designed for maximizing yields of wood decomposers. Most gourmet and medicinal mushrooms produce prolifically on this substrate. If wood is a scarce commodity and not available as a base component, please refer to Chapter 18.

The Sawdust/Bran Fruiting Formula

100 pounds sawdust
50 pounds wood chips ($^{1}/_{2}$–4 inches)
40 pounds oat, wheat, or rice bran

5–7 pounds gypsum (calcium sulfate)*

By dry weight, the fraction of bran is approximately 20% of the total mass. By volume this formula is equivalent to

64 gallons sawdust
32 gallons wood chips
8 gallons bran
1 gallon gypsum (calcium sulfate)*

The above-mentioned mixture fills 160 to 180 bags of moist sawdust/bran to a wet mass between 5.0 and 5.5 pounds. By using larger chips mixed with sawdust, aeration increases, metabolism is accelerated, and yields are enhanced. Royce (2000) noted that when sawdust less than .85 mm was

---

* Raaska (1990) found that the use of calcium sulfate (gypsum) stimulated mycelial growth of Shiitake in a liquid media supplemented with sawdust. The calcium sulfate did not, by itself, significantly affect pH at makeup. However, mycelial growth was stimulated by its addition, and there was a corresponding precipitous decline of pH and a fourfold increase in biomass versus the controls. Leatham and Stahlman (1989) showed that the presence of calcium sulfate potentiated the photosensitivity of the Shiitake mycelium, affecting fruitbody formation and development. A beneficial effect of calcium sulfate on the growth rate of other wood decomposers is strongly suspected.

Adding 2- to 3-inch diameter wood chips on top, which builds the matrix.

The ingredients are thoroughly mixed.

removed from the mix, yields were enhanced. I recommend using a standardized volumetric unit for ease of handling, anything from a plastic 4-gallon bucket to the scoop bucket of a front-end loader. In either case, simply scale up or down the afore-mentioned proportions to meet individual needs. Thorough mixing is essential.

The above weights of the sawdust and chips are approximate, based on their ambient, air-dried state. (The wood used, in this case, is red alder, *Alnus rubra*, and is highly recommended.) Bran should be stored indoors, away from moisture, off the ground, to prevent souring. (Rice bran readily contaminates and must be carefully handled. Molds and bacteria flourish in most nitrogen-rich supplements soon after milling and exposure to moisture.)

Using a 4-gallon bucket as a measurement unit, 16 buckets sawdust, 8 buckets chips, and 2 buckets bran lie ready for use. All three are mixed thoroughly together in dry form, then gypsum is added, and the final mixture is moistened to 60–65%. This formula makes 70 to 80 bags weigh-ing 5.5 pounds. Directly after makeup the bags are loaded into the autoclave for sterilization. Should the mixture sit for more than a few hours, fermentation reactions begin. Once bacteria and molds flourish, the mixture is rendered unsuitable.

If alder is unavailable, I strongly encourage substituting other rapidly decomposing hardwoods, such as cottonwood, poplar, willow, aspen, sweetgum, and similar wood types from riparian ecosystems. Although oak is the wood most widely used in the cultivation of Shiitake, Maitake, and Enokitake, its inherent, slower rate of decomposi-tion sets back fruiting schedules compared to the above-mentioned hardwoods. Sycamore, mahogany, ironwood, the fruit trees, and other denser woods require a longer gestation period, although subsequent fruitings may benefit from the increased wood density.* (Kruger, 1992) Here, a little experimentation on the part of the cultivator could have far-reaching, profound results. Mini-trials matching the strain with the wood type *must* be conducted before expanding into commercial cultivation.

By laying out sawdust first in a 10 × 10-foot square, the chips can be thrown evenly upon the sawdust, and topped by broadcasting of bran evenly over the top. This mass is mixed thoroughly together by whatever means available (flat shovel, cement or soil mixer, tractor). A mixer of less than a cubic yard in capacity is probably not more efficient than one person mixing these three ingredients by hand with a shovel. Pockets of discoloration, mold, or "clumps" should be avoided during the making up of this composition. The more competitors at makeup mean the more that are likely to survive the "sterilization" cycle.

Mixing the above components by hand becomes functionally impractical beyond 300 bags per day. At this level of production and above, automated mixing machines and bag fillers, adapted from the packaging and nursery industry, are far more efficient in terms of both time and money invested.

Three times the mixture featured on pages 150–151 created 210 6-pound (< 3 kg) bags of the fruiting formula. Once mixed and wetted, this mixture must be immediately loaded into the autoclave and sterilized.

Testing For Moisture Content

Wetting the substrate to its proper moisture content is critical to creating a habitat that encourages mycelial growth while retarding contamination. If too much water is added, exceeding the carrying capacity of the media, the excess collects at the bottoms of the bags, discouraging mycelium and stimulating bacterial blooms and anaerobic activities. Ideally, sawdust is wetted to 60–65% water. If the wetted mix can be squeezed with force by hand and water droplets fall out as a stream, then the mix is probably too wet.

The easiest way to determine moisture content is by gathering a wet sample of the mixture, weighing it, and then drying the same sample in an oven for 1 hour at 350°F (180°C). If, for instance, your sample weighed 100 grams before drying, and only 40 grams after drying, then obviously 60 grams of water were lost. The moisture content was 60%.

Once the person making the substrate obtains experience with making up a properly balanced substrate, moisture content can be fairly accurately determined by touch. Materials are measured volumetrically, correlated to weights, for ease of handling. This ensures that the mixing proceeds with speed and without unnecessary interruption.

---

* If speed of production is not the overriding issue, longer-lasting fruitings can be gained from denser hardwoods, such as the oaks, than those from the rapidly decomposing hardwoods. However, I have found Enokitake, Oyster, Reishi, Lion's Mane, and Shiitake give rise to faster fruitings of equally superior quality on alder and cottonwood, albeit briefer in duration.

Commercial double-door autoclave.

Choosing a Sterilizer, aka Retort or Autoclave

Although home-style pressure cookers are ideal for sterilizing agar media and for small-to-medium batches of grain, they have insufficient capacity for the sterilization of bulk substrates. The problems faced by the mushroom cultivator in Thailand or the United States are the essentially the same. In developing countries, the sterilizer is often a makeshift, vertical drum, heated by fire or gas. A heavy lid is placed on top to keep the contents contained. The boiling of water generates steam, which over many hours will sufficiently "sterilize" the substrate. This method works well within the model of many rural agricultural communities.

The pressurized steam autoclave is far better suited for commercial production. The most useful autoclaves for sterilizing bulk substrates are horizontal and have two doors. Since the autoclave is the centerpiece upon which the entire production process is dependent, many factors must be considered in its acquisition: size, configuration, and placement. Another important feature is its ability to hold a vacuum subsequent to the sterilization cycle. If the autoclave cannot hold a vacuum as it cools, a valve should be installed for the controlled intake of filtered air. If the influx of air is not filtered, the contents can contaminate after sterilization.

Hospital autoclaves are typically made of stainless steel and equipped with a pressurized steam jacket. These types of autoclaves are usually smaller than those needed by commercial mushroom cultivators, measuring only 2 × 3 feet or 3 × 4 feet by 3 × 6 feet deep. Furthermore, they usually have only one door, and their pressure ratings have been engineered to operate at 100 psi, far exceeding the needs of most mushroom growers. Unless obtained on the surplus market for a fraction of their original cost, most knowledgeable spawn producers avoid these types of autoclaves. The most cost-effective vessels are those developed for the canning industries. These are commonly called "retorts" and are constructed of steel pipe, $1/4$ to $3/8$ inch thick, and ideally fitted with doors *at both ends*. The doors come in a variety of configurations. Quick-opening, spider doors are popular and durable. Wing-nut, knock-off latches secure the doors that are slower to open and close. With autoclaves longer than 6 feet, steam spreader pipes are needed so that the entire mass heats up evenly. More suggestions follow for choosing an autoclave:

Recommendations for Equipping an Autoclave:

- Double-doors (i.e., doors at both ends)

- Redundant pressure/temperature gauges (at least two)

- Pressure/Vacuum Gauge (+50 psi to –50 psi) with valves

- Electrical safety interlocks with warning lights

- Hand-operated vent valve on top of autoclave for venting cold air

- 25 psi and 50 psi excess-pressure relief, safety blowout valves

- Hand-operated drain valve for drawing off condensate

- Coating of heat-resistant, anti-corrosive paint

- At least four 1-inch, and/or two 2-inch ports for inputs, exhausts, and sensors

- One-way gate valve in series with a vacuum gauge that allows the drawing in of clean-room air post autoclaving

Recommendations for the Placement of the Autoclave

- Recessed "wells" (2 feet × 3 feet × 2 inches) underneath each door with sealable drains for removing excess condensate from the auto-clave after opening

- Insulation jacket (R = 8 to R = 16)

- Length of autoclave framed in its own insulated room (R = 18 to R = 32) with active exhaust (500+ CFM)—fan activated by remote reading thermostat can redistribute excess heat to other environments, if needed

- One door of autoclave opening into clean room—fill side end of autoclave open to outside environment in moderate climates, or under protected overhang in severe climates, allowing for easy cleaning

- Escape doors remote from the door seals of the autoclave should metal fatigue suddenly release an impassable curtain of steam

Sterilization of Supplemented Substrates

Once the bags are filled, the supplemented (sawdust) substrate must be heat-treated for an extended period of time before inoculations can proceed. In a small pressure cooker, 2 to 3 hours exposure to high pressure steam at 15 psi (1 kg/cm^2) or 252°F (121°C) usually suffices for sterilizing supplemented sawdust substrates. When sterilizing more than 100 bags in a large pressure vessel, however, the thermodynamics of the entire mass must be carefully considered in choosing a successful sterilization protocol. Hundreds of bags tightly packed in an autoclave achieve different degrees of sterilization. When bags are stacked against one another, the entire mass heats up unevenly. Even so, this practice is common with those whose autoclaves must be packed to capacity in order to meet production requirements. I recommend a two-inch plenum between vertical rows of stacked bags to enhance heat penetration. Bear in mind that sawdust has high insulating properties, making heat penetration through it difficult.

Other factors affect the minimum duration of sterilization. The substrate mixture should be wetted just prior to filling. If water is added to the formula and allowed to sit for more than 6 hours, legions of contaminants spring to life. The more contaminants at makeup, the more that are likely to survive the "sterilization" cycle.

Fresh hardwood sawdust needs 2 to 3 hours of sterilization at 15 psi (1 kg/cm²). The same mass of sawdust supplemented with rice bran needs 4 to 5 hours of sterilization. Hence, one of the cardinal rules of mushroom culture: *As the percentage of nitrogen-supplements increases relative to the base substrate, the greater the likelihood of contamination, and thus the greater the need for full and thorough sterilization.*

I prefer the aforementioned formula using alder sawdust, alder chips, and oat bran. An autoclave filled tightly 5 bags high, 6 bags wide, and 8 bags deep (240 bags) *requires exposure to steam pressure for 8 hours at 15 psi to assure full sterilization.* The lower, central core is the slowest to heat up. (See page 156.) By placing, heat-sensitive "sterilization indicator strips" throughout the mass of sawdust-filled bags, a profile of sterilization can be outlined. Each cultivator must learn the intricacies of their system. The combination of variables is too complex to allow universal judgments. Each cultivator must fine-tune his or her techniques. Even the type of wood being used can influence the duration of the sterilization cycle. Woods of higher density, such as oak, have greater thermal inertia per scoop than, say, alder. Each run through the autoclave is uniquely affected by changes in the substrate formulation.

Those with ample space in their autoclaves separate the layers so thermal penetration is uniform. This is ideal. The sterilization cycle can be shortened, again best reaffirmed by sterilization sensitive markers. However, few individuals find themselves in the luxurious position of having an autoclave capable of running several hundred bags with 1 or 2 inches of separation between the layers of bags. These 1 or 2 inches could be used to increase the capacity of the run by approximately 20%. Many small-scale cultivators are soon forced to maximum capacity as their production expands with market demand. In the long run, dense packing is generally more cost efficient compared to loose packing. Hence dense packing, although not the best method, is usually the norm, not the exception with the small

to midsized cultivator. Thus, the manager of the autoclave cycle operates from a precarious decision-making position, constantly juxtaposing the needs of production versus the dangers of uneven sterilization due to heavy loading.

Packing an autoclave deeper than four bags (32 inches deep) runs other contradictory risks. In the attempt to achieve full colonization, sterilization time is typically extended, potentially causing other problems: oversterilization of the outer zones and bag fatigue. Oversterilization usually occurs when wood substrates are subjected to steam pressure (15 to 18 psi) for more than 10 hours. The sawdust takes on a dark brown color, has a distinctly different odor signature, and, most importantly, resists decomposition by mushroom mycelium. Prolonged steam sterilization results in complex chemical transformations. (I have yet to find a chemist who can adequately explain what happens from prolonged exposure.) Suffice to say that turpentines, changes in volatile oils, and toxic by-products are responsible for this radical shift in sawdust's myco-receptivity. In the end, the substrate is rendered entirely inhospitable to mushroom mycelium.

After the autoclave has been packed, the displacement of the cold air by introducing steam and top venting is absolutely critical. The cold air, if not vented, gives a false temperature/pressure reading. At 15 psi, the temperature within the autoclave should be 252°F (121°C). This arithmetic relationship between temperature and pressure is known as Boyle's Law. When a cold mass is introduced into an autoclave or pressure cooker, Boyle's Law does not come into play until the thermal inertia of the affected mass is overcome. In other words, as hot steam is being forcibly injected into the vessel, a lag time ensues as the heat is being absorbed by the cold mass, giving a false pressure/temperature reading. Eventually, thermal inertia is soon overcome, and Boyle's Law becomes operative.

Many autoclaves not only have a combined pressure/temperature gauge but also sport a separate, remote bulb sensor that records temperature deep within the autoclaved mass. This combination enables the laboratory personnel to compare

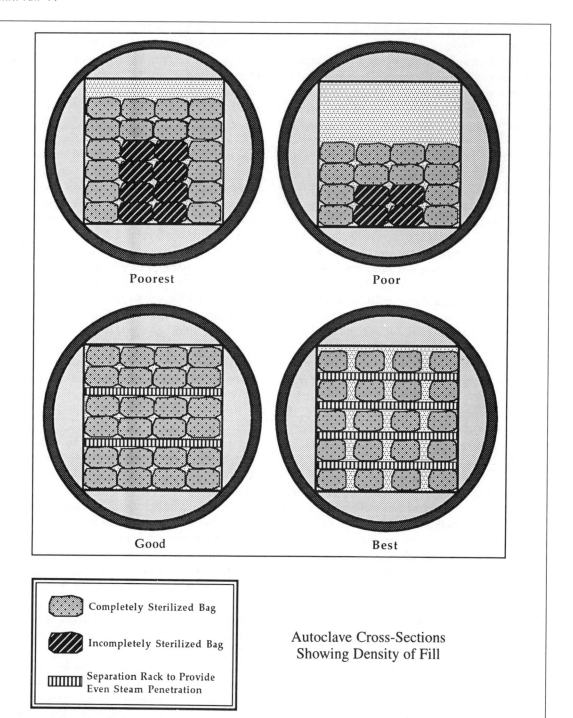

Poorest

Poor

Good

Best

Completely Sterilized Bag

Incompletely Sterilized Bag

Separation Rack to Provide
Even Steam Penetration

Autoclave Cross-Sections
Showing Density of Fill

Four profiles depicting the degrees of sterilization as affected by the method of packing the bags into the autoclave. Note central, lower core is most resistant to deep steam penetration.

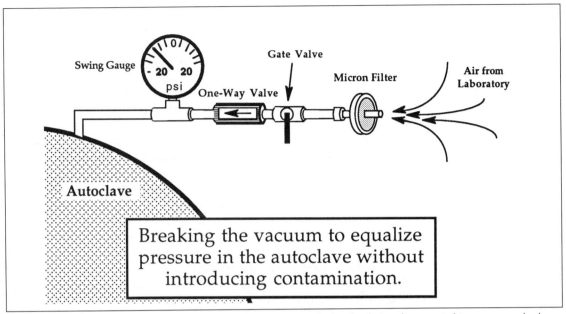

Breaking the vacuum to equalize pressure in the autoclave without introducing contamination.

A microporous filter canister is attached to a pipe equipped with a gate valve that in turn is connected to a one-way check valve leading directly into the autoclave. Located on the clean-room side, this design allows pressure to be equalized in the autoclave after the sterilization cycle without introducing contaminants.

readings between the two gauges. The duration of the autoclave run should not be timed until these differentials have been largely eliminated. (A differential of 10°F should be considered negligible.) In real terms, the differential is normally eliminated within 2 hours of start-up. Obviously smaller vessels have reduced differentials while the most massive autoclaves have substantial contradictions between pressure and temperature readings. Since the duration of "sterilization" is critical, careful consideration of these temperature trends cannot be underemphasized. Cultivators often mistakenly believe the mass has been autoclaved sufficiently when only partial sterilization has been effected. Discarding several hundred bags due to insufficient sterilization is a strong incentive for cultivators to understand the nuances of autoclave cycling. Redundant gauges are recommended since these types of devices repeatedly fail over time.

Post-Autoclaving

When the steam supply to the autoclave is cut off, pressure and temperature precipitously decline. Ideally, your autoclave should achieve a vacuum as it cools. If your autoclave or steam box does not have a tight seal and cannot form a vacuum, provisions must be made so that the air drawn in is free of airborne contamination. This usually means the timely opening of the autoclave into the clean-room air just as atmospheric pressure is attained. Commonly, an autoclave can swing in pressure from 20 psi to –20 psi several hours after steam injection has stopped. This radical fluctuation in pressure further enhances the quality of the sterilization cycle. A 40-psi pressure swing is devastating at the cellular level, disabling any surviving endospores of bacteria or conidia of contaminating molds.

Unloading the Autoclave

Once the autoclave has achieved a vacuum, the pressure must be returned to atmospheric before the door can be opened. Ideally, a gate valve has been installed on the clean room side, on a pipe connected to the combination pressure/vacuum gauge. A microporous filter canister can be attached for further insurance that the rush of air into the autoclave does not introduce contaminant spores. (See page 157.) When the pressure has equalized, the next step is to open the drain valve to draw off excess condensate, which often can be several gallons. After a few minutes, the autoclave door on the clean-room side can be opened.

If the mass has just been autoclaved, the containers will be too hot to unload by hand unless protective gloves are worn. With the door ajar, several hours of cooling are necessary before the bags can be handled freely. Bear in mind that, as the mass cools, air is being drawn in. If that air is full of dust, contamination is likely. I like to thoroughly clean my laboratory while the autoclave is running. I remove any suspicious cultures, vacuum and mop the floors, and wipe the countertops with alcohol. In a separate pressure cooker, I autoclave towels, extra water, and other equipment essential to the impending inoculation cycle. Selected personnel for laboratory work adorn laboratory garments or freshly laundered clothes.

Many spawn producers autoclave on one day, allow the vessel to cool overnight, and open the vessel the next morning. Depending on the mass of the autoclaved material, 12 to 24 hours may pass before the internal temperatures have fallen below 100°F (38°C), the minimum plateau for successful inoculations. Some of the better-equipped spawn producers have large laminar flow hoods, even laminar flow "walls" in whose airstream the sterilized mass is rapidly cooled prior to inoculation.

Atmospheric Steam Sterilization of Sawdust Substrates

Many cultivators cannot afford nor have access to large-production-style autoclaves. The size of the sterilization vessel is the primary limiting factor preventing home cultivators from becoming large-scale producers. Fortunately, alternative methods are available. Whereas straw is pasteurized for 1 to 2 hours at 160°F (70°C), supplemented sawdust is sterilized only when exposed to steam for a prolonged period of time. Many cultivators retrofit the cargo-style containers used in shipping in a fashion similar to a Phase II chamber. Large-capacity commercial laundry washers, cement mixers, cheese-making vats, beer fermentation vessels, railroad cars, semi-truck trailers, grain hoppers, and even large-diameter galvanized drain pipe can be retrofitted into functional steam chambers for the bulk processing of wood or straw-like substrates.

Once filled to capacity with bags of supplemented sawdust, steam is forcibly injected bringing the mass of the substrate to 190°F (90°C) for a minimum of 12 hours. Since water boils at 212°F (100°C) at sea

A portable laminar flow wall creates a localized clean zone in which newly autoclaved substrate can safely cool without danger of contamination.

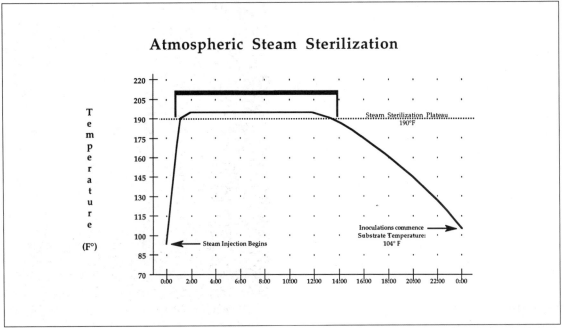

Atmospheric Steam Sterilization

Profile of atmospheric steam sterilization, also known as super-pasteurization, an alternative method for sterilizing sawdust-based substrates.

level, the mass of sawdust cannot be elevated beyond 212°F unless the pressure within the vessel is raised above 1 psi. I call this method *atmospheric sterilization* or *super-pasteurization.*

Most competitor organisms are easily killed with steam heat, with the exception of some thermo-tolerant black pin molds and endospore-forming bacteria. Every microcosm, every microscopic niche, must be subjected to 250°F (121°C) for at least 15 minutes to effect true sterilization. When processing tons of sawdust, true sterilization is rarely achieved. The cultivator must constantly compromise the ideal in favor of the practical. To this end, temperature-sensitive indicator strips help the cultivator determine sterilization profiles. If sawdust is treated in bulk and not separated into individual bags, the danger of cross-contamination is likely during the unloading and spawning process.

After 12 hours of heat-treatment, the steam is shut off. As the mass cools, air will be drawn into the sawdust. The cultivator must take precautions so that contaminants are not introduced. The best alternative is to design the inoculation room with a positive-pressurized HEPA filtration system. Many cultivators use bags or bottles fitted with a filter—either plugged cotton or a specially designed filter disk that prevents the introduction of airborne contaminants. Oftentimes, 1 to 2 days must pass until the mass naturally falls below 100°F (38°C), at which time inoculations can begin.

Super-pasteurization of supplemented oak sawdust substrates, although effective, often results in less total yield than from the same substrate sterilized. Comparative studies by Badham (1988) showed that there are no appreciable differences in yields of Shiitake between supplemented sawdust blocks subjected to *high-pressure* autoclaving versus *atmospheric* steam sterilization for the first flush. In the comparison of total yields, however, more mushrooms can be grown per pound of sawdust if pressure sterilization is employed. The greater yield from sterilized sawdust, according to Royse et al.

(1985), is not due to the survival of contaminants, but a function of the rendering of the sawdust into a form more readily digestible to the Shiitake mycelium.* Pressurized steam essentially softens the sawdust.

Sawdust can be just as easily oversterilized when subjected to pressurized steam for prolonged periods. In tightly packed autoclaves, the outer layers can become oversterilized while the inner core can remain understerilized. Spacing the bags apart so even steam penetration occurs offsets this effect. The cultivator must walk a narrow line in choosing the proper combinations of steam, pressure, duration of treatment, and most importantly, density of fill. Beginning cultivators often see more failure than success for the first several production runs until these parameters can be more specifically refined.

Inoculation of Supplemented Sawdust: Creating the Production Block

The best path for the inoculation of supplemented sawdust is via sawdust spawn. However, the direct path of grain spawn-to-supplemented sawdust is also successful, provided several precautions are taken. Inoculations of supplemented sawdust via grain spawn are prone to self-heating, a phenomenon leading to contamination. *This is especially true with Shiitake.* As supplemented sawdust is consumed by the mycelium of Lion's Mane, Enokitake, Maitake, Nameko, and the Black Poplar mushroom, exothermic reactions emerge at various rates. Regardless of the species, the incubating bags must be spaced apart to preclude thermogenesis.

---

* Those using the more rapidly decomposing hardwoods as a substrate base, such as alder, have not found yields on super-pasteurized sawdust to be depressed compared to sterilized sawdust. Moreover, the density of the wood and moisture content are major factors affecting heat penetration. The addition of buffers, calcium carbonate and calcium sulfate are recommended for the more acidic woods. For more information, see Badham (1988) and Miller and Jong (1986).

Bags of Shiitake mycelia incubating on sterilized sawdust. Note that bags are narrowly separated, but are not touching, which aids in the loss of heat.

(Open wire shelves are recommended over solid shelves.) Once inoculated, the internal temperatures of the bags soon climb more than 20°F over the ambient air temperature of the laboratory. Once the 95–100°F (35–38°C) temperature threshold is surpassed, dormant thermophiles spring to life, threatening the mushroom mycelium's hold on the substrate. For cultivators in warm climates, these temperature spirals may be difficult to control.

Since the risk of contamination is greater with supplemented sawdust, each step must be executed with acute attention to detail. The lab personnel must work as a well-coordinated team. The slightest failure by any individual makes the efforts of others useless. The same general guidelines previously described for the inoculation of sterilized agar, grain, and sawdust media parallel the inoculation steps necessary for inoculating sawdust bran.

Automatic inoculation machines have been built in the attempt to eliminate the "human factor" in causing contamination during inoculation. I have yet

Three people can inoculate 500 to 700 bags in one shift by hand.

Person	Duties	Experience Level
Lab Manager	Spawn Selector Primary Inoculator	+++
First Assistant	Sealer Product Stream Coordinator	++
Second Assistant	Shaker ($^1/_2$ time) Bag Mover Bag Labeler	+

to see a fully automatic spawning machine that out-performs a highly skilled crew. When the human factor is removed from this process, a valuable channel of information is lost. The human factor steers the course of inoculation and allows quick response to every set of circumstances. Every unit of spawn is *sensed* for any sign of impurity or undesirability. The spawn manager develops a ken for choosing spawn based as much on intuition, as on appearance, fragrance, and mycelial integrity.

Inoculations by hand require that either gloves are worn or that hands are washed frequently. With repetition, manual dexterity develops, and success rates in inoculations improve dramatically. Answering the telephone, touching your eyes, picking up a scalpel off the floor, and making contact with another person are causes for immediate remedial action.

Although one person alone can inoculate the sawdust bran bags, a well-coordinated team of three to four expedites the process with the shortest intervals of "down time" and the highest outflow of production. The process can be further accelerated by premarking bags, preshaking spawn, using agitators to evenly disperse the spawn, using gravity or belt conveyors, etc.

Shiitake mycelia on supplemented sawdust one week, three weeks, and six weeks after inoculation.

Steps and Duties for the Personnel Inoculating Supplemented Sawdust

Before proceeding, the lab must be thoroughly cleaned after the autoclave is emptied. The enriched sawdust blocks are positioned in front of the laminar flow bench. Additional blocks are stacked on movable pushcarts that can be quickly moved and unloaded in and out of the inoculation area. The laboratory personnel have prepared the inoculation site by supplying alcohol squirt bottles, paper towels, and garbage bags, marking pens, and drinking water. If only a two-person lab crew is available, the duties of the Lab Manager and the First Assistant are often combined.

Step 1

Lab Manager

Select pure spawn. Avoid any units of spawn showing the slightest disparity in growth. Be suspicious of spawn units adjacent to partially contaminated ones. Usually contamination outbreaks run through a series of consecutive inoculations, to greater and lesser degrees. Individual units of spawn that look pure, but are neighbors to contaminated units, should only be used as a last recourse. Shake the spawn, thoroughly breaking it up into its finest particles. Place the spawn immediately downstream from the bag sealer. Wipe your hands with 80% isopropanol (rubbing alcohol).

First Assistant

The First Assistant works with the Lab Manager in shaking the spawn, readying it for use.

Second Assistant

The Second Assistant positions the bags and begins prelabeling. If more than one strain or species is used, prelabeling must be done carefully, lest confusion between strains occur.

Step 2

Lab Manager

The Lab Manager holds a bag of sawdust spawn and, using a pair of aseptically cleaned scissors, cuts at a 45-degree upward angle towards the opposite corner, cutting across the previous seal. (See page 163.) This results in a "spout," facilitating the transfer of spawn from one bag to another. If you hold a bottom corner with one hand and raise the bag with the other, grasping above the newly created spout, the transferring of spawn from one bag to another container is simple and fast. (If inoculating supplemented sawdust with grain spawn, follow the techniques described on page 160 for Inoculating Sterilized Sawdust.)

As the inoculations progress, care is taken not to touch the inside walls of the bags with your hands. The bags can be pulled apart by grasping the outside plastic, expanding the opening, so that sawdust spawn can be received without hindrance. At times the First Assistant may be called upon to make sure the bags are fully opened.

First Assistant

The First Assistant closes the bags on the precleaned thermal bag sealer on the uppermost, opened portion. The sealer is activated (depending on type) and the panels of plastic meld together, capturing a volume of air in the process. Ideally, a "domed" bag can be created. (See pages 147–148.) Sometimes, multiple seals are necessary before full closure is achieved.

Second Assistant

As soon as the bag has been hermetically sealed, it is removed from the position behind the sealer and passed to the Second Assistant. The Second Assistant first ascertains the bag has been sealed. With gentle squeezing, leaks are detected by either the slow collapse of the bag, the sound of air escaping, or visual observation. (If the leak still can not be detected, I bring the bag to my face. When squeezing, my face can usually find the location of the hole as a jet of air escapes.) If the leak is due to a puncture, the hole is covered with plastic packaging tape. If the seal is imperfect, the bag is returned to the First Assistant for a second try at heat sealing. Once each bag has been properly sealed and assured of proper labeling, thorough shaking is in order. Using a combination of agitation and rotation, the sawdust spawn is mixed through the supplemented sawdust. The sawdust spawn has a lighter color than the supple-

Cutting a bag of pure sawdust spawn for use as inoculum into supplemented sawdust.

mented sawdust and hence it is easy to see when the spawn has been evenly dispersed.

The Second Assistant gently slams the sealed, domed bag on a tabletop to close any open spaces, and increase the density of the mass. (In shaking, the mixture becomes quite loose.) The bag is positioned on a wire shelf where daily observations are made for the next few weeks. The bags are kept at least a finger breadth apart. *These newly inoculated, incubating bags must not touch one another.*

Now that the duties of the three laboratory personnel are clearly defined, working in concert becomes the foremost priority. A well-organized laboratory team achieves a furious pace. Conversation is kept to a minimum. (The person doing the inoculations can't talk anyhow, except between sets of inoculations, lest his or her breath spreads bacteria. Wearing a filter mask reduces this risk.) However, times arise when one or more of the three-person production team is interrupted. During these down times, countertops should be vacuumed and wiped clean with alcohol. Garbage is consolidated.

Pouring sawdust spawn into bag of sterilized, supplemented sawdust.

Finally one's hands are washed prior to any more inoculation activity. The pace during the inoculation process should be both rhythmic and fast. Inoculations are purposely interrupted after every fifty or so bags so periodic cleaning can occur. Depending on the equipment, design of the facility, and experience of the laboratory personnel, better ways of organizing the labor during inoculations will naturally evolve. However, this method works well. Hundreds of bags can be inoculated during a single shift. Greater efficiency is realized if two or more sealers are employed simultaneously.

After inoculation, each bag can be shaken by hand. Larger farms place the bags onto a gravity conveyor leading to a multiple-bag, automatic shaking machine. As the crew's performance improves with experience, block production soars into the thousands per shift. With each bag yielding US $10 to $50+, every doubling of production over a baseline level realizes proportionally greater profits for the owners.

Once shaken, the inoculated bags are placed on open wire shelves and spaced about 1 inch apart for incubation. Be forewarned that bags contacting one another are likely to contaminate with black pin or other thermophilic molds.

Incubation of the Production Blocks

The first 2 weeks of incubation after inoculation are the most critical. If the supplemented sawdust is not fully colonized during that time period, contamination usually arises soon thereafter. Within several days of inoculation, out-gassing of volatile by-products causes a distinctly noticeable fragrance. As soon as the laboratory is entered, the atmosphere imparts a unique "odor signature." The smell is generally described as sweet, pleasant, and refreshing.

The laboratory should be maintained at 75°F (24°C) and have an ambient humidity between 30 and 50%. Since the internal temperatures of the incubating blocks are often 20°F higher than ambient air temperatures, keeping the laboratory warmer

than recommended is likely to cause the internal incubation temperatures to rise to dangerous levels. Carbon dioxide levels within the laboratory should never exceed 1,000 ppm, although 20,000 to 40,000 ppm of CO_2 is typical within the bags as they incubate. This steep slope of high CO_2 within the bag to the low CO_2 in the atmosphere of the laboratory is helpful in controlling the evolution of metabolic processes. Should the gradient be less severe, CO_2 levels can easily exceed 50,000 ppm within the incubating bags. At this and higher levels, mycelial growth lessens and contaminants are encouraged. To compensate, adjust the laboratory air handling system for the proper mixing of fresh versus recirculated air. (See Appendix 2: Designing and Building a Spawn Laboratory.)

Three days after inoculation, the mycelium becomes clearly visible, often appearing as fuzzy spots of growth. Second shaking, although essential for insuring full colonization of grain spawn, is not usually advisable in the incubation of supplemented sawdust. If complete sterilization has not been achieved, second shaking can result in a contamination bloom. If one is certain that sterilization has been achieved, second shaking *helps* colonization, especially around days 4 and 5, but most cultivators find this extra step unnecessary.

Each species uniquely colonizes supplemented sawdust. Oyster mycelium is notoriously fast, as is Morel mycelium. "Good growth" can be generally described as *fans of mycelium rapidly radiating outwards from the points of inoculation. Growth is noticeable on a daily, and in some cases, an hourly basis.* When the mycelium loses its finger-like outer edges, forming circular dials, or distinct zones of demarcation, this is often a sign that contaminants have been encountered, although they may not yet be visible. The behavior of the mycelium constantly gives the spawn manager clues about the potential success of each run.

The cultivator should not operate under the belief that the supplemented sawdust blocks are absolutely free of contamination after autoclaving. In fact, they should be viewed as having a greatly reduced population of contaminants, which pose little problem as long as quick colonization by the mycelium occurs. Large runs of supplemented sawdust, however, are more likely to host minute pockets of unsterilized substrate than smaller runs would host. Should colonization be inhibited, or encouraged by any number of factors—poor strain vigor, a dilute inoculation rate, elevated internal thermal or carbon dioxide levels—contaminants are to be expected. *This race between the mycelium and legions of competitors is a central theme operating throughout every stage of the cultivation process.*

Achieving Full Colonization on Supplemented Sawdust

Prior to the mycelium densely colonizing the blocks with a thick and tenacious mycelial mat, the supplemented sawdust appears to be grown through with a fine, but not fully articulated, mycelial network. The once brown sawdust mixture takes on a grayish white appearance (with most species). With Shiitake mycelium, this is usually between days 3 and 7. During this state, the mycelium has yet to reach its peak penetration through the substrate. Although the substrate has been captured as a geological niche, the mycelial network continues to grow furiously, exponentially increasing in its micro-netting capacity. The bags feel warm to the touch and carbon dioxide evolution is great.

Within hours, a sudden transformation occurs: The once-gray appearance of the bags flush to snow-white. The fully articulated, thick mycelial network achieves a remarkable tenacity, holding fast onto the substrate. Now when each block is grasped, the substrate holds together without falling apart, feeling solid to the touch.

The blocks can be further incubated until needed, within certain time restraints. (Refer to Chapter 21 for the particular time requirements of each species.) With a one-room laboratory, inoculations and incubation can occur in the same space. If a multiroom laboratory is being used, then the supplemented sawdust blocks are furthest downstream from the precious petri dish cultures. In fact, they should be nearest the door for ease of removal. Ideally, this sequence of prioritizing cultures should follow each step in the exponential expansion of the mycelial mass:

- Petri dish cultures are furthest upstream, i.e., are given the highest priority.

- Grain spawn is organized in rank; downstream from the cultures maintained on malt agar media. First-generation, second-generation, and third-generation spawn are prioritized accordingly. Grain spawn, incubated in jars, is best stored at angles in vertical racks.

- Sawdust spawn, being created from grain spawn, is next in line. Second generation sawdust spawn is kept next downstream.

- Supplemented sawdust blocks designed for mushroom cropping, along with any other units destined for fruiting or implantation outdoors, are incubated closest to the exit.

As the mycelium is expanded with each generation of cultures, contamination is increasingly likely. This specified flow pattern prevents reverse contamination of upstream cultures from those downstream.

Once the mycelium achieves the above-described "grip" on the supplemented sawdust, the nature of the mycelium changes entirely. The blocks cease to generate heat, and carbon dioxide evolution abruptly declines. With most species, the blocks no longer need to be treated so delicately. They can be moved to secondary storage rooms, even thrown through the air from one person to another! (A new sport!?) This state of "mycelial fortitude" greatly facilitates the handling process. The blocks should be moved out of the laboratory environment and either taken to a staging room for later distribution to the growing room or a dark refrigeration room until needed. This resilient state persists until mushrooms form within the bags, either from the response of environmental change or not. Many strains of *Lentinula edodes* (Shiitake), *Hericium erinaceus* (Lion's Mane), *Grifola frondosa* (Hen-of-the-Woods), *Agrocybe aegerita* (Black Poplar mushroom), and *Pleurotus* spp. produce volunteer crops of mushrooms within the bags as they incubate in the laboratory, without any environmental shift to stimulate them.

For many of the species listed in this book, volunteer fruitings begin 3 to 6 weeks after inoculation. Just prior to the formation of visible mushrooms, the topography of the mycelium changes. With Shiitake, "blistering" occurs. The smooth surface of the outer layers of mycelium roughens, forming miniature mountains and valleys. (See Chapter 21 for a complete description of this phenomenon.) With Lion's Mane (*Hericium erinaceus*), dense, star-like zones form. These are the immediate precursors to true primordia. If these ripe bags are not taken to the growing room in time, the newly forming mushrooms soon malform: most frequently, with long stems and small caps. (These features are in response to high CO_2, lack of light, or both.) The young mushrooms at this stage are truly embryonic and must be treated with the utmost care. The slightest damage at this stage will be seen later—at the full maturity—as gross deformations: dimpled or incomplete caps, squirrelly stems, etc. Shiitake are particularly fragile at this stage whereas Oyster mushrooms can return to near-normal forms once exposed to the conducive climate of the growing room.

Handling the Bags Post-Full Colonization

Depending upon the species, 3 to 6 weeks pass before the bags are to be placed in the growing room. Before the blocks are moved in, the growing room has been aseptically cleaned, having been washed down with a bleach solution. After washing with bleach, I tightly close up the room for 24 hours and turn off all fans. The residual chlorine becomes a disinfecting gas permeating throughout the room, effectively killing flies and reducing mold contaminants. A day after chlorine treatment, fans are activated to displace any residual gas before filling. Additional measures prior to bleaching include replacing old air ducting with new, the changing of air filters, etc.

If you space the bags at least 4–5 inches apart, the developing mushrooms will mature without crowding. Sufficient air space around each block also limits mold growth. Galvanized stainless steel and/or epoxy-coated wire mesh shelves are preferred over solid shelves. Wood shelves should not be used because they will eventually, no matter how well treated, become sites for mold growth. Farms that do use wood trays either chemically treat them with an antifungal preservative to retard mold growth or construct them from redwood or cedar. I know of no studies determining the transference of toxins from chemically treated trays into mushroom fruitbodies. However, this concern is logically based, as mushroom mycelium absorbs some toxins, especially heavy metals.

CHAPTER 18

Cultivating Gourmet Mushrooms on Agricultural Waste Products

Many wood decomposers can be grown on nonwood-based substrates such as cereal straws, cornstalks, sugarcane bagasse, coffee pulp, banana fronds, seed hulls, and a wide variety of other agricultural waste products. Since sources for hardwood by-products are becoming scarce due to deforestation, alternative substrates are in increasing demand by mushroom cultivators. However, not all wood decomposers adapt readily to these wood-free substrates. New mushroom strains that perform well on these types of alternative substrates are being selectively developed.

The more hearty and adaptive *Pleurotus* species are the best examples of mushrooms that have evolved on wood but readily produce on corncobs, cottonseed hulls, sugarcane bagasse, coffee waste, palm leaves, and cereal straws. When these materials are supplemented with a high nitrogen additive (rice bran, for instance), simple pasteurization inadequately treats the substrate, and sterilization is called for. (Without supplementation, pasteurization usually suffices.) Each cultivator must consider his or her unique circumstance juxtaposing the available substrate components, species, facilities, and market niches—in their overall system design.

Growing the Oyster Mushroom, *Pleurotus ostreatus*, on straw is less expensive than growing on sterilized sawdust. In contrast, Shiitake, *Lentinula edodes*, which barely produces on wheat straw is best grown on wood-based substrates.* When both straw and sawdust are difficult to acquire, alternative substrates are called for. Mini-trials are encouraged before substantial resources are dedicated to any commercial enterprise. I encourage readers to formulate new blends of components that could lead to a breakthrough in gourmet and medicinal mushroom cultivation.

---

* Several patents have been awarded in the cultivation of Shiitake on composted, wood-free substrates. Although these are fruitful, wood-based substrates are still preferred by most Shiitake cultivators.

Alternative Fruiting Formulas

Here is a basic formula, not using wood or straw, for the cultivation of wood-decomposing mushrooms. A nitrogen supplement, in this case, rice bran, is added to boost yields. As discussed, the substrate must be heat-treated by any one of a number of methods to effect sufficient sterilization.

    85 lbs. (39 kg) ground corncobs, peanut shells,
        chopped roughage from sugarcane bagasse,
        tea leaves, coffee, banana, saguaro cactus, etc.
    10 lbs. (4.6 kg) bran, grape pumice or approxi-
        mately 2.5 lbs. (1 kg) extracted soybean oil
    4 lbs. (1.8 kg) gypsum (calcium sulfate)
    1 lb. (.45 kg) calcium carbonate
    100–140 lbs. (45–64 kg) water or as required

The amount of calcium carbonate can be altered to effectively raise pH, offsetting any inherent acidity. The components are mixed in dry form and wetted until a 70–75% moisture content is achieved. The mixture is loaded into bags and immediately heat-treated. Should the bags sit overnight, and not be autoclaved, contaminants proliferate, making the mixture unsuitable for mushroom cultivation.

The methods described here for the cultivation of mushrooms indoors on straw can be extrapolated for cultivating mushrooms on chopped cornstalks, sugarcane bagasse, and many other agricultural waste products. In contrast to wood wastes, which should be sterilized, I believe most agricultural by-products are better pasteurized using steam or hot water baths. Pasteurization selectively kills the majority of competitors, preserving populations of noncompetitive thermotolerant fungi, resulting in a substrate with a limited window of opportunity for contamination, and in turn, favors the dominance of mushroom mycelium. Pasteurization typically occurs between 140 and 180°F (60 and 82°C) at atmospheric pressure (1 psi). Sterilization is by definition above the boiling point of water, >212°F (100°C), and above atmospheric pressure, i.e., >1 psi. A hybrid treatment, which I call atmospheric sterilization or "super-pasteurization" calls for the exposure of substrates to prolonged, elevated temperatures exceeding 190°F (88°C) for at least

12 hours. In any case, a carefully balanced aerobic environment must prevail throughout the incubation process or competitors will flourish.

Readily available, inexpensive, and needing only a quick run through a shredder (and sometimes not even this), wheat straw is ideal for both the home cultivator and commercial cultivator. Straw is a "forgiving" substrate for the small to midsized cultivator, accepting a limited number of contaminants and selectively favoring mushroom mycelium. Growing on straw is far less expensive than growing on sawdust. Many cottage growers enter the gourmet mushroom industry by first cultivating Oyster mushrooms on straw. Wheat, rye, rice, oat, and sorghum straws are the best. Hay, resplendent with abundant bunches of seed kernels, should not be used as the grain kernels tend to contaminate. However, limited numbers of grain kernels generally boost yields. Royse (1988) found that yields of Oyster mushrooms from wheat straw are enhanced by the addition of 20% alfalfa without increasing the risk of contamination. Alfalfa, by itself, is "too

Coffee grounds, if they are inoculated soon after expressing, are a ready-made substrate for Oyster mushrooms, *Pleurotus ostreatus* and *Pleurotus pulmonarius*, and for Reishi, *Ganoderma lucidum*.

hot" to use because of its elevated nitrogen content. Straw supports all the gourmet Oyster mushrooms, including *Pleurotus citrinopileatus*, *P. cystidiosus*, *P. djamor*, *P. eryngii*, *P. euosmus*, *P. ostreatus*, *P. pulmonarius*, and *Pleurotus tuberregium*. Other mushrooms, King Stropharia (*Stropharia rugosoannulata*), Shaggy Manes (*Coprinus comatus*), the Paddy Straw (*Volvariella volvacea*), and Button (*Agaricus* spp.) mushrooms also thrive on straw-based substrates, often benefiting from modest supplementation.

Coffee waste in Mexico and Central America has become a major ecological catastrophe, polluting rivers and aquifers. Coffee pulp is an ideal substrate for growing Oyster mushrooms. Not only have yields in excess of 159% biological efficiency been reported, but residual material, post-harvesting, has largely been denatured of caffeine and is suitable as feed for many farm animals (Martinez-Carrera, 1987, 1988; Leon-Chocoo, 1988; Thielke, 1989). Coffee plantations and Oyster mushroom farms are complimentary industries. This is yet another

A simple and easy method for pasteurizing straw (and other bulk materials). The drum is filled with water and heated with the propane burner at 160°F (71°C) for 1 to 2 hours.

example that mushroom cultivation is the missing link in the integration of complex systems of human enterprises within a sustainable environment.

The specifics for cultivation of each of these species are discussed further on, in Chapter 21.

Heat-Treating the Bulk Substrate

Bulk substrates like straw are generally pasteurized (as opposed to sterilized) and upon cooling, inoculated with grain spawn. Pasteurization selectively kills off populations of temperature-sensitive microorganisms. The population left intact presents little competition to the mushroom mycelium for approximately 2 weeks, giving ample opportunity for the mushroom mycelium to colonize. If not colonized within 2 weeks, the straw naturally contaminates with other fungi, irrespective of the degree of pasteurization. Straw is first chopped into

Shredding and moistening straw.

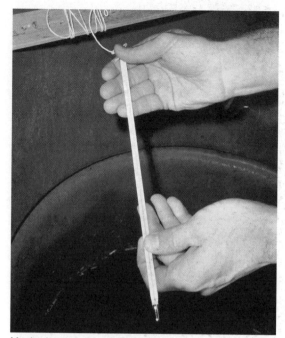

Monitoring water temperature.

Straw is stuffed into a wire basket and then placed into the hot water. A weight is placed to keep the straw submerged.

1-inch to 4-inch lengths and can be prepared via several methods.

The Hot Water Bath Method: Submerged Pasteurization

The first method is the hot water bath. Straw is stuffed into a wire basket and submerged in a cauldron of 150–180°F (65–82°C) water for 1 hour.* The soaking cauldron is usually heated from underneath by a portable propane gas burner. The straw basket is forcibly pushed down into the steaming water and held in place by whatever means necessary. A probe thermometer, at least 12 inches in length, is inserted deep into the mass, with string attached for convenient retrieval. When water temperature reaches 160°F, the straw is submerged for at least 1 hour and no longer than 2.

Upon removal, the straw is well drained and laid out in a shallow layer onto cleaned surfaces (such as a countertop) to rapidly cool. Most cultivators broadcast grain spawn over the straw by hand. Gloves *should* be worn but often are not, and yet success is the norm. In either case, the hands are thoroughly and periodically washed, every 15 minutes, to limit cross-contamination. The spawn and straw are then mixed thoroughly together and placed directly into trays, columns, wire racks, or similarly suitable containers.

Another basket of chopped straw can be immersed into the still hot water from the previous batch. However, after two soakings, the hot water must be discarded. The discolored water, often referred to as "straw tea," becomes toxic to the mushroom mycelium upon the third soaking, retarding or preventing further mycelial growth. Therefore, the water must be changed after the second soak. (Interestingly, this tea is toxic to most vegetation and could be used as a natural herbicide.)

* Stainless steel 55-gallon drums from the food/fermentation industry are preferred. If stainless steel drums are unavailable, only those designed for food storage/processing should be used. Metal drums that stored juices, oils, and glycerin are good candidates.

A brick keeps the straw submerged during pasteurization.

After 1 to 2 hours of submerged pasteurization, the basket is lifted out. After draining excess water, the straw is cooled, and grain or sawdust spawn is broadcast over the surface and mixed throughout the straw.

The "Phase II" Chamber:
Steam Pasteurization

A second method calls for the placement of straw in a highly-insulated room into which steam is injected. This room is known as the Phase II chamber. Before the straw is loaded into the Phase II chamber, it must be moistened. This can be done simply by spreading the chopped straw over a large surface area, a cement slab or plastic tarpaulin to a depth no greater than 12 inches. Water is sprayed on the straw via sprinklers over a 2- to 4-day period. The straw is turned every day to expose dry zones to the sprinkling water. After several turns, the straw becomes homogeneous in its water content, approaching 75% moisture, and is reduced to about half its original volume. Once evenly moistened, the straw is now ready for loading into the steam chamber. Short-stacking the straw is not intended to accomplish composting, but rather a way of tendering the straw fiber, especially the waxy, outer cuticle. In contrast to composting, the straw is not allowed to self-heat.

An alternative method calls for the construction of a large vat into which straw is dunked. This tank is usually fitted with high-pressure water jets and rotating mixing blades to assure full moisture penetration into the straw. If given sufficient agitation, finely chopped straw gains 75% moisture in the matter of minutes. Once moistened, the straw is loaded directly into the Phase II chamber.

One ton of wheat straw, chopped and soaked, occupies approximately 250 cubic feet of space, equivalent to 10 × 10 × 2.5 feet. This figure is helpful in sizing a pasteurization chamber. An additional 25% allowance should be made for variation in the chop size of straw, air plenums, and handling needs. Five dry tons of wheat straw functionally fills a 1,000-square-foot growing room. Most growers fill growing rooms to no more than one-fourth of total air volume. I prefer to fill to only one-eighth of capacity. This means that for every eight air spaces, one space is occupied by substrate. (In other words, the ratio of air-to-substrate space is 7:1.)

The classic Phase II room has a raised, false floor, screened several inches above the true floor upon which a latticework of steam pipes, are situated. The walls and ceiling are well insulated. The interior panels are made of heat-resistant, waterproof materials. Many convert shipping containers used to ferry cargo on ships into Phase II chambers. Others custom-build their own steam rooms.

Another important feature is a floor drain fitted with a gate valve. This valve prevents contamination from being drawn in during and after pasteurization.

Boilers provide live steam, dispersed through the pipes, and into the straw, which can be filled to a depth as great as 8 feet. The greater the depth, the longer heat takes to penetrate to the center. Heat penetration can be enhanced with high-pressure blowers. At least three remote reading temperature probes are inserted at various locations: low (within 4–6 inches), midway, and high (within 12–24 inches of the top surface). These temperature probes should be monitored periodically to gather data for the generation of a pasteurization profile specific to each run. Over time, the temperature points of each successful batch are accumulated for establishing a baseline for future operations.

When steam is injected, the outer edges of the straw mass heat up first. An outer zone of high temperature forms, and over time increasingly enlarges towards the center. Early on in the Phase II process, three thermometers can read a range from room temperature to 160°F (71°C) simultaneously. This temperature differential must be monitored carefully. The core of a densely packed Phase II steam chamber remains below 100°F (38°C) for several hours, lagging behind the hot outer shell, and then suddenly races upwards. If steam output from the boilers is not reduced in time, the entire mass continues to heat at an uncontrollable rate. Without the cold core, which in effect is a heat-sink, to deflect the spiraling increase in temperature, thermal momentum continues for an hour or two beyond the time steam injection is shut off. The entire mass must not exceed 200°F (93°C) The minimum recommended time for pasteurization is *2 hours above 160°F (71°C).*

A common oversight in steaming a mass of straw or compost is the failure to chart, every 30 minutes, the temperature profile of the mass. With measuring and charting, trend analysis is possible. If the climb in temperature is not anticipated or reduced at

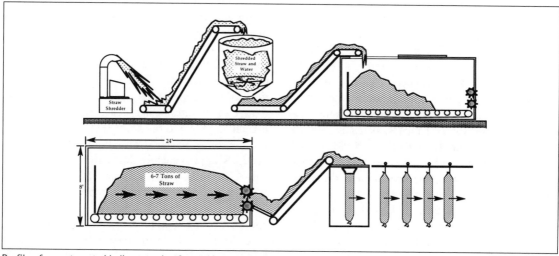

Profile of an automated bulk processing for soaking and pasteurizing straw. After pasteurization, the steamed mass is pulled via a netted floor into two outwardly rotating teethed cylinders that throw the straw onto conveyors. In this case columns are filled, although trays or other fruiting vessels could just as well be used.

Screened floor racks allow the passage of steam underneath the bulk material being pasteurized.

Inside a large Phase II designed for growing Button mushrooms.

the right time, the thermal momentum of the hot outer shell not only overwhelms the ever-shrinking cold core, but cause the entire mass to skyrocket to 200+°F (93°C), a temperature above which disaster awaits. Above this temperature plateau, noncompetitive, beneficial organisms are killed, and the substrate becomes an open habitat for many competitors that would otherwise be held in abeyance.

When the steam output from the boiler is turned off, the Phase II box should be immediately positive-pressurized with contaminant-free air. By forcing air through a HEPA filter and ducting the air directly into the Phase II chamber, you prevent contaminants from being sucked in as the mass cools. For a steam box measuring 10 cubic feet, a $1/8$-HP (horsepower) blower pushing 200 CFM (cubie feet per minute) through a 12 × 12 × 6-inch HEPA filter (99.99% @ .3 μ) adequately positive-pressurizes the chamber. (Larger Phase II rooms will require correspondingly higher pressure fans able to push air over 2–4 inches

of static pressure.) The substrate mass slowly cools in 12 to 24 hours to temperatures tolerable for inoculation, generally below 105°F (38°C).

Before the pasteurization chamber is opened, the inoculation area is intensively clean with a 10% bleach solution.* To aid the cleaning process, venturi siphon mixers are ideal for drawing bleach directly into a hose line at the faucet connection. Conveyor belts, countertops, funnels, ceilings, and walls are all cleansed with torrents of chlorinated water. Spraying down the room with such a solution is colloquially termed "bleach bombing" in the industry. Protective clothing is advised for the sanitation personnel.

---

*Most brand-name bleaches have 5.25% sodium hypochlorite. One tablespoon of bleach in a gallon of water is roughly equivalent to 200 ppm chlorine. A cup of bleach per gallon of water is equivalent to 3,200 ppm. Most mushroom mycelia are harmed above 500 ppm of chlorine.

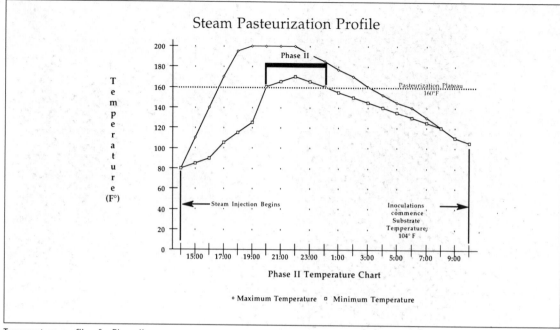

Temperature profile of a Phase II run.

Methods vary for the unloading of the Phase II chamber. Some remove the straw by hand with clean pitchforks and throw the straw onto stainless steel tables whereupon the inoculation occurs. Conveyors are favored for substrate handling by many growers. The largest Phase II chambers utilize a netted or "walking" floor, which pulls the substrate mass into two outwardly rotating, horizontally positioned, teethed cylinders. (See page 172.) As the substrate mass is forced into the space between the two outwardly rotating, spiked cylinders, the straw is separated and ejected onto a depressed platform in whose center is a funnel or ramp that leads to conveyors. (Warning: The danger of personal injury here at this juncture is notorious. Special precautions must be implemented to prevent accidents.) After the straw is thrown into the conveyor belt, grain spawn is automatically fed or hand-broadcast onto the straw as it is being ferried away. Foot-activated switches are helpful in controlling the off-loading of the substrate from the Phase II box with the conveyor.

When spawn is placed directly upon the surface of pasteurized straw, mixing is strongly advised. Cement and soil mixers, specially adapted funnels, ribbon blenders, and "Archimedes screws" suffice. If the spawn is laid upon straw and not mixed through, growth layers form resulting in uneven colonization. The advantage of removing the straw and inoculating by hand is that the process can be interrupted and recurrent cleaning can occur. With intermittent disinfecting, cross-contamination can be prevented. With automated, continuous-loop systems, the likelihood that contamination can travel throughout the facility unchecked is greater. Special attention to detailed disinfection is necessary with these systems to prevent disastrous results should pasteurization be incomplete. Once spawn has been sown throughout the straw, the inoculated substrate is placed directly into the "fruiting" containers, usually columns, trays, or bags. Each container must be vented so the mycelium can respire as it colonizes the substrate.

Alternative Methods for Rendering Straw and Other Bulk Substrates for Mushroom Cultivation

Several inexpensive, alternative methods can be used for treating straw (and other bulk materials) that do not involve heat-treatment. The first three are chemical; the last is biological. Surely other alternative methods will be developed as imaginative entrepreneurs experiment. By sequencing a substrate through a combination of biological and chemical treatments, you can avoid heat pasteurization entirely. Small pilot-scale experimentation is strongly encouraged before cultivators attempt these techniques commercially. The future use of such methods appears extremely promising.

The Hydrated Lime Bath Method

Hydrated lime (calcium hydroxide) is extremely alkaline and water-soluble. By immersing straw into water baths high in hydrated lime, competitor fungi and bacteria are largely rendered inactive from the drastic change in pH. The preparation is quite simple.

Two to four pounds of lime is added for every 50 gallons of water. (Since a gallon of water weighs 8.3 pounds this ratio is equivalent to 2–4 pounds lime/415 pounds water or about .5–1.0%.) The pH of the water skyrockets to 9.5 or higher. Once dissolved, chopped straw is immersed into this highly alkaline bath. Under these caustic conditions, pH-sensitive microorganisms soon die. Subsequent to an overnight soaking, the water is drained and discarded. (Note that this highly alkaline water kills many plants and should be prevented from entering any watershed.) The straw is then drained and inoculated using standard methods. It is not unusual for the straw to achieve a pH of 8.5 or higher after soaking. Oyster mushroom mycelia can tolerate this alkaline environment better than most competitors. After 3 or 4 days of initial growth, pH slowly falls as the mycelium races through the straw, secreting acids and enzymes. One week after inoculation, the straw should be fully colonized.

If colonization is not complete within 7 to 10 days, competitors usually arise. Optimizing the parameters for the species being cultivated greatly influences the success or failure of this simple method. Please consider that the starting pH of makeup water affects the final outcome. Each cultivator must compensate accordingly.

The Bleach Bath Method

This is similar to the hydrated lime bath method, but household bleach (5.25% sodium hypochlorite) is used as a disinfectant. I recommend adding 3–4 cups of household bleach to 50 gallons of water. A basketful of chopped wheat straw is immersed. The straw is kept submerged for a minimum of 4 and no more than 12 hours. The bleach leachate is drained off. The straw is immediately inoculated with an aggressive species such as *Pleurotus pulmonarius*. Should colonization not be complete within a week, contaminants naturally occur. Cultivators should be careful where the toxic leachate is drained. Gloves should be worn to protect the skin from chemical burning.

The Hydrogen Peroxide Technique

Hydrogen peroxide has been a traditional method for disinfection for more than a century. Recently, this method was, in a sense, rediscovered by chemist Rush Wayne, who, having become frustrated with the difficulty and expense of creating a sterile environment in his home, refined this technique to a practical level. A full description of this technique can be found at *http://members.aol.com/_ht_a/RushWayne/Perox_Mushrooms.html*. He has also published a booklet on this technique (Wayne, 1999). The usefulness of this technique on a commercial scale has yet to be proven but this method is well worth a careful economic study. Its main advantage is that it offers an alternative method to those who do not have access to many of the more expensive tools-of-the-trade. The readers should note, however, that much resident contamination can survive this process, and it is not as effective as many other methods. Nevertheless, Rush Wayne has provided an excellent service by continually refining this technique.

Hydrogen peroxide (H_2O_2) works to kill many fungal spores, yeasts, and bacteria by producing a reactive form of oxygen, which destroys cell walls. However, because fungal mycelium has evolved to decompose organic compounds in the environment using peroxides, the mushroom mycelium and the mycelia of contaminant molds are protected from its oxidizing effects. If colonies of mycelium from contaminant fungi have already developed, this method will be of limited advantage. However, most contaminant spores will be neutralized when in contact with peroxide. The advantage of hydrogen peroxide is that it topically destroys resident spores within many of the substrates useful for cultivation, especially those present in dry form, or in the media. It also provides some protection from airborne vectors when the media is exposed. However, since hydrogen peroxide is so reactive, it does not provide much protection from airborne contamination after absorption by the host substrate.

Although not thorough enough to neutralize most of the natural fungal contaminants resident in raw sawdust, straw, or composts, hydrogen peroxide can help complete the process started with many preheated substrates. When wood is baked in an oven at 300°F (149°C) for 3 hours, compounds are destroyed in the wood that would otherwise neutralize the peroxide. This technique is an effective heat pretreatment. Or, the heat-treated, hardwood fuel pellets used for wood stoves can first be boiled in water, drained, and then allowed to cool. Hydrogen peroxide can be diluted 100-fold, from 3% to .03%, into water (less than 140°F or 60°C). This water can then drench the substrate to further reduce the likelihood of competitors. When diluted hydrogen peroxide is added to wood and allowed to sit for 8 hours, the substrate can then be inoculated with a modicum of success. However, this technique is good for only one generation of transfers at most, which means one unit of pure culture spawn can inoculate five to twenty units of peroxide-treated sawdust, but these units cannot easily be used for further expansion without likelihood of a major

contamination bloom. This limitation means that this technique is more applicable to home cultivators than to commercial cultivators.

This is but one of many examples. Any technique that leaves a window of opportunity open long enough for the mushroom mycelium to dominate, generally 7 to 10 days, is worth pursuing.

The High-Pressure Extrusion Method

Another pass-through method for treating straw and sawdust utilizes the heat generated from the extrusion of a substrate from a large orifice through a smaller one. This methodology was first developed for the commercial feed industry in the creation of pelletized rabbit, chicken, and cattle feed. The effective reduction of the substrate causes frictional heat to escalate. For instance a 6:1 reduction of straw into a 10-millimeter-sized pellet creates a thermal impaction zone where temperatures exceed 176°F (80°C), temperatures sufficient for pasteurization. Large commercial units can process up to 2,000 pounds per hour. (One supplier has a home page at *www.amandus-kahl-group.de*.) Pellets between 50 and 100 millimeters in diameter are better suited for fruitbody production than the smaller pellets. However, the smaller pellets might be more useful as spawn in the outdoor introduction of mycorrhizae or for the introduction of mycelia for purposes of mycoremediation of toxic habitats. Another technology being tested uses a roller mechanism rather than a narrow orifice, and can process much more substrate mass.

The Detergent Bath Method

This method simply utilizes biodegradable detergents containing fatty oils to treat bulk substrates. Coupled with surfactants that allow thorough penetration, these detergents kill a majority of the contaminants competitive to mushroom mycelium. The substrate is submerged into and washed with a detergent solution. The environmentally benign wastewater is discarded, leaving the substrate ready for inoculation. Recently, many environmentally safe soaps have been developed, especially in Europe. Cultivators are encouraged to experiment to match the best detergents to their substrate materials. Here again, the goal is to create a process that is both simple and applicable for small- and large-scale cultivators.

The Yeast Fermentation Method

Another alternative method for rendering straw is biological. Straw can be biologically treated using yeast cultures, specifically strains of beer yeast, *Saccharomyces cerevisiae*. This method, by itself, is not as effective as those previously described one but has achieved limited success. Research into the sequencing of multiple microorganisms may improve success.

First a strain of beer yeast is propagated in 50 gallons (200 liters) warm water to which malt sugar has also been added. Recommended rates vary. Usually a 1–5% sugar broth is concocted. Fermentation proceeds for 2 to 3 days undisturbed in a sealed drum at room temperature (75°F, 24°C). Another yeast culture can be introduced for secondary, booster fermentation that lasts for another 24 hours. After this period of fermentation, chopped straw is then forcibly submerged into the yeast broth for no more than 48 hours. Not only do these yeasts multiply, absorbing readily available nutrients, which can then be consumed by the mushroom mycelium, but metabolites such as alcohol and antibacterial by-products are generated in the process, killing competitors. Upon draining, the straw is inoculated using standard procedures.

Another method of submerged fermentation uses the natural resident microflora from the bulk substrate. After 3 to 4 days of room-temperature fermentation, a microbial soup of great biological complexity evolves. The broth is now discarded and the substrate is inoculated. Although highly odoriferous for the first 2 days, the offensive smell soon disappears and is replaced by the sweet fragrance of actively growing mycelium. I hesitate to recommend it over the other procedures described here.

The outcome of any one of these methods greatly depends on the cleanliness of the substrate being used, the water quality, the spawn rate, and the aerobic state of the medium during colonization. These methods generally do not result in the high consistency of success (>95%) typical with heat treatment techniques. However, with refinement, these simple and cheap alternatives may prove practical wherever steam is unavailable.

CHAPTER 19

Cropping Containers

C hoosing the "best" type of cropping container depends upon a number of variables: the mushroom species, the cultivator, and the equipment/facility at hand. White Button growers typically grow in trays, made of either wood or metal. Facilities designed for growing Button mushrooms (*Agaricus* species) encounter many difficulties in their attempts to adapt to the cultivation of the so-called exotic mushrooms. For instance, most Oyster mushrooms have evolved on the vertical surfaces of trees, and readily form eccentrically attached stems. Because Oyster mushrooms require healthy exposure to light, the darkened, dense-packed tray system gives rise to unnatural-looking, trumpet-shaped Oyster mushrooms. This is not to say that Oyster strains cannot be grown *en masse* in trays. However, many Oyster strains perform better, in my opinion, in columns, vertical racks, or bags. After taking into account all the variables, cultivators must decide for themselves the best marriage between the species and the cropping container. Please consult Chapter 21 for specific recommendations for the cultivation of each species.

Tray Culture

Growing mushrooms in trays is the traditional method of cultivation, first developed by the Button mushroom (*Agaricus*) industry. Trays range in size from small 2 feet × 3 feet × 6 inches deep, which can be handled by one person to trays 6 feet × 10 feet × 12 inches deep, which are usually moved into place by electric or propane-powered forklifts. For years, trays have been constructed of treated or rot-resistant wood. More recently, polycarbonate and aluminum trays have been introduced with obvious advantages. Both types are designed to stack upon each other without additional structural supports.

Trays allow for the dense filling of growing rooms (up to 25% of the volume) and because *Agaricus brunnescens*, the Button mushroom, is

Wooden tray cultures of the Button mushroom (*Agaricus brunnescens*) at a commercial mushroom farm.

not photosensitive, no provisions are made for the equal illumination of the beds' surfaces. The main advantage of tray culture is in the handling of substrate mass-filling, transporting, and dumping. The Dutch are currently using tray culture for the cultivation of Button, Shiitake, Oyster, and other mushroom species.

Tray culture easily accepts a casing layer. Casing layers are usually composed of peat moss and vermiculite, and buffered with calcium carbonate and applied directly to the surface of a myceliated substrate. Button mushroom production excels from the application of a casing layer whereas it is debatable whether yields from the wood decomposers are substantially affected.* Those using trays and *not* applying a casing must take extra precaution to ensure the necessary microclimate for primordia

formation. This can be accomplished by covering the trays with either a perforated layer of plastic or breathable, anti-condensate films. The plastic is stripped off, depending upon the species, at the time of, or soon after, primordia formation. High-humidity, but less than fog-like, environments typically ensue until the primordia have firmly set. The transpiration of water, "wicking of moisture," is an active process benefiting the raising of nutrients to the primordia sites.

In North America, Davel Brooke-Webster (1987) first perfected tray culture for Oyster mushrooms. (See page 186.) This method utilizes a perforated plastic covering over the surface of trays. Since many Button mushroom farms are centered on tray technology, the replacement of the casing layer with a sheet of perforated plastic allows the cultivation of

---

* The King Oyster (*Pleurotus eryngii*) produces better second and third flushes when a casing layer is applied. However, advantages realized by the application of casing are often offset by the associated contaminants they encourage. After the first flush, I find that burying blocks outside into sandy soil provides additional flushes without further risk of contamination.

both species at the same facility. Holes (1- to 2-inches in diameter) are punched evenly through an 8-foot roll of plastic, before application. The plastic sheeting is stretched over the trays, directly after inoculation of Oyster spawn into pasteurized wheat straw. The plastic barrier prevents 98% of the evaporation that would otherwise occur had the inoculated straw remained exposed. Even with the plastic covering, humidity within the growing room should remain relatively high so that the straw exposed to the air directly below the holes does not "pan" or die back. A sure sign that the growing room humidity is too low is when brown zones of dry straw form around each puncture site while the remainder of the substrate is white with mycelium.

An advantage of the Brooke-Webster technique is bouquets of equal weight produced simultaneously on the same horizontal trays so widely used by the Button mushroom (*Agaricus brunnescens*) industry. This adaptation has been put into practice by Himematsutake (*Agaricus blazei*) growers. A disadvantage of tray culture is that equal exposure

of light over the surface of each tray, when tightly stacked upon one another, is difficult. (When providing light to tightly packed trays, cultivators usually mount the fixtures on the underside of the tray immediately above the fruiting surface. Most cultivators remove the heat-generating ballast to a remote location and recapture the heat into their air circulation system.) Another method is to line the bottom of wire mesh trays with perforated white plastic. Lights can be mounted vertically on the walls of the growing room. The white undersides of the trays help disperse light across the tops of the beds. When lighting is insufficient in Oyster mushroom cultivation, stems elongate while caps remain underdeveloped, causing abnormal, fluted, or trumpet-shaped mushrooms.

The key to the success of this method lies with strains that form bouquets of mushrooms *site-specifically* at the holes in the plastic. Ideally, primordial clusters hosting multiple mushrooms form at each locus. (See Chapter 14: Evaluating a Mushroom Strain.)

The author in front of aluminum trays filled via the Dutch pulled-net system.

Horizontal trays covered with perforated plastic allow for the simultaneous emergence of Oyster mushroom bouquets from pasteurized wheat straw. (In this case the variety being cultivated is the "sajor-caju" variety of *Pleurotus pulmonarius*. This technique can be adapted for encouraging large Portobellos (*Agaricus brunnescens*) and Himematsutakes (*Agaricus blazei*).

The Phoenix Oyster, *Pleurotus pulmonarius* fruiting from a wall formation of stacked bags.

Vertical Wall Culture

In the evolution of techniques, the *Agaricus* tray has been modified for growing Oyster mushrooms by turning it vertically so mushrooms could fruit out both faces. Usually these vertical surfaces are screened with tight wire or plastic mesh. Perforated plastic positioned between the substrate and the wire mesh allows the formation and development of mushrooms while retaining moisture. Alternately, a plastic curtain is used to envelope the container until the time of fruiting.

Racks having a breadth of 12–16 inches support full flushes and generally do not become anaerobic near the core, a problem seen when racks are 20 inches and more in breadth. If properly designed, individual 4 × 4-foot to 4 × 8-foot rack frames can be stacked upon each other in the construction of continuous mushroom walls. With one side of the frame hinged for opening, filling is made easy.

Another variation of wall culture is the building of walls by stacking polypropylene bags, sideways, on top of one another. Only the ends of the bags have an opening, causing mushrooms to form on the exposed outer surface of the constructed wall. Here, forming the bags into a square shape during incubation facilitates wall construction. Some cultivators hang walls of substrate from roof trusses.

Slanted Wall or A-Frame Culture

Slanted walls are constructed by stacking bags of inoculated substrate to build sloped faces. The advantage of the slanted wall is that a higher density of fill can be achieved within a given growing room space and harvesting is easier. The disadvantage is that mushrooms are limited to forming on only one plane—the outwardly exposed surface. In comparison, blocks that are spaced apart give rise to fruitings on five planes: the four sides *and* the top, a response seen especially with Shiitake and Oyster mushrooms.

Black Reishi, *Ganoderma lucidum*, fruiting from wall formation of stacked bags.

Many cultivators find that a wall composed of individually wrapped blocks limits cross-contamination from infected units. Isolation of the blocks, or for that matter any fruiting container, has distinct advantages, both in terms of yield enhancement and contamination containment.

Bag Culture

In the search for inexpensive, portable, and disposable containers, plastic bags have became the logical choice. High-temperature-tolerant polypropylene bags are primarily used for processing wood-based substrates, which require higher temperature treatment than the cereal straws. Once cooled and inoculated, sterilized substrates are usually filled, directly into heat-sensitive polyethylene bags. Mushrooms fruit from the top or sides.

For the cultivation of most wood decomposers, a biodegradable, heat-tolerant, and breathable plastic for bag culture is sorely needed. "Cello-phane," a wood cellulose-based plastic-like material used commonly in the mid-1900s, has some of these features (heat tolerance, gas porosity), but lacks durability. Despite this shortfall, one of the most appealing aspects of cellophane is that mushroom mycelium digests it in the course of its use. With mushrooms having such great potential for recycling wastes, it is ironic that plastics so fulfill a critical need among cultivators. Some ingenious restructuring of cellulose could satisfy this increasing market for environmentally sensitive cultivators. This technology would be highly patentable and serve a wide variety of industries—from mushroom centers to medical laboratories.

Bag culture first became popular for the Button mushroom industry and is still used to this day. Inoculated compost is filled into the bags topped with a soil-like casing layer. Individual bags are grouped on horizontal shelves. By preventing contact between the mycelium and the wooden shelves, contamination, especially from the wood-loving *Trichoderma* and *Botrytis* molds, is minimized.

A white Oyster mushroom, *Pleurotus ostreatus*, fruiting from custom made plastic grid-frame.

The hinged wall-frame—two trays latched together in the center. Mushrooms fruit out both sides. (Frame has been turned 90 degrees on side for photograph.)

Slanted wall culture of Shiitake, constructed of stacked sawdust blocks.

Primordial cluster of Oyster mushroom emerging through hole in plastic.

Many growers favor plastic bags for their ease of use. When holes are punched through the plastic, mushrooms emerge soon thereafter. Opaque, black bags encourage photosensitive strains of mushrooms to form only where the holes allow light exposure. Clear bags stimulate maximum populations of mushroom primordia but often, primordia form all over. Cultivators must remove the appressing plastic or selectively release colonies of primordia with sharp blades to insure crop maturity. However, many cultivators overcome this problem by using select Oyster strains that localize primordial clusters at exactly the puncture sites. By punching dozens of holes with stainless steel, four-bladed arrowheads, large bouquets of mushrooms are encouraged to form at the hole sites. From the force of the enlarging mushrooms, the flaps are pushed opened. This method has many advantages and is highly recommended.

Oyster mushrooms fruiting from suspended bags. Elongation of stems is due to elevated CO_2 levels.

Bag culture of the Button mushroom (*Agaricus brunnescens*).

Bouquets of Golden Oyster mushrooms (*Pleurotus citrinopileatus*) fruiting from columns of pasteurized wheat straw within a growing room.

Bag culture of the Oyster mushroom (*Pleurotus ostreatus*).

Column Culture

Growing Oyster mushrooms in columns gives rise to natural-looking fruitbodies. Having evolved on the vertical surfaces of hardwood trees, Oyster mushrooms, with their off-centered stems, grow out horizontally at first, turn, and grow upright at maturity.* This often results in the formation of highly desirable clusters, or "bouquets," of Oyster mushrooms. The advantage of cropping clusters from columns is that many young mushrooms form from a common site, allowing 1/4- to 1-pound clusters to be picked with no further need for trimming. Yields of succulent young mushrooms are maximized while spore load is minimized; harvesting is far faster than picking mushrooms individually;

---

* This response—to grow against gravity—is called "negative geotropism." See Badham (1985).

Cross-section of 15-inch-diameter column of Oyster mushroom mycelium contaminated with an anaerobic core.

Inoculating columns. First plastic ducting is cut to length and secured to the stainless steel funnel. (I had a spring-activated collar custom-made for this purpose.) A knot is tied several inches off the ground.

clusters store far better under refrigeration than individual mushrooms.

Automated column-packing machines have been developed and tested in North America and Europe with varied results. Typically the machines rely on an auger or "Archimedes screw," which forces the straw through a cylinder with considerable force. Straw can be pasteurized, cooled, and inoculated along a single production line. Columns are packed, usually horizontally, in a sleeve that can be removed for the vertical placement of the column in the growing room. Several inventive engineers-turned-mushroom-growers are currently developing production systems based on this concept. For many, separating each activity—pasteurization, inoculation, filling, and positioning—simplifies the process.

Vertical cylinders can be made of a variety of materials. The least expensive is the flexible polyethylene ducting designed for air distribution in greenhouses, available in rolls as long as 5,000 feet and in diameters from 6–24 inches.* Drain-field pipe, polycarbonate columns, and similar "hard" column materials have also been employed with varying degrees of success. The more extravagant systems utilize inner rotating, perforated, hard columns equipped with centrally located air or water capillaries that double as support frames, but I have yet to see such a system perfected. Many farms even use overhead trolleys for ferrying the columns into and out of the growing rooms. (See page 172.) Of the many variations of column culture, the ease and inexpensiveness of the flexible polyethylene tubing has yet to be surpassed.

---

* Readers should note that many suppliers sell ducting in "lay-flat" diameter, which is actually ¹/₂ of circumference. Simply divide the "lay-flat" measurement by 1.6 for true inflated diameter.

Another factor in choosing the type of column culture is greatly determined by the mushroom strain. Strains of Oyster mushrooms, which produce clusters of many mushrooms and which are site-specific to the perforations, work better in the perforated column model than in the fully exposed one. Mushroom strains that produce only one, two, or three mushrooms per cluster—as with some *Pleurotus pulmonarius* cultures—do not demonstrate an obvious advantage with the perforated column culture method. Hence, the benefits of perforated column culture can be easily overlooked unless you test many Oyster mushroom strains.

From extensive trials, I have determined functional limits in the cultivation of Oyster mushrooms in columns. Columns less than 8 inches give meager fruitings and dry out quickly. Columns whose diameters exceed 14 inches are in danger of becoming anaerobic at the core. Anaerobic cores create sites of contamination, which emanate outwards, often overwhelming the outer layer of mycelium.

As the substrate fills the column, spawn is added and the plastic elongates.

Some growers strip the columns after colonization to expose the greatest surface area. The columns are held together with two to four vertically running lengths of twine. Although the intention is to maximize yield, the massive loss of moisture, combined with the dieback of exposed mycelium, can cancel any advantage contemplated. El-Kattan (1991) and others who have conducted extensive studies have found the accumulation of carbon dioxide during colonization has an enhancing effect on subsequent yields. Studies prove the partially perforated plastic gives rise to larger fruitings of Oyster mushrooms sooner than substrates fully exposed. Exposed columns not only lose more moisture, but they also allow the sudden escape of carbon dioxide, resulting in a substantial reduction of the total mass of the substrate. Zadrazil (1976) showed that fully 50% of the mass of wheat straw evolves into gaseous carbon dioxide during the course of Oyster mushroom production! (See page 57.)

When the column is slammed to the floor during filling, the straw packs densely.

Once filled, the collar is released and the plastic is tied into a knot.

Columns are best filled with bulk, pasteurized substrates (such as straw) via conveyors leading to a stainless steel funnel. If conveyors are unavailable, then the pasteurized straw can be moved from the steam room (Phase II box) to a smooth-surface tabletop via a pitchfork. Either on the conveyor or on the tabletop, grain spawn is evenly distributed. The inoculated substrate is then directed to the recessed funnel. The funnel should be positioned 10–14 feet above the floor. Plastic ducting is precut into 12-foot lengths and tied into a tight knot at one end. The open end of the plastic tube is pulled over the cylindrical downspout of the funnel and secured via bungee cords or, preferably, a spring-activated, locking collar.

Since the tensile strength of a 12-inch-diameter, 4-millimeter-thick, polyethylene tube is insufficient to suspend the mass of moist straw tightly packed into an 8-foot-long column, care must be taken in filling. After securing the empty plastic column 4–6 inches above the floor, the plastic column slowly elongates with substrate filling. When a few air-

release holes are punched near the bottom of the column, air escapes during filling, facilitating loading. Most importantly, cavities—air pockets—must be eliminated. As the column is filled, the poly tubing stretches until partially supported by the floor. As the column fills with substrate, a worker hugs the column, gently lifts it several inches off the floor, and forcibly slams it downwards. The impact against the floor increases substrate density and eliminates cavities. This ritual is repeated until the column is filled to a height of approximately 10 feet.

Once the column has been filled to capacity, a person standing on a small stepladder removes the securing collar. The column is tied off at the top using whatever means deemed most efficient (a twist-tie, knot, collar, etc.). The column is carefully taken away from the inoculation station and another tube is immediately secured for the next fill of inoculated substrate. An 8-foot-long column, 12 inches in diameter, tightly packed can weigh 120–150 pounds depending upon moisture content,

Once ferried into the growing room, the column is inverted so that the loose straw at the top is compressed. Stainless steel arrowheads are mounted on a board used to puncture 200 to 400 holes in each plastic column. Note that 75% of the weight of the column is floor-supported.

The same column 12 days later, after inoculation with grain spawn of the Pink Oyster mushroom (*Pleurotus djamor*). Twenty-seven pounds of fresh mushrooms were harvested on the first flush.

density (determined largely by particle or "chop-size"), and spawn rate.

After the top folds of the column have been secured, the column should be inverted, upside down. The loose-filled substrate that was in the top 2 feet of the column is now at the bottom and the dense substrate at the bottom of the column is now switched to the top. This, in effect, packs the column tightly. If the straw is not tightly pressed against the plastic, causing substantial cavities, mushrooms will form behind the plastic and develop abnormally. In contrast, tight fills cause the substrate and plastic to be forcibly in contact with one another. When an arrowhead puncture is made, the plastic bursts. The mycelium is exposed to the growing room's oxygen-rich, humidified atmosphere. Given proper lighting and temperature conditions, a population of

primordia forms specific to each puncture site. Growers using this technique develop a particular fondness for strains that are site-specific to the punctures in their response to standard initiation strategies.

Once the column is removed from the inoculation station, two people carry or trolley it to the growing room, where it is hung and allowed to incubate. Overhead trolley systems similar to those employed in slaughterhouses, cold storage rooms, or even clothes dry-cleaner companies can be adapted for this purpose. If the columns are carried, care must be taken so that the columns do not sag in the middle, lest they break. Modified hand trucks fitted with a slant board are helpful in this regard.

The columns are hung to approximately 4 inches above the floor. After hanging, they noticeably stretch within a few seconds to become substantially, approximately 75% floor-supported. Columns (10 inches in diameter and greater) suspended in the air after inoculation will probably fall before the cropping cycle is completed.

Within 1 hour of the columns settling to the floor, numerous holes must be punched for aeration. If holes are not punched until the next day, substantial loss of spawn viability occurs, and a bacterial bloom ensues in the stagnant, air-deprived column. However, if the column is suspended and holes are punched too soon, the punctures elongate. The column soon is in danger of splitting apart. For an 8-foot-high column, 12 inches in diameter, at least 200 and no more than 400 $^1/_8$-inch holes should be punched for maximum yield. Stainless steel, four-bladed arrowheads mounted on a board are recommended for this purpose. Although the puncture hole is only $^1/_8$ inch in diameter, four slits 1–2 inches in length are also made. These flaps open as the mushrooms push through.

Bottle Culture

Bottle culture is an effective means for growing a variety of gourmet and medicinal mushrooms on sterilized substrates. However, bottle culture is impractical for the cultivation of mushrooms on pasteurized bulk substrates such as straw or compost. San Antonio (1971) first published a method for growing *Agaricus brunnescens*, the Button mushroom, on cased, sterilized grain from bottles. This article became a template for the cultivation of many other mushrooms. A counter-culture book on psilocybin mushroom cultivation by Oss and Oeric (aka Dennis and Terence McKenna, 1976) brought the concept of bottle culture to the forefront of small-scale mushroom cultivators. Currently, Asian growers have adapted bottle culture, originally designed for the easy cropping of Enoki mushrooms (*Flammulina velutipes*), to the cultivation of many other gourmet and medicinal mushrooms, including Lion's Mane *(Hericium erinaceus)*, Buna-shimeji (*Hypsizygus tessulatus*), Reishi (*Ganoderma*

Bottle culture of a dark gray strain of *Hypsizygus tessulatus*, known in Japan as Buna-shimeji or Yamabiko Hon-shimeji. This mushroom is currently being marketed in America as just "Shimeji." The Japanese prefer to use the Latin name *Hypsizygus marmoreus* for this mushroom. For more information, please consult the growth parameters for this species.

lucidum), Wood Ears (*Auricularia polytricha*), and some varieties of Oyster mushrooms.

The advantage of bottle culture is that the process can be highly compartmentalized and easily incorporated into the many high-speed production systems adapted from other industries. The disadvantage of bottle culture is that the substrate must be top-spawned and grain spawn cannot be mixed thoroughly through bottles containing sawdust substrates. The bottles are filled to within 2 inches of the brim with moistened supplemented sawdust and then sterilized for 2 to 4 hours at 15 to 20 psi. (The formula is the same for bag culture of Shiitake and Enokitake. Please refer to Chapter 21.) When grain spawn is added at inoculation, and the bottles are shaken, the spawn descends to a depth of only

Bottle (jar) culture (cased grain) of a Magic Mushroom known as *Psilocybe cubensis*.

Sterilized sawdust inoculated via top-spawning.

Downwardly growing mycelium a week after inoculation into sterilized sawdust.

a few inches. Hence, the mycelium quickly covers the top surface layer and then grows slowly downwards into the sterilized sawdust. Top-spawning results in imbalanced ages with the mycelium, with the highest zones being oldest and the lowest regions being youngest. The newly growing mycelium near the bottom inhibits the formation of mushrooms in the top layer of mycelium as the life cycles are engaged in two distinct and potentially contradictory activities. The discrepancy in age in zones of the mycelial mat inhibits maximum mushroom formation. When the mycelium is actively growing out, the total mycelial colony cannot easily shift from colonization to primordia formation. Nevertheless, top-spawning will result in the entire substrate becoming fully colonized, although fruitings are delayed.

An advantage of this method is that mushrooms, when they do form, arise from the maturest mycelium, at the top of the bottles. Side and bottom fruitings are rare. Top-spawning is preferred by cultivators in tropical climates where air conditioning is not an economical option. With top-spawning the substrate is slowly colonized from above, and head generation is much less than from through-spawning techniques. In contrast, through-spawning gives colder climate cultivators benefits from heat generation associated with faster spawn run. If the cultivator can afford to make the initial investment of incubating thousands of bottles until the first cycle starts, then the drawbacks are primarily that of lost time and delay in the initial production cycle, but not overall yield. Top-spawning is fast and convenient for bottle and small bag culture, although I see more benefits from through-spawning. Many cultivators in Japan accelerate the colonization process by inoculating the bottles with pressurized liquid spawn.

With the natural evolution of techniques, Asian cultivators have replaced bottles with similarly shaped, cylindrical bags. Many growers in Thailand, Taiwan, and Japan prefer this hybrid method.

Liquid-inoculation of sterilized, supplemented sawdust allows for inoculation methods resembling the high-production systems seen in a soda pop factory. With reengineering, such high-speed assembly-line machinery could be retrofitted for commercial bottle and bag cultivation.

Unless an aggregate-slurry is used, liquid spawn settles near the bottom of the bottles. For a complete discussion of liquid fermentation and inoculation techniques, please refer to Chapter 15. Bottles can be arranged horizontally in walls or fruited vertically.

Bottles of various sizes can be used. The most common are between 1 quart (1 liter) and 1 gallon (4 liters). The openings are usually between 50 and 100 millimeters in diameter. Glass bottles are not as

Bottle culture of Enoki mushrooms (Flammulina velutipes). Exposure to light caused caps to mature. The mushrooms at this stage are not marketable. They were—24 hours earlier.

Compacted compost formed to create self-supporting "megablocks" for the cultivation of the Button mushroom, *Agaricus brunnescens*. This ingenious British method of cultivation is being patented (U.K. Patent #953006987.1). The cavities allow transpiration and prevent anaerobic cores.

popular as those made from polypropylene-like materials. Each bottle is fitted with foiled cotton or an autoclavable lid equipped with a microporous filter disk. After full colonization, the lids are removed, and the surface mycelium is exposed to the growing room environment. Enoki growers often insert a coil of paper or clear plastic that encourages stem elongation.

Oyster mushrooms (*Pleurotus ostreatus*) fruiting from suspended, perforated plastic containers. A Russian cultivator, Vladimir Nesterov, is refining this method.

Applying a peat moss casing.

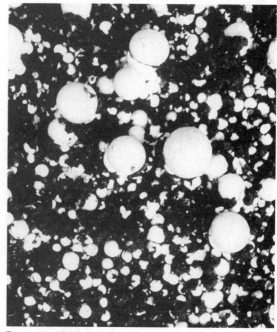

One week later, mycelium grows close to the surface.

Two weeks later, primordia form in the casing layer.

Casing: A Topsoil Promoting Mushroom Formation

Button growers long ago discovered that, by placing a layer of peat moss over compost grown through with mushroom mycelium, yields were greatly enhanced. The casing serves several functions. Foremost, the casing layer acts as a moisture bank where water reserves can be replenished through the course of each crop. The casing layer also limits damage to the mycelium from fluctuations in relative humidity. Besides moisture, the casing provides stimulatory microorganisms and essential salts and minerals. These combined properties make casing a perfect environment for the formation and development of primordia.

In the cultivation of gourmet and medicinal mushrooms, casing soils have limited applications. Cultivators should be forewarned that green-mold contamination often occurs with soil-based casing layers, especially when air circulation is poor and coupled with contact with wood framing. The possible benefits of casing are often outweighed by the risks they pose. Few woodland species are absolutely dependent upon casing soils, with the exception of the King Stropharia (*Stropharia rugosoannulata*).

A dozen or so casing soils have been used successfully in the commercial cultivation of mushrooms. They all revolve around a central set of components: peat moss, vermiculite, calcium carbonate (chalk), and calcium sulfate (gypsum). Two brands called Sunshine #2 and Black Magic peat moss are preferred by many Canadian and American growers. Recently, "water crystals," a water-capturing plastic, have been tried as a casing component with varying results. These crystals can absorb up to 400 times their weight in water and do not support contaminants, two highly desirable characteristics. Unfortunately, the fact that water crystals are not fully biodegradable and cannot be easily recovered from the spent substrate greatly limits their acceptance by environmentally inclined growers. Starch-based water absorbents tend to clump and must be added with an aggregate. Cultivators must weigh and balance these factors when designing the casing mixture.

For many years, cultivators have used the following casing formula.

Casing Formula (by volume)

10 units peat moss
.5 unit calcium sulfate (gypsum)
.5 unit calcium carbonate (chalk)

Calcium carbonate is used to offset the acidity of the peat moss and should be adjusted according to desired pH levels. Calcium sulfate, a non-pH affecting salt,* provides looseness (particle separation) and mineral salts, especially sulfur and calcium, essential elements for mushroom metabolism. Peat moss, although lacking in nutrition, is resplendent with mushroom-stimulating bacteria and yeasts. The above-described formula depends greatly on the starting pH of the peat moss. (A pH neutral peat moss I like is Sunshine brand. It seems less prone to Trichoderma molds.) Generally, the pH of the resultant mixture is 7.5 to 8.5 after makeup. As the mushroom mycelium colonizes the casing layer, its pH gradually falls. For some of the more acid-loving species mentioned in this book, calcium carbonate should be excluded. Typically, this chalk-free mixture gives pH readings from 5.5 to 6.5. Adding 10% coca nut fiber and .5–1% salt enhaces osmosis, aeration, and transportation of water to lower regions of casing. Some European growers spawn the casing at the time of application to assure dominance, prevent contamination, and set primordia evenly.

Mix the dry components together in a clean bucket or wheelbarrow. Add water slowly and evenly. When water can be squeezed out to form brief rivulets, then proper moisture has probably been achieved. A 75% moisture content is ideal and can be tested by measuring the moisture lost from a sample dried in a hot oven.

Once wetted, the casing is applied to the top of a substrate. Casing soils can be used with tray, bag, or outdoor mound culture. Although some of the following mushroom species are not absolutely dependent upon a casing soil, many benefit from it. Those species dependent upon soil microorganisms for fruitbody formation are listed below with an asterisk. In absence of soil bacteria, the "*" species will not fruit well, or at all. Typically, a 1- to 2-inch layer of casing soil is a placed onto 4–10 inches of colonized substrate.

*Agaricus bitorquis,** the Warm-Weather Button Mushroom
*Agaricus blazei,** Himematsutake
*Agaricus brunnescens,** the Portobello or Button Mushroom
Agrocybe aegerita, the Black Poplar Mushroom
*Coprinus comatus,** the Shaggy Mane
Ganoderma lucidum, Reishi or Ling Chi
*Polyporus umbellatus,** Zhu Ling
Pholiota nameko, Nameko
Pleurotus eryngii, the King Oyster
*Stropharia rugosoannulata,** the King Stropharia
Volvariella volvacea, the Paddy Straw Mushroom

(Techniques for the cultivation of *Agaricus bitorquis* are covered in detail in *The Mushroom Cultivator* by Stamets and Chilton, 1983.)

If the microclimate lacks enough air circulation and carbon dioxide builds up, contaminants (especially green molds) will flourish in peat moss–based casings. Some cultivators are discouraged from using casing soils considering the contamination potential they pose. Others, in the effort to defeat mold contaminants and kill nematodes, pasteurize or sterilize the casing mixture. By a quirk of nature, casings often encourage more contamination than they prevent.

A one- to two-inch layer of moist casing is evenly placed across the top surface of the substrate, and lightly patted down. Relative humidity is maintained at 90–95% until fans of mycelium break through the

---

* Gypsum (calcium sulfate) may affect pH by half of a point initially. Its pH-altering ability is minor until the sulfur evolves into sulfuric acid. Calcium and sulfur are essential elements in mushroom metabolic processes. If the substrate is lacking in these essential elements, yields are adversely affected. Shiitake particularly benefits from the addition of calcium sulfate to sawdust substrates.

The Button mushroom (*Agaricus brunnescens*) fruiting from cased grain.

A Magic Mushroom (*Psilocybe cubensis*) fruiting from cased grain.

surface. At this time, a classic initiation strategy is implemented: Watering is increased; rH is raised to 95–100%; air exchange is accelerated; and temperature is dropped. (See Chapter 21.) If the casing is being placed onto substrates enveloped in plastic bags, holes must be poked on the undersides for drainage.

Here is an alternative formula, containing no contaminant-supporting ingredients.

Soil-less Casing Formula (by volume)

1 unit vermiculite
1 unit "water crystals"

Mix vermiculite and water crystals when dry. Add water until fully saturated. This mixture can hold up to 90% moisture until its carrying capacity for water is exceeded. Apply a $1/4$- to $1/2$-inch layer as a casing layer to the surface of the mushroom mycelium.

CHAPTER 21

Growth Parameters for Gourmet and Medicinal Mushroom Species

The parameters outlined here are based on the author's experiences over many years of cultivation. Each mushroom species thrives on a limited range of substrates. However, strains within a species are even more specific in their habitat requirements, temperature preferences, and their flushing intervals. Wherever possible, I have identified individual strains so that readers can achieve similar yields. These strains are being kept in perpetuity in the Stamets Culture Collection. Some strains are proprietary and can be obtained by contacting the author through his business Fungi Perfecti, LLC. Most species can be obtained from a number of culture libraries such as the American Type Culture Collection. These sources are listed in the Resource Directory section in Appendix 4.

To remain competitive, cultivators must continuously search out and develop new strains from wild stocks. Although specific temperature parameters are outlined in this section, some strains will perform better outside of these prescribed limits. In general, rapid cycling strains prefer higher temperatures. The cold-weather strains require a longer gestation period before fruiting. The cultivator must customize initiation strategies to each strain, a process fine-tuned with experience.

Of the many factors already described for producing successful crops, the misapplication of only one can result in poor fruitings or absolute failure. *Each grower is strongly encouraged to conduct mini-trials before attempting to grow mushrooms commercially.* Optimization of yields is realized only if the grower becomes keenly sensitive to and satisfies the unique needs of each mushroom strain. Therefore, the following parameters should be used as a general guide, to be refined in time and with experience. The first set of parameters is centered on the incubation period, called *spawn run,* the second set is for initiating mushrooms, called *primordia formation,* and the third is for cropping or *fruitbody development.* In essence, each stage of mushroom growth has a different environment ideal for its growth. As each factor is changed, secondary effects are seen. The skill of a mushroom cultivator is measured by the ability to compensate for fluctuations in this complex mosaic of variables.

Spawn Run: Colonizing the Substrate

Spawn run spans the period of time when the mycelium is colonizing the substrate. Other than the factors described below, the amount of spawn inoculated into the substrate can greatly affect the duration of colonization, and therefore, the time to fruiting.

Moisture: Substrate moisture contents should be between 60 and 75%. Moisture contents below 40% promote slow and wispy mycelial growth. Unless a casing layer is used, the moisture content of the substrate gradually declines from initial inoculation. For instance, the water content of straw at inoculation is nearly 75%, precipitously dropping after the first flush to the 60% range, and continuing to steadily decline through the remainder of the cropping cycle. The cultivator's prime responsibility during this period is to manage the moisture reservoir as if it were a bank. Moisture loss must be limited before initiation or else the mycelium will fail in its efforts to generate mushrooms, which are themselves about 90% water! The solution: Retard the loss of substrate moisture by maintaining high humidity during spawn run.

Air Exchange: Mushroom mycelium is remarkable in its tolerance for carbon dioxide. At levels snuffing out the life of a human, the mycelium thrives. Some Oyster mushrooms' growth rates peak at 20% carbon dioxide, or 200,000 ppm. However, this CO_2 environment is equally stimulatory to competitor molds. The best levels vary with the strain, and whether one is working with pasteurized or sterilized substrates.

To reduce carbon dioxide, fresh outside air is introduced. Consequently, several other phenomena occur: Evaporation is increased; humidity drops; temperature changes; and the net number of contaminant particles entering the growing room rises as air exchanges are increased.

Temperature: As a general rule, incubation temperature runs higher than the temperature for primordia formation. Internal temperatures should not exceed 95°F (35°C) or black pin molds and other thermophilic competitors will awaken, especially under the rich CO_2 conditions created as a by-product of spawn running.

Lighting: For the species described in this book, moderate lighting has no effect, adverse or advantageous, on the mycelium during spawn run. Bright, unfiltered, direct sunlight is damaging. Light is especially harmful when intensities exceed 10,000 lux. From my experiences, the mycelial mat only becomes photosensitive after it has achieved a threshold critical mass, usually coincident with full colonization, and after carbon dioxide evolution has steeply declined.

Primordia Formation: The Initiation Strategy

By far the most critical step is that of primordia formation, called the *initiation strategy*. An initiation strategy can be best described as a shift in environmental variables, triggering the formation of mushrooms. The four major environmental factors operative in an initiation strategy are *moisture, air exchange, temperature*, and *light*. These are adjusted accordingly:

Moisture: High humidity 95–100%. Direct watering coupled with a constant, controlled rate of evaporation. Fog-like conditions are important when aerial mycelium is first exposed to the growing room environment. Once primordia form, a gradual reduction of humidity from 100% to 90–95% usually is beneficial. Humidity should be measured in at least three locations in the free air spaces directly above the mycelium-permeated substrate.

Air Exchange: By introducing air, carbon dioxide precipitously declines with a corresponding increase in oxygen. CO_2 levels should be below 1,000 ppm, ideally below 500 ppm for maximum mushroom formation. Air exchange should be adjusted specifically to lower CO_2 to the specified levels outlined for each species in the following growth parameters.

Temperature: Many strains will not form mushrooms unless temperature is dropped or raised to a critical plateau. For most strains, a temperature drop is required. Since mushroom formation is primarily a surface phenomenon, the atmosphere of the growing room has to be altered to affect a temperature change in the substrate. As a substrate is being colonized with mycelium, heat is released as a by-product. After colonization is complete, heat generation abates, and internal temperatures naturally decline to nearly equal with air temperature. This is the ideal time to synchronize the other factors favorable to mushroom formation. Note that the temperature thresholds listed for each species are what cultivators call the bed, or substrate temperature. Air temperature is adjusted upwards or downwards to affect the desired change. When air temperature is changed, a lag time follows, often for 24 to 72 hours, before the substrate temperature acclimates to the new level. In most cases, the critical temperature plateau occurs within 2–4 inches of the surface, the region supporting the creation of primordia.

Lighting: In nature, light acts as a signal alerting the mycelium to an open-air environment where, should mushrooms form, spores can be spread into the air. Light controls stem elongation and cap development. Ideal light conditions—intensity and wavelength—vary with each species and strain. Indirect natural light, or the dappled light filtering through a forest canopy, is considered ideal for woodland mushrooms. Specific photoperiods and spectral frequencies have not yet been established for all mushroom species. In these cases, cultivators resort to providing the lighting necessary to the most sensitive of the gourmet mushrooms, the Pleurotus species. Modest light is not harmful to developing mushroom mycelium; it seems unaffected by its presence. *Direct sunlight or high-intensity exposure is harmful.* The fluorescent lights used in indoor facilities do not inhibit mycelial growth, and in some circumstances may stimulate early primordia formation. For most species, light levels between 50 and 1,000 lux and 380 and 480 nanometers (green to blue) seem most stimulatory to primordia formation. (I use six 8-foot-long, "Daylight" 6,500 Kelvin

Lack of light causes Oyster mushrooms to malform into coral-like structures.

fluorescent lights to light each 1,000-square-foot growing room that also gets supplemental natural light through a row of diffusion panels.) For specific light requirements, please consult the growth parameters for the species being cultivated.

Leatham and Stahlman (1987, 1989) conducted trials with Shiitake on chemically defined media, which showed that the absence of calcium made the mycelium unresponsive to light stimulation and primordia failed to form. At low calcium levels (<40 µg/mg) Shiitake mycelium formed mushrooms when stimulated by light between 600 and 680 nanometers of red light. At high calcium levels (>130 µg/mg), the wavelengths most stimulatory for primordia formation were between 400 and 500 nanometers, what we know as blue to ultraviolet light. Calcium is naturally present in woods in sufficient quantities to allow fruiting. Just as strains of Shiitake differ in their fruiting cycles, I suspect that the "calcium factor" in triggering Shiitake formation may be strain specific. Nevertheless, the interplay between light and calcium concentrations continues to be a subject of great interest. Further studies are needed to compare the many strains, on various woods, with varying levels of calcium, and at different wavelengths.

The body of mycelium is not as sensitive to these environmental stimuli until the substrate, its impending food source, has become fully captured by it. Where there are zones of *uncolonized* substrate, the mycelium continues on its conquest of nutrients, and fruitings are delayed until colonization is complete. Substrates that are carefully and evenly inoculated colonize faster, responding readily to the four environmental stimuli described above. When the rapidly growing mycelium is forced to stop, because of natural borders or contact with competitors, the mycelium shifts gears, biologically speaking, from conquest to consolidation. The mycelium consolidates its hold of the substrate by the infinite microscopic branching of hyphae. Concurrent with this phase change, the mycelium and the substrate cool. For Oyster mushroom cultivators, this period of declining temperature leads directly to primordia formation. At this juncture, the cultivator adjusts the surrounding environment—introducing light, dropping temperature, exchanging air, and increasing moisture—to stimulate the greatest number of primordia.

Some mycologists take a different view of what causes this shift to fruiting in the mushroom life cycle. They describe the sudden lack of food for the mycelium as *nutrient deprivation*. The best example of this is the Morel. Once the sclerotia have formed remote from the nutrient base, the nutrient base is physically separated from the sclerotial colonies, and this loss of nutrition is one of the triggers stimulating fruitbody formation. (This is basically the pivotal technique upon which the patent was awarded for

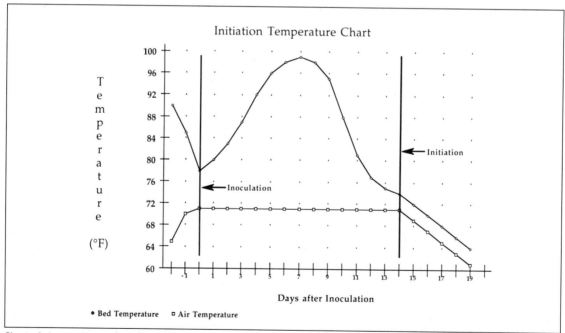

Chart of air vs. substrate ("bed") temperature during colonization. Note that bed temperature naturally declines as colonization is completed while air temperature remains constant. An initiation strategy (i.e., dropping temperature, adding moisture, increasing light, and exchanging air) is instigated to augment the mycelium's natural progression to fruiting.

growing Morels—see Ower et al. 1978.) In my mind, this is a clear case of true nutrient deprivation. However, substrate separation techniques are not generally used in the cultivation of Oyster, Shiitake, Enoki, Lion's Mane, Maitake, Wood Ear, and many of the other gourmet and medicinal mushrooms. More accurately, I would describe these fruitings as being triggered by *nutrient limitation*, not deprivation.

Fruitbody (Mushroom) Development

Moisture: Atmospheric moisture must be carefully managed to allow mushroom development but not to the advantage of competitors. While relative humidity approaches 100% during primordia formation, it should be lowered to levels whereby a constant rate of evaporation is drawn from the fruitbodies. The crop should be sprayed several times a day, as long as the mushrooms, the substrate, or the air soon reabsorbs the excess water. This dynamic process, of replenishment and loss, encourages the best crops of mushrooms. The humidity in the growing room is often reduced several hours prior to picking, extending the shelf life of the crop. This is where the "art" of cultivation plays a critical role in affecting quality.

Air Exchange: Air exchange and turbulence are managed for maximum benefit of the mushrooms, in terms of reducing carbon dioxide levels, elevating oxygen concentration, and effecting the constant evaporation of moisture from the surfaces of the maturing mushrooms.

Temperature: Temperature levels either remain the same or are raised. Typically after primordia formation, temperature controls the speed of development of the fruitbody. Naturally, warmer temperatures result in faster growth while colder temperatures slow development. One advantage of fruiting at a cooler temperature is that a firmer-fleshed, higher-quality mushroom forms at the time of harvest.

Lighting: Without adequate light, stem elongation and malformation of the cap occurs. Oyster and Enoki mushrooms are especially sensitive. Also strong light alters the pigment of the developing mushrooms. Some strains of Oyster mushrooms darken under bright light conditions; others pale. This light-sensitive response is also affected by temperature.

Duration: The timing of crops—their first appearance, the duration of harvest, and the period of time between crops—are strain and process dependent. With Shiitake on sterilized sawdust/chips/bran, I go for five crops. With Oyster mushrooms grown in columns on pasteurized straw, two to three flushes seem most efficient. Approximately 1 to 2 weeks separate the end of the first flush to the beginning of the second. Within time cycles, there are "windows of opportunity" in which fruitings occur. A period of dormancy is required between crops so nutrients can be accumulated as the mycelium prepares for the next production cycle. During these windows of opportunity, the cultivator must actively signal the mushroom mycelium with as many environmental stimuli as possible. Synchronizing this combination of events gives rise to the best possible fruitings.

Agaricus blazei, Himematsutake, is a classic Agaricus.

The Gilled Mushrooms

The gilled mushrooms are the archetypal forms we all know. They are typically umbrella shaped, with a cylindrical stem from which plates radiate outwards from the underside of the cap. Oyster, Shiitake, and Enoki are classic gilled mushrooms. Taxonomically, these mushrooms fall into the order Agaricales. Microscopically, they all reproduce spores by means of club-shaped cells called basidia.*

The most extensive treatise on the taxonomy of gilled mushrooms has to be Rolf Singer's *Agaricales in Modern Taxonomy* (1986). This massive work is the pivotal reference text on the overall systematics of fungal taxonomy. Ongoing studies use PCR-DNA profiling to further expand our understanding of interrelationships among species. As the science of mycology progresses, monographs on an individual genus or a section of a genus delve more precisely into inter-species relationships. Alexander Smith's *Mushrooms in Their Natural Habitat* (1949) still stands as the template for describing mushrooms macroscopically and microscopically, and in their relationship to the natural environment. These two works, in combination, have had the greatest influence on the course of American mycology.

---

*For a discussion of basidia, consult Chapter 9.

The Himematsutake Mushroom of the Genus *Agaricus*
Agaricus blazei Murrill

Introduction: A rising star in the lexicon of medicinal mushrooms, this unique *Agaricus* was first recognized as a novel species by an American mycologist, W. A. Murrill, who found it on the lawn of a Mr. R. W. Blaze in Gainesville, Florida (Murrill 1945). Associates of Japanese coffee growers in Brazil rediscovered this mushroom, which was well known to the locals. Upon obtaining specimens, Japanese mycologists pioneered its cultivation and are credited for bringing this species to the forefront. Cultivation centers in China and Brazil are now well established, but the primary market for *Agaricus blazei* is Japan, where it is called Himematsutake and has an excellent reputation as one of the most expensive of all edible medicinal mushrooms.

Commercial cultivation in the United States has just recently begun. Because of its preference for warmer temperatures, outdoor cultivation is practical only in the southern United States, or during the summer months in the temperate regions of the world. Under controlled conditions, this mushroom can easily be grown in a fashion similar to *Agaricus brunnescens*. Responsive to light stimulation to a degree yet to be determined, *Agaricus blazei* has

Classic, aggressive *Agaricus blazei* mycelia at 10 days.

the general appearance of an oversized Portobello (brown *Agaricus brunnescens*) Button mushroom, but with a beguiling almond fragrance and flavor that is comparable to The Prince, *Agaricus augustus* and The Almond Agaricus, *Agaricus subrufescens*.

Common Names: Royal Sun Agaricus
Himematsutake
Kawariharatake
Cogmelo de Deus
  (Mushroom of God)
Murrill's Agaricus or ABM
King Agaricus
Almond Portobello

Taxonomic Synonyms and Considerations: The mushroom that shares the closest resemblance to *Agaricus blazei* is the slender, but almond-flavored

Agaricus subrufescens Peck, its taxonomic cousin, differing slightly in the shape of the spores. The spores of *Agaricus blazei* tend to be more ovoid whereas the spores of *Agaricus subrufescens* are more ellipsoid. The spores of the similar *Agaricus augustus* Fries are much larger, 7.5–10 × 5–6 µ compared to the smaller spores, 5–4 µ, seen in *Agaricus blazei*. Freshly picked, *Agaricus blazei* usually bruise bright yellowish when cut, while *A. subrufescens* bruises only along the outer cuticle, if at all. Once *Agaricus blazei* has been harvested the tendency for yellow staining diminishes and is replaced by a dull browning reaction. Differences in the staining reac-

Spawn of *Agaricus blazei* on rye grain—note rhizomorphs and stellar centers, nexus points of growth, which can give rise to primordia.

Ideal stage for harvesting Himematsutake. Note abundant rhizomorphs at base of stipe. Himematsutake easily pulls from the casing layer and care must be taken to prevent the casing from soiling the white stems.

tion between these two species may or may not be taxonomically significant. From this author's experience, the staining reaction in *Agaricus blazei* is not a dependable taxonomic character and appears to be associated with an unstable chemical reaction of damaged tissue activated by exposure to air.

In culture, the mycelia of *Agaricus blazei* and *Agaricus subrufescens* share many similarities, in contrast to the mycelium of *Agaricus augustus*, which is anemic in its growth when compared on standard PDA or MEA media. Careful comparative DNA analysis is needed for delineating the phylogeny of *Agaricus blazei* from populations of *Agaricus subrufescens*.

Description: A classic large *Agaricus* species, this mushroom is grander in culture than most wild forms. Cap 7–25 cm broad, convex at first, soon hemispheric, then broadly convex, eventually flattening. Often cap margin is smooth, white, and splitting only in age. Cap surface covered with brownish fibrillose patches. Partial veil membranous, floccose with patches of the veil, typically tearing to form a median membranous annulus, but sometimes with remnants attached to the margin at maturity. Gills pallid at first, soon gray, and then chocolate brown when mature. Stem cylindrical, solid, tall, whitish, smooth, flesh thick, often quickly staining yellowish (ochraceous) when bruised. Growing singly or in clusters, arising from stellar sites of dense rhizomorphs leading to and often attached to the stem bases. Imparting strong scent of almonds, especially during cooking. Spore deposit dark chocolate brown en masse.

Distribution: First collected in Florida and thought to be scattered throughout the southeastern United States, this mushroom is more common in southeastern North America than most realize. In Brazil, in the Sal Hose do Rio Preto district northwest of São Paulo (the city), this mushroom is common in the fields and mountainous regions. Probably more widely distributed than the literature presently indicates.

Natural Habitat: Grows in soils rich in lignicolous debris, in mixed woods, well-composted soils, and along forest edges. This mushroom is a complex saprophyte and prefers composting soils rich in plant debris. Also grows in well-manured grasslands.

Microscopic Features: Spores chocolate brown in deposit, nearly ovoid, 5 × 4 μ. Cystidia few. Basidia four-spored, heterothallic mating system.

Available Strains: Strains are circulated through private collections and are generally not widely available. Most strains originate in Brazil, and have been further refined in Japanese and Chinese laboratories. Strains that sporulate late in their development are advantageous for maintaining air quality in the growing room, limiting cross contamination and fly infestation, and extending shelf life of the crop after harvest.

Mycelial Characteristics: Longitudinally striate mycelium, with radiating rhizomorphs overlaying a cottony mycelial undergrowth. Rhizomorphic mycelia in culture produces hyphal aggregates and pseudo-primordia after one month of incubation on 2% MEA, which fail to enlarge to maturity. Becoming loosely aerial in age, mycelia often exude a yellowish, almond-smelling metabolite. When subculturing this strain, some cultures abort in growth due to unknown factors. Maintaining strains closest to its genetic origins is strongly recommended. Repeated subculturing can lead to mutations and senescence.

Fragrance Signature: Musty grain with almond overtones.

Natural Method of Cultivation: Mound culture of this mushroom has been practiced in Brazil for the past twenty years. This mushroom benefits from soil microflora and warm temperatures, making it an ideal candidate for outdoor cultivation in the tropics and subtropics during the warmest months. Himematsutake is a prime candidate for the ever-expanding model of mycopermaculture as first described by Stamets (1993), and could be incorporated into the recycling systems of agriculturally based communities.

Recommended Courses for Expansion of Mycelial Mass to Achieve Fruiting: Standard techniques as for many other species: agar culture to grain, then grain-to-grain transfers. Mushrooms can be generated on rye grain, laid into shallow trays, and cased with unpasteurized soils, although compost-based substrates generate higher yields. Fermentation of straw, supplemented with corn meal and/or urea and/or ammonium nitrate is the basis of a standard formula used in Japan. Sterilized sawdust cultivation is shown here for the first time.

Suggested Agar Culture Media: MYPA

1st, 2nd, and 3rd Generation Spawn Media: Cereal grain spawn throughout, with excellent growth on whole rye grains.

Substrates for Fruiting: Enriched composts or pasteurized substrates supplemented with nitrogenous additives (bran, urea, chicken manure, ammonium nitrate, etc.). Like the button mushroom, *Agaricus blazei* is a secondary decomposer and a lover of well-composted, nitrogen-rich substrates that are then pasteurized and inoculated with grain spawn. In Brazil, Thailand, and elsewhere in the tropics, the composting process is aided by naturally high temperatures, and Phase II is sometimes aided by utilizing solar energy. The composting methods are similar to those described for *Agaricus brunnescens* as outlined in *The Mushroom Cultivator* (Agarikon Press, 1983), co-authored by myself and Jeff Chilton, and *Modern Mushroom Growing* by P. J. C. Vedder (Educaboek, 1978). Composts achieving a 1.5–2% nitrogen level, post Phase II, are ideal. A simple substrate, described in Stamets (1978), is fresh, week-old leached cow manure, commonly available at dairies for a few dollars per truckload. This ready-made substrate is pasteurized at 140–150°F (60–66°C) for a day, and then conditioned at 120–125°F (49–52°C) for two days until the ammonia dissipates. Upon cooling below 95°F (35°C), spawn is mixed through at rates varying from 1 to 2 cups per square foot.

Himematsutake growing in Japan on sterilized substrates.

Iwade and Mizuno (1997) proposed a compost formula suited for indoor cultivation that consists of mixing together 375 kg of chopped straw, 10 kg rice bran, 15 kg chicken manure, 8 kg lime, 10 kg ammonium sulfate, and 5 kg super-phosphate, to which 700–800 kg water is added. These ingredients are mixed together with temper-atures rapidly escalating. The baseline temperatures should not exceed 140° F (60°C), and the pile is turned every few days to insure controlled thermo-genesis between 131–140° F (55–60° C). After one week, when the temperature of the compost plateaus and descends to 122° F (55° C), the com-post is brought indoors and the temperature elevated again in a well-sealed room to 140°F for 48 hours as a Phase II fermentation. This second elevation in temperature insures the destruction of pest organisms and the promotion of thermotoler-ant microbial allies. The room is then well ventilated, and when the compost temperature drops to below 86° F (30° C), inoculation with grain spawn can commence. Many of the standard formulas for the

Button or Portobello mushroom will work well for *Agaricus blazei*.

Since heat-sensitive bacteria are essential for fruiting, untreated casing soils are recommended. Canadian sphagnum peat moss amended with 10% gypsum (by volume) produced a good casing with natural resistance to Trichoderma contamination. A method to promote primordia is to add a simple water extract of microbially enriched soil directly to the casing layer. Take 2 pounds (1 kg) of soil, add one gallon (4 liters) of water, and then screen through a coarse filter (i.e., a coffee filter or sock), diluting this extract by a factor of ten. Sprinkle 1 gallon of this diluted soil extract evenly over 10–20 square feet of casing layer one week after application. A novel bacterium, yet to be typed, (Bacterium '864': see Chen et al. 1999), appears necessary for primordia formation, similar to the beneficial activity of *Pseudomonas putida* in *Agaricus brunnescens*. When casing soils are heat-treated, *Agaricus blazei* will yield few if any mushrooms.

Recommended Containers for Fruiting: Plastic "mega-bags" weighing from 22–50 pounds (10–23 kg) and/or plastic or wooden trays. Trays should have adequate drainage and a layer of porous, non-organic material for aeration (i.e., perforated plastic) to prevent anaerobic activity in the depths of the substrate, especially since this mushroom is grown at comparatively higher tem-peratures than, for instance, *Agaricus brunnescens*.

Yield Potentials: Current biological efficiencies are estimated to average 50–75%. Iwade and Mizuno (1997) reported yields of 30–50 kg sq. meters (= 1.8–3.1 pounds/sq. feet). On supplemented sawdust, 1 pound (.5 kg) of mushrooms can be harvested from 5 pounds (2.5 kg) of substrate. Mushrooms more often form in clusters than singly. The first mushrooms formed by *Agaricus blazei* are much larger in stature than those seen in the first flush of *Agaricus brunnescens*. Outdoor cultivation biolog-ical efficiencies may be less, but given the expanse of compost beds that can be laid out, inexpensively, in humid, subtropical and tropical environments, outdoor cultivation can be proportionately more profitable. The window of opportunity for outdoor

GROWTH PARAMETERS

Spawn Run:

Incubation Temperature: 70–80°F (21–27°C)
Relative Humidity: 90–100%
Duration: 28–40 days on compost;
    60–90 days on sawdust
CO_2: >5,000 ppm
Fresh Air Exchanges: 1 per hour
Light Requirements: n/a

Primordia Formation
(after applying $1^1/_2$–2 in. casing):

Initiation Temperature: 70–75°F (21–24°C)
Relative Humidity: 80–90%
Duration: 18–24 days
CO_2: 400–800 ppm
Fresh Air Exchanges: 5–7 per hour
Light Requirements: minimal, 100–200
    foot-candles

Fruitbody Development:

Temperature: 75–80°F (24–27°C)
Relative Humidity: 75–85%
Duration: 4–8 days
CO_2: <2,000 ppm
Fresh Air Exchanges: 5–7 per hour.
Light Requirements: minimal, 100–200
    foot-candles

Cropping Cycle:

Every 2 to 3 weeks for 2–3 flushes

Harvest Hints: The best time to harvest *Agaricus blazei* is when the mushrooms still have an intact partial veil covering the gills. (See page 209). The convex caps become box-like in form, just prior to quickly expanding. Once the partial veil tears and falls to become a membranous annulus, the shelf life and the culinary excellence of this mushroom diminishes, but only slightly. This mushroom is one of the few that is considered an excellent gourmet mushroom through the range of fruitbody maturity. Once harvested, cultivators must drop temperature precipitously to refrigeration temperatures for storage. The closed veil and thick-fleshed characteristics of this mushroom further complicate the descent into the desired cold storage range of 35°F (2°C). Since the stems are whitish and often feature cottony mycelium, care in harvesting the mushrooms is essential to minimize cleaning of the clinging soil. If drying, mushrooms are sliced lengthwise.

Agaricus blazei fruiting from enriched alder *(Alnus rubra)* sawdust/chips mixture supplemented with 20% oat bran. Incubation of 3 months preceded casing, with primordia forming in 20 days.

cultivation is limited by seasonal temperatures, even in the tropics, as the heat-sink effects of the ground cools the substrate below optimum temperature plateaus. On composts, the second flush is often equal or larger than the first flush. The third flush markedly diminishes in yield. If growing Himematsutake indoors for two flushes, removal of the substrate after the second flush to an outdoor setting for subsequent fruitings allows for further cropping with minimum risk of contamination.

Fruitbodies arising from the same culture dish produce different colored pilei on compost and wood substrates. With this (but not all) strains, pasteurized composts give rise to darker forms whereas the lighter forms come from sterilized sawdust. Such variability due to substrate composition casts doubt on the taxonomic status that pileal pigmention has been given in the genus *Agaricus*. More studies on the influence that substrate has on pigmentation are needed.

Grower David Sumerlin with emerging flush of *Agaricus blazei*.

Form of Product Sold to Market: Fresh, dried, powdered, extracts, capsules, teas, and tablets. In quantity, the current price as listed from a variety of Internet sources is between $ .50 and $ 1.00 per gram.

Nutritional Content: On a dry weight basis, protein content can range from 37–48%, making this species one of the most protein-rich of all cultivated mushrooms. Analysis by this author showed, on the basis of a 100 gram sample, 9.88% moisture, 39.3% protein, 1.8% fat, 25.6% fiber, 10.1% ash, and 38.9% carbohydrate. Another uncredited analysis posted on the Internet shows that this mushroom has 36.7% protein, 3.4% fat, 6.8% fiber, 7.3% ash, 38.3% sugar, 939 mg/100 g of phosphorus, 18.2 mg/100 g iron, 41.6 mg/100 g calcium, .48 mg/100 g vitamin B1, 2.84 mg/100 g vitamin B2, 345 mg/100 g ergosterol, and 40.9 mg/100 g niacin.

Medicinal Properties: This mushroom produces 1-3 and 1-6 D-fractions of beta glucans, polysaccharides currently under investigation for immunopotentiation. The literature reports beta glucan levels up to 14%. (A recent analysis of the author's Himematsutake showed 9% beta glucans.) Its unique polysaccharides promote natural killer cells that are selectively cytotoxic on tumor cells. This mushroom has been the subject of numerous analyses for isolating constituents, both tumoricidal and immunomodulatory, for the treatment of cancers (Fujimiya et al. 1998, 1999; Ito et al. 1997; Itoh et al. 1994). The cultured mycelium also produces anti-tumor compounds (Mizuno et al. 1999). That this mushroom produces compounds specifically increasing apoptosis in cancerous cells (but not in healthy cells) and also triggers an immune response, is notable. A yellowish metabolite exuded by the mycelium apparently has bactericidal properties.

A contradition not yet reconciled at the time of this writing is a report by Stijve et al. (2000) that specimens of *A. blazei* from Brazil contained 1,000–3,200 mg/kg (.10–.32%) agaritines. In comparison, another almond flavored *Agaricus, Agaricus augustus,* has up to 2.2% agaritine content (Toth 2000), while Button mushrooms have up to .87%. (See also pages 221–223 and 313.)

Flavor, Preparation, and Cooking: This mushroom imparts a sweet almond flavor, delicate but distinct, a symphony of flavors that linger long after consumption. The aromatics of this mushroom are especially potent in the hours directly after harvest, and the flavor can be potent. When very fresh mushrooms are cooked, the slices of mushrooms undergo a mesmerizing color change from white to strong golden yellow. This color reaction is lost within a day of harvest. The thick stems are especially appealing in their crunchy texture. I recommend cooking them simply on high heat in a bit of olive oil, then seasoning with salt. Cooked mushrooms develop a wonderfully slippery texture, making it one of the most tactilely interesting of all succulent mushrooms. This mushroom is best eaten infrequently, as with daily consumption the novelty of its flavor soon fades. Some people feel the flavor is too strong and prefer the mushrooms be cooked separately so as not to overwhelm the flavors of other dishes. The golden exudate secreted from the spawn also has a very strong almond flavor. After mushrooms are harvested, even at the button stage, the gills continue to mature with spores for days while under refrigeration, becoming pink and eventually dark brown with maturity. (See page 71.) Mature specimens rate almost as good as young fruitbodies in their flavor profile.

Comments: A mushroom rapidly increasing in popularity, especially with Japanese, Brazilian, and Chinese cultivators who are eagerly exporting dried fruitbodies and fresh mushrooms. This mushroom has become the center of a $600 million (U.S.) industry in Japan since 1995. In response to demand, cultivation centers are being built in numerous countries, including the United States, Denmark, the Netherlands, and throughout the Far East.

Agaricus blazei has moved to the uppermost ranks of the best of all gourmet and medicinal mushroom species. Inevitably, the commercial Button (*Agaricus bisporus* = *brunnescens*) industry in North America and Europe will awaken to the potential of this species. Since this mushroom is phototropic, Button growers must face the difficult task of installing sufficient lighting if they want to switch to this species. Although not absolutely critical for primordia formation, exposure to light enhances yields.

This species can be grown on sterilized sawdust and manure based composts. Like *Agaricus brunnescens*, the activity of soil-borne bacteria appear critical for fruitbody maturation. Casing soils should have a pH of 7, not be sterilized, and in most cases, not be pasteurized, as the beneficial heat-sensitive microflora can easily be killed. If using a heat-treated peat moss–based casing, taking a few cups (200 ml) of native soil, dissolving into 5 gallons (15–16 liters) of water, and then using this microbially rich soil to moisten the casing at makeup provides the essential microflora necessary for fruiting. One method contours the casing layer with alternating mounds of 2^1/$_2$ inches (6 cm) of casing soil—a ridge bed—to enhance yields and inoculate egg-sized plugs of grain spawn 8 inches (20 cm) apart, and to a depth of 4 inches (10 cm) (Iwade and Mizuno 1997). From this author's experience, methods that rely on this type of inoculation overcome barriers that are otherwise encountered from through-mixing of spawn, allowing for concentrated "island" colonies of mycelium, which can readily become the launching pads for primordia. This approach resembles an early technique first employed by Button mushroom growers but later abandoned in favor of through-spawning (Atkins 1966).

Rough casing layers varying in particular size, but devoid of undecomposed wood fragments are recommended. A 1^1/$_2$- to 2-inch (2.5–5 cm) casing layer should be applied after 3 to 4 weeks of colonization. A thick casing soil appears important for adequate primordial formation and enlargement. Cultivators in Brazil elect to use native soils.

This author prefers Canadian sphagnum peat moss. The casing layer should be thick enough that mycelia shows in the valleys of casing layer three weeks after application. If showing sooner, watering can be increased to discourage overlay. At warmer temperatures, the rate of transpiration/evaporation is greater, and cultivators must frequently monitor moisture levels to prevent damaging dehydration. The recommended humidity is not as high as with many other mushrooms, allowing for active water-transport to the surface. Primordia tend to form deep within the casing layer, as opposed to the surface. Young primordia are massive in comparison to other mushrooms, and channel water from afar.

A patent (#5,048,227) has been issued for using a lattice-like separation layer between the spawned substrate and within the casing soil thus increasing, according to the claim, yields by 1.4 times or greater.

The separation lattice is nearly 3 cm thick, placed directly within 5 cm of casing soil laid directly upon the colonized compost beds. This technique limits the number of up-channeling rhizomorphs, reduces surface area, and in doing so, increases the likelihood that any primordium forming will have greater access to nutrients drawn from the mother mycelium. This interruption layer centralizes and distances the loci for primordia formation, resulting in an increase in the size and earlier formation of mushrooms. As *Agaricus blazei* fruitbodies tend to be large and have a high protein content, each erupting primordium has high demands for nutritional resources from the surrounding substrate. A first flush of *Agaricus blazei* is analogous to the second and third flushes that create the Portobello form of *Agaricus brunnescens*.

When dense clusters of primordia first form, the more rapidly developing fruitbodies can fuse and pull adjacent mushrooms from the substrate, a feature which can be avoided by surgically separating touching mushrooms while they are at the primordial stage. Mushrooms grown from immature composts can show slight brownish stains, or fissures on the stems, hiding hollowed portions filled with amber, almond fragrant liquids. This disease recedes as the compost matures during the flushing cycle.

Agaricus blazei can be grown on the recycled sawdust blocks from the end of the cultivation cycles of Shiitake (*Lentinula edodes*), Maitake (*Grifoloa frondosa*), Reishi (*Ganoderma lucidum*), and other primary saprophytes. Turning the compost piles made from the above for several weeks and mixing with wheat straw has provided a satisfactory compost medium for growing Himematsutake. The net nitrogen of composted Shiitake blocks approaches 2%, near the target nitrogen levels for classic Portobello composts. The use of this mushroom on "spent" substrates from the cultivation of primary saprophytes is on-going for idealizing a sequence of mushroom species growing on the same medium.

Peculiar to *Agaricus blazei* is that molds are not as likely to contaminate a peat moss casing layer once the mycelium emerges from below, a sure sign that the mushroom has predominated the microsphere. This dominance is so pronounced as to set Himematsutake apart from other species needing casing layers to encourage fruitings. This phenomenon also allows for a window of time in which experiments can be conducted without becoming victimized by mold infestation. The bottom of the trays must have adequate drainage and aeration or the mycelium will go into stasis and fruitings will be retarded. Growing Himematsutake indoors for the first two crops and then making ridge rows of mounded substrate excavated from trays for outdoor cultivation may be a good paradigm for warmer climates.

Cultivators in Hawaii and Florida have been growing this mushroom outdoors in mound culture. One common observation is that it does not attract the common insects that plague Oyster mushroom cultivation. The almond scent may repel insect pests. Also noteworthy is that the mushrooms are slow to rot, which means shelf life of the crop is good, provided the mushrooms are chilled directly after harvest. By all comparisons, *Agaricus blazei* is an exceptional mushroom. As with many species, pioneering Japanese mycologists first "blazed" the path for its cultivation and deserve credit for bringing this species forward, ironically one that was originally discovered in the United States but never recognized for its superb qualities. *Agaricus blazei* is an important new cultivar, ranking high in both culinary value and medicinal properties (Stamets and Yao 1999).

The Portobello Mushroom of the Genus *Agaricus*
Agaricus brunnescens Peck

Tray fruiting of the Portobello mushroom.

Introduction: First cultivated by an agronomist to Louis XIV, Olivier de Serres, in the seventeenth century, who found that transplanting mycelium from horse stable manure into new compost gave rise to more mushrooms. This insight eventually led to the multibillion-dollar industry of today. Early cultivators found that the constant temperatures of caves near Paris benefited production, and this began the age of indoor cultivation. In the 1980s, this mushroom was "reinvented" with the marketing of a large, brown form touting a made-up Italian sounding name—"Portobello," newly revised to "Portabella." In a marketing coup, the North American Button industry has made millions from this large brown form, more characteristic of second and sometimes third flushes. Previously these mushrooms were viewed as unsalable because of their unusual size but favored by workers who took them home to their families.

Common Names: Portobello
Portabella
Ports
Brown Button Mushroom
Champignon, Button
 Mushroom, or Blanco
 Bello (white form)
Crimini or Baby Portobello
 (small form)

Taxonomic Synonyms and Considerations: *Agaricus brunnescens* is synonymous with *Agaricus bisporus* (Lge.) Singer, and with its white form, formerly called *Agaricus hortensis*. Through selection of wild forms, cultivators have isolated stout-statured strains having thick pileal flesh, short gills, and short stems. *Agaricus bitorquis* is very similar but has four-spored basidia, and prefers warmer temperatures.

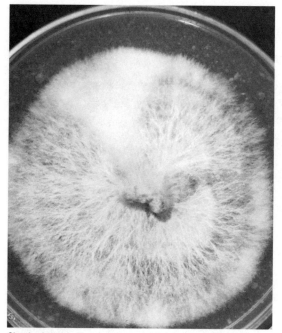

Classic *Agaricus brunnescens* at 14 days on malt extract agar. Note cottony sector.

Test tray of *Agaricus brunnescens*. By casing grain spawn with peat moss, strains can be quickly tested for their temperature preferences and inspected for overall appearance.

Description: Young mushrooms globose at first, robust, soon convex, expanding with age to broadly convex, eventually plane with age. Young mushrooms, depending upon strains, can be whitish, off-white beige to brown; smooth but can be covered with fine brown fibrils, which sometimes group to form scales. Partial veil thick, often doubled, breaking to form a median annulus. Cap flesh whitish, thick, with thin gills, not attached to the stem. Gills pallid when young, then pinkish, and soon darkening to chocolate brown with spore maturity. Stem short, stout, firm, swelling towards the base to which rhizomorphs are often attached.

Distribution: Widely distributed through the temperate regions of North America and Europe. A unique tetrapolar population (Callac et al. 1993) is localized around Palm Springs.

Natural Habitat: In well-manured grounds, in rich soils, in meadows, grasslands, and along forest edges, roadsides, and according to Arora (1986) under cypress trees in western coastal California.

Microscopic Features: Spores ellipsoid, 5.5–8.5 × 4–6.5 μ, chocolate brown in deposit, binucleate. Basidia predominately two-spored, homothallic, with minor populations four-spored, heterothallic.

Available Strains: The current collections of commercial strains are largely descendants of a pool of only one half dozen strains originating from Europe in the early 1900s. Of those, U-1 and U-3 are most well known and gave rise to the progeny most widely in use today. Royse et al. (1982) and Kerrigan et al. (1993) noted that most commercially cultivated strains originated from the same genetic population in Europe. Strains are temperate in nature (fruiting at (55) 60–70°F or (14) 16–21°C), have thick pileal flesh, short stems, and thin gills. Agaricus researchers see the need to continually diversify its strain bank in order to protect against virus and/or senescence. Few wild strains compare to the commercial strains currently in use in terms of productivity, uniformity of the crop, and flavor. The American Type Culture Collection lists several strains, notably Lambert's original #10892, San Antonio's #24558, Kneebone's #36416 and #36425, and Horst U-1 #62462. Brown strains such as those isolated from San Antonio's #56128, and Kneebone's #36416 make for good-forming portobellos, and earlier this century were more popular until the white varieties became the choice of U.S. consumers in the 1950s. The trend has now reversed, and brown forms are becoming increasingly popular (for instance, the Crimini varieties.) Mixing dikaryons originating from early culture collections created many of the newer strains. Most laboratories rename or renumber strains in order to maintain a proprietary advantage and to confuse competitors. Tissue culturing of strains from store-bought mushrooms

usually results in poorly growing strains—a trait that inherently protects the spawn industry.

Rick Kerrigan, a noted Agaricus expert, is credited with making the largest recent deposit of wild strains to any culture library since the early works of pioneers like Kneebone and San Antonio. He created the Agaricus Resource Program or ARP, offering a bounty for any new wild strain of *Agaricus brunnescens*, successfully acquiring hundreds of new isolates from nature. Kerrigan's work revealed that the California wild native populations are being slowly supplanted by European varieties, which have escaped into the environment from cultivated specimens. He fears extinction of native varieties is possible in the near future (Kerrigan et al. 1996, 1998).

The susceptibility of the Button mushroom to virus and blotch diseases is a major concern worldwide. Hence, the pursuit of virus-resistant strains and the preservation of genetic diversity are top priorities of knowledgeable breeders. For the industry to continue to thrive, collecting and banking wild strains must be a priority.

Mycelial Characteristics: Off-white, dingy with age, struck with brownish hues, divergently rhizomorphic, with an overlayering of cottony, aerial mycelium developing in age or in response to carbon dioxide. Strains can easily senescence, often first seen with the loss of rhizomorphs, an increase in cottony forms lacking feather-like outer edges and an overall decline in the speed of growth. Grain spawn exudes a yellowish metabolite and becomes mealy with age.

Fragrance Signature: Grain spawn musty. Not particularly pleasant.

Natural Method of Cultivation: Making rows of compost impregnated with clumps of grain spawn, or using compost spawn, has met with unpredictable success and is rarely practiced anymore as mushroom farms have moved to control the environments better with indoor cultivation.

Recommended Courses for Expansion of Mycelial Mass to Achieve Fruiting: Grain spawn is mixed into formulated manure-enriched composts, and then cased with soil.

The Portobello strain is a brown variety of the common Button mushroom, *Agaricus brunnescens* otherwise known as *Agaricus bisporus*. Portobello mushrooms' amazing success is due to ingenious marketing and adapting a method of cultivation to produce a high profit product.

Isolated Portobellos become large when not competing for nutrients from nearby mushrooms.

a quick way to determine if the strain will fruit, and to gain insights into its form without having to resort to compost making (see page 218).

Substrates for Fruiting: This mushroom benefits from a thorough composting of manure-enriched straw that is turned every other day and then undergoes a secondary high-temperature fermentation (Phase II) to encourage the proliferation of thermotolerant allies (*Actinomyces* spp.). Numerous compost formulas succeed, using chicken, horse, and cow manures as a source of nitrogen and microbes. A complete list of compost formulas using materials readily available in most agricultural regions of the world are described in the list of recommended reading at the end of this description. For small batches, a simple formula is posted on the Internet at *www.mushroomadventures.com/compost.html.* Another simple, no-turn method, which was described in my first book of 1978, is to use leached cow manure available from dairies, buffered with gypsum and/or other amendments, pasteurized, and then conditioned with a Phase II strategy. Once inoculated, a casing layer hosting active bacteria, especially Pseudomonads, is critical for primordia formation. The positive benefits from the non-sporulating bacterium *Pseudomonas putida* was first identified by Hayes et al. (1969) as being critical for fruiting in *Agaricus brunnescens*. Visscher (1978) determined that pasteurization of the casing layer at 140°F (60°C) for 4 hours significantly retarded and reduced fruitings due to the loss of stimulatory bacteria, but did not prevent mushroom formation. Hence, high heat-treatment of casing soils is not recommended. When composing a casing mix, the ecology of the peat moss (absence of nematodes, insects, undecomposed woody tissue, etc.) has a dramatic influence on the cultivation success. When stored for prolonged periods, basic microflora rebuild in peat mosses that have otherwise been heat-treated by the supplier.

Suggested Agar Culture Media: PDYA, OMYA, and/or DFA. I find that this species likes PDYA better than most other media.

1st, 2nd, and 3rd Generation Spawn Media: Standard agar-to-grain for the creation of masters, and then grain-to-grain transfers for three or more generations. Some laboratories use Popsicle-like sticks that are inserted directly into bags of sterilized grain. The actual expansion of this mushroom as provided by most spawn laboratories is many orders of magnitude beyond what I would recommend. My personal preference for spawn is rye. Grain used for spawn production is usually buffered with calcium carbonate and calcium sulfate, submerged in water, boiled, and then removed for placement into jars and/or polypropylene bags before sterilization. If millet is used, parboiling prior to sterilization is absolutely essential so the grain kernels adequately saturate with water, greatly aiding in the colonization of the mycelium. Casing the grain with soil is

Recommended Containers for Fruiting: Since this mushroom best grows from horizontal surfaces, trays 6–12 inches in depth or plastic bags weighing 40–60 pounds are the standard for the industry.

GROWTH PARAMETERS

Spawn Run:

Incubation Temperature: 70–77°F (23–25°C)
Relative Humidity: 90–100%
Duration: 18–20 days
CO_2: > 5,000 ppm
Fresh Air Exchanges: 1 per hour
Light Requirements: n/a

Primordia Formation:

Initiation Temperature: 60–65°F (16–18°C)
Relative Humidity: 95–100%
Duration: 12–18 days
CO_2: 400–800 ppm.
Fresh Air Exchanges: 5–7 per hour
Light Requirements: none needed

Fruitbody Development:

Temperature: 60–65°F (16–18°C)
Relative Humidity: 85–90% (95%)
Duration: 4–7 days
CO_2: <1,000 ppm
Fresh Air Exchanges: 5–7 per hour
Light Requirements: none needed

Cropping Cycle:

Every other week for 3 flushes

Yield Potentials: 5–8 pounds per square foot of tray, filled to a depth of 7–10 inches. Dutch Agaricus growers report up to 35 kilograms per square meter in three flushes, which is equivalent to approximately 7.7 pounds per square foot. Yields are accumulated over three flushes, with either uneven first flushes giving rise to the Portobello forms, or more typically larger mushrooms forming during the second and third flushes. Some cultivators place a perforated "interrupter" to isolate fruitings so that the first flush gives rise to the more expensive Portobello forms.

Harvest Hints: For the Portobello form, the mushrooms are harvested well after the veil has fallen, when the mushrooms are broadly convex but not plane. Mushrooms are twisted from the bed, trimmed and then set, stems down, into the collecting basket. This mushroom succeeds where most fail: *Agaricus brunnescens* is one of the few species whose spores and gill tissue are not unsavory. Portobellos are tasty when the cap fully expands, and the gills have enlarged to permit massive sporulation. With most mushroom species, I have found that adult mushrooms are slightly less tasty than young ones, with the progressive loss of flesh to gill formation and sporulation as they age. Good pickers can harvest up to 40 pounds per person per hour, at a cost of $.50–.60 per pound in labor costs.

Form of Product Sold to Market: Fresh, canned whole or sliced, in soups and sauces, the Button mushroom is as American as apple pie. The recent use of Portobellos in the making of "mushroom burgers," a meat sandwich substitute is yet another venue for consumption. Young, close capped Portobello mushrooms are being marketed under the name of 'Crimini'. New products being touted in Japan from this mushroom uses an isolated constituent to combat bad breath and another as a "skin whitener."

Nutritional Content: Crude protein 25–33%; 9% fiber; 10% ash. The FDA has officially designated Button mushrooms as "healthy foods" because they are low in fat and have no cholesterol.

Medicinal Properties: This mushroom contains compounds that inhibit the enzyme aromatase. Aromatase is associated with tumor growth. Compounds inhibiting aromatase have potential for the treatment or prevention of breast cancer (Bankhead, 1999). A diet of mushrooms in mice with implanted tumors showed a decrease in aromatase as mushroom consumption increased. However, *Agaricus brunnescens* contains hydrazines, carcinogenic compounds that have been thought to dissipate only from prolonged, high-temperature heating. More than 80 percent of known hydrazines are carcinogenic. The most notable carcinogenic hydrazine from this mushroom is agaritine, a powerful mutagen, which is activated by the mushroom enzyme tyrosinase, making it heat stable. Enzymes in the digestive system convert agaritine into carcinogenic by-products. The chemical culprits worthy of concern are: 4-(hydroxymethyl)phenyl-hydrazines and 4 (hydroxymethyl)benzene diazonium ions (Walton et al. 1997). Free radicals can also activate *Agaricus* hydrazines into highly carcinogenic subconstituents (Tomasi et al. 1987) as well as catalytic processes in the kidneys (Price et al. 1996). Hence, there are several modes of activating agaritine into highly carcinogenic derivatives.

The damaging effects of agaritine's derivatives may be partially suppressed by the mushroom's antioxidants, which, in turn help create host-generated superoxide dismutases (SODs), and the activity of aromatase inhibitors. Walton et al. 1998 asserted, however, that the mutagenic and pre-mutagenic compounds are not affected by quick cooking (10 minutes at 437°F/225°C) but are only slightly reduced by prolonged heat treatment in boiling water for 4 hours at 212°F (100°C). A study of blanched, canned mushrooms showed that the agaritine content was reduced tenfold in comparison to fresh mushrooms, from 229 mg/kg to 15-18 mg/kg (Andersson et al. 1999). However, this reduction may have been due to leaching of the hydrazines into the surrounding water used for blanching in combination with the prolonged, high pressure steaming processing used for canning. Another report by Sharman et al. 1990, found most fresh samples of this mushroom had agaritine levels within the range of 80–250 mg/kg but with one dried sample having 6,520 mg/kg, a comparatively high level. This result suggests that agaritine production may be a strain-specific trait, as this one dried, sliced sample had more than 8 times the agaritine content of other samples in this same study. In contrast, dried Shiitake mushrooms, *Lentinula edodes,* have either undetectable or extremely low levels of agaritines, in the 0.082 mg/kg range (Stijve et al. 1986; Hashida et al. 1990). Hashida's study reported marked reduction of agaritines from boiling in water at 212°F

(100°C) for 10 minutes, a report in direct contradiction to Walton's 1998 study.

A Swiss report estimated that with the average consumption of 4 grams per day of *Agaricus bisporus* (=*Agaricus brunnescens*) the lifetime increase in cancer risk would be approximately two cases per hundred thousand lives (Shepard et al. 1995). In a metropolitan area of twenty million residents, approximately the size of Los Angeles, two hundred people would be expected to get cancer in their lifetime from eating *Agaricus brunnescens* mushrooms, all other factors being equal.

However, other investigations have questioned the cause and effect relationship of agaritine in Button mushrooms and its mutagenic properties (Pilegaard et al. 1997; Matsumoto et al. 1991; Papaparaskeva et al. 1991; and Pool-Kobel 1990). Benjamin (1995) noted that early studies are controversial and potentially flawed. One study had the intravenous introduction of mushrooms into mice. Another study showed that mice implanted with cancer cells (Sarcoma 180) and then fed dried mushrooms showed inhibited tumor growth (Mori et al. 1986). More recent studies reconfirm that a diet of this mushroom, both raw and baked, induced tumors in mice (Toth et al. 1998).

The cited research is highly controversial and raises concerns about the human consumption of *Agaricus brunnescens* as a health food. For years, the conventional wisdom was that hydrazines would be destroyed with cooking. Anti-cancer polysaccharides, aromatase-inhibiting compounds, and antioxidants known from *Agaricus brunnescens* (Kweon 1998), may neutralize the carcinogenic effects of the hydrazines, but, in my opinion, the jury is still out on this issue. Eating this mushroom raw, especially with free radical inducing foods, is definitely not recommended. And yet, in the United States, up to 80 percent of all Button mushrooms consumed are eaten uncooked. I am disturbed that the most commonly cultivated mushroom in the world has few studies authenticating its beneficial medicinal properties, in stark contrast to the numerous studies on Shiitake, Maitake, Reishi, Yun Zhi, and others. The Portobello mushroom may be gourmet, but in the absence of scientific studies, I doubt that, at this time, *Agaricus brunnescens* can be considered medicinally beneficial.

What to do? The financial future of the Button/Portobello industry may well depend on recognizing the risks, and aggressively developing low agaritine or agaritine-free strains. As analyses have shown more than an eight-fold difference in the concentrations of agaritine in *Agaricus brunnescens* mushrooms, clearly some strains already in cultivation are much lower in agaritine content than others. Pursuing low agaritine strains should be a top research priority within the *Agaricus* industry, especially within the venue of the spawn producers. Given variations in agaritine levels in existing strains, a breeding program for creating agaritine-free strains is a task preeminently achievable in the near future. Certainly the Button mushroom industry has clear economic and ethical incentives for doing so.

Flavor, Preparation, and Cooking: Probably the mushroom with the most recipes, this tasty mushroom is stir-fried, cooked into soups, and placed onto pizzas, and is the primary mushroom available in the markets and restaurants of North America and Europe. Button mushrooms have an exceptionally good texture and a much thicker flesh than most. Although it is popular, eating this mushroom raw is not recommended as the nutritional and medicinal benefits are not realized, and there is the danger for negative effects from heat sensitive compounds. Recipes are on page 443.

Comments: The simplest way to encourage the Portobello forms is to shorten the primordia formation period by prematurely raising temperatures so fewer mushrooms are set, allowing for the creation of larger fruitbodies. To further encourage the Portobello forms on subsequent flushes, mushrooms are selectively harvested to encourage dispersed formation of larger mushrooms, which then mature first and cause adjacent primordia to go dormant. Preferred strategies are developed by individual growers through trial and error.

Dutch method of fruiting the White Button variety of *Agaricus brunnescens*.

When mushrooms form close to one another, the potential for the spread of diseases increases. Some growers apply ample water at initiation—essentially saturating the casing layer—at a rate of 2–3 liters per square meter. No additional water is added to stimulate subsequent flushes. By using lightly chlorinated water (200 parts per million) disease is limited while not adversely affecting mushroom formation and development. Most cultivators water only as needed during cropping.

In 1998, sales of this species represented 98% of the United States mushroom market, with more than 861 million pounds sold. Per capita consumption is now at a record 4.10 pounds The continuing trend emphasizes increasing consumption of fresh mushrooms. Worldwide, this mushroom is responsible for more than a million metric tons per year! Probably the most well-researched of all cultivated mushrooms, the common Button mushroom has become the center of a multibillion-dollar industry. The *Agaricus* industry has increasingly become dominated by fewer and fewer players, with a corresponding reduction in the number of commercial spawn producers. This trend has concentrated enormous financial resources in a handful of companies who largely control the industry. Few are organically certified. Few care to be.

Because of the attractiveness of the compost and the mushrooms to insect parasites, the Button mushroom industry has been notorious for its use of pesticides, most of which have become banned in the United States and Europe due to their carcinogenic and ground-water damaging properties. The reliance on chemical solutions to pest problems has slowly waned with improved composting practices, especially indoor composting techniques. Nevertheless, I encourage consumers who love this mushroom, but cannot grow their own, to support organic producers who have not developed a dependency on chemical solutions to their pest management strategies.

Compost formulas are crude in form but not in design. Phase I composting is for the proliferation of microorganisms, which convert cellulose compounds into cellular proteins. Phase II is the elevation of temperatures of the microbially mature compost to neutralize pests by selectively favoring thermotolerant fungi and actinomyces, which are then put into stasis upon the return of the compost to just above ambient temperatures. Thus, the composts become selectively favorable to secondary decomposers, such as *Agaricus brunnescens*. Recent advances in composting practices, including refluxing leachate from Phase I composting back into the compost minimizes odor, and potentiates nutritional conversion, and consequently yields. Another advancement is the ample aeration of the compost to prevent the evolution of volatile sulfur compound in the anaerobic core (Perrin and Maculey, 1995). Short composting methods have evolved using various biological "activators" (such as *Bacillus subtilis*) and claim to enhance yields by denaturing cellulosic ingredients. The number of agricultural wastes that can be incorporated into composts is remarkable. For more information on compost preparations, please consult the recommended references listed below.

Modern Mushroom Growing by P. Vedder, 1978, Educaboek, Culemborg, Netherlands.

The Mushroom Cultivator by P. Stamets and J. Chilton, 1983, Agarikon Press, Olympia, Washington.

"Agaricus: The Leader in Production and Technology," Chapter III: Compost Materials and Composting in *Edible Mushrooms and Their Cultivation* by S. T. Chang and P. Miles, 1989, CRC Press, Boca Raton, Florida.

The Black Poplar Mushroom of the Genus *Agrocybe*
Agrocybe aegerita (Brigantini) Singer

Mycelium of *A. aegerita* 3 and 7 days after inoculation.

Introduction: This mushroom has a mellow and attractive flavor when young. *Agrocybe aegerita* grows prolifically on deciduous wood debris, often forming large clusters—both in nature and in the controlled environment of the growing room. This species is an excellent candidate for stump recycling, especially in the southeastern United States.

Common Names: The Black Poplar Mushroom
The Swordbelt Agrocybe
Pioppino (Italian)
Yanagi-matsutake (Japanese)
South Poplar Mushroom or
Zhuzhuang-Tiantougu
(Chinese)

Taxonomic Synonyms and Considerations: A very variable fungus which may well be split into several distinct taxa with more research, this mushroom was once called a *Pholiota, P. aegerita*. Other synonyms are *Pholiota cylindracea* Gillet (Singer, 1986), or *Agrocybe cylindracea* (DC. ex Fr.) Maire, a name still preferred by Asian mycologists. Watling (1982) prefers *Agrocybe cylindrica* (De Candolle ex Fries) Maire. *Agrocybe molesta* (Lasch) Singer and *A. praecox* (Pers. ex Fr.) Fayod are related species and can be cultivated using the same methods described here. Their flavor, in the opinion of many mycophagists, is not as good as *A. aegerita*. More work on the taxa of the southeastern *Agrocybes* is needed.

Description: A substantial mushroom, often up to 12 inches in diameter. Cap convex to hemispheric, expanding to plane at maturity, smooth, yellowish

Twenty days after inoculation, primordia voluntarily form on malt extract agar.

GROWTH PARAMETERS

Spawn Run:
Incubation Temperature: 70–80°F (21–27°C)
Relative Humidity: 95–100%
Duration: 20–28 days
CO_2: > 20,000 ppm
Fresh Air Exchanges: 0–1 per hour
Light Requirements: n/a

Primordia Formation:
Initiation Temperature: 50–60°F (10–16°C)
Relative Humidity: 95–100%
Duration: 7–14 days
CO_2: <2,000 ppm
Fresh Air Exchanges: 4–8 per hour
Light Requirements: 500–1,000 lux

Fruitbody Development:
Temperature: 55–65°F (13–18°C)
Relative Humidity: 90–95%
Duration: 4–6 days
CO_2: <2,000 ppm.
Fresh Air Exchanges: 4–8 per hour
Light Requirements: 500–1,000 lux

Cropping Cycle:
Two flushes, 10–14 days apart

gray to grayish brown to tan to dingy brown, darker towards the center. Gills gray at first, becoming chocolate brown with spore maturity. Stem white, adorned with a well-developed membranous ring, usually colored brown from spore fall.

Natural Habitat: Growing saprophytically, often in clusters, on stumps in the southeastern United States and southern Europe. Preferring hardwoods, especially cottonwoods, willows, poplars, maples, box elders, and in China on tea-oil trees.

Distribution: Not known to occur in North America outside of the southeastern states of Mississippi, Louisiana, and Georgia. Common across southern Europe and in similar climatic zones of the Far East.

Microscopic Features: Spores smooth, ovoid to slightly ellipsoid, brown, 9–11 × 5–6.5 (7) μ, lacking a distinct germ pore. Clamp connections present.

Available Strains: Strains are commonly available from culture libraries. Most strains cloned from wild specimens produce on hardwoods.

Mycelial Characteristics: Longitudinally linear, becoming cottony, usually not aerial. White at first, soon becoming spotted brown, and eventually tan brown. Primordia usually form on malt extract agar media.

Fragrance Signature: Mealy, farinaceous, but not pleasant.

Natural Method of Cultivation: Stumps of the above-mentioned trees. Outdoor wood-chip beds also produce, much in the same manner as for the cultivation of *Stropharia rugosoannulata*.

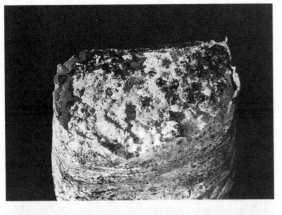

The Black Poplar mushroom (*Agrocybe aegerita*) fruiting on supplemented alder sawdust/chips 32, 33, 34, and 35 days after inoculation.

Recommended Courses for Expansion of Mycelial Mass to Achieve Fruiting: Petri dish cultures blended via Eberbach stirrers to create liquid inoculum, which is, in turn, injected into sterilized grain. This grain spawn can be used to make sawdust spawn for outdoor inoculations or for inoculating directly into supplemented, sterilized hardwood sawdust.

Suggested Agar Culture Media: MYPA, PDYA, OMYA, and DFA.

1st, 2nd, and 3rd Generation Spawn Media: Cereal grains (rye, wheat, milo, sorghum, etc.) for first and second generation spawn. Sawdust is recommended for the third generation, which can

then be used as spawn for inoculation directly into slices cut into stumps. Sawdust spawn is also recommended for inoculation into sterilized, supplemented hardwood sawdust.

Substrates for Fruiting: I have fruited this species on supplemented oak and alder sawdust/chips. Willow, poplar, cottonwood, and maple are just as likely to support substantial fruitings.

Recommended Containers for Fruiting: Polypropylene bags and trays. This mushroom is better grown from horizontal surfaces than from vertical ones.

Yield Potentials: Up to 1 pound of fresh mushrooms per 5–6 pound block of sterilized sawdust/chips/bran. Given the size of fruitings occurring naturally, large-diameter willow, poplar, and cottonwood stumps could sustain massive fruitings for many years.

Harvest Hints: A more fragile mushroom than it at first appears, this mushroom should be encouraged to grow in clusters. If mushrooms are harvested before the veils break, shelf life is prolonged.

Form of Product Sold to Market: Fresh; because of this mushroom's resemblance to the Button mushroom (*Agaricus brunnescens*) marketing is not as difficult as with many "new" species.

Nutritional Content: Not known to this author.

Medicinal Properties: None known, although closely related species produce unique antibiotics.

Flavor, Preparation, and Cooking: Finely chopped and stir-fried, cooked in a white sauce and poured onto a fish or chicken, or baked in a stuffing, this species imparts a mild but satisfying, pork-like flavor.

Comments: This mushroom benefits from the application of a $1/2$-inch casing directly onto the top surface layer of mycelium. However, if a condensing fog environment is provided, combined with high turbulence, an even plane of primordia can form absent any casing layer. (See page 227.) For cluster formation, primordia should be encouraged to appear in groups of 4 to 20.

This species figures as one of the best for recycling stumps in the humid southeastern United States. The willow-populated swamps of Louisiana seem like an ideal setting for the deliberate cultivation of *A. aegerita*. Regions of Chile, Japan, and the Far East, as well as southern Europe, have coincident weather patterns that should support growth.

The Shaggy Mane of the Genus *Coprinus*
Coprinus comatus (Muller: Fries) S. F. Gray

Introduction: Shaggy Manes have long been a favorite among mushroomers in North America and Europe. Easy to identify, often growing in massive quantities, this brilliantly white mushroom is hard to miss and difficult to confuse with poisonous species. Their fragile constitution and unique method of self-destruction, combined with their mild but excellent flavor, has made the Shaggy Mane a popular mushroom among hikers and hunters.

After experimenting with its cultivation, I am pleasantly surprised at how well this species adapts to a wide variety of indoor and outdoor substrates. Although the commercial cultivation of this mushroom is limited by its predisposition to disintegrate into an inky mess, this mushroom is fantastic for those who can consume it within 2 days of picking.

Classic, cottony Shaggy Mane mycelium on malt extract agar.

Common Names: The Shaggy Mane
Lawyer's Wig
Maotou-Guisan (Chinese)

Taxonomic Synonyms and Considerations: *Coprinus comatus* is considered a taxonomically "clean" species by most mycologists and can be accurately identified by sight.

Description: Cap 4–10 (15) cm high by 3–4 (5) cm thick, vertically oblong, dingy brown at first, soon white, and decorated with ascending scales. Gills crowded, white to pale, long, broad and slightly attached or free to the stem. Stem 6–12 (15) cm long by 1–2 cm thick, equal, hollow, and bulbous at the base, and adorned with a movable, membranous collar-like ring, that separates from the cap margin as the mushrooms enlarge. As the mushrooms mature, the gills blacken, or "deliquesce," transforming into a black, spore-ladened fluid, which drips from the rapidly receding cap margin. The cap eventually totally recedes, leaving only the stem.

Distribution: Growing in the late summer and fall throughout the temperate regions of the world.

Natural Habitat: In lawns, meadows, around barnyards, in wood chips, along roadsides, and in enriched soils.

Microscopic Features: Spores black, 11–15 × 6.0–8.5 µ, ellipsoid, with a germ pore at one end. Subhymenium cellular.

Available Strains: This mushroom is easy to clone. Cultures are widely available from libraries throughout the world. However, a sporeless strain is needed before commercial cultivation will become practical. I have isolated a late-sporulating strain from a fruiting on manure-enriched sawdust. Sporeless strains are needed as the maturing spores trigger deliquescence, causing the mushrooms to disintegrate into a black spore-enriched fluid. This spore liquid presents unique problems to cultivators in their attempts to isolate spores away from contaminants.

Mycelial Characteristics: Mycelium usually white, cottony, aerial, often develops "tufts" (hyphal aggregates) with maturity. Most strains form mycelial mats asymmetrically shaped along the outer edge.

Fragrance Signature: Farinaceous and mildly sweet.

Natural Method of Cultivation: Inoculation of spawn directly into manure-enriched soils or 4- to 6-inch deep beds of hardwood sawdust. Newly laid or fertilized lawns that are frequently watered are perfect habitats for Shaggy Manes. Cow or horse manure, mixed with straw or sawdust, is also ideal. Hardwood sawdust spawn should be used as inoculum for establishing outdoor patches.

Recommended Courses for Expansion of Mycelial Mass to Achieve Fruiting: Adaptive to liquid fermentation, grain spawn can be created with ease. Grain spawn is directly inoculated into pasteurized, supplemented straw-based composts. I have had poor experiences with bran-supplemented sawdust compared to manure-enriched straw-based substrates. Primordial dials form on sterilized sawdust but often fail to develop.

Suggested Agar Culture Media: PDA, PDYA, MEA, MPYA, DFA, or OMYA.

1st, 2nd, and 3rd Generation Spawn Media: Grain (rye, wheat, and sorghum) throughout.

Substrates for Fruiting: The straw/manure compost formulas described by Stamets and Chilton (1983) support substantial fruitings of this mushroom. Another excellent medium can be made from the manure- and urea-enriched sawdust discharged from horse stables. Paper and pulp wastes have also been proven to support fruitings.* Like most coprophiles, this mushroom greatly benefits from the placement of a peat moss–based casing soil.

---

* Mushroom production on pulp waste stopped in British Columbia in the late 1980s after concerns about residual, heavy metal contamination. Please check with paper manufacturers before using their products as a substrate for mushroom production.

GROWTH PARAMETERS

Spawn Run:
Incubation Temperature: 70–80°F (21–27°C)
Relative Humidity: 95–100%
Duration: 12–14 days
CO_2: 5,000–20,000 ppm
Fresh Air Exchanges: 0–1 per hour
Light Requirements: n/a

Primordia Formation:
Initiation Temperature: 60–70°F (16–21°C)
Relative Humidity: 95–100%
Duration: 12–15 days after casing
CO_2: 500–1,000 ppm
Fresh Air Exchanges: 4–8 per hour
Light Requirements: 500–1,000 lux for 8
  hours per day

Fruitbody Development:
Temperature: 65–75°F (18–24°C)
Relative Humidity: 80–90%
Duration: 5–7 days
CO_2: 500–1,000 ppm
Fresh Air Exchanges: 4–8 per hour
Light Requirements: 500–1,000 lux for 8
  hours per day

Cropping Cycle:
Two to three flushes, 4–10 days apart

Recommended Containers for Fruiting: Trays and bags. This mushroom is not inclined to grow from anything but horizontal surfaces.

Yield Potentials: When soya flour was added at a rate of 2% to paper pulp fiber (giving the substrate a 0.5% nitrogen content), yields approached 80% B.E. However, these yields were lower than that generated from manure compost, which often exceed 100% B.E. (See Mueller et al., 1985.)

Harvest Hints: Since this mushroom deliquesces from the end of the gills upward to the stipe, mushrooms should be picked before the slightest hint of the gills turning black. If picked when no basidia have matured, mushrooms can be kept in cold storage for 4 to 5 days. Any mushrooms that begin to deliquesce should be removed from the fresher fruitbodies since the enzymes secreted by one deliquescing mushroom will decompose adjacent mushrooms, regardless of age.

Form of Product Sold to Market: If this mushroom ever gets to market, it is there all too briefly due to self-deliquescence. In the matter of 2 days from harvest, the mushrooms turn into a black ink-like slurry unless precautions are not taken. Harvesting immature mushrooms, submerging the mushrooms upside-down in cold water, and storing them under refrigeration can preserve them. Packing Shaggy Manes in refrigerated, nitrogen gas–filled containers also extends shelf life. Currently, Shaggy Manes that are sold at farmer's markets are usually from wild collections. Young shaggy manes can be thinly sliced and quickly dried for storage. Freeze-drying is also an option.

Nutritional Content: 25–29% protein (N × 4.38); 3% fat; 59% carbohydrates; 3–7% fiber; and 1.18% ash (Crisan and Sands, 1978; Samajpati, 1979).

Medicinal Properties: A novel antibiotic has been isolated from this species and is currently being characterized by American researchers. Ying (1987, page 313) reports that the "inhibition rates against Sarcoma 180 and Ehrlich carcinoma are 100% and 90% respectively." This author knows no other research on the antitumor properties of this mushroom.

Flavor, Preparation, and Cooking: Shaggy Manes were the first mushrooms that seduced me into the art of mycophagy. It may seem odd, but I prefer this mushroom for breakfast. I like to prepare the mushrooms by frying thinly cut dials (stem included) in a frying pan with onions and light oil. Once they are slightly browned, the mushroom can be added to an omelet. Or, fry the mushrooms in butter at medium heat and serve on whole-wheat toast. Many of the recipes listed in this book can incorporate Shaggy

Shaggy Manes fruiting from cased horse manure-enriched straw.

Kelly Chadwick with Shaggy Manes fruiting from cased, leached cow manure.

Manes. Since this mushroom has considerably more moisture than Shiitake, for instance, the water can be used for steaming vegetables or other foods.

Comments: This is a great mushroom to grow in your yard and in compost piles. Once an outdoor patch is established, Shaggy Manes can fruit for many years. For impatient cultivators, indoor cultivation is recommended. For mycological landscapers not concerned about territorial confinement of their mushroom patch, the Shaggy Mane is an excellent companion to garden plants.

After pasteurized compost is inoculated, the substrate is completely colonized in 2 weeks with a cottony, non-rhizomorphic mycelium. When colonization is complete, a moist casing (peat moss/gypsum) layer is applied. After 10 days, the mycelium can be seen reaching through the upper surface of the casing. At this stage, lower the temperature, increase watering, and introduce light to stimulate fruiting. Yields can be substantially increased if the casing layer is vigorously raked just as the mycelium beings to show on the surface of the casing.

The primordia form as circular dials, between the size of a dime and a quarter. The primordia are unique in that they are wide and flat. An inner collar forms within the dial and arises to form a dome. This dome soon shoots up to form a recognizable

"Wild" Shaggy Manes fruiting in yard.

mushroom. The circular zone visible at the primordial stage becomes the movable ring resting on the stem of the mature mushroom.

For more information, consult Van de Bogart (1976), Stamets and Chilton (1983), and Mueller et al. (1985).

The Enoki Mushroom of the Genus *Flammulina*
Flammulina velutipes (Curtis ex Fries) Singer

Mycelia of Enokitake.

Introduction: The Japanese lead in popularizing this mushroom. In the wild, *Flammulina velutipes* is a short, furry-footed mushroom. Enoki morphs while seeking light and oxygen. Usually cultured in chilled growing rooms, the mushroom has abnormally small caps and long stems that are achieved by elevating carbon dioxide levels and limiting light exposure. This unnatural shape makes the harvesting of Enoki easy.

Common Names: Enokitake (Japanese for "The Snow Peak Mushroom." Enokidake is an alternative spelling)
Nametake ("Slimy Mushroom")
Yuki-motase ("Snow Mushroom")

The Winter Mushroom
The Velvet or Furry Foot Collybia
The Golden Mushroom (Thailand)

Taxonomic Synonyms and Considerations: Formerly known as *Collybia velutipes* (Fr.) Quel.

Enoki culture producing abundant primordia on malt agar medium.

Wild fruiting of *Flammulina velutipes* from stump. Note shortness of stems and breadth of caps of wild vs. cultivated fruitings.

Description: Cap 1–5 cm in diameter, convex to plane to upturned in age, smooth, viscid when wet, bright to dull yellowish to yellowish brown to orangish brown. Gills white to yellow, attached to the stem. Stem usually short 1–3 inches, yellow to yellowish brown, darkening with age and covered with a dense coat of velvety fine brown hairs near the base. In culture, the morphology of this mushroom is highly mutable, being extremely sensitive to carbon dioxide and light levels. Cultivated specimens usually have long yellowish stems, small white to yellowish caps. When spores mature, the caps darken to brown.

Distribution: Widespread throughout the temperate regions of the world, growing from sea level to tree line.

Natural Habitat: Primarily on hardwoods, occasionally on conifers, commonly growing in the late fall through early winter. This mushroom can freeze, thaw, and continue to grow. One definition of Enoki

in Japanese is "huckleberry," implying that this mushroom grows on that shrub in Japan.

Microscopic Features: Spores white, 6–8 × 3–4 µ, ellipsoid. Hongo (1988) lists a spore size of Japanese varieties of 5–7.5 × 3–4 µ. Moser (1983) reports spores of European collections measuring 8–9 × 4.5–6 µ. These differences may underscore the wide range of varieties included within this complex. Clamp connections present.

Available Strains: Vast populations of this species thrive in the wild, offering cultivators a rich resource for new strains. Most strains require a cold shock and/or growth in temperature ranges from 40–60°F (4.4–13°C) Strains of this mushroom vary in their sensitivity to light and carbon dioxide levels. White or yellow capped strains are common.

Mycelial Characteristics: White, longitudinally linear, becoming finely appressed and tinged light brown to spotted with golden yellow-brown zones with age. The surface roughens, resembling fine sandpaper, and sometimes becomes beaded at the earliest stage of primordia formation. Long-stemmed, small-capped mushrooms commonly form along the inside periphery of the petri dish or during cold storage of culture slants.

Fragrance Signature: Grain spawn musty smelling, not pleasant.

Natural Method of Cultivation: Stump culture is possible, as evidenced by the penchant that this species has for logs, stumps, and wood debris in the wild. However, stump culture should not be encouraged to those who cannot distinguish Enoki mushrooms from the small, wood-decomposing poisonous mushrooms such as the deadly members of the genus *Galerina* or *Conocybe*.

Recommended Courses for Expansion of Mycelial Mass to Achieve Fruiting: Liquid-inoculated grain spawn mixed directly through sterilized, supplemented sawdust at a rate of 10–15%. The rapidly decomposing hardwoods such as alder, cottonwood, willow, aspen, and poplar are recommended. If selecting cultures actively forming primordia, and placing the mycelium into the Eberbach stirrer for the generation of liquid-inoculated grain spawn, the duration from colonization to fruiting can be shortened by a week. The timely disturbance of developing primordia often results in bursts of regrowth.

Suggested Agar Culture Media: PDYA, MYA, OMA, or DFA.

1st, 2nd, and 3rd Generation Spawn Media: Grain spawn throughout.

Substrates for Fruiting: A wide variety of hardwoods (oak, alder, poplar, cottonwood, aspen, willow, birch, beech, etc.), and some softwoods (Douglas fir) although the latter are, in general, less productive. The supplemented sawdust formula on page 150 works well for many varieties of Enoki mushrooms. One advantage, however, of growing this mushroom on Douglas fir is that this wood has a natural resistance to green mold, especially *Trichoderma*. The pH range for fruiting falls between 4 and 5. Enokitake also grows on a wide variety of paper products.

GROWTH PARAMETERS

Spawn Run:
Incubation Temperature: 70–75°F (21–24°C)
Relative Humidity: 95–100%
Duration: 14–18 days
CO_2: >5,000 ppm
Fresh Air Exchanges: 0–1 per hour
Light Requirements: n/a

Primordia Formation:
Initiation Temperature: 40–50°F (4–10°C)
Relative Humidity: 95–100%
Duration: 3–5 days
CO_2: 2,000–4,000 ppm
Fresh Air Exchanges: 2–4 per hour
Light Requirements: 20–50 lux

Fruitbody Development:
Temperature: 50–60°F (10–16°C)
Relative Humidity: 90–95%
Duration: 5–8 days
CO_2: 2,000–4,000 ppm
Fresh Air Exchanges: 2–4 per hour
Light Requirements: 20–50 lux

Cropping Cycle:
Two to three crops, 10–12 days apart

Recommended Containers for Fruiting: Most automated Enoki farms utilize polypropylene bottles for ease of handling and speed of harvesting. A cylinder of plastic or paper is formed into a cylinder fitted within the open top of each bottle. (This causes the stems to grow long and facilitates harvesting.) If Enoki are grown in bags, the sidewalls of the bags should extend 6 inches above the plane of the fruiting surface to encourage the desirable elongated stems. The plastic walls are stripped down just prior to harvest.

Yield Potentials: Biological efficiency rating to 150%, the preponderance of which is stem mass. If

Enoki mushrooms fruiting from the book *The Mushroom Cultivator.*

grown in 1-liter bottles, yields of 3–5 ounces are standard for the first flush.

Harvest Hints: The difficulty of picking several hundred mushrooms, one by one, is daunting. If you stimulate the elongation of the stem through CO_2 elevation, cropping can be quickly accomplished. If culturing in bottles, firmly grasp the cluster and pull. With either method, trim any residual substrate debris off with a knife or a pair of scissors. Some strains of Enoki re-assimilate the damaged stem butts and form more primordia upon them for the second flush. If only minor cap development is allowed, and the mushrooms are picked before the gills mature, shelf life is greatly extended. However, some connoisseurs favor the flavor of the tender cap over the tougher, stringy stem.

Form of Product Sold to Market: Bunches of whitish to yellowish, long-stemmed mushrooms are usually shrink-wrapped and marketed in 3- or 5-oz. packages. Most of the Enoki available is brought into this country from Japan.

Bottle culture of Enokitake in Japan.

Nutritional Content: Variable, influenced by substrate components. Crude protein 17–31%; fat 1.9–5.8%; fiber 3.7%; ash 7.4%.

Medicinal Properties: The water-soluble polysaccharide *flammulin*, is 80–100% effective against Sarcoma 180 and Ehrlich carcinoma according to Ying (1987). An epidemiological study in Japan found a community of Enoki growers near the city of Nagano that had unusually low cancer rates. Frequent Enoki consumption was thought to be the cause. Ikekawa et al. (1968) reported anticancer activity from extracts of this mushroom. Zeng et al. (1990) and Qingtian et al. (1991) have published studies of the anti-tumor properties of this mushroom. The polysaccharide thought to be active is commonly referred to as FVP for *"Flammulina velutipes* polysaccharide."

Flavor, Preparation, and Cooking: This mushroom is surprisingly flavorful, *including* the stems, an opinion not shared by Singer (1986). Traditionally, Enoki is lightly cooked, served in soups or in stir-fries with vegetables, fish, and chicken. The stems are often left long, thus posing some interesting problems in swallowing. I prefer Enoki finely cut, almost diced, and then cooked at high heat for a short period of time. At a recent mycological society gathering, the addition of finely chopped Enoki to a cream sauce, stems and all, resulted in a *crème supérieur*.

Comments: This mushroom is the classic example of the influences light and carbon dioxide have on fruitbody formation. Like Oyster mushrooms, this mushroom's appearance is contingent upon the environment in which it was grown. The growing room environment can be tuned to elicit the perfect crop. Over time, experienced growers can orchestrate flushes with precision and generate cluster-bouquets of golden mushrooms. Properly managed, each bundle achieves a remarkably similar weight.

Under outdoor conditions (moderate light/low CO_2), this mushroom is short-stemmed with caps as wide as the stems are long. The lower regions of the stem develop darkened fuzz, hence the common name "The Velvet Foot." Under the lighted, high carbon dioxide conditions, the stems greatly

A Japanese Enokitake grower.

elongate and are yellow to white in color. The caps remain relatively small. While CO_2 determines the length of the stem, light is an overriding factor in influencing the formation and development of the cap. Thus under high CO_2 and no light conditions, thin stems may form usually without any caps. Most strains behave in this fashion but responses vary. Depending on the surrounding environment, the stems can be as short as 1 inch to as long as 12 inches. The cap-to-stem ratio varies from 1:1 to 1:100. The range in the shape of the fruitbody as a response to extreme environmental conditions is remarkable.

The surface mycelium undergoes a radical transformation during the period of pre-primordia formation. The mycelium yellows, and then forms dingy, blemished brown and white zones, which soon evolve into a roughened, beaded surface. From this micro-landscape, a high population of minute, squat, yellow primordia emerge. The mushrooms appear virtually stemless. If carbon dioxide levels are kept elevated, above 5,000 ppm, significant stem elongation continues. Japanese cultivators have invented the technique of fruiting in bottles that are topped with a cylindrical insert of clear plastic or paper. The cylinder pools carbon dioxide and the stems elongate. This technique encourages the formation of highly uniform flushes of mushrooms in each bottle.

For more information on the development of the mushroom strains in response to humidity levels, see McKnight (1985, 1990, 1992).

The task of picking Enoki mushrooms can be daunting.

The Clustered Woodlovers of the Genus *Hypholoma*

For cultivators, the genus *Hypholoma* (Fries) Kummer includes several interesting species, all of which thrive in cold weather, not producing when temperatures exceed 60–65°F (15–18°C). Aggressive wood decomposers, they share similar cultural requirements. And, they all produce a type of mycelium that is quite distinct from other saprophytes I have cultivated. Their uniquely beautiful mycelium is not only fantastically rhizomorphic, but luxuriously satin-like. After the mycelium has captured a substrate, a several-week resting period precedes primordia formation. If this resting period can be shortened, indoor cultivation may prove more commercially feasible. With current methods, the Hypholomas endure and proliferate in outdoor settings and fit perfectly within the Natural Culture models described in this book. I believe this genus of mushroom holds great promise for decomposing woody debris from logging and for reducing the risk of fuel load in the forest.

Hypholoma means "mushrooms with threads" because of the thread-like veil that connects the cap to the stem when young and for the bundles of rhizomorphs radiating outwards from the stem base. In North America, the name *Naematoloma* was used for years to delimit the fleshier species of this genus, following usage by Singer and Smith. However, *Hypholoma* has been officially conserved against *Naematoloma*, which means that only the name *Hypholoma* is proper to use.

Species of *Hypholoma* are closely related to *Psilocybe* and *Stropharia*. These genera belong to the family Strophariaceae (or subfamily Stropharioideae sensu Singer). They are distinguished from one another on the basis of microscopic features, features so subtle that many researchers have remarked on the usefulness of representing this group as one enveloping macro-genus. Since *Psilocybe* was published first, this name would officially take precedence.*

---

*For more information on these genera and their species, refer to my first book, *Psilocybe Mushrooms and Their Allies* (1978).

Hypholoma capnoides, the Smokey Gilled Woodlover, on conifer stump.

The Brown-Gilled Woodlover
Hypholoma capnoides (Fries) Quelet

Cultured mycelium of *H. capnoides* at 7 and 14 days. Note growth zones.

Introduction: Before the publication of my first book, I had a long-term affection for this mushroom but never attempted eating it until Elsie Coulter of Hayden Lake, Idaho, first told me that *H. capnoides* was her favorite edible mushroom. When a person with the depth of knowledge of an Elsie Coulter tells you a mushroom is her choice edible, you better listen! Another seasoned mushroom hunter rated this species as a 7 on a scale of 10.

A true saprophyte, *H. capnoides* is an aggressive conifer stump decomposer. One precaution is in order. *Hypholoma capnoides* is not a mushroom for those unskilled in mushroom identification. Several poisonous mushrooms resemble this mushroom and inhabit the same ecological niche. I can imagine how overly enthusiastic mycophiles, in their lust for delectable fungi, could mistake *Hypholoma fasciculare,* a toxic species, or *Galerina autumnalis,* a deadly poisonous mushroom sharing the same habitat, for *H. capnoides.* Cultivators should be forewarned that several mushroom species, inoculated or not, can inhabit a single stump or log. This danger is entirely avoided by honing identification skills, or by growing *H. capnoides* indoors on sterilized sawdust/chips.

Common Names: The Brown-Gilled Clustered
Woodlover
Smokey Gilled Hypholoma
Elsie's Edible

Taxonomic Synonyms and Considerations:
Hypholoma capnoides is known by many as
Naematoloma capnoides (Fr.) Karst. A sister species
to *H. capnoides* is *Hypholoma fasciculare* (Hudson
ex Fr.) Kummer [=*Naematoloma fasciculare* (Fr.)
Quelet], well worth knowing since it is poisonous!
These two mushrooms are sometimes difficult to tell
apart until the mushrooms are upturned and the
gills are examined. *H. capnoides* has smoky brown
gills whereas *H. fasciculare* has gills that are bright
greenish yellow to dingy yellow brown in age.
Furthermore, *H. fasciculare* is extremely bitter flavored whereas *H. capnoides* is mild. Once studied,
these two species can be separated without difficulty.

Description: Cap orange to orangish yellow to
orangish brown to dull brown, 2–7 cm broad at
maturity. Convex with an incurved margin, soon
expanding to broadly convex to almost flattened,
occasionally possessing an obtuse umbo. Cap margin often adorned with fine remnants of the partial
veil, soon disappearing, pale yellowish, becoming
buff yellow in age. Surface smooth, moist, and
lacking a separable gelatinous skin (pellicle). Gills
attached, soon seceding, close, white at first, soon
grayish, and eventually smokey grayish purple
brown in age. Stem 5–9 cm long, enlarged at the
base, covered with fine hairs. Partial veil cortinate,
sometimes leaving a faint annular zone, becoming
dusted purple brown with spores on the upper
regions of the stem. Usually growing in clusters.

Distribution: Widely distributed across North
America, particularly common in the western
United States. Also found throughout the temperate
regions of Europe, probably widely distributed
throughout similar ecological zones of the world.

Natural Habitat: A lover of conifer wood, especially Douglas fir, this mushroom is frequently found

on stumps or logs. I often find this mushroom, along
with other interesting relatives, in "beauty bark"
used for landscaping around suburban and urban
buildings. Although not reported on alder in the
wild, I have successfully grown this species on sterilized wood chips of *Alnus rubra*.

Microscopic Features: Spores purple brown in
mass, 6–7 × 4.0–4.5 µ, ellipsoid, smooth, with a
germ pore at one end. Cheilocystidia, pleurocystidia,
and clamp connections present. Context monomitic.

Available Strains: Strains are easily acquired by
cloning the cap context or the flesh adjacent to the
pith located at the stem base. Some culture laboratories list this mushroom, often under the name of
Hypholoma capnoides.

Mycelial Characteristics: Producing a white, silky,
rhizomorphic mycelium, usually exquisitely formed,
and growing out in distinct zonations. Mycelium
becomes overlaid with yellow tones in age but not
the rusty brown colorations that are typical of
Hypholoma sublateritium.

Fragrance Signature: A fresh, sweet, forest-like, pleasant fragrance, similar to *Stropharia rugosoannulata*.

Natural Method of Cultivation: Using nature as
a guide and applying the methods used for *H.
sublateritium*, I recommend cultivating this mushroom on stumps. This aggressive species is one of
the best for recycling millions of conifer stumps left
in the aftermath of logging. Clusters hosting dozens
of fruitbodies and weighing up to 4 pounds have
been collected. *Hypholoma capnoides* and *H. sublateritium* fruit for more than 6 years on the same
stumps and logs, even when the wood has lost its
bark, has become highly pulped, and falls apart
when kicked. Both species are ideal candidates for
mycological landscapes.

**Recommended Courses for Expansion of Mycelial
Mass to Achieve Fruiting:** Nutrified agar into liquid
fermentation for 48 hours. (The broth used for
fermentation should be fortified with 2–5 grams

of sawdust per liter.) Once fermented, the liquid-inoculum is transferred into sterilized grain that can be expanded two or three more generations. The grain spawn can inoculate sterilized sawdust/chips, but the spawn rate should not exceed 10% (moist spawn/moist sawdust). The sawdust/chip blocks can be fruited after a resting period, or be used for outdoor planting. The yield of this mushroom is not enhanced by bran-like supplements.

Suggested Agar Culture Media: MEA, PDYA, DFA, or OMYA.

1st, 2nd, and 3rd Generation Spawn Media: Grain spawn for the first two generations. Sterilized sawdust is best for the third generation. The sawdust can be fruited or used as spawn for outdoor inoculations.

Substrates for Fruiting: Unsupplemented alder, oak, or conifer (Douglas fir) sawdust, chips, logs, or stumps.

Recommended Containers for Fruiting: Bags, bottles, or trays.

Yield Potentials: .10–.25 pounds of mushrooms per 5 pounds sawdust. Yields should substantially improve with strain development and further experimentation. I have not seen reports on the cultivation of this mushroom anywhere in the literature.

Harvest Hints: Mushrooms should be harvested when the caps are convex. Since the stems elongate much in the same manner as Enokitake (*Flammulina velutipes*), the harvest method is similar. Outdoors, this mushroom forms clusters, often with several dozen mushrooms arising from a common base.

Form of Product Sold to Market: Not yet marketed.

Nutritional Content: Not known to this author.

Medicinal Properties: Given this species' woodland habitat and success in combating competitors, I think *H. capnoides* should be carefully examined for its antibacterial and medicinal properties.

Flavor, Preparation, and Cooking: Nutty and excellent in stir-fries. See recipes in Chapter 24.

GROWTH PARAMETERS

Spawn Run:
Incubation Temperature: 70–75°F (21–24°C)
Relative Humidity: 95–100%
Duration: 20–28 days (+ 20-day resting period)
CO_2: >10,000 ppm
Fresh Air Exchanges: 0–1 per hour
Light Requirements: n/a

Primordia Formation:
Initiation Temperature: 45–55°F (7–13°C)
Relative Humidity: 98–100%
Duration: 10–14 days
CO_2: 1,000–2,000 ppm
Fresh Air Exchanges: 1–2 per hour
Light Requirements: 200–500 lux

Fruitbody Development:
Temperature: 50–60°F (10–16°C)
Relative Humidity: 90–95%
Duration: 10–14 days
CO_2: 1,000–5,000 ppm
Fresh Air Exchanges: 1–2 per hour, or as required
Light Requirements: 200–500 lux

Cropping Cycle:
Two crops, 4 weeks apart

Comments: High-yielding strains of *Hypholoma capnoides* for indoor cultivation have not yet been developed. I hope readers will clone wild specimens, especially those forming unusually large clusters and screen for commercially viable strains. At present, this mushroom is better grown outdoors than indoors. *Hypholoma capnoides* is one of the few gourmet mushrooms adaptive to cultivation on conifer stumps and logs. Once sawdust spawn is implanted into fresh cuts via wedge or sandwich inoculation techniques, rhizomorphs soon form.

The Clustered Woodlover, *Hypholoma capnoides*, growing from a Douglas fir stump.

H. capnoides fruiting from sterilized hardwood sawdust.

Large-diameter stumps and logs have been known to produce crops every season for more than a decade. Agroforesters should carefully consider the judicious use of this fungus in designing polyculture models. This mushroom is one of the best candidates for decomposition of conifer stumps.

Indoor strategies closely mimic the methods used for Enokitake: Stems elongate in response to elevated carbon dioxide levels and cap development is influenced both by light and carbon dioxide. As a baseline, and until more information is accumulated, I recommend pursuing a parallel cultivation strategy.

Kuritake (The Chestnut Mushroom)
Hypholoma sublateritium (Fries) Quelet

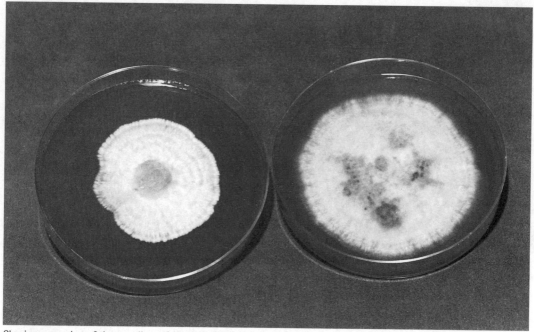

Classic progression of the mycelium of *H. sublateritium*. The mycelium is whitish at first, soon speckled, and eventually rusty brown in color.

Introduction: A favorite of Midwest mushroom hunters, great clusters of this species are often found on dead hardwoods, especially stumps, logs, and soils rich in wood debris. The Japanese have pioneered its cultivation using techniques similar to Shiitake. *Hypholoma sublateritium* is another excellent candidate for recycling stumps. In Japan, a preferred outdoor method is to inoculate hardwood logs that are then partially buried in shady, natural settings. This method has worked especially well for me.

Common Names: Kuritake (Japanese for "The Chestnut Mushroom")
Brick Top
Cinnamon Caps
Red Woodlover

Taxonomic Synonyms and Considerations: Synonymous with *Naematoloma sublateritium* (Fr.) Karsten.

Description: Cap 2–7 cm broad, hemispheric to convex, expanding with age to broadly convex, and eventually plane in age. Cap tan to brown to brick red, darker towards the center, and a lighter yellow near the margin. Margin incurved at first and covered with floccose remnants of the partial veil, soon straightening, and eventually uplifted at maturity. Flesh relatively thick, bruising yellowish.

GROWTH PARAMETERS

Spawn Run:

Incubation Temperature: 70–75°F (21–24°C)
Relative Humidity: 95–100%
Duration: 20–28 days (+ 14–28 day resting period)
CO_2: >10,000 ppm
Fresh Air Exchanges: 0–1 per hour
Light Requirements: n/a

Primordia Formation:

Initiation Temperature: 50–60°F (10–16°C)
Relative Humidity: 98–100%
Duration: 8–12 days
CO_2: 1,000–2,000 ppm.
Fresh Air Exchanges: 1–2 per hour
Light Requirements: 100–200 lux

Fruitbody Development:

Temperature: 50–60°F (10–16°C)
Relative Humidity: 90–95%
Duration: 7–14 days
CO_2: 1,000–5,000 ppm
Fresh Air Exchanges: 1–2 per hour or as required
Light Requirements: 100–200 lux

Cropping Cycle:

Two crops, 2 weeks apart

Gills close, bluntly attached to the stem, pallid at first, becoming dark purple gray when mature. Stem 5–10 cm long × 5–10 mm thick, solid, and covered with fibrillose veil remnants below the annular zone. This species often forms large clusters.

Distribution: Growing in eastern North America, Europe, and temperate regions of Asia (Japan and Korea).

Natural Habitat: Fruiting in the late summer and fall, primarily on the stumps of oaks, occasionally chestnut. Dr. Alexander Smith's comments that the largest specimens he found were "at the edge of an old sawdust pile in an oak-hickory woods" and that this species "is a highly prized esculent" should encourage cultivators in their pursuit of outdoor cultivation methods (Smith, 1949, pages 509–510).

Microscopic Features: Spores dark gray brown, 6.0–7.5 × 3.5–4.0 μ, smooth, ellipsoid, with a faint germ pore. Cheilocystidia, pleurocystidia, and clamp connections present. Context monomitic.

Available Strains: Strains are available from most culture libraries, including several from the American Type Culture Collection. ATCC #64244 is a good fruiting strain and is featured in this book.

Mycelial Characteristics: Mycelium white, cottony at first, soon linearly rhizomorphic with a silkish sheen. Soon after colonization of a 100 × 15-mm MYA petri dish, the mycelium develops zones of tawny or rusty brown discolorations emanating from the site of inoculation, which predominate in age. On sterilized sawdust this discoloration immediately precedes primordia formation.

Fragrance Signature: Pleasant, sweet, reminiscent of the refreshing fragrance from a newly rained-upon forest, similar to the scent from the King Stropharia (*Stropharia rugosoannulata*).

Natural Method of Cultivation: Hardwood logs can be pegged or sawdust-inoculated and laid horizontally side by side. Untreated sawdust can be used to bury the logs to $^1/_3$ their diameter. Oak, chestnut, and perhaps the stumps of similar hardwoods can be inoculated using any one of the methods described in this book.

Recommended Courses for Expansion of Mycelial Mass to Achieve Fruiting: When a sample of hardwood sawdust is added to the liquid fermentation broth, the mycelium grows more vigorously than without. After 3 to 4 days of fermentation, the liquid-inoculum should be distributed directly into sterilized grain. This grain spawn can be expanded several more generations or used directly for implantation into sterilized, supplemented sawdust. Oak sawdust is preferred. After 50 to 60 days

Inoculated logs of Kuritake, *Hypholoma sublateritium*, fruiting outdoors in a shaded location.

Kuritake fruiting on supplemented alder sawdust/chips.

from inoculation, the substrate can be initiated by lowering temperature to the prescribed levels.

Suggested Agar Culture Media: MYPA, OMYA, PDYA, or DFA.

1st, 2nd, and 3rd Generation Spawn Media: Grain spawn throughout, or alternatively sawdust spawn for the third generation. Plug spawn can be created from sawdust spawn.

Substrates for Fruiting: Supplemented alder, chestnut, poplar, hickory, cottonwood, or oak sawdust, logs, and stumps.

Recommended Containers for Fruiting: Bottles, bags, or trays.

Yield Potentials: .25 to .50 pound of fresh mushrooms per 5 pounds of moist, supplemented sawdust.

Harvest Hints: Cluster formation should be encouraged as well as stem elongation to facilitate

harvesting. Clusters can be firmly grasped and the base substrate trimmed off. Mushrooms are better presented as a "cluster-bouquet" than individually separated. The brilliant yellowish zone around the cap margin, in contrast to the reddish brown color, makes this mushroom highly appealing and aesthetically pleasing.

Form of Product Sold to Market: Fresh and dried. Some mycophiles in Midwest prefer to pickle this mushroom. This mushroom, known as Kuritake (the Chestnut Mushroom) in Japan, is primarily marketed in fresh form.

Nutritional Content: Not known to this author.

Medicinal Properties: Like *H. capnoides*, this mushroom is worthy of investigation for its potential medicinal properties. The only reference I have found is a short note in *Icons of Medicinal Fungi*, which states that "the inhibition rates of *Hypholoma sublateritium* (Fr.) Quel. against sarcoma 180 and

Ehrlich carcinoma (sic) is 60% and 70% respectively" (Ying 1987, page 325). The reference supporting this claim is in Chinese and lacks further elaboration.

Flavor, Preparation, and Cooking: Excellent in stir-fries or baked. Many of the recipes listed in this book can incorporate Kuritake.

Comments: The cultivation of this mushroom was pioneered at the Mori Mushroom Institute in Kiryu, Japan. Chestnut, oak, or similar logs are inoculated with sawdust or plug spawn and partially buried parallel to one another in a shady, moist location.

This mushroom should figure as one of the premier candidates for incorporation into model for sustainable mycopermaculture. It can be grown indoors on blocks of sterilized sawdust. Once these blocks cease production, they can be buried outside for additional fruitings. Another alternative is that the expired fruiting blocks can be broken apart and the resident mycelium can be used as sawdust spawn for implantation into stumps and logs.

Alexander H. Smith in *Mushrooms in Their Natural Habitats* (1959) notes that some strains of this species from Europe have been implicated in poisonings. I have seen no other reference to this phenomenon in North America or Japan, where this mushroom has long been enjoyed as a favorite edible. Nor have I heard of any recent reports that would suggest alarm.

Hypholoma sublateritium, Kuritake or the Cinnamon Cap, fruiting on alder logs inoculated with recycled sawdust spawn blocks.

The Beech Mushrooms of the Genus *Hypsizygus*

The rich flavor of the Shimeji mushroom can be summed up by the Japanese phrase "for fragrance, Matsutake; for flavor, Shimeji." However, the name Shimeji is widely used to describe some of the best Japanese gourmet mushrooms that grow "on wet ground." For years the general name Shimeji has been assigned to about twenty mushroom species, causing widespread confusion among amateur and professional mycologists.

A number of scientific articles have attempted to clarify what is the "true Shimeji," which the Japanese call Hon-shimeji. (See Clemencon and Moncalvo, 1990; Nagasawa and Arita, 1988.) The Hon-shimeji of Japan is actually a *Lyophyllum*, i.e., *Lyophyllum shimeji* (Kawam.) Hongo. This species is not commercially cultivated, and according to researchers at the Mori Mushroom Institute, may even be a mycorrhizal species (Motohashi, 1993).* The confusion is understandable because young specimens of *Lyophyllum shimeji* look very similar to *Hypsizygus tessulatus*, known in Japan as Buna-shimeji or the Beech Mushroom.

The genus *Hypsizygus* was first described by Rolf Singer and contains two excellent, edible and choice mushrooms. Collected in the wild by Native Americans (Singer, 1986), these mushrooms are otherwise not well known to other North Americans. Species in this genus are generally saprophytes but can become "facultative parasites" when trees, particularly elms and beeches, are dying from other diseases. These mushrooms have a tendency to grow high up on the trunks of trees, making the collecting of wild specimens difficult for the unprepared or unathletic. (*Hypsi*- means "on high or aloft" and -*zygus* means "yoke.") Only two species are known in this genus, *H. tessulatus* (Bull.:Fries) Singer and *H. ulmarius* (Bull.:Fries) Redhead. Both cause a brown rot of hardwoods.

Firmer fleshed and better flavored than most *Pleurotus* species, the *Hypsizygus* mushroom group out-classes the Oyster varieties commonly cultivated by North American and European growers in terms of flavor and texture. Here again, the Japanese are credited for first commercially cultivating *Hypsizygus*. Recent Japanese research shows that *H. tessulatus* is active in retarding tumor growth when consumed. Studies are ongoing to more precisely determine their medicinal properties.

The genus *Hypsizygus* most closely resembles *Lyophyllum* and to a lesser degree *Pleurotus*. These taxa are separated by the following combination of features: First, *Hypsizygus* and *Pleurotus* typically grow on wood, above ground level. *Lyophyllum* grows on the ground, in soils rich in woody debris. Furthermore, *Hypsizygus* species lack numerous granules within the basidia, a feature that is characteristic of members in the genus *Lyophyllum*. (These granules are siderophilous, i.e., the granules become apparent in acetocarmine, a stain used by mycologists to bring out internal cell features.) *Hypsizygus* spores are small, generally less than 7 μ, and more ovoid in shape compared to the spores of species in the genus *Pleurotus*, whose spores are more cylindrical and greater than 7 μ in length. Otherwise, mushrooms in these three genera closely resemble each other macroscopically, making cultivated mushrooms difficult to differentiate without the use of a microscope. To my eye, *Hypsizygus ulmarius* looks much more like an Oyster mushroom while *Hypsizygus tessulatus* resembles the terrestrially bound *Lyophyllum* species, especially when young.

For cultivators, another notable advantage of *Hypsizygus tessulatus* and *H. ulmarius* over *Pleurotus* species is their much reduced spore load. Oyster mushroom growers in this country might want to follow the lead of the Japanese in switching over to *Hypsizygus* cultivation for the many clear advantages.

---

* I find it more likely that *L. shimeji* depends upon soil microflora for fruitbody formation in the same manner as *Stropharia rugosoannulata* rather than it being a mycorrhizal species.

Buna-shimeji (The Beech Mushroom)
Hypsizygus tessulatus (Bulliard: Fries) Singer

H. tessulatus mycelium 4 and 8 days after inoculation onto malt extract agar.

Introduction: A delicious species, *H. tessulatus* falls under the umbrella concept of the Japanese Shimeji mushrooms. Firm textured, this mushroom is considered one of the most "gourmet" of the Oyster-like mushrooms. Recently, this mushroom has been attributed to having anti-cancer properties. Increasingly better known, this obscure mushroom compares favorably to *P. ostreatus* and *P. pulmonarius* in North American, European, and Japanese markets.

Common Names: The Beech Mushroom
Buna-shimeji (Japanese for
   "Beech Mushroom")
Yamabiko Hon-shimeji
   ("Mountain Echo
   Mushroom")
Tamo-motashi ("The Elm
   Oyster Mushroom")

Taxonomic Synonyms and Considerations: The delineation of this species has struggled through a taxonomic quagmire, resulting in a terribly confused history that has only recently been resolved. Originally published by Bulliard as *Agaricus tessulatus* in 1791 from Europe, this collection became the type for a new genus, the genus *Hypsizygus* as defined by Singer. In 1872, Peck described a paler

H. tessulatus 6 weeks after inoculation onto malt extract agar.

form as *Agaricus marmoreus* and Bigelow transferred this species to the genus *Hypsizygus*, naming it *Hypsizygus marmoreus* (Peck) Bigelow.

Redhead (1984) initially proposed synonymy between *H. tessulatus* (Bull. ex Fries) Singer and *H. ulmarius* (Bull.:Fr.) Redhead, which he has since reconciled into two separate, discrete taxa. *Hypsizygus tessulatus* is a synonym of both *Hypsizygus marmoreus* and *Pleurotus elongatipes*, the Long-Footed Oyster Mushroom (Redhead, 1986) and is *not* a synonym of *Hypsizygus ulmarius*. Because of the confusion associated with the application of the species name *H. tessulatus*, Japanese mycologists prefer to use the taxonomically "clean" name of *H. marmoreus* (Peck) Bigelow.

The name *tessulatus* refers to the water spots on the caps. *Marmoreus* means marbled, again in reference to the markings on the cap surface. *Elongatipes* refers to the ability of this mushroom to form long stems, especially promoted when the mushroom arises deeply from a cleft or wound in a tree. All these features are clearly expressed when this mushroom is grown indoors.

In Japan, *H. tessulatus* is marketed under the name of "Yamabiko Hon-shimeji" or just "Hon-shimeji." In the United States, this species is simply marketed under the name "Shimeji." Adding to the confusion, the name "Buna-shimeji" was first applied by Imazeki to *Lyophyllum ulmarius* (= *H. ulmarius*) but now is used in *exclusive* reference to *Hypsizygus marmoreus* (= *H. tessulatus*). (See Nagasawa and Arita, 1988.)

Description: Mushrooms hemispheric to plane, spotted to marbled with "water spots" on the cap, measuring 2–7 cm. Cap margin inrolled to incurved when young. Mushrooms dark tan, becoming gray tawny brown to creamy brown when mature. Gills bluntly attached to the stem, close, firm, and "wax-like." Stem thick, centrally or eccentrically attached, often tapering towards the base, with variable lengths depending on the method of cultivation.

Wild fruiting of *H. tessulatus* from cleft in a cottonwood tree.

Distribution: Throughout the temperate hardwood forests of Europe, Asia, and North America.

Natural Habitat: A saprophyte on elms, beech, cottonwoods, maple, willow, oak, and other hardwoods. I have found this mushroom arising from clefts in the trunks of dying cottonwoods. Like many Oyster mushrooms, this species can behave either as a facultative parasite on dying hardwoods or a true saprophyte on dead trees. (The wild fruiting featured to the left is from a dying cottonwood.)

Microscopic Features: Spores white, spherical to egg-shaped, relatively small, 4.0–6.5 × 3.5–5.0 µ. Clamp connections numerous. Hyphal system monomitic.

Available Strains: Strains are available from Japanese, American, and Canadian culture libraries. Strains range from white to gray to black varieties and are typically darker when young and/or when grown under cool conditions. Those who have patented strains of Hon-shimeji could be referring to *H. tessulatus*, *H. ulmarius*, or a *Lyophyllum* species, depending on the taxonomic system they were following.

Strains are easily cloned from wild specimens. DAOM #190991 is a light-colored Canadian strain, one of several strains featured in this book. (See page 253, top.) Strains vary in their duration to fruiting, in the color of the mushrooms at maturity, and in their response to carbon dioxide levels.

Mycelial Characteristics: Mycelium white, cottony, resembling *P. ostreatus* mycelium but not as aerial. Also, the mycelium of *H. tessulatus* does not exude the yellowish orange metabolite nor does it form the

Bottle culture of *H. tessulatus* (as *H. marmoreus*) in Japan. Note marbled caps, sometimes referred to as "water spots" by mycologists.

classically thick, peelable mycelium, two features that are characteristic of *Pleurotus* species.

Fragrance Signature: Sweet, rich, not anise-like, but oyster-esque, and pleasing.

Natural Method of Cultivation: Inoculation of logs or stumps. This mushroom can sometimes be grown outdoors in deep beds (>6 inches) of wood chips. Sawdust spawn is best used for inoculating outdoor beds. When sterilized, supplemented sawdust blocks have finished fruiting indoors, I recommend burying the blocks outdoors into sawdust, in shady settings, to promote fall fruitings.

Recommended Courses for Expansion of Mycelial Mass to Achieve Fruiting: The path of mycelial expansion that I recommend is to go from agar cultures into liquid fermentation, then into sterilized grain, then to sawdust and finally into supplemented sawdust.

Suggested Agar Culture Media: MYPA, PDYA, DFA, or OMYA

1st, 2nd, and 3rd Generation Spawn Media: The first two generations of spawn can be grain. The third generation can be sawdust or grain.

Substrates for Fruiting: Supplemented sawdust. Good wood types are oak, alder, beech, or elm. The effectiveness of other woods has not yet been established. From my experience, straw does not provide commercially viable crops unless inoculated up to 25% of its weight with sawdust spawn.

Recommended Containers for Fruiting: Bottles or narrowly opened bags are generally preferred so that stem elongation can be encouraged. (See page 257, bottom.) Open bag culture results in squat-looking mushrooms.

Yield Potentials: $1/2$ pound of fresh mushrooms per 5-pound blocks (wet weight) of supplemented hardwood sawdust/chips.

Harvest Hints: This mushroom is "waxy" when young, firm, and deliciously edible. The firmness of the flesh is gradually lost as the mushrooms enlarge. Mushrooms are best picked when the caps are still convex and the margin remains incurved.

GROWTH PARAMETERS

Spawn Run:
Incubation Temperature: 70–75°F (21–24°C)
Relative Humidity: 95–100%
Duration: 30–45 days
CO_2: >5,000 ppm
Fresh Air Exchanges: 0–1 per hour
Light Requirements: n/a

Primordia Formation:
Initiation Temperature: 50–60°F (10–15°C)
Relative Humidity: 98–100%
Duration: 7–12 days
CO_2: 500–1,000 ppm
Fresh Air Exchanges: 4–8 per hour
Light Requirements: 400–600 lux

Fruitbody Development:
Temperature: 55–65°F (13–18°C)
Relative Humidity: (85) 90–95%
Duration: 5–10 days
CO_2: 2,000–4,000 ppm
Fresh Air Exchanges: 2–4 per hour
Light Requirements: 400–600 lux

Cropping Cycle:
Two crops, 3 weeks apart

Form of Product Sold to Market: Fresh, dried, and powdered. Extracted fractions for cancer treatment may be available in the near future.

Nutritional Content: Not known.

Medicinal Properties: An unpublished report from the National Cancer Institute of Japan showed strong anti-tumor activity. Mice were implanted with Lewis Lung carcinoma and given aqueous extracts of the fresh mushrooms. At a dose equivalent to 1 gram/kilogram of body weight per day, tumors were 100% inhibited, resulting in total regression of the tumors. Control groups confirmed that, in absence of the mushroom extract, tumors were uninhibited

Bag culture of Buna-shimeji on supplemented alder sawdust and chips.

in their growth. No studies with human subjects have been reported.

Flavor, Preparation, and Cooking: A superior edible mushroom with a firm, crunchy texture and mildly sweet nutty flavor, this mushroom is highly esteemed in Japan. Versatile and delicious in any meal, *H. tessulatus* enhances vegetable, meat, or seafood stir-fries and can be added to soups or sauces at the last minute of cooking. The flavor dimensions of this mushroom undergo transformations difficult to describe. Buna-shimeji can be incorporated into a broad range of recipes.

Comments: A quality mushroom, Buna-shimeji is popular in Japan and is being intensively cultivated in the Nagano Prefecture. The only two mushrooms that come close to this species in overall quality are *H. ulmarius* or *Pleurotus eryngii*.

In the same environment ideal for Shiitake (i.e., normal light, CO_2 less than 1,000 ppm), my strains of *H. tessulatus* produce a stem less than 2 inches tall and a cap many times broader than the stem is long. When I reduce light and elevate carbon dioxide

Commercial cultivation of *H. tessulatus* (=*H. marmoreus*) in Japan using standard bottle technique.

levels, mushrooms metamorphosize into the form preferred by the Japanese. Here again, the Japanese have set the standard for quality.

In the growing room, abbreviated caps and stem elongation is encouraged so that forking bouquets emerge from narrow-mouthed bottles. Modest light levels are maintained (400 lux) with higher than normal carbon dioxide levels (>2,000 ppm) to promote this form of product. From a cultivator's point of view, this cultivation strategy is well merited, although the mushrooms look quite different from those found in nature. This cultivation strategy is probably the primary reason for the confused identifications. When visiting Japan, American mycologists viewed abnormal forms of *H. tessulatus*, a mushroom they had previously seen only in the wild, and suspected it to be a *Lyophyllum*.

Many of the strains of *H. marmoreus* cultivated in Japan produce dark gray brown primordia with speckled caps. These mushrooms lighten in color as the mushrooms mature, becoming tawny or pale woody brown at maturity. All the strains I have obtained from cloning wild specimens of *H. tessulatus* are creamy brown when young, fading to a light tan at maturity. Both have distinct water markings on the caps. The differences I see may only be regional in nature.

Although we now know that Hon-shimeji (i.e., true Shimeji) is not *H. tessulatus*, but is *Lyophyllum shimeji* (Kwam.) Hongo, habits in identification are hard to break. Many Japanese, when referring to cultivated Hon-shimeji, are in fact thinking of *H. tessulatus*.

This mushroom does not exude a yellowish metabolite from the mycelium typical of *Pleurotus* species. Petersen (1993) found that *H. tessulatus* produces a mycelium-bound toxin to nematodes, similar to that present in the droplets of *P. ostreatus* mycelium. This discovery may explain why I have never experienced a nematode infestation in the course of growing *Hypsizygus tessulatus*.

Given the number of potentially valuable by-products from cultivating this mushroom, entrepreneurs might want to extract the water-soluble anti-cancer compounds and/or nematicides before discarding the waste substrate.

For more information, consult Zhuliang and Chonglin (1992) and Nagasawa and Arita (1988).

Shirotamogitake (The White Elm Mushroom)
Hypsizygus ulmarius (Bulliard: Fries) Redhead

H. ulmarius mycelia at 4 and 8 days on malt yeast agar media.

Introduction: A relatively rare mushroom, which usually grows singly or in small groups on elms and beech, *Hypsizygus ulmarius* closely parallels the morphology of an Oyster mushroom but is far better in flavor and texture. Unique microscopic features qualify its placement into the genus *Hypsizygus*. Increasingly popular in Japan, *H. ulmarius* has yet to be cultivated commercially in North America where I believe it would be well received by discriminating markets. This mushroom can become quite large.

Common Names: The Elm Oyster Mushroom
Shirotamogitake (Japanese for "White Elm Mushroom")

Taxonomic Synonyms and Considerations: An Oyster-like mushroom, *Pleurotus ulmarius* (Bull. ex Fr.) Kummer, became *Lyophyllum ulmarium* (Bull.:Fr.) Kuhner and is most recently placed as *Hypsizygus ulmarius* (Bull.: Fr.) Redhead. *H. ulmarius* and *H. tessulatus* are closely related, living in the same ecological niche. *Hypsizygus ulmarius* is not as common, much larger, lighter in color, and has a flared, thin, uneven, and wavy margin at maturity. *H. ulmarius* is similar to an Oyster mushroom in form. *H. tessulatus* is smaller, stouter, with a thicker stem and a cap that is speckled with dark "water

H. ulmarius, Day 35 and 36 after inoculation into a gallon jar containing sterilized, supplemented alder sawdust/chips.

markings." To my eye, *H. tessulatus* is closer in form to a *Lyophyllum* while *H. ulmarius* looks more like a *Pleurotus*. (Please refer to the taxonomic discussion of *H. tessulatus* on page 249.)

Description: Mushrooms hemispheric to plane, sometimes umbellicate, uniformly tan, beige, grayish brown, to gray in color, sometimes with faint streaks, and measuring 4–15 cm. (This mushroom can become quite large.) Cap margin inrolled to incurved when young, expanding with age, even to slightly undulating. Gills decurrent, close, often running down the stem. Stem eccentrically attached, thick, tapering, and curved at the base. Usually found singly, sometimes in groups of two or three, rarely more.

Distribution: Throughout the temperate forests of eastern North America, Europe, and Japan. Probably widespread throughout similar climatic zones of the world.

Natural Habitat: A saprophyte on elms, cottonwoods, beech, maple, willow, oak, and occasionally on other hardwoods.

Microscopic Features: Spores white, spherical to egg-shaped, 3–5 µ. Clamp connections numerous. Hyphal system monomitic.

Available Strains: Strains are available from Japanese, American, and Canadian culture libraries. The strain featured in this book originated from Agriculture Canada's Culture Collection, denoted as DAOM #189249, produces comparatively large fruitbodies. (See above.) Patents have been awarded, both in Japan and in the United States, to a Japanese group for a particular strain of *H. ulmarius*, which produces a "convex cap."* (See Kawaano et al.,

---

* A mushroom strain producing a convex cap is hardly unusual, let alone patentable. During their life cycle, most gilled mushrooms progress from a hemispheric cap, to one that is convex, and eventually to one that is plane with age.

GROWTH PARAMETERS

Spawn Run:
Incubation Temperature: 70–80°F (21–27°C)
Relative Humidity: 95–100%
Duration: 14–21 days, + 7-day rest
CO_2: >10,000 ppm
Fresh Air Exchanges: 0–1 per hour
Light Requirements: n/a

Primordia Formation:
Initiation Temperature: 50–55°F (10–13°C)
Relative Humidity: 98–100%
Duration: 5–10 days
CO_2: <1,000 ppm
Fresh Air Exchanges: 4–8 per hour or as needed
Light Requirements: 500–1,000 lux

Fruitbody Development:
Temperature: 55–65°F (13–18°C)
Relative Humidity: (90) 94–98%
Duration: 4–7 days
CO_2: 600–1,500 ppm
Fresh Air Exchanges: 4–8 per hour or as needed.
Light Requirements: 500–1,000 lux

Cropping Cycle:
Two crops, 7-10 days apart

Fragrance Signature: Sweetly oysteresque with a floury overtone, not anise-like, but pleasant.

Natural Method of Cultivation: Inoculation of partially buried logs or stumps. I suspect that this mushroom will probably grow in outdoor beds consistently of a 50:50 mixture of hardwood sawdust and chips, much like the King Stropharia.

Recommended Courses for Expansion of Mycelial Mass to Achieve Fruiting: This mushroom adapts well to the liquid fermentation methods described in this book.

Suggested Agar Culture Media: MYPA, PDYA, OMYA, or DFA.

1st, 2nd, and 3rd Generation Spawn Media: Grain spawn throughout. For outdoor cultivation, sawdust spawn is recommended. Supplementing sterilized sawdust definitely enhances mycelial integrity and yields.

Substrates for Fruiting: Hardwood oak or alder sawdust supplemented with rye, rice, or wheat bran is preferred. I have had good results using the production formula of sawdust/chips/bran using alder as the wood base.

Recommended Containers for Fruiting: Polypropylene, autoclavable bags, bottles, and/or trays. This mushroom grows well horizontally or vertically.

Yield Potentials: My yields consistently equal $^1/_2$–1 pound of fresh mushrooms from 5 pounds of moist sawdust/chips/bran.

Harvest Hints: I especially like this mushroom when it grows to a fairly large size, but prior to maximum sporulation. I gauge the thickness at the disk as the criterion for harvesting, preferring at least 2 inches of flesh.

Form of Product Sold to Market: Fresh, sold only in Japan and Taiwan at present.

Nutritional Content: Not known to this author.

Medicinal Properties: I know of no published studies on the medicinal properties of *Hypsizygus ulmarius*. Anecdotal reports, unpublished, suggest

1990.) The method of cultivation described in the patent appears the same as for the cultivation of *H. tessulatus, Flammulina velutipes,* and *Pleurotus* species. Patenting a mushroom strain because it produces a convex cap would be like trying to patent a tree for having a vertical trunk, or a corn variety for producing a cylindrical cob.

Mycelial Characteristics: Mycelium white, cottony, closely resembling *P. ostreatus* mycelium. A few strains produce primordia on 2% MYA media.

H. ulmarius fruiting from bag of sterilized sawdust.

this mushroom is highly anticarcinogenic. Much of the research has been done in Japan. Mushrooms in this group, from the viewpoint of traditional Chinese medicine, are recommended for treating stomach and intestinal diseases.

Flavor, Preparation, and Cooking: The same as for most Oyster mushrooms. See recipes in Chapter 24.

Comments: *Hypsizygus ulmarius* is an excellent edible, ranking, in my opinion, above all other Oyster-like mushrooms. I find the texture and flavor of a fully developed *H. ulmarius* mushroom far surpasses that of the even the youngest specimens of *Pleurotus ostreatus* or *Pleurotus pulmonarius*. Oyster mushroom cultivators throughout the world would do well to experiment with this mushroom and popularize it as an esculent.

This mushroom is extraordinary for many reasons. When the caps grow to the broadly convex stage, lateral growth continues, with an appreciable increase in mass. (In most cultivated mushrooms, when the cap becomes broadly convex, this period signifies an end to an increase in biomass, and marks the beginning of reapportionment of tissue for final feature development, such as gill extension, etc.) If cultivators pick the mushrooms too early, substantial loss in yields results. Not only is *H. ulmarius*'s spore load substantially less than most *Pleurotus* species, but the quality of its fruitbody far exceeds *P. ostreatus, P. pulmonarius, P. djamor, P. cornucopiae,* and allies. In my opinion only *P. eryngii* and *H. tessulatus* compare favorably with *H. ulmarius* in flavor and texture.

The Shiitake Mushroom of the Genus *Lentinula*
Lentinula edodes (Berkeley) Pegler

Shiitake mushrooms (pronounced *shee ta' kay*) are a traditional delicacy in Japan, Korea, and China. For at least a thousand years, Shiitake mushrooms have been grown on logs, outdoors, in the temperate mountainous regions of Asia. To this day, Shiitakes figure as the most popular of all the gourmet mushrooms. Only in the past several decades have techniques evolved for its rapid cycle cultivation indoors, on supplemented, heat-treated, sawdust-based substrates.

Cultivation of this mushroom is a centerpiece of Asian culture, having employed thousands of people for centuries. We may never know who first cultivated Shiitake. The first written record of Shiitake cultivation can be traced to Wu Sang Kwuang who was born in China during the Sung Dynasty (A.D. 960–1127). He observed that, by cutting logs from trees that harbored this mushroom, more mushrooms grew when the logs were "soaked and striked." (See page 34.) In 1904, the Japanese researcher Dr. Shozaburo Mimura published the first studies of inoculating logs with cultured mycelium (Mimura, 1904, 1915). Once inoculated, logs produce 6 months to a year later. With the modern methods described here, the time period from inoculation to fruiting is reduced to only a few weeks.

Introduction: Log culture, although traditional in Asia, has yet to become highly profitable in

Mycelium of *L. edodes* 2, 10, 20, and 40 days after inoculation onto a malt extract agar (MEA) media.

North America—despite the hopes of many wood-lot owners. However, log culture does generate modest supplementary income and fits well within the emerging concept of mycopermaculture.

In contrast, indoor cultivation on sterilized sawdust-based substrates is proving to be highly profitable for those who perfect the technique. Most successful American growers have adapted the methods originating in Asia for the cultivation of this mushroom on sterilized substrates by doubling or tripling the mass of each fruiting block and by "through-spawning." The Japanese, Chinese, Taiwanese, and Thai production systems typically utilize cylindrically shaped bags filled with 1 kilogram of supplemented sawdust that are top-inoculated. This method gives a maximum of two flushes whereas the more massive blocks (2–3 kilograms apiece) provide four or five flushes before expiring.

The method I have developed, and which is illustrated in this book, gives rise to fruitings within 20 to 35 days of inoculation, two to three times faster than most cultivators achieve on sterilized substrates. This technique is fully described in the ensuing instructions.

Common Names: Shiitake (Japanese for "Shii Mushroom")
Golden Oak Mushroom
Black Forest Mushroom
Black Mushroom
Oakwood Mushroom
Chinese Mushroom
Shiangu-gu or Shiang Ku (Chinese for "Fragrant Mushroom")
Donku
Pasania

Taxonomic Synonyms and Considerations: Berkeley originally described Shiitake mushrooms as *Agaricus edodes* in 1877. Thereupon the mushroom has been variously placed in the genera *Collybia,*

When Shiitake mycelium appears at the end of a log, fruiting can be induced if the log is submerged in water for 24 hours, and then kept in a shaded, moist environment.

Oak logs incubating after inoculation with Shiitake plug spawn in uninsulated, shade-clothed greenhouses in Japan and in United States.

Armillaria, Lepiota, Pleurotus, and *Lentinus.* Most cultivators are familiar with Shiitake as *Lentinus edodes* (Berk.) Singer. Shiitake has recently been moved to the genus *Lentinula* by Pegler. A tropical *Lentinus* that forms large clusters, *Lentinus squarrosulus* grows on sterilized sawdust in a fashion similar to Shiitake. (See page 275.)

The genus *Lentinula* was originally conceived by Earle in the early 1900s and resurrected by Pegler in the 1970s to better define members formerly placed in *Lentinus.* White spores, centrally to eccentrically attached stems, gill edges that are often serrated, and distinct preferences for woodland environments characterize both genera. The genera differ primarily in microscopic features. The genus *Lentinula* is monomitic, i.e., lacking dimitic hyphae in the flesh, and have cells fairly parallel and descending in their arrangement within the gill trama. Members in the genus *Lentinus* have flesh composed of dimitic hyphae and have highly irregular or interwoven cells in the gill trama.

In 1975, Pegler proposed this species be transferred to *Lentinula.* Although Singer has disagreed with this designation, many taxonomists concur with Pegler (Redhead, 1993). Pegler believes Shiitake is more closely allied to the genera like *Collybia* of the Tricholomataceae family than to mushrooms like *Lentinus tigrinus,* the type species of the genus *Lentinus.* Furthermore, *Lentinus* shares greater affinities to genera of the Polyporaceae family where it is now placed rather than to other gilled mushrooms. This allegiance bewilders most amateur mycologists until the mushrooms are compared microscopically. Recent DNA studies support this delineation. Cultivators must keep abreast of the most recent advances in taxonomy so that archaic names can be retired, and cultures are correctly identified. For more information, consult Hibbett and Vilgalys (1991), Singer (1986), Redhead (1985), Pegler (1975, 1983).

Description: Cap 5–25 cm broad, hemispheric, expanding to convex and eventually plane at maturity. Cap dark brown to nearly black at first, becoming lighter brown in age, or upon drying. Cap margin even to irregular, inrolled at first, then incurved, flattening with maturity and often undulating with age. Gills white, even at first, becoming serrated or irregular with age*, white, bruising brown when damaged. Stem fibrous, centrally to eccentrically attached, fibrous, and tough in texture. Flesh bruises brownish.

Distribution: Limited to the Far East, native to Japan, Korea, and China. Until recently, not known from North America. Two sitings of wild Shiitake have been confirmed from Washington and California (Ammirati, 1997; Desjardin, 1998). These naturalized races have apparently escaped from cultivated Shiitakes and are adapting. Considering the continued deforestation of the Far East, where the genome of this mushroom appears increasingly endangered, these sitings may serve to protect the Shiitake genome.

Natural Habitat: This mushroom grows naturally on dead or dying broadleaf trees, particularly the Shii tree (*Castanopsis cuspidata*), *Pasania* spp., *Quercus* spp., and other Asian oaks and beeches. Although occasionally found on dying trees, Shiitake is a true saprophyte exploiting only necrotic tissue.

Microscopic Features: Spores white 5–6.5 (7) × 3–3.5 μ, ovoid to oblong ellipsoid. Basidia four-spored. Hyphal system monomitic. Pleurocystidia absent. Clamp connections and cheilocystidia present.

Available Strains: Many strains are available and vary considerably in their duration to fruiting, the formation of the fruitbodies and their adaptability to different wood types. ATCC #58742, ATCC #62987, and Mori #465 are all strains that fruit in 10 to 14 weeks after inoculation onto supplemented sawdust.

---

* Although taxonomists frequently refer to the even margin of Shiitake gills, as a grower, I see the Shiitake gill margins progressing from even to irregular edges as the mushrooms mature, especially on the first flush. I would hesitate to rely on this feature as one of taxonomic significance. It may be a strain-determined phenomenon.

After soaking, logs are placed in a young conifer forest which provides shade and helps retain humidity. Logs are watered via sprinklers twice a day. Mushrooms appear 1 to 2 weeks after removal from the soaking pond.

Shiitake forming on oak logs.

Agaricus blazei, Himematsutake or the Almond Portobello, at ideal stage for harvest. (Above)

Agaricus blazei fruiting on cased, leached cow manure compost. (Left)

Fully mature *Agaricus blazei*, also known as the Royal Sun Agaricus. (Below)

White form of the most widely cultivated mushroom in the world, *Agaricus brunnescens,* the Button Mushroom or "Blanco-Bellos," fruiting from cased compost. At this Dutch farm, the substrate is pulled onto each of the 200+ foot long trays via a net dragged by hydraulic winches. (Above)

Agaricus brunnescens, the Portobello Mushroom, fruiting from cased compost. (Left)

One initiation strategy sets the stage for three consecutive flushes of these Button mushrooms. Note rapidly developing "first-flush" mushrooms surrounded by resting primordia. Once these larger mushrooms mature, the young primordia awaken in waves for consecutive flushes over a period of a few weeks. Watering is minimized as touching mushrooms tend to get bacterial blotch. (Below)

Agrocybe aegerita, the Poplar or Pioppino Mushroom, growing from supplemented sawdust.

Flammulina populicola, a close relative of *Flammulina velutipes*, is native to the Rocky Mountains of North America, but prefers warmer temperatures.

Flammulina velutipes, the Winter Mushroom or Enokitake, growing from bottles of sterilized sawdust in Japan.

Hypholoma capnoides, the Smokey Gilled Woodlover, growing on sterilized sawdust. This aggressive mushroom fruits on many pines and hardwoods and is an excellent edible species. (Above)

Hypholoma sublateritium, the Red Brick Top or Kuritake, fruiting on sterilized sawdust. (Above right)

Hypholoma sublateritium fruiting from partially buried logs inoculated with sawdust spawn. (Below and right)

Harvesting *Lentinula edodes*, the Shiitake Mushroom. (Left)

Shiitake Mushrooms fruiting from blocks of supplemented, alder sawdust. (Above and below)

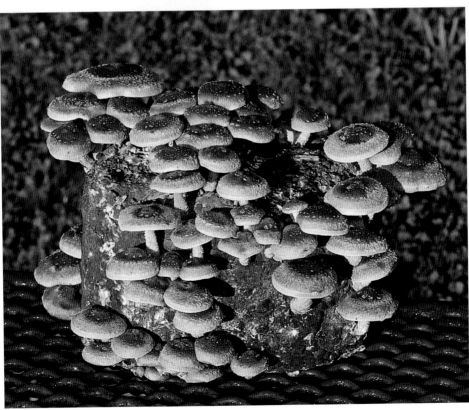

Commercial Shiitake *(Lentinula edodes)* cultivation at Stamets Farm in the United States.

Undersides of Shiitake mushrooms showing serrated gill edges.

Shiitake mushrooms fruiting on eucalyptus logs in Brazil.

Shiitake mushrooms fruiting on oak logs in the Pacific Northwest of the United States.

Pholiota nameko, at ideal stage for harvest, fruiting on hardwood sawdust. (Above)

Nameko growing from logs inoculated with sawdust spawn. (Right)

Nameko fruiting from maple and alder sections. Sawdust spawn is laid in between sections as they are stacked. (Below)

Column culture of the Golden Oyster Mushroom fruiting from pasteurized wheat straw in the United States. (Above)

Bottle culture of the Golden Oyster Mushroom *(Pleurotus citrinopileatus,* aka *Pleurotus cornucopiae)* fruiting from sterilized sawdust in Japan. (Top left)

Mature Golden Oyster Mushrooms. (Center and bottom left)

Pleurotus eryngii, the King Oyster Mushroom. (Top left)

Pleurotus djamor, the Pink Oyster Mushroom, fruiting from pasteurized wheat straw. (Above and below left)

The Pink Oyster Mushroom and its albino form. (Below)

The Old World European Blue Oyster, *Pleurotus ostreatus*, fruiting from pasteurized wheat straw. (Above left)

The New World American Beige Oyster, *Pleurotus ostreatus*, fruiting from espresso coffee grounds. (Above right)

Pleurotus tuberregium, the King Tuber Oyster Mushroom, a pantropical mushroom, fruits from a large sclerotium. (Below and right)

The type collection of *Psilocybe cyanofibrillosa* fruiting wildly from Douglas fir sawdust.

Psilocybe azurescens cultivated outdoors in hardwood chips.

Psilocybe azurescens ready for harvest. (Above left)

Sacred mushroom patches of *Psilocybe cyanescens* sensu lato, sometimes called Fantasi-takes, cultivated outdoors. (Above right and below)

LaDena Stamets with 5 pound specimens of *Stropharia rugoso-annulata*, the King Stropharia or Garden Giant Mushroom. (Above)

Typical fruiting. (Above right)

Young buttons of King Stropharia fruiting from cased, pasteurized wheat straw. (Right)

LaDena Stamets with King Stropharia at ideal stage for consumption. (Below)

Two pound specimen of King Stropharia. (Below right)

Fruitings of *Volvariella volvacea*, the Paddy Straw Mushroom, from rice straw.

The antler form of Reishi.

LaDena Stamets with organic Reishi *(Ganoderma lucidum)* fruiting from sterilized sawdust in the United States. (Above left)

Reishi grown from suspended bags in Thailand. (Above right)

Reishi grown in Japan on logs. (Below left)

Reishi grown in the southern United States from logs buried into pots and topped with soil. (Below right)

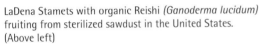

Three-quarter-pound and 1-pound clusters of *Grifola frondosa*, Maitake or Hen-of-the-Woods, fruiting from blocks of supplemented alder sawdust.

Outdoor fruitings of Maitake arising from blocks buried into soil.

Succulent bouquet of Maitake ready for consumption. (Above)

Mycologist Bill Chapman holding giant, wild Maitake. (Right)

Wild fruitings of Maitake from a stately oak tree in a colonial graveyard in upstate New York. (Below)

Mycologist David Arora, crazy about Maitake. (Below right)

Wild *Trametes versicolor* (synonymous with *Coriolus versicolor*), called Turkey Tail or Yun Zhi, the Cloud Mushroom or Kawaritake, growing on a conifer log in the Olympic National Park of Washington state in the United States. (Above)

Trametes versicolor is easily cultivated on sawdust—both hardwood and conifers. This mushroom is one of the most well-studied of all medicinal mushrooms. I cultivated this strain from my apple tree. (Left)

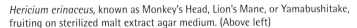

Hericium erinaceus, known as Monkey's Head, Lion's Mane, or Yamabushitake, fruiting on sterilized malt extract agar medium. (Above left)

Hericium erinaceus with characteristic spines. This mushroom possesses unique nerve-growth stimulant factors. (Above right)

Hericium erinaceus produces prodigiously on sterilized sawdust. In one week, a mushroom arises from a tiny hole (1–2 mm in diameter) in the bag to become a one-pound (1/2 kilogram) specimen. (Below right)

Yamabushitake growing in Japan using the bottle method. (Below left)

Auricularia polytricha, The Wood Ear Mushroom, is at the center of a complex of closely-related, easy-to-grow species.

Tremella fuciformis, White Jelly Mushroom, fruits when it parasitizes another fungus, *Hypoxylon archerii*, a wood saprophyte.

Polyporus tuberaster, the Canadian Tuckahoe Mushroom, can sprout from a star-like sclerotium to form edible mushrooms.

Lentinus squarrosulus is a fast growing cousin to Shiitake and fruits only 14 days from inoculation at 77–95°F (25–35°C).

Macrocybe crassa, fruiting from a sawdust block in Thailand, prefers 80–90°F (28–32°C).

Termitomyces robustus is one of the best of the edible mushrooms but defies human attempts at cultivation. So far only ants know the secret to growing this delicacy.

Nature is the source of all strains: collecting wild mushrooms in the Old Growth Rainforest, Upper Elwa River Valley, Washington State, U.S.A.

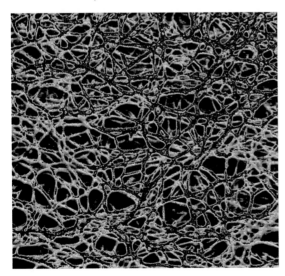

The interconnecting structure of the mushroom mycelium acts as a mycofiltration membrane to capture nutrients.

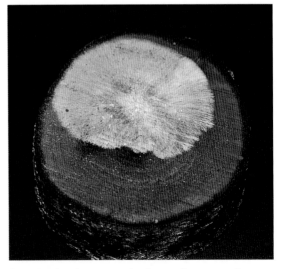

From the site of spore germination, mushroom mycelium spreads over the cut face of an alder "round."

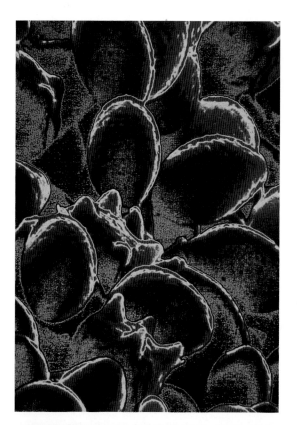

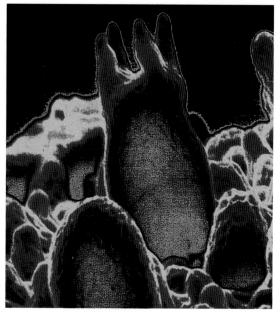

Young, developing basidia, before spore formation. (Above and left)

After meiosis, the fully mature, tetrapolar, mushroom basidium is ready to release spores in paired opposites with a force of 25,000 Gs. (Below)

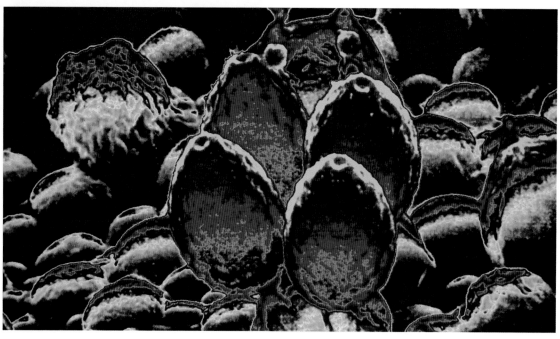

Shiitake mushrooms growing on alder log.

Shiitake mushrooms growing on oak logs.

I have developed a strain, designated Stamets CS-2, which originally came from China and produces in as quickly as 14 days from inoculation onto supplemented alder sawdust! I have found, however, that by forestalling primordia formation until 45 to 50 days after inoculation, the mycelium achieves greater tenacity, giving rise to better-quality fruitbodies. Using my methods in combination with this fast-fruiting strain, the Shiitake fruiting cycle is completed in 90 days from the date of inoculation, yielding three flushes and averaging 1.5–2 pounds of mushrooms per block. The first flush arises from blocks that are totally white in color. Volunteer fruitings on MEA media, on grain, and on sawdust in 3 weeks are characteristic of this unusual strain.

Mycelial Characteristics: Mushroom mycelium white at first, becoming longitudinally linear and cottony-aerial in age, rarely, if ever truly rhizomorphic. In age, or in response to damage, the mycelium becomes dark brown. Some strains develop hyphal aggregates—soft, cottony ball-like structures—that may or may not develop into primordia. Many mycologists classify this species as a white rot fungus for the appearance of the wood after colonization. However, the mycelium of Shiitake is initially white, soon becoming chocolate brown with maturity, leaving a white-pulped wood.

Fragrance Signature: Grain spawn has a smell similar to crushed fresh Shiitake, sometimes slightly astringent and musty. Sawdust spawn has a sweeter, fresh, and pleasing odor.

Natural Method of Cultivation: On hardwood logs, especially oak, sweetgum, poplar, cottonwood, eucalyptus, alder, ironwood, beech, birch, willow, and many other non-aromatic, broadleaf woods. The denser hardwoods produce for as long as 6 years. The more rapidly decomposing hardwoods have approximately half the life span, or about 3 years. The fruitwoods are notoriously poor for

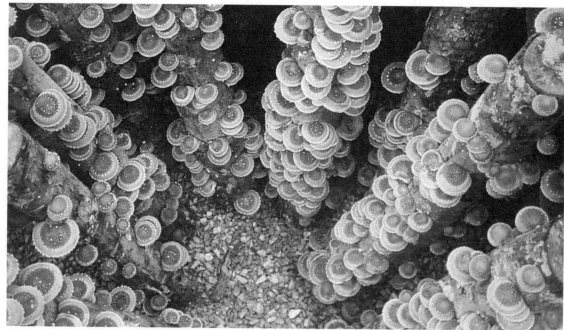

Shiitake growing on eucalyptus logs in Brazil.

growing Shiitake. Although Shiitake naturally occurs on oaks and beeches, the purposeful cultivation of this mushroom on hardwood stumps in North America has had poor success thus far.

For information on the cultivation of Shiitake on logs, see Kruger (1992), Fujimoto (1989), Przybylowicz and Donoghue (1988), Leatham (1982), Komatsu (1980, 1982), Kuo and Kuo (1983), and Harris (1986). Several studies on the economics of log cultivation have been published to date. Kerrigan (1982) published a short booklet on the economics of Shiitake cultivation on logs, which sought to show the profitability of Shiitake log culture. Gormanson and Baughman (1987) published an extensive study and concluded that profitability of growing Shiitake outdoors, as in Japan, was marginal at best. Roberts (1988) reviewed their statistical models and concluded that Shiitake cultivation on logs was not profitable. In the most recent study on log Shiitake industry in the United States, Rathke and Baughman (1993) concluded that when a production threshold of 4,000 logs/year was

Approximately 20 days after inoculation into sterilized sawdust, the surface topography of Shiitake blisters, a phenomenon many cultivators call "popcorning."

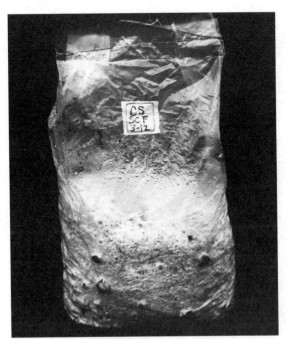

Twenty-five days after inoculation onto sterilized, supplemented alder sawdust, brown primordia form on the peaks of the blisters. When ten mushrooms form, I remove the plastic to allow unencumbered development. Note: Most methods for growing Shiitake on sterilized sawdust require 70 to 90 days before primordia are visible, and usually only after the blocks have become brown in color.

achieved by an *experienced* grower, the net profit on money invested, after costs, was a mere 5.76% return after-tax equivalent. Increasing production to 8,000 logs had no appreciable increase in profitability. Joe Deden, who once was the mushroom project manager at the Forest Resource Center outside of Lanesboro, Minnesota, has been instrumental in matching Shiitake strains with various woods, and bringing Shiitake log cultivation to the forefront of public awareness. I am cautiously optimistic that, with continued refinement of technology, profitability should increase, but believe log cultivation better fits economies of developing countries.

Presently, I believe a stand-alone Shiitake log operation in the United States can only be profitable if substantial subsidies are provided to offset the costs of materials and labor. However, this is not to say that Shiitake cultivation on logs is not attractive for those seeking a minor supplemental income, or for those, who simply enjoy cultivating mushrooms at home for their friends and family. Shiitake cultivation on logs may be a perfect example that "smaller is better."

Recommended Courses for Expansion of Mycelial Mass to Achieve Fruiting: Modern methods use a hardwood-based substrate, amended with a nitrogen-rich supplement (rice bran, wheat bran, rye bran, oatmeal, wheat flour, etc.). The mixture is moistened and packed into heat-tolerant plastic bags, sterilized, and inoculated with either grain or sawdust spawn. The Shiitake bags or "blocks" as they are commonly called are then placed into indoor, controlled-environment growing rooms. The first flush is initiated by simply elevating humidity and watering. Subsequent flushes are initiated by submerging the blocks into cold water for 24 to 48 hours. The fruiting cycle usually spans no more than 3 months.

This species adapts well to liquid culture techniques. I prefer to use malt agar media supplemented with 3–5 grams of alder sawdust. Once the cultures are grown out, they are blended in a stirrer, sub-proportioned into Erlenmeyer flasks containing malt sugar, yeast, and alder (2%, .1%, .5%, respectively), fermented for 2 days, and injected into sterilized rye grain. These liquid-inoculated grain masters are then used to inoculate sawdust for the creation of sawdust spawn which can then be used to inoculate the fruiting substrate: sawdust/chips/bran. For further information on Shiitake under liquid culture conditions, see Raaska (1990), Yang and Jong (1987), and Leatham (1983).

Suggested Agar Media: PDYA, MYA, and OMYA with the addition of .2% of the hardwood sawdust used in the production block.

1st, 2nd, and 3rd Generation Spawn Media: Rye, wheat, sorghum, or corn throughout for the first two generations. I recommend sawdust spawn for the third and final generation.

Blocks of Shiitake can be synchronized to produce a crop that matures simultaneously.

First-flush Shiitake forming from white blocks 35 days after inoculation.

GROWTH PARAMETERS:

Spawn Run:
Incubation Temperature: 70–80°F (21–27°C)
Relative Humidity: 95–100%
Duration: 35–70 days (strain-dependent)
CO_2: >10,000 ppm
Fresh Air Exchanges: 0–1 per hour
Light Requirements: 50–100 lux

Primordia Formation:
Initiation Temperature: 50–60°F
    (10–16°C)*
60–70°F (16–21°C)**
Relative Humidity: 95–100%
Duration: 5–7 days
CO_2: <1,000 ppm
Fresh Air Exchanges: 4–7 per hour
Light Requirements: 500–2,000 lux at
    370–420 nm

Fruitbody Development:
Temperature: 50–70°F (16–18°C)*
60–80°F (21–27°C)**
Relative Humidity: 60–80%
Duration: 5–8 days
CO_2: <1,000 ppm
Fresh Air Exchanges: 4–8 per hour
Light Requirements: 500–2,000 lux at
    370–420 nm***

Cropping Cycle:
Every 2–3 weeks for 8–12 (16) weeks

---

* Cold-weather strains; ** Warm-weather strains. Fluctuations of temperatures within these ranges are beneficial to the development of the mushroom crop. *** Light levels below 500 lux cause noticeable elongation of the stem.

Substrates for Fruiting: Broadleaf hardwoods such as oak, ironwood, sweetgum, beech, poplar, cottonwood, and alder.

The formula described on page 150 utilizing sawdust, chips, rice, or rye bran, and buffered with gypsum is ideal for high-yield, indoor production. At makeup this substrate hovers around 5.5 to 6.0. Prior to fruiting, the pH drops to 3.5 to 4.5. (The optimal range for fruiting, according to Chang and Miles, 1989, falls between 4.2 and 4.6.) Other recipes utilize a variety of supplements, including various grains. The cereal brans, most flours, tea leaves, yeast, molasses, etc., are widely used. For further information on formulating sawdust based media, consult Jong (1989), Royse and Bahler (1986), San Antonio (1981), and Ando (1974). The Forest Research Institute of New Zealand published one of the first studies exploring the usefulness of pines (*Pinus radiata*—the Monterey Pine), which produced satisfactorily yields when combined with a hardwood such as beech or poplar and supplemented with barley grain. (The ratio was 6 parts pine: 3 parts hardwood: 1 part grain.)

Recommended Containers for Fruiting: Polypropylene, high-density, thermotolerant polyethylene bags, usually fitted with a microporous filter patch, or stuffed with a cotton plug. Bottle fruitings are impractical. Tray fruitings *à la* Button mushroom culture have been employed with some success in Europe. However, the advantage of bag culture is that contaminants can be isolated, limiting cross-contamination of adjoining substrates.

Yield Potentials: 1.5–3 pounds of fresh mushrooms from 6 pounds of sawdust/chip/bran. Biological efficiency rating of 100–200% using the methods described herein.

Harvest Hints: Humidity should be constantly fluctuated during fruitbody development and then lowered to 60% RH for 6 to 12 hours before the crop is harvested. This causes the cap's leathery, outer skin to toughen, substantially extending shelf life. I prefer to pick the mushrooms when the margins are still inrolled, at a mid-adolescent stage. However, greater yields are realized if the fruitbodies

After each flush, the blocks sit dormant for 7 to 10 days, after which they are placed into a soaking tank and submerged in water for 24 to 36 hours.

After submersion, the blocks are spaced well apart and placed onto open-wire shelves. These same blocks on solid shelves would contaminate with green molds.

are allowed to enlarge. For best results, the growing room manager must carefully balance the interests of quality vs. yield throughout the cropping process.

Although these mushrooms can withstand a more forceful water spray than Oyster and other mushrooms, Shiitake gills readily bruise brownish, reducing quality. (Outdoor-grown Shiitake commonly has brown spots caused by insects. These damaged zones later become sites for bacterial blotch.) Mushrooms should be trimmed flush from the surface of the blocks with a sharp knife so no stem butts remain. Dead stems are sites for mold and attract insects. Thumbs should be wrapped with tape, or protected in some manner, as the pressure needed to cut through Shiitake stems is substantially greater than that of most fleshy mushrooms.

Form of Product Sold to Market: Fresh mushrooms, dried, powdered, and extracts. In Japan Shiitake wine, Shiitake cookies, and even Shiitake candies are marketed.

Nutritional Content: Protein 13–18%; niacin (mg/100 g): 55; thiamin (mg/100 g): 7.8; riboflavin (mg/100 g): 5.0. Ash: 3.5–6.5%. Fiber: 6–15%. Fat: 2–5%. Vetter (1995) found that the caps had 15.24% protein while stems had 11%.

Medicinal Properties: Lentinan, a water-soluble polysaccharide (ß-1,3 glucan with ß-1,6 and ß-1,3 glucopyranoside branchings) extracted from the mushrooms, is approved as an anticancer drug in Japan. The Japanese researcher Chihara was one of the first to publish on the anticancer properties of Shiitake, stating that lentinan "was found to almost completely regress the solid type tumors of Sarcoma 180 and several kinds of tumors including methylchloranthrene-induced fibrosarcoma in synergic host-tumour system" (Chihara, 1978, p. 809). The mode of activity appears to be the activation of killer and helper "T" cells.

Another heavyweight polysaccharide, called KS-2, isolated by Fujii et al. (1978), also suppressed

Sarcoma 180 and Ehrlich ascotes carcinoma in mice. Other protein-bound fractions have shown differing degrees of antitumor activity. Clearly, a number of anti-tumor compounds are produced in Shiitake besides the well-known lentinan.

In the past twenty years, more than a hundred research papers have been published on the chemical constituents of Shiitake and their health stimulating properties. In an early study (Sarkar, 1993), an extract from the cultured mycelium of Shiitake interrupted the replication of the type 1, herpes simplex virus. Ghoneum (1998) has shown that an arabinoxylane derivative from fermenting Shiitake is effective in slowing the HIV virus. In clinical trials at the San Francisco General Hospital, Gordon (1998) found that a combination of lentinan with didanosine (ddI) showed a mean increase of 142 CD-4 cells/mm3 over a 12-month period compared to a decrease in CD4 cells in patients treated with ddI alone. Odani et al. (1999) has isolated a novel serine proteinase inhibitor from the fruitbodies with a molecular mass of 15,999. In a study on the effect of a novel low molecular polysaccharide fraction on human cells, interleukin1 and apoptosis on human neutrophils decreased while increasing interleukin 1 and apoptosis in the monocytic (U937) human leukemia cells. (Sia et al., 1999). Yamamoto et al. (1997) found that Shiitake's mycelium produces a water soluble lignin-polysaccharide fraction, unique

from lentinan, which has potent anti-viral and immunopotentiating activities *in vitro* and *in vivo*. These studies confirm the medical significance of this species, and encourage further research.

In a series of clinical studies by Ghoneum et al (1994, 1995, 1996), patients afflicted with a variety of cancers, some with advanced malignancies, were treated with mycelially derived hemicellulose compounds and showed significant improvement. Ghoneum (1998) found that arabinoxylane, a fraction from the fermentation of Shiitake, Turkey Tail (*Trametes versicolor*), and the split gill mushroom (*Schizophyllum commune*), increased human NK activity by a factor of 5 in 2 months. Ghoneum's studies are the first clinical trials with cancer and mushrooms in the United States. Shiitake has also shown promise in lowering blood pressure. (Kabir and Yamaguchi, 1987; Jong et al., 1991.) Novel antibiotics have recently been isolated from Shiitake. (Hirasawa, 1999). The cholesterol-lowing compound was identified as eritadenine, an adenine derivative.

A very small percentage of individuals are allergic to Shiitake mushrooms, and a rare form of dermatitis, exacerbated by sunlight, has been reported in Japan (Hanada et al. 1998).

For more information on the medicinal properties of Shiitake, consult Mori et al. (1987), Fujii et al. (1978), Jong (1991), Ladanyi et al. (1993) and Jones (1995).

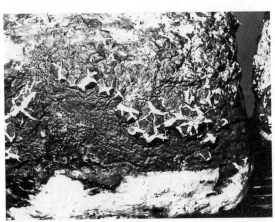

Two to three days after submersion, mushrooms form below the surface, cracking the outer brown skin.

Four days from the soak tank, mushrooms visibly extend beyond the surface plane of the sawdust block.

Flavor, Preparation, and Cooking: Shiitake can be enjoyed in a wide variety of dishes. A traditional Japanese soup recipe calls for slicing the mushrooms and placing them in a preheated chicken broth complemented with chopped green onions. The addition of miso (a vegetarian substitute for chicken) is also delicious. The Shiitake are steeped in this soup broth for a few minutes and served hot. The flavor and fragrance of slightly cooked Shiitake is tart and totally different than the flavor imparted from thorough cooking. Chinese restaurants usually rehydrate Shiitake and simmer them in the broth of stir-fries.

Our family regularly consumes Shiitake. Our favorite and standard method is to baste the gills with tamari and sesame oil, and then to bake or grill the mushrooms whole. Not only does this save time, but also the flavor of the Shiitake is potentiated through the absorption of the sauces in the gill tissue. Simply scrumptious! A standard way to cook Shiitake is to sauté the mushrooms. The stems are first cut off and the mushrooms are pulled apart, starting from the cut stem base. By tearing the mush-

rooms, cells are pulled apart along cell walls, preserving the flavor within. Canola or olive oil is added to the wok or frying pan that is then brought to high heat. Once hot, the torn mushrooms are added, stirred frequently, and cooked until the majority of water has been evaporated. While the mushrooms are being cooked, a distinct change in fragrance occurs, becoming more meat-like. Chopped onions, sliced walnuts or shaved almonds, and other condiments can be added as desired. This preparation can be used as a base in many dishes. By adding stir-fired Shiitake to steamed rice, fish, pasta, chicken, and vegetables, a culinary dish of extraordinary dimensions is created. Other dishes using this species includes Shiitake Quiche, Shiitake Pâté, etc. Please refer to the recipes in Chapter 24.

When I was an impoverished, near-starving student living in a remote, unheated A-frame house in the boondocks, Jeff Chilton generously delivered flats of fresh Shiitake, which I eagerly consumed, cooked and raw. (An act of generosity I will never forget.) To this day it is the only mushroom I enjoy

Seven days after submersion, crop is ready for harvest. This process can be repeated three to four times before the blocks cease production.

The "Donko" form of Shiitake can be induced by fluctuating humidity during primordia formation, especially during cool, dry conditions.

Occasionally, 1-pound mushrooms can be harvested from a 4-pound block of sterilized, supplemented sawdust.

without the benefit of cooking, fully aware that their potential nutritional contribution is largely untapped.

Comments: When we compare Shiitake to Oyster (*P. ostreatus*) mushrooms, several notable similarities and differences in their growth requirements are unveiled. Shiitake cannot be grown on the wide range of substrates that the highly adaptive Oyster mushrooms can exploit. Both are phototropic, with Shiitake primordia most stimulated by light exposure of 100–200 lux of green to ultraviolet at 370–420 nanometers (Ishikawa, 1967) while Oyster mushrooms maximally produce mushrooms at 2,000 lux at 440–495 nanometers (Eger et al., 1974). I find that although Shiitake primordia are stimulated into formation at this low light level, the development of the fruitbody is retarded unless light levels are increased. Since primordia formation can span a week, I prefer to give the blocks the higher exposure of light initially rather than risking malformation later on. Furthermore, Shiitake produces fairly normal-looking mushrooms under high carbon dioxide conditions (>2,000 ppm) while Oyster mushrooms deform with exaggerated stems and underdeveloped caps. Other notable distinctions are that Shiitake have a thicker cap, a distinct cap cuticle, a lower spore load, and a markedly longer shelflife than the Oyster mushroom.

The cultivation of Shiitake on sterilized, supplemented sawdust calls for a set of techniques very different than for most other mushrooms. (The formula for production is described on page 150.) Shiitake strains are abundant, most will produce, but a few are remarkably more aggressive than others. Exceptionally aggressive strains of Shiitake tend to be warm-weather races, tolerant of temperatures up to 90°F (32°C). *Employing a super-aggressive strain of Shiitake, propagating the mycelium according to the procedure outlined above, inoculating at a high rate, and using as the base medium a rapidly decomposing hardwood (red alder—Alnus rubra) have allowed me to accelerate the Shiitake life cycle*

Dr. Andrew Weil holding 5 pounds of freshly picked, organically grown Shiitake.

far faster than any that has been published to date. If the supplemented bags of sawdust are agitated 7 to 10 days after inoculation, fruitbody formation is triggered soon thereafter. This method causes fruitbody formation in as short as 14 days from inoculation.*

Early formation of Shiitake has disadvantages. If the network of mycelium is insufficiently formed, lacking both density and tenacity, high-quality mushrooms cannot be supported. If allowed 4–6 weeks of colonization, top-grade Shiitake is produced.

When the first crop is picked from the white blocks, they must be carefully cut flush with the

---

* Agitation of partially sterilized bags often results in a contamination bloom. These same bags would otherwise be completely colonized by the mushroom mycelium if left undisturbed. With sufficiently high spawning rates (10–20% wet weight spawn/dry weight substrate) secondary shaking post inoculation is unnecessary.

outer surface with a sharp knife or chunks of the sawdust substrate will be pulled off. I prefer to hold back the fruitings until 28 to 35 days after inoculation, allowing less than a dozen mushrooms on the first flush, and then exposing the substrate to the conditions recommended for crop development.

The first flush from white blocks is unique and calls for a strategy totally different than for subsequent flushes. Timing is critical. If one is not attentive, the window of opportunity can pass. During incubation, the outer surface of the myceliated sawdust appears as a smooth flat plane, pressed flush to the surface walls of the polypropylene bags. After a wait of 20 to 25 days until the blocks start "buckling"—an irregular, blister-like surface topography forms.* These formations are the precursors to primordia. (See page 264, bottom.) Several days after this surface topography forms, temperatures are dropped and small brown spots form at the peaks of the blisters. Often appressing against the interior plastic walls, the primordia can form overnight, measuring 1–3 mm in diameter. Should more than a dozen mushrooms form, or if they develop underneath the plastic, the crop quality greatly suffers. The cultivator must assess the maturity of the primordia population and expose the sensitive mycelium to the air precisely at the right time by stripping the plastic bags from the blocks.

The mycelium is suddenly thrust into the highly aerobic environment of the growing room. Massive evaporation begins from the newly exposed, aerial, fluffy white mycelium. For this first flush, which forms topically on the outer surface of the sawdust block, the humidity must be maintained at 100% under fog-like conditions until the desired number of primordia form. The sudden shift from the CO_2-rich environment within the bag to the highly aerobic environment of the growing room signals the block to bear fruit. At this stage, the Shiitake blocks are snow-white in color and dotted with several brown-headed primordia.

Since these events occur rapidly and the window of opportunity is so narrow, all the skills of the cultivator come into play. Allowing too many

---

* Some cultivators call this "blistering" or "popcorning."

primordia to form is a real problem. The more mushrooms that are set, the smaller they will be, increasing the labor at harvest. The fewer mushrooms set, generally the larger they will become. Despite the number of mushrooms that form, the yield remains constant! The first flush from a moist 6-pound alder sawdust/chip/bran is usually ³/₄ to 1 pound of mushrooms per block.

Once six to a dozen mushrooms form, relative humidity is lowered, and air turbulence is increased to affect greater evaporation. The aerial mycelium collapses, or in mushroom lingo, "pans." This flattened mycelium becomes the thickened coat of dead cells, eventually giving rise to the brown skin so characteristic of the remainder of the block's lifespan.

After the first flush, the fruiting blocks must dry out. The humidity in the growing room is lowered to 30–50% and maintained around 70°F (21°C).* After 7 to 10 days of dormancy, the now browning blocks weigh only 3–4 pounds of their original weight. The blocks are submerged in water (non-chlorinated) for 24 to 48 hours. [If the water temperature is 45–55°F (7–13°C), 48 hours is recommended. If the water temperature is above 60°F (15–16°C) then the blocks should not be submerged for more than 24 hours.] At our farm, the blocks are so buoyant as to necessitate extraordinary efforts to keep them submerged. When the number of blocks exceed 500, the process of handling becomes too labor-intensive. Some large-scale cultivators use winch-driven trolley cars on tracks that drive into the depths of soaking ponds, only to be ferried out the next day. These trolley cars then become the growing racks during the fruiting cycle.

When the blocks are removed from the soak tank, they should be placed directly back into the growing room onto open-wire shelves. During transport, a forceful spray of water removes any extraneous debris *and* cleans the outer surfaces. If the humidity is raised to 100% at this point in time, disaster soon results. Green molds (*Trichoderma* species) flourish. The constant, and at times, drastic

---

* This fruiting strategy is specific to warm-weather strains of Shiitake.

A Shiitake wine marketed in Japan.

fluctuations in humidity improve crop quality but discourage contamination. When Shiitake growers visit me, the most frequent remark I hear is that green molds are totally absent from the thousand or more blocks in my growing rooms. The absence of green molds is largely a function of how the growing rooms are operated on a daily basis; they are minimally influenced by air filtration. The key is to encourage Shiitake growth and discourage competitors by fluctuating humidity several times per day from 70–90%. The rapid evaporation off the surfaces of the blocks retards green mold contamination. Shiitake grows better at lower humidities than most mushrooms.

At least once, preferably twice a day, the blocks are washed with a moderately forceful spray of water. (Humidifiers are turned off.) Once the crop is watered, the floor is cleaned by hosing all dirt and debris into the central gutter where it is collected and removed. After this regimen, the room feels "fresh." Three days from soaking, white, star-shaped fissures break through the outer brown surface of the

Shiitake block. (See page 269.) The blocks are wafted with water every 8 to 12 hours. Since the humidifiers are set at 70–75%, they infrequently come on compared to the initiation strategy used with Oyster mushrooms.

One week after soaking, the crop cycle begins with the picking of the first mushrooms. Daily watering schedules are dictated by the crop's appearance. At maturity, the moisture content of the mushrooms must be lowered before picking, a technique that will greatly extend shelf life. After the harvest is completed, the blocks are dried out for 7 to 10 days, after which the resoaking process begins anew.

Harvesting Shiitake mushrooms from supplemented alder sawdust blocks.

This cycle can be repeated several more times. After five flushes, with an accumulated yield of 2–3 pounds of fresh mushrooms per 6–7 pounds of sawdust/chip/bran block, the Shiitake mycelium can produce, at most, rapidly maturing miniature mushrooms—few and far between. This is a sure sign that maximum yields have been achieved. Another way of determining whether the block is incapable of producing more mushrooms is to drop the block from waist level to a cement floor. A Shiitake block with good yield potential will strike the surface and not break apart. An expired block will burst upon impact. (A direct measure of mycelial fortitude!) As the mycelium loses vitality, the tenacity of the mycelial mat is also lost. At completion, the blocks are 1/2 to 1/3 of their original size and are often blackish brown in color. The "spent" blocks can now be recycled by pulverizing them back into a sawdust-like form. The expired Shiitake substrate is then resterilized for the sequential cultivation of Oyster, Maitake, Zhu Ling, or Reishi mushrooms. See Chapter 22 for more information.

Many supplements can be used to enhance Shiitake production from sterilized sawdust. Tan and Chang (1989) examined the effect of seventeen formulations on yield and found that a formula (dry weight) consisting of 71% sawdust, 18% used tea leaves, 7% wheat bran, and 1.4% calcium carbonate gave the highest yields. When calcium sulfate was substituted for calcium carbonate, no effect on yield was seen. In their opinion, spent tea leaves proved to be an excellent supplement for enhancing yields. Notably, only two strains of Shiitake were used, with only four replicates of each formula for the data trials. Another study by Morales et al. (1991) found that the addition of cotton waste to a sawdust/bran (12.5%) formula significantly improved yields of Shiitake.

These examples can be used as guidelines for Shiitake cultivators who will, inevitably, develop precise formulas and strategies that maximize yields according to their unique circumstances. My method consistently gives first flushes of Shiitake in 25 to 35 days, with subsequent flushes 10 to 14 days apart. My final flushes end 3 months after the first begin. For most other Shiitake growers on sterilized sawdust, their first flushes are just beginning after 3 months of incubation. (Apparently I am not alone in growing Shiitake so rapidly. Others have achieved similarly fast fruitings although no studies have been published.) (See Jong, 1993.)

The "secret" of this method is the culmination of a combination of factors: the use of an aggressive strain sustained on a unique agar media formula; a high spawning rate; a rapidly decomposing hardwood (alder); water rich in minerals; and the sensitive care of a good cultivator.

Similar to Shiitake, *Lentinus squarrosulus* shows promise as a new cultivar.

The Nameko Mushroom of the Genus *Pholiota*
Pholiota nameko (T. Ito) S. Ito et Imai in Imai

P. nameko mycelia 5 and 10 days after inoculation onto MEA media.

The genus *Pholiota* is not known for its abundance of deliciously edible species. Many species are characterized by a glutinous, slimy veil, which coats the surface of the cap making the mushrooms quite unappealing. However, the Japanese have discovered that one species in particular, *Pholiota nameko*, is a superior gourmet mushroom.

Introduction: *Pholiota nameko* is one of the most popular cultivated mushrooms in Japan, closely ranked behind Shiitake and Enokitake. This mushroom has an excellent flavor and texture. *P. nameko* would be well received by North Americans and Europeans if it were not for the thick, translucent, glutinous slime covering the cap. (This mucilaginous coat is common with many species of *Pholiota*.) Although unappealing to most, this slime soon disappears upon cooking and is undetectable when the mushrooms are served. My son and I have engaged in more than one culinary battle to get the last tasty morsels of this mushroom! This mushroom is a superb edible, which can be grown easily on sterilized, supplemented sawdust and/or on logs.

Nameko primordia form 21 days after inoculation on supplemented alder sawdust/chips.

The same block 48 hours later.

Common Names: Nameko or Namerako (Japanese for "Slimy Mushroom")
Slime Pholiota
Viscid Mushroom

Taxonomic Synonyms and Considerations: *P. nameko* is synonymous with *Pholiota glutinosa* Kawamura. Formerly placed in *Collybia* and *Kuehneromyces*, this mushroom is uniquely recognizable for its smooth cap and glutinous veil covering the mushroom. The type specimen of *Collybia nameko* T. Ito was found, upon reexamination, to be none other than *Flammulina velutipes*, although its original description by Ito obviously conformed to the mushroom we now call Nameko. Hence, the Latin name is burdened by interpretations by several mycologists. The reader should note that Nameko is a common name, once applied to many Japanese mushrooms with a viscid or glutinous cap. The common name has since become restricted to one species, i.e., *Pholiota nameko*.

Description: Cap 3–8 cm, hemispheric to convex, and eventually plane. Surface covered with an orangish, glutinous slime, thickly encapsulating the mushroom primordia, thinning as the mushrooms mature. The slime quickly collapses, leaving a viscid cap. Cap surface smooth. Gills white to yellow, becoming brown with maturity. Partial veil glutinous/membranous, yellowish, adhering to the upper regions of the stem or along the inside peripheral margin. Stem 5–8 cm long, equal, covered with fibrils and swelled near the base.

Distribution: Common in the cool, temperate highlands of China and Taiwan, and throughout the islands of northern Japan. Not known from Europe or North America.

Natural Habitat: On broadleaf hardwood stumps and logs in the temperate forests of Asia, especially deciduous oaks and beech (*Fagus crenata*).

Microscopic Features: Spores cinnamon brown, ellipsoid, small, 4–7 µ × 2.5–3.0 µ, smooth, and lacking a distinct germ pore. Pleurocystidia absent.

GROWTH PARAMETERS

Spawn Run:
Incubation Temperature: 75–85°F (24–29°C)
Relative Humidity: 95–100%
Duration: 2 weeks
CO_2: >5,000 ppm
Fresh Air Exchanges: 0–1 per hour
Light Requirements: n/a

Primordia Formation:
Initiation Temperature: 50–60°F (10–15.6°C)
Relative Humidity: 98–100%
Duration: 7–10 days
CO_2: 500–1,000 ppm
Fresh Air Exchanges: 4–8 per hour
Light Requirements: 500–1,000 lux

Fruitbody Development:
Temperature: 55–65°F (13–18°C)
Relative Humidity: 90–95%
Duration: 5–8 days
CO_2: 800–1,200 ppm
Fresh Air Exchanges: 4–8 per hour
Light Requirements: 500–1,000 lux

Cropping Cycle:
Two crops in 60 days, 10–14 days apart

Clamp connections are present. This mushroom also has a conidial stage that allows the formation of spores directly from the mycelium. Nameko is unique (in contrast to the other species listed in this book) in that a single spore can project a homokaryotic mycelium, and generate mushrooms with homokaryotic spores.

Available Strains: Strains are available from most Asian culture libraries.

Mycelial Characteristics: Whitish, longitudinally radial, becoming light orangish or tawny from

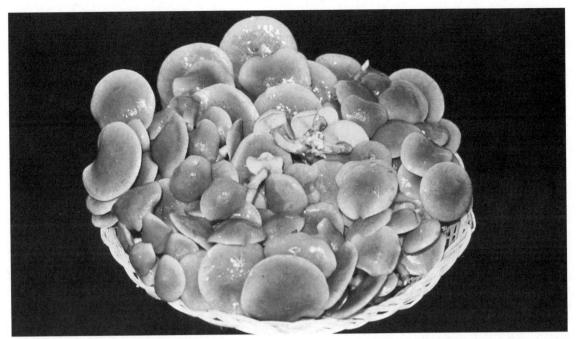

The harvested crop ready for cooking.

the center as the mycelium ages. On sterilized grain the mycelium is densely cottony white and becomes speckled with yellowish to orangish zones at maturity.

Fragrance Signature: Musty, farinaceous, not pleasant.

Natural Method of Outdoor Cultivation: On logs of broadleaf hardwoods, especially beech, poplar, and assorted deciduous oaks, *à la* the method for Shiitake. Because of the high moisture requirement for this mushroom, partially burying the logs in a moist soil base is recommended. Amazingly, I have had fruitings of this mushroom for more than 4 years on alder logs nestled into wood chips in the forest. This mushroom is an excellent decomposer of stumps, especially maple and oak.

Recommended Courses for Expansion of Mycelial Mass to Achieve Fruiting: Spawn is quickly generated with liquid-inoculation of grain from petri dish cultures. The grain spawn can be exponentially

expanded two generations using standard grain-to-grain inoculations. Intermediate sawdust spawn can then be produced from the grain spawn for final inoculation into supplemented hardwood sawdust. For many cultivators, going from grain spawn to supplemented sawdust is an easier and more direct approach.

Suggested Agar Culture Media: MEA, MYA, PDYA, and DFA.

1st, 2nd, and 3rd Generation Spawn Media: Grain spawn throughout.

Substrates for Fruiting: The supplemented sawdust formula described in Chapter 17 is recommended. Arita (1969) recommends that no more than 10% rice bran should be used as a supplement for oak hardwood formulations. However, I have found that 20% rice bran supports a more massive first flush and second flush when using *Alnus rubra* (red alder). Arita also found that the addition of 15% rice bran was the optimum if using conifer sawdust (*Pinus* *densiflora* or Asian Pine, and *Cryptomeria japonica* or Japanese cedar) as the base substrate. This is one of the few gourmet mushrooms that will give rise to substantial fruitings on a conifer wood.

Recommended Containers for Fruiting: Autoclavable bottles and polypropylene bags.

Yield Potentials: The fruitings pictured in this book yielded, on the first flush, an average of slightly more than 1 pound of mushrooms from 5-pound blocks of hardwood sawdust supplemented with rice bran.

Form of Product Sold to Market: Primarily fresh mushrooms are sold. An interesting, tasteful, if not elegant mushroom, *Pholiota nameko* is a mushroom well worth cultivating. Whether or not its marketing in North America will be successful is another question. *P. nameko* is *very* popular in Japan. Its flavor is so distinct and appealing as to win over the squeamishly skeptical.

Nutritional Content: Crude protein (N x 4.38): 20.8%; fat: 4.2%; carbohydrate: 66.7%; ash: 8.3%; and fiber: 6.3%. Vitamins (in milligrams per 100 grams dry weight): thiamin 18.8; riboflavin 14.6; niacin 72.9. Minerals (in milligrams per 100 grams dry weight): calcium 42; potassium 2,083; iron 22.9; sodium 63.

Medicinal Properties: According to Ying (1987), water and sodium hydroxide extracts of this mushroom are 60% and 90% effective, respectively, against Sarcoma 180 implanted in white mice. Furthermore, resistance to infection by *Staphylococcus* bacteria is substantially improved. This author knows of no parallel studies by Western researchers. The references making these medical claims are in Chinese.

Flavor, Preparation, and Cooking: This is a very slimy mushroom—a feature that has caused less consternation in Asia than in North America. Nameko is easily diced into miniature cubes and can be used imaginatively in a wide variety of menus, from stir-fries to miso soups. Nameko has a wonderful crunchy texture. Although pleasantly satisfying when lightly cooked, I prefer the strong nutty flavor that thorough cooking evokes. Once the glutinous slime is cooked away, the mushroom becomes very appetizing.

In Japan one can often find fresh Nameko for sale, while in the United States only canned mushrooms have been available, and only infrequently. Every fall, select Japanese restaurants in the United States feature this mushroom in a traditional autumn soup, which is clear and subtly seasoned with tiny bits of coriander leaf and other herbs.

Comments: This mushroom is more sensitive to moisture and carbon dioxide levels than most. For indoor cultivation, a precise initiation strategy is called for. I prefer not to use a casing layer as it promotes contamination, makes the cleaning of mushrooms tedious, and is unnecessary with good environmental controls in the growing room.

Should a casing layer not be applied, the block of supplemented sawdust must be exposed to a "condensing fog" environment during the primordia formation period. If the aerial mycelium suddenly dehydrates and dies back, surface primordia will be prevented and no crops will form. In this event, the cultivator must either roughen the surface of the block and/or apply a moist casing layer, two second-choice alternatives.

To initiate mushroom formation, lower temperatures to the 50–60°F (10–15°C) range, lower carbon dioxide levels, increase relative humidity to 98–100% rH, increase light levels to >500 lux, and mist the surface mycelium frequently with a fine spray of water. Approximately a week after initiating, orange streaks of slime form across the exposed surface of the mycelium. It is essential that the cultivator encourage the formation of this marmalade-looking goop. Soon thereafter, populations of primordia form and emerge within this overlaying, glutinous mass. So elastic is this material that it can be stretched more than 6 inches with each pull. This glutinous layer acts as a moisture bank promoting mushroom formation and development. Should this layer collapse due to dehydration, the primordia are at risk of aborting.

Rather than removing the entire polypropylene bag, I recommend that most cultivators cut off the top portion of the incubation bag, leaving 3- to 4-inch sidewalls of plastic surrounding the exposed, upper surface of the sensitive mushroom mycelium. These plastic walls will help collect moisture, enhancing primordia formation. If done properly, the mushroom stems will elongate to exactly the height of these walls, facilitating harvest.

The second flush is comparatively poor unless the surface is disturbed by roughening to expose viable mycelium below. A paddle with extruding nails or a wire brush serves this purpose well. Once the surface layer is torn apart, humidity is again raised to achieve the condensing-fog atmosphere. Soon thereafter (4 to 7 days), the mycelium becomes aerial, fuzzy, regenerates the orange slime layer, giving rise to another break of mushrooms. To achieve

a third flush, I recommend turning the block upside down, roughening its surface, and following a similar strategy to that described above. Fourth and fifth flushes are usually not substantial. However, I have put Nameko blocks into my "mushroom graveyard," letting nature prevail, and have been pleasantly surprised at getting more mushrooms. Be forewarned! For a slug, there can be no better feast than this slimy *Pholiota*!

The advantages of not using a casing layer are less work and less risk of green mold (*Trichoderma*) contamination, and most importantly, the harvested mushrooms are free of debris. Because of the glutinous nature of the *P. nameko* fruitbodies, casing debris readily adheres to and is difficult to remove from the harvested mushrooms and your fingers.

Similar to *Nameko* in its growth requirements, *Pholiota limonella* (= *Pholiota squarrosa-adiposa*) is a choice edible. This species is native to maple and alders. Some cultivators have mistakenly called this mushroom a *Hypholoma* species.

The Oyster Mushrooms of the genus *Pleurotus*

Oyster mushrooms are by far the easiest and least expensive to grow. For small cultivators with limited budgets, Oyster mushrooms are the clear choice for gaining entry into the gourmet mushroom industry. Few other mushrooms demonstrate such adaptability, aggressiveness, and productivity as these species of *Pleurotus*. Preeminent wood decomposers, *Pleurotus* species grow on a wider array of forest and agricultural wastes than any other mushroom group. They thrive on most all hardwoods, on wood by-products (sawdust, paper, pulp sludge), all the cereal straws, corn and corncobs, sugarcane bagasse, coffee residues (coffee grounds, hulls, stalks, and leaves), banana fronds, cottonseed hulls, agave waste, soy pulp, and numerous other materials containing lignin and cellulose. More than any other group of mushrooms, *Pleurotus* species can best serve to reduce hunger in developing nations, and to revitalize rural economies. To this end, worldwide Oyster mushroom production has surged in recent years, from 169,000 metric tons in 1987 to 909,000 in 1990.

What is most extraordinary about Oyster mushrooms is their conversion of substrate mass into mushrooms. Biological efficiencies often exceed 100%, some of the greatest, if not *the* greatest, in the world of cultivated mushrooms. In the course of decomposing dry straw, nearly 50% of the mass is liberated as gaseous carbon dioxide, 20% is lost as residual water, 20% remains as "spent" compost, and 10% is converted into dry mushrooms. (See Chapter 7 for an explanation of biological efficiency.) This yield can be also expressed as a 25% conversion of the wet mass of the substrate into fresh mushrooms. This formula is greatly affected by the stage at which the mushrooms are harvested.

On a dry-weight basis, Oyster mushrooms have substantial protein, ranging from 15–35% and con-

Growing Oyster mushrooms is not only a commercially viable enterprise—it can also be an art form. "Living" art by Zeger Reyers of the Netherlands.

tain significant quantities of free amino acids. They are replete with assorted vitamins such as vitamin C (30–144 mg per 100 grams), vitamin B, and niacin (109 mg per 100 grams). The variation in the reported nutritional analysis of Oyster mushrooms is due to several factors. Not only is the protein content affected by the type of substrate but also by the spawning medium and rate. Finally strains of *Pleurotus* vary in their nutritional composition and yield performances.

For more information on the nutritional properties of Oyster mushrooms, refer to the articles by El Kattan (1991), Rai et al. (1988), Bano and Rajarathnam (1982), and Gunde-Cimerman (1999).

Three notable disadvantages persist in the cultivation of Oyster mushrooms. Foremost is that the mushrooms are quick to spoil, presentable to the market for only a few days. (This can also be a good argument that local producers supply local markets.) Second, the spore load generated within the growing room can become a potential health hazard to workers. Sporeless strains, which tend to have short gills and are thicker fleshed, prolonging storage, are highly sought after by Oyster growers. Third, the grower must wage a constant battle against the intrusion of flies. Oyster mushrooms attract Sciarid and Phorid flies to a far greater degree than any other group of mushrooms described in this book. The flies swirl in a frenzied aerial dance around mature Oyster mushrooms aroused by spore release.

New strains of Oyster mushrooms are easy to acquire by cloning wild specimens. Most clones will grow to fruition in culture, with deformity of the fruitbody and excessive spore load being the most commonly encountered negative characteristics. Often, wild clones of Oyster mushrooms result in frenetically growing mycelium, replete with multiple sectors, and readily produce mushrooms on malt sugar agar media.

The colors of the Oyster mushrooms span the rainbow: white, blue, gray, brown, gold, and pink! Of all these, the high-temperature-tolerant *Pleurotus pulmonarius* is the easiest to grow. For flavor, the King Oyster, *Pleurotus eryngii*, reigns supreme. The Golden Oyster, *Pleurotus citrinopileatus*, and the Pink Oyster, *Pleurotus djamor*, are the most brilliantly colored. The Tree Oyster, *Pleurotus ostreatus*, is the most widespread throughout the hardwood forests of the world, which host the most diverse varieties from temperate climates. In sterile culture, the dimorphic *Pleurotus cystidiosus* is by far the most unique.

In growing Oyster mushrooms, several valuable by-products are generated. After the crop cycle is complete, the remaining substrate is rendered into a form that can be used as feed for cattle, chickens, and pigs. Using the spent straw as a nutritious food source could help replace the wasteful practice of feeding grain in the dairy and cattle industry. For more information on the applicability of "spent" straw from Oyster mushroom cultivation as fodder, please consult Zadrazil, 1976, 1977, 1980; Streeter et al., 1981; Sharma and Jandlak, 1985; Bano et al.,1986; and Calzada et al., 1987.

Feed is but one use of myceliated straw. In the end, the remaining myceliated substrate mass is an excellent ingredient for building composts and new soils. The waste straw may yield another by-product of economic importance: environmentally safe but potent nematicides. At least five Oyster mushroom species secrete metabolites toxic to nematodes (Thorn and Barron, 1984; Hibbett and Thorn, 1994). Lastly, the waste straw remains sufficiently nutritious to support the growth of *Stropharia rugosoannulata* outdoors. Additional products could include the recapturing of considerable quantities of enzymes secreted in the course of straw decomposition.

From a taxonomic point of view, the genus *Pleurotus* has been hard to place. Singer (1986) throws the genus into Polyporaceae family along with *Lentinus*. Others have suggested the genus belongs to the Tricholomataceae. However, the more the genus *Pleurotus* is studied, the more discrete this group appears. Until DNA studies indicate otherwise, I am following Watling and Gregory (1989) who place the genus *Pleurotus* into their own family, the Pleurotaceae. Hilber (1997) and Thom et al. (2000) recently published the most updated treatises on the taxonomy of *Pleurotus*.

The Golden Oyster Mushroom
Pleurotus citrinopileatus Singer

Introduction: Few mushrooms are as spectacular as this one. Its brilliant yellow color astonishes all who first see it. This species forms clusters hosting a high number of individual mushrooms, whose stems often diverge from a single base. Its extreme fragility post-harvest limits its distribution to faraway markets. Spicy and bitter at first, this mushroom imparts a strong nutty flavor upon thorough cooking. *Pleurotus citrinopileatus* grows quickly through pasteurized straw and sterilized sawdust, and thrives at high temperatures.

Common Names: The Golden Oyster
Mushroom
Tamogitake (Japanese)
Il'mak (Russian term for
"Elm Mushroom")

P. citrinopileatus mycelium 5 days after inoculation onto malt extract agar medium.

Taxonomic Synonyms and Considerations: *Pleurotus citrinopileatus* is closely allied to *Pleurotus cornucopiae* (Paulet) Roll. and is often considered a variety of it. Moser (1978) and Singer (1986) described *P. cornucopiae* var. *cornucopiae* as having a tawny brown cap whereas *P. citrinopileatus* has an unmistakably brilliant yellow pileus.

Singer (1986) separated *P. citrinopileatus* Singer from *P. cornucopiae* (Paulet ex Fr.) Rolland sensu Kuhn. and Rom. (=*P. macropus* Bagl.) on the basis of the arrangement of the contextual hyphae. According to Singer *P. citrinopileatus* has monomitic hyphae whereas *P. cornucopiae* has dimitic hyphae, a designation that has caused considerable confusion since he used this feature as a delineating, subgeneric

distinction.* Upon more careful examination, Parmatso (1987) found that the context was distinctly dimitic, especially evident in the flesh at the stem base. This observation concurs with Watling and Gregory's (1989) microscopic observations of *P. cornucopiae*.

Hongo (1976) describes the Golden Oyster mushroom as a variety of *P. cornucopiae*, i.e., *Pleurotus cornucopiae* (Paulet ex Fries) Rolland var. *citrinopileatus* Singer. Petersen's (1993) interfertility studies showed a culture of *P. citrinopileatus* from China was indeed sexually compatible with *P. cornucopiae* from Europe. From my own experiences, the golden color of *P. citrinopileatus* can be cultured out, resulting in a grayish brown mushroom closely

---

*Singer first collected *P. citrinopileatus* when fleeing German forces during World War II. He traveled east, across Asia, and during his travels found the Golden Oyster mushroom. Dried samples were brought to the United States for study years later. This contradiction in the arrangement of the contextual hyphae may simply be a result of poor specimen quality. Contextual hyphae are more easily compared from tissue originating near the stem base than from the cap. Hence, such confusion is not uncommon when examining old and tattered herbarium specimens.

P. citrinopileatus mycelium 2 and 10 days after inoculation onto malt extract agar media.

conforming, macroscopically, to *P. cornucopiae* var. *cornucopiae*.* Geographically, *P. citrinopileatus* is limited to Asia whereas *P. cornucopiae* occurs in Europe. Neither has yet been found growing wildly in North America. With the onset of commercial cultivation of these mushrooms adjacent to woodlands in North America, it will be interesting to see if these exotic varieties escape. In this book, I am deferring to the use of *P. citrinopileatus* rather than *P. cornucopiae* var. *citrinopileatus*.

Description: Caps golden to bright yellow, 2–5 cm, convex to plane at maturity, often depressed in the center, thin fleshed, with decurrent gills that show through the partially translucent cap flesh. Stems white, centrally attached to the caps. Usually growing in large clusters arising from a single, joined base. Clusters are often composed of fifty to one hundred

or more mushrooms. As strains of this species senesce, the yellow cap color is lost, becoming beige, and fewer mushrooms are produced in each primordial cluster.

Distribution: Native to the forested, subtropics of China, southern Japan, and adjacent regions.

Natural Habitat: A saprophyte of Asian hardwoods, especially oaks, elm, beech, and poplars.

Microscopic Features: Spores pale pinkish buff, 7.5–9.0 × 3.0–3.5 μ. Clamp connections present. Hyphal system dimitic.

Available Strains: Strains of this mushroom have been difficult to acquire in North America. While traveling through China in 1983, I made clones of Chinese mushrooms. Using a Bic lighter and a small scalpel, I inoculated ten test tube slants without the

---

*Curiously, when the strain loses its golden color through continued propagation, the bitter flavor is also lost.

Bottle culture of *P. citrinopileatus* in Hokkaido, Japan.

benefit of any laboratory facility. Only one or two were pure. One of those clones survived the return trip. This is the strain prominently featured here.

Mycelial Characteristics: Cottony, whitish mycelium, often with tufts of dense growth, sometimes with yellowish tones, and occasionally run through with underlying rhizomorphic strands. Primordia are yellow at first, especially from strains kept close to their natural origins. Mycelium dense on grain. Colonization of bulk substrates at first wispy, becoming denser with time. This mushroom casts a much finer mycelial mat at first than, for instance, *Pleurotus ostreatus* or *P. pulmonarius* on wheat straw.

Fragrance Signature: Grain spawn smells astringent, acrid, nutty, and sometimes "fishy," with a scent that, in time, is distinctly recognizable to this species.

Natural Method of Cultivation: This species grows on logs and stumps, especially of *Ulmus* and *Carpinus* species much like *P. ostreatus*. Hilber

(1982) reported that, per cubic meter of elm wood, the yield from one season averaged 17–22 kilograms! Also grown on cottonseed hulls, sugarcane bagasse, straw, and sawdust in China. In the United States wheat straw or hardwood sawdust are most frequently employed for substrate composition.

Recommended Courses for Expansion of Mycelial Mass to Achieve Fruiting: Grain spawn sown directly into sterilized sawdust or pasteurized substrates. The generation of intermediate sawdust spawn is not deemed necessary. Straw inoculated with grain spawn has substantially greater yields than straw inoculated with sawdust spawn.

Suggested Agar Culture Media: MYA, MYPA, or PDYA.

1st, 2nd, and 3rd Generation Spawn Media: Rye, wheat, sorghum, milo, or millet.

Substrates for Fruiting: Pasteurized wheat, cottonseed hulls, chopped corncobs, and hardwood

GROWTH PARAMETERS

Spawn Run:

Incubation Temperature: 75–85°F (24–29°C)
Relative Humidity: 90–100%
Duration: 10–14 days
CO_2: 5,000–20,000 ppm.
Fresh Air Exchanges: 1–2 per hour
Light Requirements: n/a

Primordia Formation:

Initiation Temperature: 70–80 (90)°F
  (21–27 (32)°C)
Relative Humidity: 98–100%
Duration: 3–5 days
CO_2: <1,000 ppm
Fresh Air Exchanges: 4–8
Light Requirements: 500–1,000 lux

Fruitbody Development:

Temperature: 70–85°F (21–29°C)
Relative Humidity: 90–95%
Duration: 3–5 days
CO_2: <1,000 ppm
Fresh Air Exchanges: 4–8
Light Requirements: 500–1,000 lux

Cropping Cycle:

Two crops, 10–14 days apart

sawdusts. Alternative substrates being developed commercially are sugarcane bagasse, paper by-products, banana fronds, and peanut hulls. Every part of the coffee plant can be recycled growing Oyster mushrooms—from the coffee grounds, hulls, stalks, limbs, and leaves!

Recommended Containers for Fruiting: Perforated plastic columns, bags, trays, and bottles.

Yield Potentials: This species is not as prolific as the more commonly cultivated *P. ostreatus* and *P. pulmonarius* in the conversion of substrate mass to mushrooms. After the second flush, comparatively few mushrooms form. Biological efficiency rating:

25–75% indoors on wheat straw. Yield efficiencies are higher on cottonseed-amended substrates.

Harvest Hints: Since picking individual mushrooms is tedious and often damages the fragile fruitbodies, cultivators should pursue strategies that encourage clusters hosting large numbers of young mushrooms. Marketing of clustered bouquets is far easier than selling individual mushrooms.

Form of Product Sold to Market: Fresh and dried mushrooms. (The golden color fades in drying.) This mushroom is especially popular in Asia.

Nutritional Content: Not known to this author. This mushroom is likely to have a similar nutritional profile as *P. ostreatus*.

Medicinal Properties: According to Ying (1987) in *Icons of Medicinal Fungi*, *P. citrinopileatus* potentially cures pulmonary emphysema. The supportive references are in Chinese. This mushroom is likely to be similar to *P. ostreatus* in its cholesterol-reducing properties.

Flavor, Preparation, and Cooking: Mushrooms are better broken into small pieces and stir-fried, at high heat for at least 15 to 20 minutes. This mushroom is extremely bitter and tangy when lightly cooked, flavor sensations pleasant to few and disdained by most. However, when they are crisply cooked, a strong, appealing cashew-like flavor eventually develops. This progression of flavors, primarily affected by the duration of cooking, underscores the Golden Oyster mushroom's versatility as an esculent. This mushroom is becoming increasingly popular as a garnish in salads.

Comments: An eye-stopper, the Golden Oyster Mushroom is one of the most spectacular of all gourmet mushrooms. When strains of this mushroom are overcultured, the golden color is one of the first features to be lost. When fruiting this mushroom, the brightness of the gold cap color is directly related to the intensity of light in the growing room. A high temperature tolerant mushroom, primordia will not form below 60–65°F (16–18°C). Coupled with the brevity of time between spawning and fruiting, is its fondness for cottonseed hulls; this

Column culture of *P. citrinopileatus*.

A beautiful clustered bouquet of the *P. citrinopileatus*.

Harvested bouquets of *P. citrinopileatus* ready for retail packaging.

mushroom is better suited for cultivation in warmer climates of Asia, the southern United States, or Mexico, or during the summer months in temperate regions. Its penchant for forming clusters, which I call "Golden Gourmet Bouquets," makes harvesting easy and prevents damage to individual mushrooms.

P. citrinopileatus does, however, have some limitations that should be carefully considered before embarking on large-scale commercial cultivation. *Pleurotus citrinopileatus* is extremely fragile, easily breaking if mishandled, especially along the thin cap margin, complicating long-distance shipping. The fruitbodies quickly lose their bright yellow luster subsequent to harvest. Higher spawning rates (15–20% fresh spawn/dry substrate) are required to assure the full colonization of most pasteurized materials. And, cropping yields are not nearly as good compared to other *Pleurotus* species. However, its rarity, striking gold color, and broad range of flavors make this species uniquely marketable and pleasurable to grow.

Fungi Perfecti's Golden Oyster Mushroom Kit.

The Abalone Mushroom
Pleurotus cystidiosus O. K. Miller

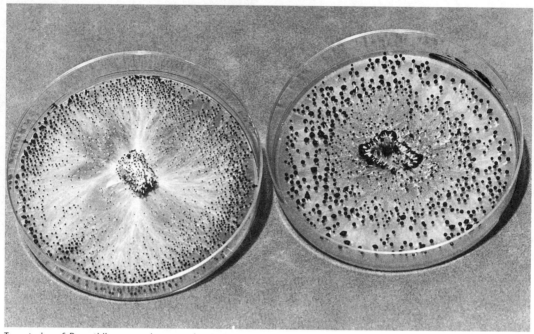

Two strains of *P. cystidiosus* growing on malt extract agar media.

Introduction: This mushroom and its close allies have a unique biology and are by far the most interesting of all the Oyster mushrooms. The asexual stage appears to be a "contaminant" to most cultivators. In fact, species in this group are dimorphic—having a sexual and asexual life cycle. A nearly identical species, *Pleurotus abalonus*, is commercially cultivated in Asia, particularly Taiwan and Thailand. Dr. Orson K. Miller first described *Pleurotus cystidiosus* in 1969 from a maple in Indiana.

Common Names: The Abalone Mushroom
The Maple Oyster Mushroom
Miller's Oyster Mushroom

Taxonomic Synonyms and Considerations: *Pleurotus cystidiosus* shares greatest similarity, from a cultural viewpoint, with *P. abalonus* Han, Chen and Cheng and *P. smithii* Guzman, and may well be conspecific with these two taxa. Hilber (1989) believes that a combination of features can delimit *P. abalonus* from *P. cystidiosus*. *P. abalonus* has a cap that is darker in color, white pileocystidia, and brown cheilocystidia, whereas *P. cystidiosus* has a cap lighter in color, translucent brownish pileocystidia, and thin-walled hyaline cheilocystidia. Furthermore, Hilber states these taxa can be further delineated by spore size.

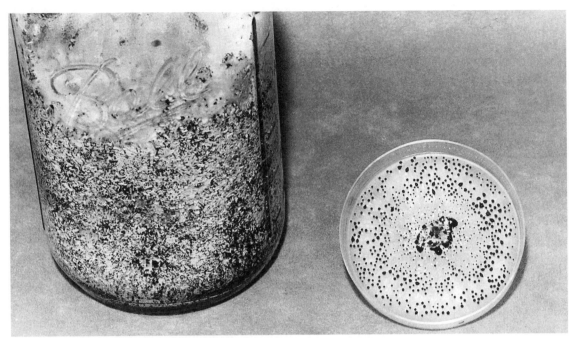

P. cystidiosus on rye grain and on malt agar medium.

At one time, *P. cystidiosus* was thought, incorrectly, to be synonymous with *P. corticatus* (Jong and Peng, 1975). This proposed synonymy led to the depositing of several mislabeled strains into international culture libraries. *P. cystidiosus* is taxonomically discrete from *P. corticatus*. Currently, *P. corticatus* (Fr.:Fr.) Kumm and *P. dryinus* (Pers.:Fr.) Kumm are considered synonyms.

Description: Cap convex to hemispheric, eventually plane, measuring 2–5 cm broad, and cream to dingy white in color. (*P. abalonus* is reportedly darker colored. See Bresinksi et al., 1987.) Cap margin often irregular. Gills broad, sometimes widely spaced, strongly decurrent, with irregular edges. Stem thick, centrally or eccentrically attached, and relatively short.

Distribution: From the eastern and southeastern United States (Louisiana, Mississippi, North Carolina), Taiwan, and South Africa. This mushroom is probably distributed throughout similar ecological zones of the world.

Natural Habitat: The type collection made by Dr. O. K. Miller came from red maple (*Acer rubrum*). Also native to eastern cottonwood (*Populus deltoides*), sweetgum (*Liquidambar styraciflua*), and Asian oaks (*Quercus nuttalli* and allies).

Microscopic Features: Spores white 11–14 × 4–5 μ. Hyphal system monomitic. Sterile cells on the cap well developed. The coremic structures on the gills can be seen with a hand lens. Dikaryotic mycelium, capable of producing mushrooms, only arises from two-celled conidia. If asexual spores (conidia) are single celled, then the strain is monokaryotic, and incapable of producing fertile mushrooms.

Available Strains: Strains are easily obtained from most culture libraries. ATCC #28599 is the type culture.

Mycelial Characteristics: *P. cystidiosus* is the most unusual Oyster mushroom I have seen in culture. At first the mycelium resembles any Oyster strain—white, racing linearly, soon fluffy white and aerial. However, as it grows outwards, black droplets of bundled spores on small stems form, radiating outwards from the center as the mycelium matures. These are coremia—stalk-like cells whose tops are fitted with liquid droplets of black spores. The spore-laden black droplets do not pose a contamination threat to other cultures in the laboratory until they dry and harden, at which time they can become airborne. If petri dishes are not handled carefully, the droplets will streak across the media, freeing them. One advantage to the cultivator of this imperfect stage is that inoculation of any substrate with pure culture spawn has an additional complement of asexual spores, effecting a simultaneous "spore-mass" inoculation. Colonization is comparatively fast.

Fragrance Signature: Musty, farinaceous, not pleasant, and not anise-like.

Natural Method of Cultivation: Dead or dying maple, cottonwood, sweetgum elms, beeches, oaks, and poplars can be inoculated via plug or sawdust spawn. Stump culture should be possible, given the success with the species' close relatives.

Recommended Courses for Expansion of Mycelial Mass to Achieve Fruiting: Cultures grown on nutrified agar media can be immersed into sterile water and chopped in a high-speed blender for several seconds. The resulting broth can inoculate sterile water fortified with malt/yeast and fermented for 48 hours using the techniques described in Chapter 15. This liquid-inoculum is then transferred directly into sterilized grain. Grain spawn should be used within one week of inoculation. No more than two generations of grain spawn are recommended. If the fermentation is continued for 5 to 7 days, asexual conidia form, facilitating the direct inoculation of bulk substrates.

Suggested Agar Culture Media: MYPA, CMYA, DFA, or PDYA.

GROWTH PARAMETERS

Spawn Run:
Incubation Temperature: 75–85°F (24–30°C)
Relative Humidity: 90–95%
Duration: 12–16 days
CO_2: 5,000–20,000 ppm
Fresh Air Exchanges: 1 per hour
Light Requirements: n/a

Primordia Formation:
Initiation Temperature: 70–80°F (18–24°C)
Relative Humidity: 95–100%
Duration: 4–5 days
CO_2: 500–1,000 ppm
Fresh Air Exchanges: 4–8 per hour
Light Requirements: 1,000–2,000 lux

Fruitbody Development:
Temperature: 70–80 (90)°F (21–27°C)
Relative Humidity: 85–90%
Duration: 4–8 days
CO_2: <2,000 ppm
Fresh Air Exchanges: 4–5 per hour
Light Requirements: 500–1,000 lux

Cropping Cycle:
Two crops, 10 days apart in 30 days

1st, 2nd, and 3rd Generation Spawn Media: Liquid or grain spawn throughout. Sawdust can be used as the final spawn medium if desired.

Substrates for Fruiting: Sterilized hardwood sawdust (maple, oak, beech, or elm), pasteurized wheat, rice, or paddy straw.

Recommended Containers for Fruiting: Bags, columns, trays, or bottles.

Yield Potential: Biological efficiency rated at 50–75%, higher on sawdust-based substrates than on straw.

Wild fruiting of *P. cystidiosus.*

P. cystidiosus fruiting from bags of sterilized sawdust.

Harvest Hints: Mushrooms should be picked before the caps expand beyond convex. Individual mushrooms can become quite large unless cluster formation is promoted. The stem is edible.

Form of Product Sold to Market: Fresh mushrooms and dried are sold in Thailand, Taiwan, China, and elsewhere in Asia.

Nutritional Content: Not known to this author.

Medicinal Properties: Not known to this author.

Flavor, Preparation, and Cooking: This mushroom can be cooked like most Oyster mushrooms in stir-fries, white sauces, or adorning lamb. Please refer to the recipes in Chapter 24.

Comments: My experience has been that cultures from Thailand and Taiwan, where this group of mushrooms is commercially cultivated, produce abundantly on rice straw and perform less productively on wheat straw. Strains are more narrowly specific in their fruiting requirements, requiring sustained warmth, and must be more carefully matched with the fruiting substrate than, for instance, *P. ostreatus*, a species more adaptive to a wider variety of materials. For more information, consult Guzman et al., 1991; Jong and Peng, 1975; Peng, 1974; and Miller, 1969.

The Pink Oyster Mushroom
Pleurotus djamor (Fries) Boedjin sensu lato

Mycelium of *P. djamor* 5 and 10 days after inoculation onto malt extract agar media.

Introduction: This species encompasses a complex of brilliantly pink Oyster mushrooms. The Pink Oyster varieties are the most common occurring wild *Pleurotus* in pan-tropical climatic zones of the world. Known for its speed to fruiting, adaptive ability to flourish on a wide variety of base materials, and high temperature tolerance, this species is so aggressive as to colonize unpasteurized bulk substrates before competitors can flourish. When growing this mushroom en masse, clusters often form lacking any pink pigment, although usually these forms are in the minority.

Common Names: The Pink Oyster Mushroom
The Salmon Oyster Mushroom
The Strawberry Oyster
The Flamingo Mushroom
Takiiro Hiratake (Japanese)
Tabang Ngungut (Dasun-
Northern Borneo)

Taxonomic Synonyms and Considerations: This mushroom has a trail of synonyms, when taken as a whole, represent a large complex of pan-tropical varieties. *Pleurotus flabellatus* (Berk. and Br.) Saccardo, *P. ostreato-roseus* Singer, and *P. salmoneostramineus* Vasil. are included within the *Pleurotus djamor* species complex. Although the mushrooms

Mycelium of *P. salmoneo-stramineus* (=*P. djamor*) 14 days after inoculation.

are usually pink, this color is usually temporal, fading as the mushrooms mature.

Originally published by Fries in 1838 as *Lentinus djamor* Fr., the Friesian concept has been amended to include many varieties. No type collection survived the passage of time. Corner (1981) proposed the pink-gilled forms should be called *P. djamor* var. *roseus*. He reports that the spores of this variety are cream colored. Guzman et al. (1993) calls "*P. djamour*" (sic) synonymous to *P. flabellatus*, describing the spores as white to gray to light honey yellow.

However, my studies reveal that the color of the fruitbody directly influences the color of the spores. Pink mushrooms give pink spores. White to beige mushrooms, from the same dikaryon that produced the pink mushrooms, gives off-white to light gray-beige spores. As the pink mushrooms fade with maturity, the spore color also changes. Redhead (1993) suspects this pigment is present in the cytoplasm and not in the outer spore coat.

One contradiction with this proposed synonymy is with the cellular arrangement of the flesh, best seen at the stem base. According to Singer (1986)

and Pegler (1983) *P. flabellatus* has a monomitic hyphal system whereas *P. salmoneo-stramineus* has dimitic hyphae. Corner (1981) considers *P. djamor* and *P. flabellatus* synonyms and states that both have dimitic hyphae, in apparent contradiction to Singer and Pegler. Guzman (1993) notes that young specimens of *P. djamor* appear to have monomitic hyphae, with dimitic forms developing in age. *P. ostreato-roseus* Singer is probably included within the *P. djamor* complex. Another Pink Oyster mushroom is the African *Pleurotus eous* (Berkeley) Saccardo. *P. eous*'s relationship to *P. djamor* should be carefully checked for synonymy. They may be the same species. (See Corner, 1981; Pegler, 1972; and Zadrazil, 1993.)

This large group of Pink Oyster varieties sorely needs further study. I would not be surprised to find that the Pink Oyster mushrooms represent a large complex of varieties in a state of rapid convergent and/or divergent evolution. Until DNA studies are completed, the taxonomy of this group is unlikely to be further resolved by macroscopic or microscopic analyses.

Description: Sharing the general shape and appearance of *P. ostreatus*, except the primordia are bright reddish ("salmon-egg") pink, becoming pinkish as mushrooms develop, eventually a dull pink to light pinkish cinnamon colored, and often fading to straw colored when overmature. (The color transitions are not only age-dependent, but vary between strains and are influenced by light conditions.) Cap convex expanding with age to broadly convex to plane. Cap margin inrolled at first, then incurved, and eventually flattening and upturning at maturity. The gills are particularly strongly pigmented with pinkish tones when young, fading to a creamy beige in age. Commonly growing in clusters of multiple mushrooms. When cultivated, variant forms often appear from the same fruiting container as the pink forms. These variants range in color from beige to cream to white, usually with white to gray gills, and often with a highly undulating, scallop-like cap margin.

Distribution: A tropical mushroom complex, widespread throughout the tropics and subtropics.

Mushrooms from this group have been collected in Thailand, Cambodia, Singapore, Vietnam, Sri Lanka, Malaysia, New Guinea, North Borneo, Japan, Brazil, Mexico, and the Antilles. Reports of a Pink Oyster mushroom from Amazonia probably belong to this species complex.

Natural Habitat: Preferring tropical and subtropical hardwoods, including palms and rubber trees, and also found on bamboo.

Microscopic Features: From the same fruiting column, I obtained pink spores from pink mushrooms and light beige spores from mushrooms that were originally pink but faded to cream beige. Spores measure 6–10 x 4–5 μ, smooth, and cylindrically shaped. Clamp connections present. Cheilocystidia present. Pleurocystidia absent. Hyphae arranged dimitically.

Available Strains: The body of strains available from this complex is mind-boggling. American Type Culture Collection has several cultures, of which ATCC #34552 (called "*P. salmoneo-stramineus*") is

The Pink Oyster mushroom fruiting from bags of sterilized sawdust in Thailand.

a fruiting strain. (See page 299.) A strain I have in my culture collection grows extraordinary quickly, producing mushrooms 10 days after inoculation onto pasteurized wheat straw. Cultivators should note that these cultures are often identified by any of the above-mentioned names. Please refer to the above discussion of the taxonomy of this mushroom complex.

Mycelial Characteristics: White at first, casting a longitudinally linear mycelium, often overrun with long, diverging rhizomorphs, eventually cottony with maturity, and aerial. Most strains soon develop strong pinkish tones, especially as the mycelium matures, at and around the sites of primordia formation. Flaming pink primordia often form as cluster colonies along the inside periphery of the petri dish and/or around the site of inoculation. As grain (rye) spawn matures, pink rhizomorphs and mycelia can predominate. A milky gray metabolic exudate collects at the bottom of the incubation containers.

Fragrance Signature: At first, the fresh mycelium is similar to *Pleurotus citrinopileatus*, the Golden Oyster mushroom, in that its fragrance is acrid and has a peculiar "bite" to it. After prolonged storage, the spawn and/or mushrooms develop a fish-like odor. The mushrooms, as they dry in bulk, give off a sickening, unpleasant odor. Once dried, the mushrooms impart a more pleasant fragrance.

Natural Method of Cultivation: This mushroom has been traditionally cultivated on hardwood stumps and logs by native peoples. Cultivators in Asia found that this species can quickly colonize unpasteurized cereal straws before contaminants emerge.

Recommended Courses for Expansion of Mycelial Mass to Achieve Fruiting: Cultures in petri dishes are cut out, fragmented in a high-speed blender, and used to generate grain spawn. Liquid-inoculated, 1/2-gallon grain jars are fully grown through in 4 to 5 days. However, the spawn must be continually expanded to preserve vigor and quality. Liquid-inoculated grain masters can be generated two more orders of magnitude. Once colonized, the grain

A variety of the Pink Oyster mushroom, which produces unusually large mushrooms from columns of wheat straw.

spawn should be implanted directly into the fruiting substrate, such as wheat straw. Grain spawn inoculated into pasteurized bulk substrates such as straw at a 10–20% ratio (wet spawn to dry substrate), results in fruitings within 2 weeks.

Suggested Agar Culture Media: MYPA, PDYA, OMYA, or DFA.

1st, 2nd, and 3rd Generation Spawn Media: Grain spawn for all three generations.

Substrates for Fruiting: Hardwood sawdust, cereal straw, corn waste, coffee residue, cotton waste, banana fronds, palm debris, and sugarcane bagasse. One formula employed by Brazilian growers calls for the proportionate mixing of 100 pounds sugarcane to 8 pounds rice bran to 3 pounds rice straw to 2 pounds calcium carbonate. The mixture is mixed, wetted, and pasteurized at 140°F (60°C) for 2 to 4

GROWTH PARAMETERS

Spawn Run:
Incubation Temperature: 75–85°F (24–30°C)
Relative Humidity: 95–100%
Duration: 7–10 days
CO_2: >5,000 ppm
Fresh Air Exchanges: 0–1 per hour
Light Requirements: n/a

Primordia Formation:
Initiation Temperature: 65–75°F (18–25°C)
Relative Humidity: 95–100%
Duration: 2–4 days
CO_2: 500–1,000 ppm
Fresh Air Exchanges 5–8 per hour
Light Requirements: 750–1,500 lux

Fruitbody Development:
Temperature: 70–85°F (20–30°C)
Relative Humidity: 85–90%
Duration: 3–5 days
CO_2: 500–1,500 ppm
Fresh Air Exchanges: 5–8 per hour
Light Requirements: 750–1,500 lux

Cropping Cycle:
Two crops, 7–10 days apart

This variety of the Pink Oyster mushroom fruits in only 10 days from inoculation into pasteurized wheat straw.

hours. Bano et al. (1978) found that this mushroom (as "*P. flabellatus*") gave the highest yields when cottonseed powder was added at 132 g per kg of dry wheat straw. The total mass of the mushrooms grown was 85% over the yields from unsupplemented wheat straw. Interestingly, the protein content of the dried mushrooms also rose to 38%!

Royse and Zaki (1991) found that the equal addition of the commercially available supplements Spawn Mate II and Fast Break at a combined rate of 168 g per kg of wheat straw substantially enhanced yields of this mushroom as "*P. flabellatus*." In these tests, biological efficiency increased from 22% to 77% in a 28-day harvest period. I would

expect that the yields would be similarly enhanced with most Oyster mushrooms.

Recommended Containers for Fruiting: Polyethylene bags or columns, trays, or racks.

Yield Potentials: Given good crop management, biological efficiency rated at 75–150%, largely dependent on the age of the fruitbody at harvest. Some strains of this species are equally as productive, in terms of biological efficiency, as the most vigorous strains of *P. pulmonarius* and *P. ostreatus*.

Harvest Hints: Mushrooms should be picked when moderately young, and handled carefully to not bruise the brilliantly colored gills. This mushroom spoils rapidly, not having a marketable shelf life of more than 4 to 5 days from the date of harvest.

Form of Product Sold to Market: Fresh or dried. Mushrooms are best presented gills up, for maximum visual impact. However, the unique pink color makes marketing an interesting challenge depending upon the market niche. In markets like New York and Los Angeles "pink is hot" and the color, as a friend once told me, "sells itself." As one can imagine, this mushroom may not sell as well as the non-pink varieties in less sophisticated rural American markets.

Nutritional Content: Not fully known to this author. Probably similar to *P. pulmonarius*. Bano et al. (1978) reported that the mushrooms grown from unsupplemented wheat straw, compared to wheat straw supplemented with cottonseed powder, had protein contents of 30% and 36%, respectively. Supplements and spawning rates have a direct impact on the protein content of the mushrooms grown.

Medicinal Properties: Not known to this author.

Flavor, Preparation, and Cooking: The flavor of this mushroom is not as appealing as many of the other Oyster species listed in this book. Many strains are tougher fleshed and more tart than other Oyster species. The pink color soon disappears upon contact with heat. Upon drying, a majority of (but not all) specimens lose their pinkish tones. Although this mushroom is not my personal favorite, some of my students prefer it to *P. ostreatus* and *P. pulmonarius*.

Comments: This complex of Pink Oyster mushrooms hosts some of the fastest growing strains of mushrooms in the genus *Pleurotus*. For those with limited access to pasteurization equipment, and living in a warm climate, strains of *P. djamor* uniquely fulfill a critical need. Its speed of colonization, short but productive fruiting cycle, and adaptability to diverse substrate materials make this species affordable to many cultivators, especially those in underdeveloped countries.

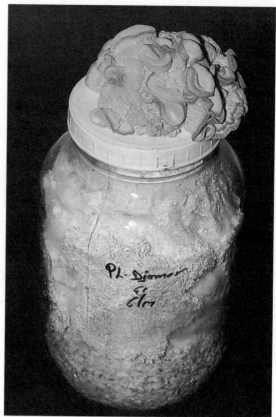

Spawn of warm weather Oyster mushrooms, such as *Pleurotus djamor*, are dangerous to keep in the laboratory for too long. Here, a fruiting of Pink Oyster mushrooms pushes through the seam of a sealed grain-spawn jar. If these mushrooms matured and sporulated in the laboratory, the room would have to be thoroughly cleaned to prevent contamination of other cultures.

Zadrazil (1979) noted that this mushroom (as "*P. flabellatus*") and *Stropharia rugosoannulata* proved to be the best at rendering straw, after fruiting, into a nutritious feed staple for ruminants, especially cattle.

For more information, please consult Bononi et al., 1991, "*Pleurotus ostreatoroseus* cultivation in Brazil." *Mushroom Science XI*, A.A. Balkema, Netherlands.

The King Oyster Mushroom
Pleurotus eryngii (De Candolle ex Fries) Quelet sensu lato

P. eryngii mycelia 5 and 10 days after inoculation onto malt extract agar media.

Introduction: *Pleurotus eryngii* is by far the best-tasting Oyster mushroom, well deserving of the title, the King Oyster. Popular in Europe, this stout, thickly flesh mushroom, is one of the largest species in the genus. Preferring hardwoods, this mushroom is easy to grow. Although this mushroom grows on the cereal (wheat) straws, the yields are not as substantial as that of *Pleurotus ostreatus* and *Pleurotus pulmonarius* on this same material, at the same rate of spawning, unless supplements are added or a unique spawning method is employed.

Common Names: The King Oyster
Boletus of the Steppes*

Taxonomic Synonyms and Considerations: Synonymous with *Pleurotus fuscus* (Batt.) Bres. Varieties specific to ecological niches have been commented upon by Bresinksy et al. (1987) and although morphologically these varieties appear identical, the distribution of ecotypes is quite distinct.

Description: Cap 3–12 cm in diameter, at first convex, expanding with age, becoming funnel-shaped, with the margin typically inrolled, extending with age. Stem 3–10 cm in length, central, thick, tapering downwards. Gills fairly distant, thin, grayish, and decurrent. Growing individually or in small groups. Cultivated mushrooms achieve a greater

---

* According to Zadrazil, Vasilkov called this mushroom "Boletus of the Steppes."

P. eryngii fruiting from supplemented alder sawdust/chips.

stature and overall size compared to ones collected in the wild.

Distribution: Throughout southern Europe, North Africa, central Asia, and southern Russia.

Natural Habitat: Terrestrial, growing on the buried roots of hardwoods. This mushroom is thought to a facultative parasite on dying *Eryngium campestre*, a member of the thistle family.

Microscopic Features: Spores white, ellipsoid, 10–14 × 4–5 μ. Clamp connections present. Context monomitic.

Available Strains: Most strains originate from Europe. Strains are commonly available from many culture libraries.

Mycelial Characteristics: Whitish, longitudinally radial at first, sometimes rhizomorphic, soon thickening and becoming cottony in age.

Fragrance Signature: Grain spawn and myceliated straw smells rich, sweet, and classically Oyster-esque but not anise-like.

Natural Method of Cultivation: Outdoors, on log sections turned vertically, and on stumps inoculated with plug spawn. This species is easily grown on straw outdoors using the mound method. Some strains are native to conifers (*Abies* spp.). If brought into culture, these races could help recycle conifer stumps throughout the world.

Recommended Courses for Expansion of Mycelial Mass to Achieve Fruiting: Traditional or liquid-inoculation of grain spawn to fruiting substrates, preferably sterilized sawdust. Pasteurized straw cultivation is comparatively less productive unless inoculated by the following method. I recommend inoculating wheat straw with equal quantities of sawdust and grain spawn. In other words, every ton of wheat straw (2,000 pounds dry weight) should simultaneously receive 100 pounds of grain spawn (wet weight) and 100 pounds of sawdust spawn (wet weight). This combination spawning method gives rise to large specimens on wheat straw.

Suggested Agar Culture Media: MYPA or PDYA.

GROWTH PARAMETERS

Spawn Run:
Incubation Temperature: 75°F (24°C)
Relative Humidity: 90–95%
Duration: 12–16 days
CO_2: 5,000–20,000 ppm
Fresh Air Exchanges: 1 per hour
Light Requirements: n/a

Primordia Formation:
Initiation Temperature: 50–60°F (10–15°C)
Relative Humidity: 95–100%
Duration: 4–5 days
CO_2: 500–1,000 ppm
Fresh Air Exchanges: 4–8 per hour
Light Requirements: 500–1,000 lux

Fruitbody Development:
Temperature: 60–70°F (15–21°C)
Relative Humidity: 85–90%
Duration: 4–8 days
CO_2: <2,000 ppm
Fresh Air Exchanges: 4–5 per hour
Light Requirements: 500–1,000 lux

Cropping Cycle:
Two crops, 14 days apart in 45 days

1st, 2nd, and 3rd Generation Spawn Media: Rye, wheat, sorghum, milo, or millet.

Substrates for Fruiting: Most hardwoods, wheat straw, and cottonseed hulls support fruitings. This mushroom is not as adaptive as *P. pulmonarius* and *P. ostreatus* to a broad range of substrates. Nevertheless, many materials can be used. I have been pleased with its performance on recycled, resterilized waste Shiitake substrate. However, I would not recommend this approach for commercial purposes unless the preferred wood type or alternative substrate materials were exceedingly scarce and cost-prohibitive. If cultivating this mushroom on wheat straw, the addition of 5–10% cottonseed meal

had the greatest effect in enhancing yield (Upadhyay and Vijay, 1991).

Recommended Containers for Fruiting: Trays, plastic bags, columns, and bottles.

Yield Potentials: 1 pound of mushrooms per 5 pounds of sterilized sawdust/chips/bran, or approximately 90% biological efficiency. Wheat straw fruitings, from my experience, have tallied approximately half of that from enriched sawdust. The stage at which the mushrooms are picked significantly affects yield efficiencies.

Harvest Hints: This mushroom can become quite large if the substrate has a sufficient nutritional base. The stage at which the fruitbody should be picked depends largely upon the strain and the cultivator's preference. I prefer harvesting the mushrooms just before the cap margin flattens out, when the cap margins are inrolled or deeply incurved, and the mushrooms are at an adolescent stage.

Form of Product Sold to Market: Mushrooms collected in the wild are sold in markets in Spain, Morocco, and other southern European countries.

Nutritional Content: Not known, although expected to be similar or exceeding *P. ostreatus*.

Medicinal Properties: Not known.

Flavor, Preparation, and Cooking: Stir-frying until edges become crispy golden brown. A chewy, nutty mushroom, this species is far superior to *P. ostreatus* and *P. pulmonarius*. This mushroom, like other Oyster mushrooms, goes well with Italian dishes, and especially with lamb, pork, and fish.

Comments: The King Oyster's stout form, short gills, and thick flesh, coupled with its pleasing flavor strongly commends this species among connoisseur growers and chefs. The short gills mean this mushroom releases comparatively fewer spores per pound of harvested mushrooms, a significant advantage over other Oyster species. *Pleurotus eryngii* has a better constitution than other Oyster species and, in many authors' opinions, is the best flavored. Gary Lincoff (1990) reported that this mushroom received the highest acclamations of any of the mushrooms

LaDena Stamets about to harvest *P. eryngii* from column of pasteurized wheat straw.

tasted during a culinary tour of mycophagists sampling the treasured mushrooms of Europe. This is the only Oyster species I know that ships well over long distances and has an extended shelf life.

Although other cultivators have recommended a casing layer, I have found its application unnecessary. My best fruitings of *Pleurotus eryngii*, in terms of both yields and quality, have been on 20% bran-enriched alder sawdust. Three weeks after inoculation with grain spawn, the fully colonized bags of sterilized sawdust/chips/bran are brought into the growing room. The tops of the bags are horizontally sliced opened, resulting in a 3- to 4-inch plastic wall around and above the surface plane of the mycelium. In effect, these sidewalls protect the supersensitive aerial mycelium from sudden dehydration. Condensation is promoted. Coupled with a descending fog environment within the growing room, the perfect microclimate for primordia formation is provided.

Zadrazil (1974) showed mycelial growth peaked when carbon dioxide levels approached 220,000 ppm or 22%. The stimulatory effect of CO_2 on mycelial growth allows this mushroom to grow under conditions that would be stifling for most other mushrooms and lifeforms. Optimum pH levels at the time of spawning should be between 7.5 and 8.5. On wheat straw, the pH naturally declines to a range of 5.5 to 6.5, a range ideal for fruiting.

The Tarragon Oyster Mushroom
Pleurotus euosmus (Berkeley apud Hussey) Saccardo

P. euosmus mycelia 3 and 10 days after inoculation onto malt extract agar media.

Introduction: The Tarragon Oyster mushroom is closely related to *Pleurotus ostreatus*. According to Watling and Gregory (1989), this mushroom is generally considered a form or variety of *P. ostreatus*, but differs in the strong smell of tarragon. It has been reported, to date, from England and Scotland. *P. euosmus* behaves, in culture, similarly to *P. ostreatus*.

Common Names: The Tarragon Oyster Mushroom

Taxonomic Synonyms and Considerations: *Pleurotus euosmus* can be distinguished from *P. ostreatus* by its odor (tarragon) and by spore size. The spore size of *P. euosmus* is 12–14 µ, substantially larger than the 7.5–11 µ spores of European *P. ostreatus* collections. However, I would not be surprised that these taxa are found to be conspecific through interfertility or DNA studies.

Hilber (1989) suggests synonymy between these two taxa without elaboration. The morphology of this mushroom—with its depressed cap at maturity and long running gills—bears strong resemblance to *Pleurotus ostreatus*.

A *Pleurotus eous* (Berkeley) Saccardo is a discretely separate species from *P. euosmus* and is more closely allied to the pink *P. djamor* varieties than to the gray brown *P. ostreatus* and allies. (See Pegler, 1972, and Corner, 1981.) Chang and Miles make reference to nutritional analysis of "*Pleurotus eous*" in *Edible Mushrooms and Their Cultivation* (1987, page 28) without further elaboration. See also the taxonomic discussions of *P. djamor* and *P. ostreatus*.

Description: Cap 5–15 cm broad, convex at first, soon broadly convex, expanding to plane, and typically deeply depressed in the center. Mushrooms beige-tan at first, becoming dingy brown with time, sometimes with a hint of blue, becoming light beige tawny in age. Margin even at first, often irregular in age. Gills dingy, decurrent, broad, running deeply down the stem. Stem short or sometimes absent.

Distribution: Limited to the British Isles, known from England and Scotland, but not yet reported from Ireland.

Natural Habitat: Preferring elm (*Ulmus* species) stumps and logs.

Microscopic Features: Spores pale pinkish lilac, oblong and narrow, measuring 12–14 × 4–5 μ. Otherwise similar to *P. ostreatus*.

Available Strains: Strains are available from some British, European, and American culture libraries.

Mycelial Characteristics: White, longitudinally linear, cottony, aerial, fast growing, and classically Oysteresque. Soon after colonizing a petri dish of MYPA, the mycelium tears off in thick sheets.

Fragrance Signature: Sweet, pleasant, slightly anise-like, and virtually identical to *P. ostreatus*.

Natural Method of Cultivation: I know of no one purposely growing this mushroom outdoors. However, given its close affinity to *P. ostreatus* and that it is native to elm stumps, this mushroom is likely to produce prodigiously using the natural culture techniques described in this book.

Recommended Courses for Expansion of Mycelial Mass to Achieve Fruiting: Transfer cultures from nutrified agar media into sterilized water and blend in a high-speed stirrer for several seconds. This liquefied mycelium should inoculate sterilized grain. Once colonized, grain spawn can be introduced directly into pasteurized straw or sterilized sawdust.

GROWTH PARAMETERS

Spawn Run:
Incubation Temperatures: 70–80°F (21–27°C)
Relative Humidity: 98–100%
Duration: 7–14 days
CO_2: >10,000 ppm
Fresh Air Exchanges: 0–1 per hour
Light Requirements: n/a

Primordia Formation:
Initiation Temperature: 65–75°F (18–24°C)
Relative Humidity: 95–100%
Duration: 7–10 days
CO_2: <2,000 ppm
Fresh Air Exchanges: 4–8 per hour
Light Requirements: 750–1,500 lux

Fruitbody Development:
Temperature: 70–80°F (21–27°C)
Relative Humidity: 90–95%
Duration: 4–8 days
CO_2: <1,000 ppm
Fresh Air Exchanges: 4–8 per hour
Light Requirements: 750–1,500 lux

Cropping Cycle:
Three crops, 2 weeks apart

Suggested Agar Culture Media: MYA, MYPA, PDYA, or OMYA.

1st, 2nd, and 3rd Generation Spawn Media: Grain spawn for the first two generations, hardwood sawdust spawn for the final stage.

Substrates for Fruiting: Hardwood sawdust, cereal straw, cottonseed hulls, sugarcane bagasse, coffee wastes, paddy straw, paper by-products, and many other materials. This mushroom will probably grow on many more substrate materials given modest experimentation.

P. euosmus fruiting from pasteurized wheat straw.

Recommended Containers for Fruiting: Bags, bottles, columns, and trays.

Yield Potentials: 75–100% biological efficiency, greatly affected by the size of the mushrooms at harvest, and the number of flushes allowed.

Harvest Hints: Mushrooms should be harvested before heavy sporulation. Since this mushroom strongly resembles *Pleurotus ostreatus* in terms of biology and appearance, the same guidelines for picking should be followed.

Form of Product Sold to Market: Fresh, dried, and powdered.

Nutritional Content: 25% crude protein, 59% carbohydrates, 12% fiber, 9% ash, and 1.1% fat.

Medicinal Properties: Not known to this author. Probably similar to *P. ostreatus*.

Flavor, Preparation, and Cooking: Versatile and flavorful, this mushroom can be incorporated into a wide variety of recipes. I prefer to sauté young mushrooms at high heat in light oil and to add cashews or almonds along with onions to adorn white fish or salmon that is then baked. Please refer to the recipes in Chapter 24.

Comments: The cultivation of *Pleurotus euosmus* parallels the cultivation of *P. ostreatus* and grows at a midlevel temperature range, not requiring a cold shock to initiate. The cultures in my collection produce uniform, medium-sized fruitbodies specific to puncture holes in the containers. Clusters of five to ten mushrooms are common, rarely numbering more, with the majority of the primordia forming reaching full maturity. These features may be strain specific. Please refer to the discussion of *P. ostreatus*, a close relative and possible future synonym of this mushroom.

The Tree Oyster Mushroom
Pleurotus ostreatus (Jacquin ex Fries) Kummer

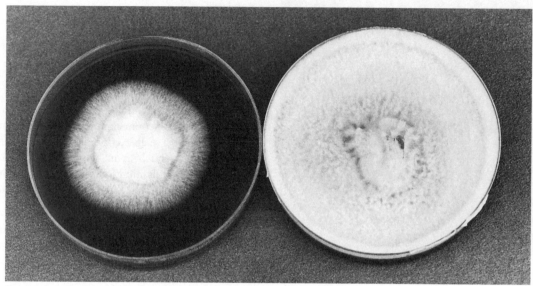

P. ostreatus mycelia 4 and 10 days after inoculation onto malt extract agar media.

Introduction: The archetypal Oyster mushroom, *Pleurotus ostreatus* has long been a favorite of mushroom hunters, especially in the springtime in lowland, hardwood forests. Prolific producers on a wide array of substrates, strains of this species are plentiful and easy to grow. Enjoying a worldwide reputation, specimens of extraordinary size have been collected from the wild. For instance, in the fall of 1988 near the north coast of Sicily, Salvatore Terracina, a farmer, collected a *P. ostreatus* nearly 8 feet in circumference, 20 inches thick, weighing 42 pounds! For the prepared and astute cultivator, cloning this monster could have resulted in some extraordinarily productive strains.

Common Names: The Oyster Mushroom
Oyster Shelf
Tree Oyster
Straw Mushroom
Hiratake (Japanese for "Flat Mushroom")
Tamogitake (Japanese)

Taxonomic Synonyms and Considerations: *Pleurotus ostreatus* is the type species for the genus *Pleurotus* and represents a huge complex of subspecies, varieties, and strains. An "old" species originally described by Fries, no collection of the original type is known. This complicates comparisons with other taxa, compounded by the fact many strains labeled as *P. ostreatus* are in fact *P. pulmonarius* and vice versa. For a mushroom so widely cultivated, I am surprised (and relieved) that only recently has the taxonomy become clearer, largely through the works of Petersen, Vilgalys, and Hilber.

Pleurotus ostreatus is so similar to *P. pulmonarius* that they are difficult to separate macroscopically. The Western collections of Oyster mushrooms on conifers usually fall into *P. pulmonarius* species concept. Furthermore, when *P. pulmonarius* is found in the Wild West, it prefers the higher altitude, drier coniferous forests to the hardwood river valleys where *P. ostreatus* dominates. Furthermore, *P. pulmonarius* is primarily found in the spring to early summer whereas *P. ostreatus* is common from the spring through late fall. A recently named species, *P. populinus* Hilber and Miller has a marked preference for black cottonwood (*Populus trichocarpa*) and aspen (*Populus tremelloides* and *P. tridentata*). Unlike *P. ostreatus*, *P. populinus* has, according to Vilgalys et al. (1993), a buff-colored, non-lilac spore print and larger spores, measuring 9–12 × 3–5 μ.

An Oyster strain from Florida, "*Pleurotus florida* Eger" is considered by this and other authors to be a synonym of *P. ostreatus* because spores from each species are cross-fertile, the mycelium forms clamp connections, and mushrooms grown from this mating produce fertile fruitbodies. The Florida variety differs primarily in its preference for warmer temperatures at fruiting, i.e., 75°F (24°C) and above. (See Li and Eger, 1978.) Guzman (1993) suggests that *P. florida* is conspecific with *P. pulmonarius*. Others believe *P. florida* is merely a variety of *P. ostreatus*. Hilber (1982) noted that the original strain of Eger's *P. florida* is, in fact, interfertile with *P. ostreatus*. Vilgalys (1993) concurs with Hilber, but solely on the basis of DNA comparisons. In our book, *The Mushroom Cultivator* (Stamets and Chilton, 1983), I incorrectly suggested synonymy between *P. florida* and *P. floridanus*, the latter being a distinctly separate species moved to the genus *Lentinus* by Pegler (1983).

Another sometimes bluish Oyster mushroom called *Pleurotus columbinus* is also in doubt as a separately valid species. Singer proposes *Pleurotus columbinus* to be a variety of *P. ostreatus*, i.e., *P. ostreatus* var. *columbinus* (Quel. apud Bres.) Quel. This placement concurs with the long-held view of many cultivators. One feature of this variety is its nearly perfect, even cap margin and broadly convex cap. The North American *Pleurotus sapidus* shares

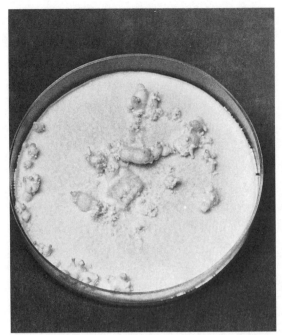

P. ostreatus mycelium 21 days after inoculation.

synonymy with *P. ostreatus*, according to Vilgalys et al. (1993).

Unless comparative DNA techniques are employed, or interfertility studies between known species are conducted, mistaken identifications between these taxa are likely. Those cloning wild specimens are therefore encouraged to retain a dried specimen for future verification of identification. For more information, please consult Hilber (1982), Kay and Vilgalys (1992), Petersen and Hughes (1992), and Vilgalys et al. (1993).

Description: Cap convex at first, expanding to broadly convex, eventually flat and even upturned in age; 5–20 cm (+) in diameter. White to yellow to grayish yellow to tan, rarely with pinkish tones, to lilac gray to gray-brown. Cap margin smooth to undulating like an oyster shell. Color varies according to the strain, lighting, and temperature conditions. Stems are typically eccentrically attached to the cap. Flesh generally thin. Some strains form clusters; others form individuals.

A sporeless strain of *P. ostreatus*. Under the microscope, the gill planes are entirely free of a hymenium and basidia.

Distribution: Distributed throughout the temperature and tropical forests of the world.

Natural Habitat: Common on broadleaf hardwoods in the spring and fall, especially cottonwoods, oaks, alders, maples, aspens, ash, beech, birch, elm, willows, and poplars. (From an evolutionary point of view, this mushroom has been very successful, given its ability to saprophytize a broad range of tree species.) Although seen on dying trees, *P. ostreatus* is thought to be primarily a saprophyte, but behaves as a facultative parasite at the earliest opportunity. Occasionally occurring on composting bales of straw, and in Mexico, on the waste pulp from coffee production. (The occurrence of *P. ostreatus* on these last two habitats might be a result of this species escaping from the woodland environment and taking advantage of a niche provided by the coffee industry.) *P. ostreatus*, and particularly *P. ostreatus* var. *columbinus*, is occasionally found on conifers, especially *Abies*. The most abundant fruitings of this species is in low valleys and riparian habitats.

Microscopic Features: Spores white* to slightly lilac to lilac gray, 7.5–9.5 × 3–4 μ. Clamp connections present. Context monomitic.

Available Strains: The genome of strains for this species is vast and increasingly explored by home and commercial cultivators. Cold- and warm-weather strains are available from numerous culture libraries. Amycel's #3001 and Penn State's #MW44, cold-weather strains, are popular. A warm-weather strain I cloned from mushrooms growing on a fallen oak in a ravine near San Diego, produces an attractive white mushroom in as short as 10 to 12 days from inoculation onto wheat straw. (See page 312.) The same strain produces a brown-capped mushroom at cold temperatures, a reaction typical of most Oyster varieties. "Sporeless" strains have obvious advantages for indoor cultivation, especially if the long gestation period before fruiting can be shortened. One sporeless strain available from the French-based Somycel Company is #3300. For more information on sporeless strains, and how to develop them, consult the article by Imbernon and Labarere (1989).

Mycelial Characteristics: Whitish, longitudinally radial, soon becoming cottony, and in age forming a thick, tenacious mycelial mat. Aged mycelium often secretes yellowish to orangish droplets of a metabolite, a toxin to nematodes. This metabolite deserves greater study.

Fragrance Signature: Sweet, rich, pleasant, distinctly anise, and almost almond-like.

---

* From my experiences, Oyster mushrooms from river-valley habitats in western Washington and Oregon produce a white to gray buff spore print, and not distinctly lilac as reported for the eastern forms. Furthermore, I have recently collected a pale rose variety of *P. ostreatus* on alder (*Alnus rubra*), which I have never encountered before. The pale rose color has been described for *P. pulmonarius*, but not for *P. ostreatus*. I suspect there are many more varieties of *P. ostreatus* unique to the western coastal regions of the Pacific Northwest than are presently described.

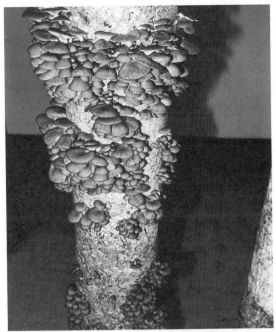

A dark, cold-weather strain of *P. ostreatus* fruiting 21 days after inoculation onto wheat straw. Note primordia form specific to punctures in plastic.

The same column 48 hours later. Mushrooms are ready for harvest.

Natural Method of Cultivation: On logs or stumps outdoors. Pagony (1973) reported that, on average, more than 1 pound of mushrooms per year was harvested from inoculated poplar stumps for more than 3 years. Of the 200 poplar stumps, ranging in size from 6–12 inches, which were inoculated in the spring, *all* produced by the fall of the following year. As expected, hardwoods of greater density, such as oak, took longer to produce but sustained yields for a longer period.

Recommended Courses for Expansion of Mycelial Mass to Achieve Fruiting: Agar-to-grain-to-cereal straw, or agar-to-grain-to-supplemented, sterilized sawdust. Since this mushroom thrives under liquid culture conditions, spawn generated by these methods is highly recommended. See Chapter 15.

Suggested Agar Culture Media: MYPA, PDYA, OMYA, or DFA. Optimal growth seen at pH 5.5 to 6.5.

1st, 2nd, and 3rd Generation Spawn Media: Rye, wheat, milo, sorghum, corn, and millet. Sawdust spawn is not needed for indoor cultivation methods. However, sawdust spawn is ideal in the inoculation of stumps and logs in outdoor settings.

Substrates for Fruiting: A wide array of agricultural and forest waste products can be used, including but not limited to straw (wheat, rye, oat, rice, and barley straw), cornstalks, sugarcane bagasse, coffee pulp, banana waste, cotton waste and cottonseed hulls, hardwood sawdusts, paper by-products, soybean waste, palm oil by-products, agave waste, and even the pulp remaining from tequila production! The pH at makeup can vary between 6.0 and 8.0 but should fall to an optimum of 5.0 at fruiting for maximum biomass production. (See El-Kattan et al., 1990.)

Martinez et al. (1985) reported yields of 132% biological efficiency (four flushes) from coffee pulp that was fermented for 5 days, pasteurized, and inoculated with wheat grain spawn. Further, they found residual caffeine from the spent substrate was reduced by more than 90%. (Caffeine represents a significant toxic waste to streams in coffee-growing regions of the world.) Martinez-Carrera (1987) validated the results with yields in excess of 100% biological efficiency on the same substrate and presented the first model for utilizing this abundant waste product.

Platt et al. (1982) published studies on the utility of cotton straw as a substrate for this mushroom. Their yields average 600–700 grams per kilogram of dry cotton straw, in other words 60–70% biological efficiency. Oyster mushrooms have been studied for recycling mango and date waste. Yields of 12% from a 1:1 ratio of mango waste to rice straw were best of series of experiments by Jwanny et al. (1995).

Recommended Containers for Fruiting: Megabags, trays, racks, and columns. (See page 182.)

Yield Potentials: 75–200% biological efficiency, greatly affected by the size of the fruitbodies harvested, and the number of flushes orchestrated.

Harvest Hints: Mushrooms should be picked when young, and preferably in clusters. Once the gills produce abundant spores, storability rapidly declines. Workers should wear filter masks effective down to 7 microns to eliminate the inhaling of spores. Mushroom surfaces should be slightly dry at harvest. Mushrooms should be chilled first to 35°F (1–2°C) and then placed into end-user containers (for restaurants or consumers) and covered with breathable, anticondensate plastic film.

Form of Product Sold to Market: Fresh and dried mushrooms. Waste straw substrate has been test marketed as a cattle feed.

Nutritional Content: By dry weight: crude protein: 10–30%; vitamin C: 30–144 mg/100 g; niacin: 109 mg/100 g; folic acid: 65 mg/100 g. High in potassium: 306 mg/100 g. For further information, see Bano and Rajarathnam (1982), Miles and Chang (1986), and Rai et al. (1988).

A white Oyster variety of *P. ostreatus* isolated from Southern California that fruits 10 days from inoculation onto wheat straw.

Medicinal Properties: Recent studies (Gunde-Cimerman et al., 1995, 1999; Bobek et al., 1998) show that *Pleurotus ostreatus* and other closely related species naturally produce Lovastatin (3-hydroxy-3-methylglutaryl-coenzyme A reductase), a drug approved by the FDA in 1987 for treating excessive blood cholesterol. More Lovastatin is present in the caps than in the stems, more concentrated on the mature gills, and especially in the spores. One model showed that plasma cholesterol turnover was significantly enhanced by 50% with a corresponding 25% decrease in the liver compared to the controls. (Bobek et al., 1995). This compound and others related to it may explain the often-reported cholesterol-reducing effects of many woodland mushrooms.

When mice were implanted with Sarcoma 180 and Oyster mushrooms constituted 20% of their

GROWTH PARAMETERS

Spawn Run:

Temperature: 75°F (24°C)
Relative Humidity: 85–95%
Duration: 12–21 days
CO_2: 5,000–20,000 ppm
Fresh Air Exchanges: 1 per hour
Light Requirements: n/a

Fruitbody Development:

Temperature: 60–70°F (10–21°C)
Relative Humidity: 85–90%
Duration: 4–7 days
CO_2: <1,000 ppm
Fresh Air Exchanges: 4–8 per hour
Light Requirements: 1,000–1,500 (2,000) lux

Primordia Formation:

Initiation Temperature: 50–60°F (10–15.6°C)
Relative Humidity: 95–100%
Duration: 3–5 days
CO_2: <1,000 ppm
Fresh Air Exchanges: 4–8 per hour
Light Requirements: 1,000–1,500 (2,000) lux*

Cropping Cycle:

Three to four crops, 7–14 days apart, over
45–55 days

---

* Eger et al. (1974) determined that *P. ostreatus* forms the most primordia in response to a light intensity of 2,000 lux or about 185 foot-candles. Light intensities exceeding 2,000 lux/hour caused a precipitous drop in the number of primordia forming. At 10,000 lux/hr. (>925 foot-candles), primordia failed to form. Their studies showed that continuous, optimal light stimulation during the primordia formation resulted in the largest population of primordia. (However, I note that if the base nutrition or the strain cannot support the development of such large population of primordia, the cultivator may actually reduce yield efficiency.) Studies by Kalberer (1974) showed that total yield was maximized (and stem mass minimized) at 300–430 lux at 12 hours per day. The critical primordia period extends for 6 to 7 days. If the total light exposure, measured in "lux hours" per day fails to exceed 2,400, equivalent to 200 lux or approx. 19 foot-candles of continuous lighting, *P. ostreatus* are triggered into "coral" formation. Continuous lighting at optimal levels prevents revegetation of developing primordia, a phenomenon occurring with many strains grown in underexposed settings. Once primordia are well formed, diurnal cycles are recommended. As insightful as the research is, strain sensitivity could sway light optima in either direction.

daily diet, the tumors were inhibited by more than 60% after one month compared to the controls (Ying, 1987). In another study, when rats were fed a diet composed of 5% Oyster mass, and administered dimethylhydrazines to induce tumors, fewer formed than the controls. In this study, Zusman et al. (1997) found that when rats were given corncobs 15% colonized by Oyster mushrooms, they were significantly protected from treatment with chemicals that otherwise induced colon cancer, reducing incidence from 47% to 26%. Corncobs without mycelium provided no protection.

Workers picking mushrooms indoors commonly report allergic reactions to the spores of *P. ostreatus*. Symptoms include fever, headache, congestion, coughing, sneezing, nausea, and general malaise (Kamm et al., 1991; Horner et al., 1993). Workers, who at first can tolerate contact with Oyster spores, often develop increased sensitivity with continued exposure. Filtration masks help but do not entirely

Paul Stamets with fruiting of *P. ostreatus* from a decomposing copy of *The Mushroom Cultivator*.

solve this workplace-related problem. The question as to whether or not spores of Oyster mushrooms can carry virus harmful to humans has not yet been satisfactorily answered. Few individuals are allergic to Oyster mushrooms after they have been cooked. For more information, consult Reshef et al. (1988) and Mori et al. (1998). Lehrer et al. (1994) found that, in a comparative study of 701 patients, approximately 10% of Americans and Europeans showed an allergenic response from extracts of *Pleurotus ostreatus*, while *Psilocybe cubensis* showed the highest allergenic response, 12% and 16%, respectively.

Flavor, Preparation, and Cooking: Stir-fry in a light oil at high heat until golden brown and then cook with other condiments. Some of the best recipes can be found in John Pisto's book *Cooking with Mushrooms: A Culinary Guide to Chef Pisto's Favorite Fungi* (1997, Pisto's Kitchen, Pacific Grove, California).

Comments: The Oyster mushrooms are the easiest to grow. The disadvantages in their cultivation are in their short shelf life post-harvest and the health problems posed by the prolific spore load generated within the confines of the growing room.

Cold- and warm-weather strains of this mushroom are widely in use. The above-described temperatures for initiating *P. ostreatus* are based on cold-weather strains. Strains evolving in warm geographical niches behave more in accordance with the parameters outlined for *Pleurotus pulmonarius*. (See page 319.)

Pleurotus ostreatus is an extraordinarily interesting mushroom from many viewpoints. It is highly tolerant and responsive to carbon dioxide levels; Zadrazil (1974) noted that mycelial growth peaks at 280,000 ppm or 28% CO_2. Unless CO_2 levels are reduced to less than 1,000 ppm (.01%), noticeable malformations of the fruitbodies occur: typically long stems and small caps. In fact, the cap-to-stem ratio is an accurate measurement of atmospheric carbon dioxide levels in the growing room and is used as a visual cue by Oyster cultivators for increasing air exchange.

This mushroom species is also super-sensitive to light levels. (See Eger, 1980.) In low light, a similar effect to that seen under elevated carbon dioxide conditions is induced. When the mushroom is exposed to high light levels, pigmentation of the cap is usually enhanced. Blue strains become bluer. Brown-capped strains become a richer brown. Similar results are also seen at lower end temperatures given constant light conditions.

Thorn and Barron (1984) first noted that *P. ostreatus* exudes a metabolite toxic to nematodes. As the nematode lies stunned, the mycelium eventually invades through one of its orifices, quickly consuming the internal organs. From an evolutionary view, this is remarkable that a saprophytic mushroom can become predatory to an animal in its quest for new sources of nitrogen. This may well explain why nematodes have never been reported as a pathogen in Oyster mushroom cultivation whereas their occurrence in the cultivation of the Button mushroom (*Agaricus brunnescens*) is economically devastating and commonplace.

Several studies have examined the usefulness of the spent substrate from Oyster mushroom production for use as feed by ruminants (Jalc et al. 1996; Bano et al. 1986; and Calzada et al. 1987). Oyster mushrooms can play a pivotal role in joining together essential components within an integrated agricultural model.

Oyster mushrooms are notorious for their massive spore release. Here, Oyster mushrooms, two days beyond the time they should have been harvested, have produced so many spores that they form strings hanging from the edges of the caps. Workers should wear masks to minimize respiratory problems during harvest.

The Phoenix or Indian Oyster Mushroom
Pleurotus pulmonarius (Fries) Quelet, *"P. sajor-caju"*

Culture of *P. pulmonarius* 2 and 4 days after inoculation onto sterilized malt extract agar media.

Introduction: According to studies recently published by Vilgalys et al. (1993), *Pleurotus pulmonarius* is virtually indistinguishable from *P. ostreatus*, differing largely in its habitat preference for conifer woods. In the western United States, *P. pulmonarius* is usually found at higher altitudes than *P. ostreatus*, which prefers the lowland, river valleys. In the western United States, *P. pulmonarius* and *P. ostreatus* grow on a variety of hardwoods, with *P. pulmonarius* primarily a spring mushroom and *P. ostreatus* growing most prevalently in the summer to fall. The North American collections show a wider range in color varieties than the European collections. Given its wide range, *P. pulmonarius* hosts a large complex of varieties, offering cultivators a rich resource for new strains. Most of these strains fruit in culture.

Common Names: The Indian Oyster
The Phoenix Mushroom
Dhingri (in northern India)
"Pleurotus sajor-caju"
(misapplied by cultivators)

Taxonomic Synonyms and Considerations: This mushroom was first published as *Agaricus pulmonarius* Fr. in 1821. Similar to *P. ostreatus* (Jacq.:Fr.) Kummer and *P. populinus* Hilber and Miller, this species can be separated from them by a combination of habitat, macroscopic, and microscopic features. (See comments in Introduction.) *P. pulmonarius* and *P. populinus* both share a preference for aspen and black poplar. *P. pulmonarius* is usually more darkly pigmented and has spores generally not longer than 10 microns long compared to the paler *P. populinus* whose spores often measure up to 15 microns in length.

Cultivator-mycologists have mistakenly called a variety of this mushroom "*Pleurotus sajor-caju*." The true *Pleurotus sajor-caju* (Fr.) Singer has been returned to the genus *Lentinus* by Pegler (1975), and is now called *Lentinus sajor-caju* (Fr.) Fries. *Pleurotus sajor-caju* (Fr.) Sing. has a distinct veil, a persistent ring on the stem, and trimitic or dimitic hyphae composing the flesh. (*P. pulmonarius* is monomitic.) In light of this new information, Singer's remark in *The Agaricales in Modern Taxonomy* (1986, p. 178) concerning the similarity of *P. sajor-caju* and the likelihood of its sharing synonymy with *Lentinus dactyliophorus* and *Lentinus leucochrous* is now understandable. He was describing a mushroom completely different from the one cultivators grow in United States and Europe, commonly called "*P. sajor-caju.*"

The name "*P. sajor-caju*" has been misapplied so frequently that confusion will likely reign for a considerable time. Many of the scientific papers published on the extraordinary yields of "*P. sajor-caju*" on straw, cotton wastes, coffee residues (ad infinitum) are undoubtedly referring to a strain of *P. pulmonarius*. (See Hilber, 1989, p. 246.) Since the name has become so entrenched by cultivators, naming a new variety, i.e., *Pleurotus pulmonarius* var. *sajor-caju*, seems like a good compromise. Until then, cultivators should refrain from calling this commercially cultivated Oyster mushroom "*Pleurotus sajor-caju*" as it is incorrect.

Description: Cap convex at first, expanding to broadly convex, eventually flat or upturned and often wavy in age; 5–20+ cm in diameter. Grayish white to beige to lilac gray to gray-brown, sometimes with pinkish or orangish tones. (At high temperatures, the cap is lighter in color. Under the same light conditions, under cold conditions, the cap becomes very dark gray to gray black.) Cap margin smooth to undulating like an Oyster. Color varies according to the strain, lighting, and temperature conditions. Stems are typically eccentrically attached to the cap. Veil absent. Flesh generally thin. Strains of this mushroom rarely form clusters of more than 5 or 6 mushrooms.

Distribution: Widely reported from North America and Europe.

Natural Habitat: In the eastern United States, this mushroom primarily decomposes hardwoods, while in the western regions, the species can be found at middle elevations (1,200-3,000 meters) on conifers (*Abies* and *Picea*). Common in the spring and summer.

Microscopic Features: Spores white to yellowish to lavender gray when dense, more or less cylindrical, 7.5–11 × 3–4 μ. Clamp connections present. Hyphal system monomitic.

Available Strains: Plentiful, available from most all culture libraries, and frequently mislabeled as "*Pleurotus sajor-caju*." A near sporeless strain, known as "3300 INRA-Somycel," produces about $1/100$th of the spore load of normal strains, but is less productive. (See Imbernon and Houdeau, 1991.) The development for high-yielding, low-sporulating strains of *Pleurotus* are essential to limit the impact spores have on the health of workers.

Mycelial Characteristics: White, linear, becoming cottony, and eventually forming a thick, peelable mycelial mat. If cultures on agar media or on grain are not transferred in a timely fashion (i.e., within 2 weeks), the mycelium becomes so dense as to make inoculations cumbersome and messy. Overincubated cultures cannot be cut, even with the sharpest surgical-grade scalpel, but is torn from the surface of the agar media.

Fragrance Signature: Grain spawn sweet, pleasant, and distinctly "Oyster-esque."

Natural Method of Cultivation: When the first log with fruiting Oyster mushrooms was brought from the forest into the camp of humans, probably during

P. pulmonarius fruiting from jar of sterilized sawdust.

the Paleolithic epoch, Oyster mushroom cultivation began. This mushroom is *exceedingly easy* to cultivate and is especially aggressive on alder, cottonwood, poplar, oak, maple, elm, aspen, and some conifers. Other materials used for natural culture include wheat, rice or cotton straw, corncobs and sugarcane bagasse.

Since this mushroom grows wildly on conifers, such as *Abies* (firs) and *Picea* (spruce), cultivators would be wise to develop strains that could help recycle the millions of acres of stumps that characterize the western forests of North America, if not the world.

Recommended Courses for Expansion of Mycelial Mass to Achieve Fruiting: Liquid-inoculated grain spawn sown directly into pasteurized straw or sterilized sawdust. This mushroom is more economically grown on pasteurized substrates, especially the cereal straws, than on wood-based substrates.

Suggested Agar Culture Media: MYPA, PDYA, OMYA, and/or DFA.

1st, 2nd, and 3rd Generation Spawn Media: Grain spawn throughout.

Substrates for Fruiting: Broadly adaptive, producing mushrooms on a great array of organic debris. The substrate materials proven to result in the greatest yields are the cereal (wheat, rice) straws, hardwood sawdusts, cornstalks, sugarcane bagasse, coffee waste (Martinez et al. 1985), pulp mill sludge (Mueller and Gawley 1983), cotton waste, and numerous other agricultural and forest waste by-products. Royse and Bahler (1988) found that the addition of 20% alfalfa hay to wheat straw increased yields substantially. In their studies, yields peaked when a combination of wheat straw, alfalfa, and delayed release nutrients were employed. (The effect of delayed release nutrients on Oyster mushroom yield is discussed in detail by Royse and Schisler, 1987a, 1987b). Alfalfa hay, as any compost maker knows, is considered "hot" because of its elevated nitrogen component. Adding these nitrogenous supplements can boost yields but the cultivator must balance whether or not this advantage is offset by the likely increase in contamination rates. (As a rule, the likelihood of competitor molds increases directly as nitrogen levels are elevated.) For more information, consult Zadrazil (1980).

Recommended Containers for Fruiting: Perforated plastic bags, columns, bags, trays, vertical racks, and bottles. (See page 182.)

Yield Potentials: Biological efficiency 100–200%, greatly affected by the size of the fruitbody at the time of harvest and whether or not a fourth or fifth flush is achieved.

Harvest Hints: Because this mushroom grows so quickly, the timing of harvest is critical to the quality of the overall crop. Mushrooms more often form individually, in twos or threes, but rarely more. New mushrooms often form where the old mushrooms have been cut, a trait not generally seen with other *Pleurotus* species. If the mushrooms are picked at full maturity, they are quick to rot, especially if kept within a container where gas exchange is limited. Under these conditions, bacteria proliferate, and hundreds of primordia form directly on the rotting fruitbodies. (See page 439.)

Form of Product Sold to Market: Mostly fresh. Some products, especially soup mixes, feature dried, powdered mushrooms.

Nutritional Content: Crude protein (N x 4.38): 14–27%; fat: 2%; carbohydrates: 51% (on a dry weight basis). The variation in the reported protein composition of *P. pulmonarius* and its close relatives is discussed by Rai et al. (1988). For additional information on the nutritional aspects of this mushroom (identified as "*P. sajor-caju*"), see Bano and Rajarathnam (1982) and El-Kattan et al. (1991). At 25% protein, this mushroom has about half of the protein represented in a hen's egg, and about one-third of most meats.

Medicinal Properties: Not known to this author.

Flavor, Preparation, and Cooking: *P. pulmonarius* enhances any meal featuring fish, lamb, and pork. Of course, it also is excellent with most vegetarian cuisines. Slicing and/or chopping the mushrooms

---

GROWTH PARAMETERS

Spawn Run:
Incubation Temperature: 75–85°F (24–29°C)
Relative Humidity: 90–100%
Duration: 8–14 days
CO_2: >5,000 ppm
Fresh Air Exchanges: 1 per hour
Light Requirements: n/a

Primordia Formation:
Initiation Temperature: 50–75 (80)°F
   (10–24 (27)°C)
Relative Humidity: 95–100%
Duration: 3–5 days
CO_2: 400–800 ppm
Fresh Air Exchanges: 5–7 per hour
Light Requirements: 1,000–1,500 (2,000)
   lux

Fruitbody Development:
Temperature: 65–75°F (18–24°C)
Relative Humidity: 85–90% (95%)
Duration: 3–5 days
CO_2: 400–800 ppm
Fresh Air Exchanges: 5–7 per hour
Light Requirements: 1,000–1,500 (2,000)
   lux

Cropping Cycle:
Every 7–10 days for three flushes

---

and adding them into a stir-fry is the most popular method of preparation. Young mushrooms are far superior to adult specimens, in texture and flavor.

Comments: This species complex hosts an enormous number of strains. The most popular are the warm weather varieties currently being marketed by spawn manufacturers, often under the name "*Pleurotus sajor-caju*." This mushroom is more widely cultivated than any other Oyster mushroom in North America and Europe.

P. pulmonarius fruiting from 25-pound bag of wheat straw.

Because it is tolerant of high temperatures, renowned for its speed to fruiting and yield efficiencies, many cultivators are initially attracted to this mushroom. However, compared to the other Oyster-like species mentioned in this book, I hesitate to call it a "gourmet" mushroom. Although high yielding, I do not hold it in high regard for numerous reasons, such as its

- Continued growth after harvest
- Lack of cluster-bouquet formation
- Premature fruiting
- Quickness to spoil
- Production of high spore loads
- Attractiveness to fungus flies

These may be merely the complaints of a critical connoisseur. Many people use and like this species. *P. pulmonarius* remains the favorite of many of the largest Oyster growers in the world, especially those located in warmer climatic zones.

Okwujiako (1990) found that the vitamin thiamin was critical for growth and fruitbody development in *P. pulmonarius*. By simply adding yeast extract to the base medium, you can provide vitamins essential for enhanced fruitbody production. For more information on the cultivation of *P. pulmonarius* refer to Bano and Raharathnam, 1991, and Azizi et al. (1990). (Please note that these authors describe *Pleurotus sajor-caju* when, in fact, they were probably cultivating a variety *P. pulmonarius*.)

For more information, consult Vilgalys et al. (1993), Hilber (1982), Kay and Vilgalys (1992), and Petersen and Hughes (1992).

The King Tuber Oyster Mushroom
Pleurotus tuberregium (Fr.) Singer

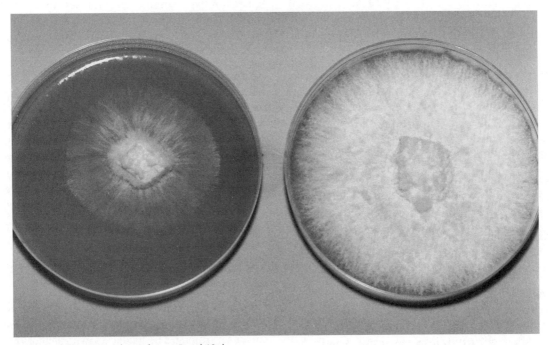

Cultures of *Pleurotus tuberregium* at 5 and 10 days.

Introduction: An unusual *Pleurotus* species, this mushroom is native to Africa and Australasia, and has only recently been cultivated. Noted for the large tuber-like sclerotium that can sprout classically shaped Oyster mushrooms. For teachers, this is an ideal species for demonstrating the role of sclerotia in the mushroom life cycle.

Common Names: King Tuber
Tiger Milk Mushroom
Omon's Oyster Mushroom

Taxonomic Synonyms and Considerations: *Pleurotus tuberregium* has been suggested to be a *Lentinus, Lentinus tuberregium* (Fr.) Fries, a view further supported by Pegler (1983) primarily due to the dimitic arrangement of the hypha and the overall tough flesh of the fruitbody. Recent molecular studies by Isikhuemhen et al. (2000) show that this species clearly belongs to the genus *Pleurotus*. Several races of this mushroom have been isolated from the subtropics and this species is likely the center of a constellation of divergent forms.

Description: Sclerotia are globose to ellipsoid, whitish to dingy beige on sterilized or pasteurized substrates, darkening with age, and dark brown to nearly black when developing in contact with soils. The initial primordia develop from the sclerotia as a spike-like proboscis, narrow at the apex and broad at the base. Once several inches high, a fluted cap emerges, and the stem develops a coating of darkened fibrils. The cap, centrally attached to the stem, is umbilicate, with decurrent gills, and with a margin, which is initially inrolled at first. Stem scabrous with brownish fibrillose remnants. Quickly maturing, this mushroom can achieve considerable size, up to 10 inches (25 cm) in diameter.

Distribution: Indigenous to Nigeria, sub-Sahara Africa, Malaysia, Papua New Guinea, Australia, New Caledonia, Indonesia, Myanmar, and the Yunnan province of China, this species is probably widely distributed throughout tropical regions of the world.

Natural Habitat: Growing on decaying logs, and more commonly on buried roots of deciduous trees in the Australasia region.

Microscopic Features: Spores whitish, ellipsoid, 6–12 × 2–4 μ, whitish in deposit. Dimitic hyphae, tetrapolar mating strategy.

Available Strains: Strains are widely available from numerous culture libraries listed in the resource section of this book. Isikhuemhen (1999) observed that strains originating from the Australasian-Pacific region could bypass the sclerotia stage whereas those from Nigeria do not.

Mycelial Characteristics: Longitudinally linear mycelium, soon tomentose and cottony in a petri dish. On sawdust, extensive rhizomorphs may fan as the mycelium climbs inside of the incubation containers. Once sclerotia have been separated from the mother mycelium and implanted into soils, the sclerotia can generate aerial, fuzzy mycelium, often a precedent for fruiting. (This method is not unlike that which is described for the patent on Morels, and is typical of most sclerotia-producing mushrooms.)

Fragrance Signature: Grain spawn sweet, souring with age.

Natural Method of Cultivation: A mass of mycelium or the sclerotia can be implanted outdoors into sandy/clay soils and lightly covered with soil (1–2 inches, or 2.5–5.0 cm). Approximately 15 to 30 days later, the sclerotial "eggs" can sprout, generating one or several trumpet-shaped oyster mushrooms over a week's time. Soil temperatures must, however, sustain temperatures above 75°F (24 °C) before fruitings can occur in earnest. Typically, in the Northern Hemisphere, May to August is ideal for outdoor fruitings.

Recommended Courses for Expansion of Mycelial Mass to Achieve Fruiting: Standard procedure for most species, except that the primary goal is to optimize the production of sclerotia. The formation of sclerotia is apparently unaffected by exposure to light. The size of the sclerotium appears to directly influence the size of the generated fruitbody. After 6 weeks of incubation on supplemented sawdust, golf-ball-sized sclerotia can be harvested.

Suggested Agar Culture Media: MYPA, PDYA, OMYA, and/or DFA.

1st, 2nd, and 3rd Generation Spawn Media: Grain spawn throughout.

Substrates for Fruiting: Highly adaptive, numerous formulas allow for sizable sclerotia formation, a precursor for fruiting. The standard sawdust/chips/bran formulation outlined in this book for Shiitake works well. Nianlai et al. (1998) recommend a formula with the ratios, by weight, of 39% sawdust, 39% cottonseed hulls, 20% rice or wheat bran, 1% calcium carbonate, and 1% sucrose, balanced to a pH of 6 to 7. Isikhuemhen (1999) noted that wheat straw could be used to produce sclerotia.

Recommended Containers for Fruiting: Trays with sclerotia buried by sand or loam. The sclerotia can sprout fruitbodies directly.

Yield Potentials: Sclerotia yield averages approximately 10% of the wet mass of the substrate, or about 5% conversion dry mass of sclerotia from dry mass of substrate. Fruitbodies from the sclerotia can be as much as 25% of the mass from which they spring. The sclerotia, when fresh, are about 50% moisture, whereas the fruitbodies are nearly 90% water.

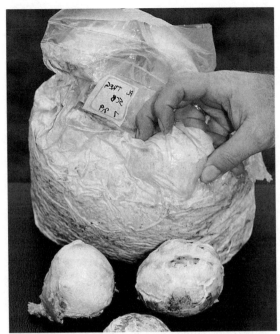

On supplemented sawdust, sclerotia begin forming in 4 weeks and sizable sclerotia can be harvested at 6 to 8 weeks from date of inoculation.

Harvest Hints: For medicine, the sclerotia can be used fresh or dried and made into a poultice to treat skin diseases. In Africa, this species is commonly combined with *Garcinia cola* (Bitter Cola), and used to treat asthma, cough, stomach pain, and many other diseases. If dried, the sclerotia can be stored for prolonged periods, and later used in teas, soups, or extracted in hot water and then preserved in alcohol.

Form of Product Sold to Market: Fresh sclerotia are sold as medicine and mushrooms are sold as food by local tribespeople in Nigeria and Ghana, and many other countries in Africa. The fresh sclerotia are inedible due to their bitterness. When dried, the bitter flavor diminishes.

Nutritional Content: Crude protein (N × 4.38): 14–27%; fat: 2%; carbohydrates: 51% (on a dry weight basis). See Isikhuemhen (1997).

GROWTH PARAMETERS

Spawn Run:
Incubation Temperature: 77–95°F (25–35°C)
Relative Humidity: 90–100%
Duration: 20–30 days
CO_2: >5,000 ppm
Fresh Air Exchanges: <1 per hour
Light Requirements: n/a

Sclerotia Formation:
Initiation Temperature: 75–80°F (24–27°C)
Relative Humidity: 95–100%
Duration: 30–90 days
CO_2: >5,000 ppm
Fresh Air Exchanges: 5–7 per hour
Light Requirements: None; incubate, if possible, in darkness

Primordia Formation:
Initiation Temperature: 75–90°F (24–32°C)
Relative Humidity: 95–100%
Duration: 14–28 days from the time of burying sclerotia
CO_2: <5,000 ppm
Fresh Air Exchanges: 5–7 per hour
Light Requirements: 1,000–1,500 (2,000) lux

Fruitbody Development:
Temperature: 86–95°F (30–35°C)
Relative Humidity: 85–90% (95%)
Duration: 4–7 days
CO_2: <2,000 ppm
Fresh Air Exchanges: 5–7 per hour
Light Requirements: lux

Cropping Cycle:
The sclerotia can form one or many fruitbodies in one flush and/or occasionally two flushes

Soon after buried in sand, a sclerotium generates a spike-like primordium.

Medicinal Properties: Traditionally, this species has been used for treating a wide range of ailments, most prominently topical treatments for skin diseases, and internally for the treatment of heart disease, diabetes, and stomach disorders (Isikhuemhen, 1999). According to Singer (1986), the sclerotia is used by native peoples for such diverse medicinal purposes as stomach pain, constipation, fever, blood pressure, and even smallpox (Singer 1986, and Oso 1977).

Flavor, Preparation, and Cooking: As an edible this mushroom is best cooked at the primordial stage—with the elongated stem and underdeveloped cap. Once the caps begin to enlarge, the texture of this mushroom becomes very tough, and is difficult to consume. Usually the sclerotia are dried first, powdered, and then used as a poultice or an admixture into a soup or tea.

Comments: The only species in the genus *Pleurotus* to produce a true sclerotium, *Pleurotus tuberregium* can become quite large. The size of the fruitbody is

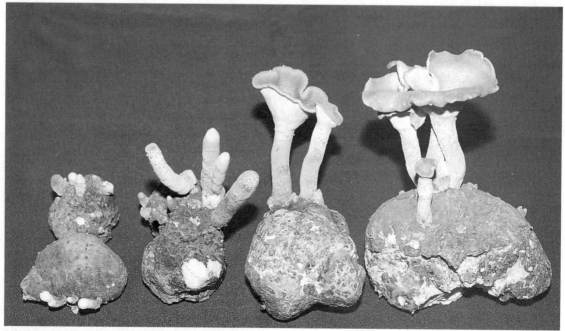

Different stages of mushroom growth from sclerotia.

Pleurotus tuberregium sclerotia, after being buried in sand for several weeks, metamorphosizing into mushrooms.

Ebikare Isikhuemhen holding *Pleurotus tuberregium* fruiting from block of sterilized sawdust. Not all strains must go through a sclerotial stage before mushrooms form.

directly related to the size of the sclerotium from which it emerges. When forming in the ground, the sclerotia can achieve a mass many orders of magnitude greater than that seen from pasteurized straw or sterilized sawdust-based substrates. *Pleurotus tuberregium* is an adaptive species like many other members in this genus, and can also be grown on wheat straw for the production of sclerotia. As with *P. ostreatus*, the mycelium traps nematodes (Hibbett and Thorn, 1994).

Much of the research on the edibility, taxonomy, and medicinal properties of this mushroom have come to light through the continuing work of Dr. Omon Isikhuemhen, a native of Nigeria, where this mushroom has had a long tradition of use. For more information I refer you to the articles written by him, which are listed in the bibliography.

The Caramel Capped Psilocybes (Pacific Coast *Teonanacatls*: Mushrooms of the Gods)* of the Genus *Psilocybe*

Psilocybe cyanescens Wakefield *sensu lato*

For millennia, Psilocybes have been used for spiritual and medicinal purposes. Cuanderos— Mesoamerican shamans—relied upon them to diagnose illness and to prognosticate the future. Through the works of R. Gordon Wasson, Jonathan Ott, Andrew Weil, Terence McKenna, and others, these mushrooms became well known to North Americans. In the mid-1970s, a group of dedicated mycophiles from the Pacific Northwest of North America pioneered the outdoor domestication of the temperate, wood-loving *Psilocybe* species. From these species, many imaginative cultivators learned techniques applicable to the cultivation of many other woodland gourmet and medicinal mushrooms. Since I have studied this group for many years and since this constellation of species has become the template for natural culture in North America, it seems fitting that this species complex be explored further.

Introduction: First cultivated in Washington and Oregon in the late 1970s, this complex of species is primarily grown outdoors in wood chip beds. Indoor cultivation is possible but pales in comparison to natural culture methods. Species in the *P. cyanescens* complex are not as high yielding per pound of substrate as some of the fleshier mushrooms in the genus and hence have little or no

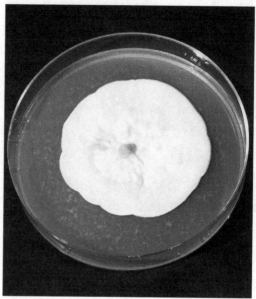

The satin-like mycelium of *Psilocybe azurescens*, a sister species to *P. cyanescens*, growing on malt extract agar.

commercial appeal. However, these mushrooms enjoy a popular reputation and are sought by thousands of eager hunters every fall. Because they are infrequently encountered in the wild, many mycophiles create a mushroom patch in the privacy of their backyards.

---

*The genus *Psilocybe* as monographed by Guzman (1983) has species that contain indole alkaloids (psilocybin, psilocin) that are known to be "psychoactive." Many species in the genus *Psilocybe* do not possess these pharmacologically active compounds However, in addition, a number of mushroom species unrelated to *Psilocybe* also contain these indoles. This section is offered for its academic value and does *not* encourage the violation of any ordinances restricting the possession or propagation of any illegal substance. Readers should further note that some individuals react negatively from the ingestion of psilocybian mushrooms. The author's research on this group was conducted under the provisions of a Drug Enforcement Administration license.

P. cyanescens complex fruiting on alder wood chips overlaid with a thin layer of straw.

Common Names: Cyans
                   Caramel Caps
                   Blue Angels
                   Potent Psilocybe
                   Fantasi-takes

Taxonomic Synonyms and Considerations: The name *Psilocybe* is Greek and means "bald head," which refers to the smooth surface texture of the cap. The genus *Psilocybe* has such close affinities to *Stropharia* and *Hypholoma* that separation of these genera continues to present unique taxonomic difficulties. These genera are clustered within the family Strophariaceae, which also includes the more distantly related genus *Pholiota*. Alexander Smith (1979) proposed that the family might best be represented by only two genera: the genus *Pholiota*

and the macro-genus *Psilocybe* that would also envelop species of *Stropharia* and *Hypholoma* (as *Naematoloma*).

Currently, the most thorough treatment of the genus can be found in Gaston Guzman's *The Genus Psilocybe: A World Monograph* (1983) and in my recent book *Psilocybin Mushrooms of the World* (1999). The monograph contains much original research and updates more than two decades of data accumulated by Roger Heim, R. Gordon Wasson, and other researchers. Wasson and Heim's beautifully illustrated monograph *Les Champignons Hallucinogenes du Mexique* (1958) revealed new Mesoamerican species, many of which were preempted by the nearly simultaneous publication of Singer and Smith's update on *Psilocybe* (1958), which listed several novel species and sections.* This

---

* According to the International Rules of Nomenclature, the names are prioritized according the date of publication. The first author to publish a description in Latin and deposit specimens into an *internationally recognized and accessible herbarium* is granted first right of use. All subsequently published names are subjugated synonyms.

The abundance of capitate pleurocystidia is in contradiction to the species concept put forth for "*P. cyanescens*" (Stamets Collection #78-34).

1978; Arora, 1979; Lincoff, 1981; Arora, 1991 and *numerous* papers published since 1958) show a mushroom that, in fact, possesses abundant, capitate pleurocystidia. Since this feature is consistent and obvious in water mounts under a microscope, and occurs in such high numbers, the mushroom in question cannot be the true *P. cyanescens*. In fact, I believe no species concept has yet been published to accurately delimit this mushroom.

A recently named species, *Psilocybe azurescens*, originates from the Columbia River basin near Astoria, Oregon is similar to the Pacific Northwest "*P. cyanescens*" (Stamets and Gartz, 1995; Stamets, 1999). This mushroom, distinguished by its comparatively great size, non-undulating cap margin is a close cousin, possibly belonging to Australian *Psilocybe subaeruginosa* Clel complex or the European *Psilocybe serbica* Moser et Horak complex. The third species in this group from the Pacific Northwest is distinguished by its forking cheilocystidia, and is called *Psilocybe cyanofibrillosa* Stamets and Guzman. (See Stamets et al., 1980.) I know of several more taxa that I have yet to publish. Despite the unusual attention these mushrooms have received, the taxonomy of this group needs further exploration. This group of new Psilocybes falls within an expanded concept of Singer and Smith's *Stirps Cyanescens* as amended by Guzman.

This species complex is fairly easy to identify. The mushrooms are generally cosmopolitan, and virtually absent from virgin forest ecosystems. They thrive in sawdust and chips from alder and Douglas firs. The mushrooms are collyboid—forming clusters that resemble the genus *Collybia* in habit only. The caps are uniquely caramel to chestnut colored and strongly hygrophanous. The cap is featured with a separable gelatinous skin and brown gills, which produce purple, brown spores. The bases of the stems radiate clusters of thick white rhizomorphs. These features separate this group of mushrooms from all others. This group can be further delimited into two subgroups: those possessing and those lacking pleurocystidia. Species having pleurocystidia can be lageniform or fusoid-ventricose with a narrow or bulbous apex.

event set the stage for a heated debate on *Psilocybe* that persisted for years. (Consult Smith, 1977; Ott, 1978; and Singer, 1986, pp. 570–571, and footnotes.)

Our limited understanding of the temperate, wood-inhabiting *Psilocybe*, particularly the *Psilocybe cyanescens* group, derives from, or more accurately suffers from, our interpretations of Singer and Smith's publications of 1958. We now know that a large constellation of species, subspecies, and races revolves around the species concept of *P. cyanescens*. Mycologists in the past have improperly misapplied species concepts from the European continent to North American candidates.

The type collection of *Psilocybe cyanescens* described by Wakefield from England lacks pleurocystidia, microscopic sterile cells on the surfaces of the gills. The photographs of a mushroom species from western North America identified in popular field guides as *Psilocybe cyanescens* (see Stamets,

Psilocybe cyanofibrillosa fruiting from Douglas fir sawdust.

Description: Caps are hemispheric at first, soon convex, expanding to broadly convex and eventually plane in age, 2–10 cm in diameter. Caps are strongly hygrophanous, sometimes chestnut especially when old or when the gills have fully matured. Cap margins are typically even at first, and straightening with age. Some varieties develop a pronounced, distinct, and undulating margin. (See page 327. Other species in this group have even margins.) Gills are colored brown to dark brown, often mottled, and bluntly attached to the stem, typically with a thin whitish margin. The stem is centrally attached to the cap, silky white to dingy brown near the base, often covered with fine fibrils, which may or may not bruise bluish. The stems are cartilaginous, even, straight to sinuous, usually swelling and curved towards the base. The base of the stem is usually fitted with a thick, radiating array of white rhizomorphs. Although mushrooms can be found individually, the majority grows in gregarious or collyboid clusters.

Distribution: The species represented in this group of mushrooms are found throughout the temperate forests of the world, including but not limited to the coastal Pacific Northwest of North America, northeastern North America, the British Isles, Eastern Europe, southern Argentina, North Africa, New Zealand, and temperate regions of Australia.

Natural Habitat: Primarily preferring deciduous woods, especially in riparian habitats, these mushrooms fruit in the fall. Possibly a saprophyte of seed cones of pines or firs, they show a particular fondness for the decorative mulch ("beauty bark") used in landscaping around newly constructed buildings. In fact, they are common in urban and suburban areas and are actually *rare* in natural settings. Ideal locations for collecting this mushroom are in the landscaped property of government facilities: courthouses, libraries, utility companies, and even police stations. Less obvious but productive locations are rhododendron, rose, and azalea gardens. Frequently

Psilocybe azurescens fruiting in the fall from a bed of alder chips implanted in the spring.

A sacred Psilocybe patch.

found along freeways, around soil mixing companies, and nurseries, these mushrooms are particularly fond of disturbed wood chip piles that have become overgrown with tall grasses. This group of mushrooms is common along the Oregon and Washington coasts in the public campgrounds in November. In fact, seasoned hunters look upon RV motor homes as an indicator species of productive habitats—Psilocybes are usually growing nearby.

Microscopic Features: Spores are purple brown, 8–12 × 5–8 μ, ellipsoid, smooth, featuring a distinct germ pore at one end. Gill margins are banded with long-throated, finger-like, nonforked or forked sterile cells. Some species have pleurocystidia while most do not. Clamp connections are present.

Available Strains: Strains are easy to obtain by joining a mycological society where mushrooms of this group are quietly exhibited during fall forays. In most countries (except Spain, Czechoslovakia,

Japan, the Netherlands, and others), it is illegal for culture companies to sell cultures except those licensed by the government.

Mycelial Characteristics: These Caramel-Capped Psilocybes behave similarly in culture, producing luxuriantly satin-like mycelia on sterilized malt agar media. The mycelium is white, cottony at first, soon silky rhizomorphic, usually radiating outwards with diverging fans from the site of inoculation. Mycelium often, but not necessarily, bruises bluish.

Fragrance Signature: Sour, unpleasant, nearly nauseating to some people, farinaceous, and reminiscent of spoiling corn.

Natural Method of Cultivation: This mushroom conforms to essentially the same strategy as does *Stropharia rugosoannulata*. (See growth parameters for that species.) Sawdust spawn is broadcast into hardwood or conifer (Douglas fir) chips that have been laid down outdoors in a partially shaded environment. The wood chips should be variable in size, ranging from $1/8$th inch in diameter to 4 inches in length. (Here again, the concept of building a wood-based matrix is essential.)

Recommended Courses for Expansion of Mycelial Mass to Achieve Fruiting: Liquid-inoculated grain spawn to 50:50 sawdust/wood chips used as spawn into outdoor beds.

Suggested Agar Culture Media: MYA, PDYA, OMYA, or DFA.

1st, 2nd, and 3rd Generation Spawn Media: Grain spawn throughout with the final stage of mycelial expansion being sawdust/chips.

Substrates for Fruiting: Hardwood sawdust and wood chips, especially alder, cottonwood, oak, birch, and beech. Douglas fir will also support fruitings.

Recommended Containers for Fruiting: Trays or framed beds outdoors.

Yield Potentials: Unavailable.

Nutritional Content: Not known to this author.

GROWTH PARAMETERS

Spawn Run:
Incubation Temperature: 65–75°F (18–24°C)
Relative Humidity: 95–100%
Duration: 45–60 days
CO_2: >5,000 ppm
Fresh Air Exchanges: 0–1 per hour
Light Requirements: n/a

Primordia Formation:
Initiation Temperature: 45–55°F (7–13°C)
Relative Humidity: 95–100%
Duration: 10–14 days
CO_2: 1,000–2,000 ppm
Fresh Air Exchanges: 2–4 per hour or as needed
Light Requirements: 400–800 lux

Fruitbody Development:
Temperature: 50–65°F (10–18°C)
Relative Humidity: 90–95%
Duration: 10–20 days
CO_2: 1,000–2,000 ppm
Fresh Air Exchanges: 2–4 per hour or as needed
Light Requirements: 400–800 lux

Cropping Cycle:
Two crops, 3–4 weeks apart

Medicinal Properties: Shamans have used these mushrooms for centuries to diagnose illness. Anecdotal reports received by this author suggest a few venues for possible medical research. An elderly friend suffering from a chronic hearing degeneration told me that small amounts of these mushrooms, too small to cause intoxication, had a remarkably positive effect on his hearing. Residual effects would carry over for several days. He urged me to tell others of his experiences.

Adorning the stem bases of *Psilocybe cyanescens* and *Psilocybe azurescens*, the rhizomorphic mycelia can regrow.

The same stem butt of *Psilocybe azurescens* regrows, projecting diverging rhizomorphs over wetted cardboard, creating spawn.

These mushrooms can enhance the mind's ability for complex visualization, and may possibly be a tool for controlling the autonomic nervous system. They should be used only under the careful supervision and guidance of a physician or shaman. Those unprepared for this experience often suffer from attacks of paranoia, fits of uncontrolled laughter, and other symptoms of psychological alteration.

Western scientists have largely ignored mushrooms of this group for their potential medicinal properties, instead focusing on their short-lived effects on the central nervous system. For a mushroom to manufacture a crystalline structure representing 1% of its mass must be for some evolutionary advantage. Yet the only advantage known is that humans are attracted to them, and thus ensure their survival. As drugs, the active compounds—psilocybin and psilocin—have virtually no addictive potential for humans. These compounds act as short-lived anti-metabolites and a substitute for serotonin, a primary neurotransmitter in the mammalian brain.

Flavor, Preparation, and Cooking: These mushrooms have an extremely bitter, revolting flavor, and can induce temporary nausea. Thorough cooking may destroy the active compounds in these mushrooms. If boiled for a prolonged period, and the excess water is discarded, the mushrooms are largely denatured of the psychoactive compounds and are edible.*

Comments: Mushrooms of this complex are rare in the wild. With the expansion of civilization, debris piles of chipped wood have accumulated around human dwellings. This rapidly emerging ecological niche has been quickly exploited by the Caramel-Capped Psilocybes.

---

* Author's note: Readers should be forewarned that cooking might not fully destroy the active compounds.

Yet, I am still mystified that in over twenty years of collecting, and in talking to dozens other collectors, that the *P. cyanescens* group has not yet been found in purely natural settings in the Pacific Northwest of North America. (A friend found one interesting collection from a seed cone of *Pinus radiata* at a college campus, suggesting that decomposing fir cones could be its original niche in nature.) Two possibilities: Either these mushrooms are exceedingly rare in nature, evolving from microniches (such as a minor player in saprophytizing fir cones), and are now undergoing a population explosion; or these species were just recently introduced from Australia or Europe. The European *P. serbica* Moser and Horak and *Psilocybe bohemica* Sebek, and the Australian *P. australiana* Guzman and Watling (=?*P. subaeruginosa* Clel.) fall into the Caramel-Capped Psilocybe complex. In any case, the sudden availability of chipped wood for use in landscaping has brought these autumnal species to the forefront of all the mycoflora found in suburban and urban settings. They all thrive on paper products, especially cardboard, quickly projecting exquisitely formed, thick white rhizomorphs.

Many gardeners unsuspectingly grow these mushrooms in the course of cultivating ornamental plants, especially rhododendrons and roses. Since mushrooms in this group are some of the few that thrive well into November in Washington and Oregon, the mycological landscaper is blessed with mushrooms at a time when few others are in their prime.

By cutting the stem butt of *Psilocybe azurescens* and "sandwiching" it into wetted, corrugated cardboard, the tissue soon regrows into a vigorous culture, in effect, creating "cardboard spawn." By placing wood chips on top of the myceliated cardboard, a renewable, perennial outdoor, sacred mushroom patch can be started, one that can last decades.

For more information on the Psilocybes and related mushrooms, consult *Psilocybin Mushrooms of the World* by Paul Stamets, 1999, Ten Speed Press, Berkeley, California.

The King Stropharia of the Genus *Stropharia*
Stropharia rugosoannulata Farlow apud Murrill

Introduction: Majestic and massive, few mushrooms are as adaptive as *Stropharia rugosoannulata* to outdoor cultivation. Popularly grown in Europe, this species is now the premier mushroom for outdoor bed culture by mycophiles in temperate climates. Known for its burgundy color when young, and its mammoth size, this mushroom is rapidly gaining popularity among mycologically astute recycling proponents.

Although *Stropharia rugosoannulata* can be cultivated in growing rooms, commercial cultivation seems uneconomical when compared to the yields achieved from other mushrooms. The time from spawning to cropping is nearly 8 to 10 weeks, slow by any standard. The casing layer cannot be fully heat-treated without significant reduction in yields. And, unpasteurized casing soils, when kept for prolonged periods of time in humidified growing rooms, tend to contaminate with a plethora of green mold and other weed fungi. With peat moss based casing, low temperature pasteurization at 130–140°F (54–60°C) for only 30 minutes may be the best course for indoor cultivation. I also recommend experimentation with soaking a synthetic casing material composed of water crystals and vermiculite with bacterially enriched water. This type of casing could stimulate fruiting by providing the

Typical *S. rugosoannulata* mycelium 10 days after inoculation onto malt extract agar medium.

essential microflora without encouraging competitor molds.

Until the gestation period can be shortened for indoor cultivation, the most practical method, given time, effort, and money, is outdoors in shaded beds of wood chips and straw.

Common Names: King Stropharia
Garden Giant or Gartenriese
Burgundy Mushrooms
The Wine Cap
Wine Red Stropharia
Godzilla Mushrooms

Taxonomic Synonyms and Considerations: At various times, called *Stropharia ferrii* Bres. or *Stropharia imaiana* Benedix.

Description: Cap 4–13 cm reddish brown at first, fading in age, broadly convex to plane at maturity. Margin incurved at first, connected by a thick, membranous veil. Veil breaking with age to form a thick membranous ring radially split with gill-like ridges, usually darkened with spores. Teeth-like veil remnants often seen at the time the ring separates from the cap. Stem thick, equal, enlarging towards the base where thick, white radiating rhizomorphs are attached.

Distribution: This mushroom is especially common in the mid-Atlantic states like New York, New Jersey, and Massachusetts. Although first described from North America in 1922, this mushroom is now found in Europe, New Zealand, and Japan. This mushroom probably became widely distributed through the export of ornamentals and wood chips. Yokoyama has deposited a strain in the culture library of the American Type Culture Collection (ATCC #42263 = IFO 30225) isolated from rice straw.

Natural Habitat: In hardwood forests and/or among hardwood debris or in soils rich in undecomposed woody matter, especially common in the wood chip mulch used in outdoor urban/suburban plantings of ornamentals.

Microscopic Features: Spores purple brown, giving a purplish black spore print, measuring 11–13 × 7.5–8 μ, smooth, ellipsoid. Clamp connections present.

Available Strains: "Vinnetou," a popular European strain; "Olympia," a widely distributed strain from the Pacific Northwest of North America.

Mycelial Characteristics: Whitish, linear longitudinally radial, not aerial, comparatively slow growing. Mycelium in culture lacks the pronounced rhizomorphs seen in nature. Under sterile conditions, the mycelium is cottony, not aerial, and often flattens with differentiated plateau-like formations. (If the culture dishes are taped with Parafilm, the mycelium can become aerial.) After prolonged incubation on grain, the mycelium secretes a clear yellow metabolic exudate. If not transferred to new media and allowed to overincubate, this highly acidic exudate jeopardizes the vitality of the host mycelium. This fluid may have interesting pharmacological, antibiotic, and/or enzymatic properties and, to my knowledge, has not yet been analyzed.

Fragrance Signature: A unique, strong, phenolic-like fragrance is imparted by the mycelium after

S. rugosoannulata fruiting from wheat straw topped ("cased") with a layer of soil.

colonizing grain. Once on sawdust, the mycelium out-gasses a rich, pleasing, and forest-like scent.

Natural Method of Outdoor Cultivation: This mushroom can be easily transplanted, a technique first used by wild collectors—whether they were aware of it or not. The trimmings from the base of the stem, resplendent with thick white rhizomorphs, quickly regrows when placed in contact with moist wood debris. Soon collectors of wild mushrooms found this mushroom growing in their backyards, aggressively seeking compost piles, sawdust, or straw-mulched soils. With the advent of commercial spawn, debris mounds were designed and constructed with *Stropharia rugosoannulata* in mind.

Our family grows *Stropharia rugosoannulata* in two ways. Our preferred method is to inoculate wood chips provided by our county utility company. (We live in a rural area with little automobile traffic.) The mixture of wood chips is mostly alder, with some Douglas fir and hemlock mixed in. Our only prerequisite is a minimum of leafy matter, which means we like to acquire our chips before mid-April, a perfect time for inoculation. Trucks dump several loads of chips into a pyramidal pile. Using metal rakes (and a tractor), we spread the pile until it is a depth of about 1 foot. (Downhill sides of the pile can get up to 3 feet in depth. The exposed surface face of the downhill slope, provides adequate aeration

Azureus Stamets holding 4-pound specimen of *S. rugosoannulata*.

and discourages activity from anaerobic organisms.) Upon this pile we use a 5-pound bag of sawdust/chip spawn per 100 square feet as our minimum inoculation rate and up to 4 units of spawn for a concentrated inoculation rate. For the first 4 days, I heavily water using a standard yard sprinkler. Subsequently, I water for ¹/₂ hour in the morning and evening, unless of course, it's been raining.

In 2 to 3 weeks, rhizomorphs can be detected in their first stages of growth. In 8 weeks, island colonies are distinct and abundant, usually separated by a few feet. The inoculated spawn creates island colonies that quickly become iceberg-like in formation, seeking the moist chips below and stimulated by the bacteria and nutrients near the wood/soil interface. These pyramidal colonies gradually expand to the surface. In 12 weeks, a large contiguous mycelial mat is formed. Provided that temperatures at ground level exceed 60°F (15–16°C), fruitings can be abundant beginning in

late July to the end of September. *Stropharia rugosoannulata*, like *Agaricus augustus*, the Prince, is a summer mushroom in the Pacific Northwest of North America. The mycelium is, however, tolerant of extreme temperature swings, thriving in 40–90°F (4–32°C) window.

Breaks or flushes of mushrooms can first be seen as smooth, reddish "stones," as if strewn into the wood chips by a playful child. As soon as they are touched, you realize that these "stones" are primordia. And for primordia, they are enormous, measuring an inch or two across. The cultivator is well advised to establish a standard walkway so that actively growing young mushrooms, unseen, are not crushed underfoot.

Hungarians have pioneered an interesting version of companion planting. King Stropharia mushrooms are grown in rows of baled wheat straw. The wheat straw is then wetted and inoculated with pasteurized, chopped wheat straw spawn. (This is called "substrate spawn.") Long rows of impregnated wheat straw are left to decompose in fields, adjacent to rows of other crops. Mushrooms usually arise from the straw/soil interface, and can be harvested during the late summer and early fall. After the growing season has ended, the straw and waste cornstalks are tilled under, enriching the soil for next year's crops.

Recommended Courses for Expansion of Mycelial Mass to Achieve Fruiting: This is a mushroom that greatly benefits from being disturbed or agitated. In the laboratory, the mycelial growth rate steadily declines unless it is frequently disrupted. To achieve the rapid colonization of mycelium on grain, liquid-inoculation is recommended. Otherwise growth is painstakingly slow, thereby allowing more opportunity for contamination. Once grain spawn is grown out, a single half gallon of spawn can functionally inoculate five to ten 5-pound bags of sterilized hardwood sawdust or finely chopped cereal straw. If liquid-inoculation is chosen for transfer into sterilized sawdust, two to four colonies of mycelium grown out in 100 × 15-mm petri dishes is recommended per 1,000 ml of sterile water. The

A happy mycophile holding a 3-pound specimen of *S. rugosoannulata*.)

LaDena Stamets among 5-pound specimens of *S. rugosoannulata*.

inoculum is then transferred at a rate of 50 ml per 5-pound bag of bulk substrate. In either case, once inoculated, this species enjoys frequent shaking, at least weekly, until full colonization is seen.

After the bag cultures have matured, they can be used for expansion into another generation of bags containing sterilized bulk (sawdust or straw) at a rate of 1:10. Again, thorough shaking on a weekly basis is critical for complete colonization.

Suggested Agar Media: MYA, PDYA, OMYA, or DFA.

1st, 2nd, and 3rd Generation Spawn Media: Grain (rye, wheat, milo, sorghum, or corn) should be inoculated from 3- to 4-day fermented mycelium. (A high inoculation rate is required if not using liquid-inoculation techniques.) Spawn jars should be shaken every 3 to 4 days after inoculation to ensure full colonization. On sterilized grain, the mycelium of this species revitalizes, bursting into new growth after being disturbed. A second generation of grain

spawn can be inoculated from the first generation and then the grain spawn can be mixed into a moistened 50:50 sawdust/chips blend. A minimum of a 10% inoculation rate should be used or preferably a rate of 15–20%.

Substrates for Fruiting: For indoor cultivation, pasteurized straw or sterilized sawdust is inoculated with grain spawn and incubated under high carbon dioxide conditions. Once colonized, a microbially rich soil (a "casing layer") is placed upon it to promote fruiting. This soil can be heat-treated to kill insects but should not be exposed to temperatures higher than 140°F (60°C) for more than 1 hour. Otherwise, the mushroom-promoting bacteria are killed, hindering or preventing fruitings. Two to three weeks after casing, long, silky white, braided rhizomorphs appear, soon giving rise to dark red-brown primordia. (See page 335.) Spent Shiitake or Oyster production blocks (sawdust/chips/bran) can be resterilized for further reduction by *Stropharia rugosoannulata*.

GROWTH PARAMETERS

Spawn Run:

Incubation Temperature: 70–80°F (21–27°C)
Relative Humidity: 95–100%
Duration: 25–45 days
CO_2: >20,000 ppm
Fresh Air Exchanges: 0–1 per hour
Light Requirements: n/a

Primordia Formation:

Initiation Temperature: 50–60°F (10–16°C)
Relative Humidity: 95–98%
Duration: 14–21 days
CO_2: <1,500 ppm
Fresh Air Exchanges: 4–8 per hour or as
   needed to affect CO_2
Light Requirements: 100–500 lux

Fruitbody Development:

Temperature: 60–70°F (16–21°C)
Relative Humidity: (85) 90–95%
Duration: 7–14 days
CO_2: <1,500 ppm
Fresh Air Exchanges: 4–8 per hour
Light Requirements: 100–500 lux

Cropping Cycle:

Two crops, 3–4 weeks apart

Recommended Containers for Fruiting: Trays or 20-gallon (or larger) bags.

Yield Potentials: 50–100% biological efficiency indoors. Outdoors, I have had woodchip beds produce hundreds of pounds.

Harvest Hints: For best flavor, this mushroom must be picked before or at the time the veil is tearing along the cap margin. As soon as the gills become dark gray, signifying the production of spores, edibility precipitously declines. The bulbous stem is particularly succulent, the presence of maggots notwithstanding.

Form of Product Sold to Market: Fresh mushrooms are sold at farmer's markets, usually by small organic farms selling other types of produce. The price ranges from $4–7 per pound. Americans are particularly attracted to this large mushroom. I have not seen this mushroom sold in dried form. It is too large to pickle.

Nutritional Content: Not known to this author.

Medicinal Properties: Not known to this author.

Flavor, Preparation, and Cooking: Young Buttons, sometimes weighing $1/2$ to 1 pound apiece, can be cut lengthwise to create King Stropharia steaks. Basted with soy sauce and herbal spices, this mushroom is superb on the barbecue during the summer, a time when it produces most prolifically. This mushroom can also be used in stir-fries. In either case, I prefer to cook this mushroom well.

King Stropharia should not be eaten for more than 2 or 3 days in a row. From European reports, some individuals who daily consumed this mushroom, failed to rebuild the enzymes necessary for digestion, an event possibly potentiated by alcohol, and resulting in a bad case of indigestion and/or nausea. I know of one, formerly enthused King Stropharia grower, who grew several hundred pounds of this mushroom, featuring it at summer garden parties. Upon his third day of imbibing, he was the only one of twenty guests to experience extreme gastrointestinal revolt. To this day, he now views King Stropharia (and me) with great suspicion.

Comments: *Stropharia rugosoannulata* is a mushroom with complex biological requirements, and yet one of great utility for gardeners and recyclers. On sterilized malt agar media, the mycelium grows anemically. On sterilized grain and sawdust, the mycelium grows out from the site of inoculation for a few inches and then radically declines in its rate of growth. Unless the mycelium is disturbed, growth continues to falter. If at this stage, the grain or sawdust is disturbed, the mycelium recoils from the concussion and bursts into a period of new growth. Often, the mycelium must be disturbed several times to assure full colonization on steril-

Representative fruiting of *S. rugosoannulata*. Note white rhizomorphs attached to base of stem.

ized substrates. When pasteurized sawdust is inoculated, the growth pattern is unhampered, unless green molds proliferate. The color of the sawdust changes from a dark brown to a light yellow brown just prior to the appearance of the white rhizomorphic mycelium.

Once the mycelium is implanted into wood chips outdoors, the mycelium undergoes a radical transformation in its pattern of growth. The mycelium is activated by microflora in soils, particularly bacteria. In response, thick cord-like and braided rhizomorphs form. This luxurious mycelium spreads from the sites of inoculation, and can travel substantial distances, generating satellite colonies, often hundreds of feet away from the mother colony. One mixed wood chip bed I inoculated had a depth of 1–2 feet, measuring approximately 20 by 30 feet in size. This patch yielded at least 200 pounds of mushrooms over its 2-year lifespan. After 3 years, the wood chips were rendered into a rich soil-like humus. (See page 28.)

LaDena Stamets with King Stropharia at the ideal stage for harvest.

Young specimens of this mushroom have an excellent flavor. The flavor quality steadily, nay, precipitously, declines as the mushroom matures, evidenced by the darkening of the gills, a sign of spores maturing. Once the thick veil ruptures and the gills throw spores, the mushroom rapidly looses any gourmet qualities. The mushrooms can weigh up to 5 pounds apiece. When these giants mature, with their huge surface areas of sporulating basidia, the spore cast is phenomenal. King Stropharia's flesh is far denser when young than when the caps are fully expanded. The stem is also edible although it is often permeated through with maggot holes, nearly invisible in young specimens. In older specimens the huge stems often become hollow carcasses in which so many fattened maggots have grown that the mushrooms spontaneously move from the rumblings of these inhabitants.

Another preferred method is to inoculate a mulched bed of moistened wheat straw in the garden and/or among shrubbery. I have found that specimens grown on wheat straw in open settings are relatively free of insect invasion compared to those grown on wood chips. For more information on the incorporation of this mushroom into an integrated farm model, please refer to Chapter 5: Permaculture with a Mycological Twist on page 39.

See also Ingle (1988), Chilton (1986), Stamets and Chilton (1983), and Steineck (1973).

Although impressive at this stage, this mushroom has passed its prime for consumption.

The rhizomorphs at the base of the stems will regrow if placed into moist cardboard (see page 332).

The Paddy Straw Mushroom of the Genus *Volvariella*
Volvariella volvacea (Bulliard: Fries) Singer

V. volvacea mycelia at 2 and 4 days after inoculation onto malt extract agar.

Introduction: Prodigiously fast-growing and one of my favorite mushrooms for the table, this mushroom thrives at warm temperatures (between 75–95°F or 24–35°C) and dies when temperatures drop below 45°F (7°C). This temperature range limits its cultivation in all but the warmest climates or months of the year. In subtropical and tropical Asia, many farmers rely on the cultivation of *V. volvacea* as a secondary source of income, making use of waste rice straw and cottonseed hulls. This mushroom has become an economic mainstay in the agricultural economies of Thailand, Cambodia, Vietnam, Taiwan, and China.

Two methods have evolved for its cultivation. The first method is outdoors, simple, and low-tech, owing its success to the rapidity of *V. volvacea*'s life cycle. The second method has been developed for intensive, indoor commercial cul-tivation, more closely resembling the composting procedures practiced by the *Agaricus* industry in the promotion of *Actinomyces* colonies, except that manure is not employed.

Once you have tasted fresh Paddy Straw mushrooms, it is easy to understand the high esteem this mushroom is given in Asia. I find fresh *V. volvacea* to be one of the best of all edible mushrooms.* The duration from inoculation to fruiting process is awesomely fast, giving competitors little opportunity to flourish. The Paddy Straw mushroom is an ideal species for the low-tech cultivation by rural people in subtropical and tropical climates. After the rice harvest, farmers mulch their rice straw into mounds, and inoculate with commercial spawn. This companion method of farming has given economic stability to a rural population, providing a much-needed supplementary income during the off-season.

---

* Canned Paddy Straw mushrooms lack the richness of the fresh fruitbodies.

Soaked straw is thrown into a tapering, trapezoid-shaped form.

After 4 to 6 layers, the frame is lifted off, and the process begins anew.

Soaked cottonseed hulls are thrown around the outer inside edge of the frame and then inoculated with grain spawn. Additional layers are built in the same fashion.

Common Names: Paddy Straw Mushroom
Straw Mushroom
Chinese Mushroom
Fukurotake (Japanese)

Taxonomic Synonyms and Considerations: *V. volvacea* and *V. bakeri* (Murr.) Shaffer are synonymous, according to Vela and Martinez-Carrera (1989).

Description: As the name implies, this mushroom's most distinctive feature is its volva, or cup, and resembles a classic *Amanita* except that an annulus is lacking. Cap is 5–15 cm broad, egg-shaped at first, soon expanding to campanulate or broadly convex with a slight umbo. Cap smoky brown to cigar-brown to blackish brown, darker when young, fading in age and/or with exposure to light. Margin edge radially ridged. Gills free, white at first, soon pinkish, close to crowded. Stem 4–20 cm long × 1.0–1.5 cm thick, white to yellowish, solid, and smooth. The stem base is encased in a thick volva.

Distribution: Thriving throughout tropical and subtropical Asia, this mushroom grows singly or in groups. Also found in eastern North America in hot houses, composts, or soils, especially in the southeastern states. Discarded experiments from the

Multiple square or rectangular mounds can be covered with plastic or a heavy layer of straw, which is kept moist.

In as short as 7 days, mushrooms form. A hoop frame covered with plastic helps maintain humidity and warmth during the primordia formation period.

The outdoor harvest can be substantial.

University of British Columbia in Vancouver, Canada, are suspected in creating a recurring patch of the Paddy Straw mushroom that persisted for nearly a decade (Kroeger, 1993).

Natural Habitat: On composting rice straw, sugarcane residue, leaf piles, and compost heaps during periods of warm weather from the spring through autumn. Several reports of this species growing in northern temperate climates have been traced to "escapees" from mushroom cultivators.

Microscopic Features: Spores pink to salmon brown, (6) 7.5–9 × 4–6 μ. Cheilocystidia, pleurocystidia, and chlamydospores are present.

Available Strains: Widely available, both from wild and developed stocks. Strains of *V. volvacea* die under cold storage. Many cultivators have found that cultures store best at ambient room temperature (i.e., above 45°F, or 7.2°C). Cold-weather strains have yet to be developed. For more information, consult Jinxia and Chang (1992) and Chang (1972).

Mycelial Characteristics: Longitudinally linear, soon aerial and disorganized, grayish white at first, soon, dingy yellowish brown, eventually becoming light gray brown to reddish brown, often with complex discolored zones.

GROWTH PARAMETERS

Spawn Run:
Incubation Temperature: 75–95°F (24–35°C)
Relative Humidity: 80–95%
Duration: Day 5–10 days
CO_2: >5,000 ppm
Fresh Air Exchanges: 1
Light Requirements: no light

Primordia Formation:
Initiation Temperature: 80–90°F (27–32°C)
Relative Humidity: 90–100%
Duration: 4–6 days
CO_2: 1,000–5,000 ppm
Fresh Air Exchanges: 4–5
Light Requirements: 250–500 lux

Fruitbody Development:
Temperature: 80–90°F (27–32°C)
Relative Humidity: 85–95%
Duration: 6–10 days
CO_2: 1,000–5,000 ppm
Fresh Air Exchanges: 4–5
Light Requirements: 500–750 lux

Cropping Cycle:
7–12 days

Fragrance Signature: Mycelium musty, not pleasant to this author.

Natural Method of Cultivation: On rice straw using a simple composting technique. Straw and cottonseed hulls are separately submerged in water. Saturated straw is laid directly on the ground in approximately a 2 × 2-ft. square; to a 2- to 3-inch depth. (See page 343.)

Recommended Courses for Expansion of Mycelial Mass to Achieve Fruiting: Traditional or liquid fermentation methods work well for spawn generation. Indoors, commercial cultivators use a single-stage composting schedule that contrasts with the methods Button mushroom growers use.

Rice straw is chopped into 4- to 5-inch (10- to 12-cm) pieces, soaked and stacked outside to tenderize the straw for no more than 2 days. A moisture content of 75% is desired. Cottonseed hulls are soaked for 2 to 7 days, allowed to ferment, and then layered onto the straw at the rate equivalent to 10–20% of the rice straw. For outdoor cultivation, spawn is added directly to the cottonseed hulls as each layer is built. The mass is covered with the goal of obtaining at least 90°F (32°C) for the next 5 to 7 days.

For indoor cultivation, the mass is mixed together and bulk-steamed at 120–140°F (49–60°C) for 2 to 4 days. The nitrogen-rich cottonseed hulls and other supplements contribute to the self-heating of the substrate. When the mass is cooled to 90–100°F (32–38°C), grain spawn is thoroughly mixed through. The bed temperature is usually 15–20°F (8–12°C) above room temperature during colonization. Four to five days after spawning, a moist, shallow casing layer can be provided to stimulate even mushroom formation, which usually occurs 4 to 6 days later.

Suggested Agar Culture Media: MYA, OMYA, and PDYA.

1st, 2nd, and 3rd Generation Spawn Media: Rice, rye, wheat, sorghum, milo, corn, or millet in liter or quart bottles. The final spawn is usually provided in convenient-to-use 5- to 10-liter plastic bags.

Substrates for Fruiting: Straw, preferably rice, hardwood sawdusts. Wheat straw also supports fruitings, although not nearly as well as supplemented, composted rice straw. One study showed that the best supplement for wheat straw is wheat bran (5%) and/or cotton hulls (10%). (See Li et al., 1988.) The pH optimum for fruiting falls between 7.5 and 8.0.

Recommended Containers for Fruiting: Trays, bags. Outdoor methods use no containers. The substrate is shaped into long rectangular mounds, narrowing at the top. The frames are covered with loose rice straw, cloth, or plastic to retain humidity.

Paddy Straw mushrooms are best in their egg form.

Yield Potentials: On average, *V. volvacea* produces two substantial flushes of mushrooms in quick succession, with the first giving 75% of the total yield and the second producing the remaining 25%.

Harvest Hints: For the best flavor as well as the best form for market, the mushrooms should be picked before the universal veil breaks, i.e., in the egg form. In the matter of hours, egg-shaped fruitbodies develop into annulate fruitbodies. Light has a governing influence on the color and overall quality of the harvestable crop.

Form of Product Sold to Market: Fresh to local markets and usually canned for export. Rarely sold in a dried form.

Nutritional Content: 26–30% protein, 45–50% carbohydrates, 9–12% fiber, and 9–13% ash. According to Ying (1987), this mushroom is rich in vitamins C and B, minerals, and assorted amino acids.

Medicinal Properties: None known to this author.

Flavor, Preparation, and Cooking: Sliced thin and stir-fried or as a condiment for soups. Used in a wide array of Asian dishes. I like to inject onion-soaked soy (or tamari) sauce via syringe into each Paddy Straw egg, cover with foil, and bake in an oven at 375°F (190°C) for 30 to 45 minutes. This mushroom, when eaten whole, explodes in your mouth creating a flavor sensation *par excellence*. Canned Paddy Straw mushrooms fail to provide a flavor experience comparable to fresh *V. volvacea*.

Comments: This mushroom is widely cultivated by farmers in China, Thailand, Vietnam, and Cambodia for supplemental income. Spawn is purchased from a central laboratory and expanded as much as possible. The egg-form is self-preserving, limiting the loss of moisture and extending shelf life.

This mushroom can be grown on uncomposted straw-based substrates although yields are substantially improved if the substrate is "fermented" or short-cycle composted. If rice straw is composted with supplements for 4 to 5 days, pasteurized, and inoculated, yields can be maximized. "Green composting" is the simultaneous inoculation of the compost while it is being formulated. The heat generated within the composting straw/cottonseed hull mass accelerates the growth of the thermophilic and heat-tolerant *Volvariella volvacea*.

In China, Paddy Straw mushrooms are sold in great quantities in outdoor markets.

Paddy Straw mushrooms fruiting on pasteurized rice straw beds in Thailand.

Cultivating the Paddy Straw mushroom is difficult in most regions of North America. Louisiana, Georgia, Alabama, and certain coastal regions of Texas have climates suitable for outdoor cultivation, as do regions of Mexico. Should temperatures fall below 70°F (21°C) for any period of time, fruitings will be limited and the critical increase in temperature caused by thermogenesis will be forestalled. Additionally, most cultures die when chilled below 45°F (7.2°C) unless they are flash-frozen in liquid nitrogen.

The most comprehensive English-language book on the cultivation of this mushroom is by S. T. Chang, published in 1972 entitled *The Chinese Mushroom (Volvariella volvacea): Morphology, Cytology, Genetics, Nutrition and Cultivation*. This book is now out of print and is much sought after. It needs to be reprinted. See also Ho (1971).

Renowned mycologist Thaithatgoon with Paddy Straw mushrooms.

The Noble Polypore, *Bridgeoporus nobilissimus,* the first mushroom to earn the title of an endangered species, growing on a stump in the Old Growth Forest, Mt. Rainier National Park, Washington State, U.S.A.

The Polypore Mushrooms

O f all the medicinal mushrooms, the polypores reign supreme, having been used for millennia. Polypores are more often incorporated into the pharmacopeia of native peoples than any other type of mushroom. Historically, cultures from tropical Amazonia to the extreme northern subpolar zones of Eurasia have discovered the power of polypores in preserving and improving human health. Polypores have also figured prominently in the cosmological view of native peoples, often being referred to as sources of eternal strength and wisdom.

The Agaria of Sarmatia, a pre-Scythian culture, used a polypore at the time of Christ to combat illness. They bestowed this polypore with the name of *agarikon*, undoubtedly to honor its value to their society. The Greek philosopher Dioscorides, recorded its name as *agaricum* circa A.D. 200. Its use persisted throughout the Middle Ages and was prescribed as one of the herbal remedies for tuberculosis. Researchers believe this mushroom was *Fomitopsis officinalis*, and/or *Fomes fomentarius*, canker parasites of conifer trees. Specimens of *Fomitopsis officinalis* have been retrieved from the graves of Pacific Northwest Coast Indian shamans. Thirteen individual carvings of this wood conk were collected in the late 1800s and mistakenly thought to be woodcarvings until Blanchette et al. (1992) studied them. Used as a poultice to relieve swellings, inflammations, and sweating, this conk was called "the bread of ghosts," and imparted supernatural powers. In Haida mythology this polypore is also directly connected to the origination of woman, and a protector of the female spirit.

Throughout the past twenty years, I have been repeatedly told by travelers to Mexico, South America, and the Middle East of Christian churches who have for centuries paid homage to crosses in whose centers were glass spheres housing what appeared to be a species of wood conk. To this day, the identity of this revered conk remains shrouded in mystery. These are but a few examples. One naturally wonders how many species and uses have not yet come to the attention of Western science, perhaps forever obscured by the passage of time.

"Drawing of an argillite plate, carved by Charles Edenshaw in approximately 1890, depicting the Haida myth of the origin of women. Fungus Man is paddling the canoe with Raven in the bow in search of female genitalia. Of all the creatures that Raven placed in the stern of the canoe only Fungus Man had the supernatural powers to breach the spiritual barriers that protected the area where women's genital parts were located...." Redrawn from a photograph, courtesy of the Field Museum of Natural History, Chicago (Blanchette et al. 1992, p. 122).

In the fall of 1991, hikers in the Italian Alps came across the well-preserved remains of man who died more than 5,300 years ago. Dubbed the "Iceman" by the news media, he was well equipped with a knapsack, flint axe, and a string of dried Birch Polypores. (Birch Polypores, *Polyporus betulinus*, are now known as *Piptoporus betulinus*.) These polypores, like many others, can be used as tinder, for starting fires, and medicinally, in the treatment of wounds. Further, by boiling the mushrooms, a rich tea with anti-fatiguing, immunoenhancing, and soothing properties can be prepared. Of the essentials needed for travel into the wilderness, this intrepid adventurer had discovered the value of the noble polypores. Ironically, as a group, the polypores remain largely unexplored. Their use may had been better known a thousand years ago than today.

Reishi or Ling Chi, *Ganoderma lucidum*, is the best known of all the polypores. The pattern of two outward spirals seen on top of the cap is an oft-repeated artistic motif in China since the Ming Dynasty. Its frequent use in art is a tribute to the esteem in which this mushroom has been held. So extensive are the medicinal claims for this fungus, that this mushroom is called the "Panacea Polypore." Claimed to cure cancer, heart disease, diabetes, arthritis, high-altitude sickness, sexual impotency, and even chronic fatigue syndrome, it is no wonder that this mushroom has been for centuries heralded as "The Mushroom of Immortality."

Three other notable polypores enjoying reputations as medicinal fungi are Maitake, *Grifola frondosa*; Zhu Ling, *Polyporus umbellatus*; and Yun Zhi (Turkey Tail), *Trametes versicolor*. Maitake has been found to be effective, *in vitro*, against the HIV

The spiraling insignia often adorned the breastplate of nobility in China, denoting their rank and social status, and was associated with long life and good fortune.

virus by the National Cancer Institute of the National Institute of Health's anti-HIV drug screening program. Turkey Tail and other mushrooms have also shown antiviral activity (Collins and Ng, 1997; Piraino,1999). During a visit to the Institute of Materia Medica in Beijing, *Polyporus umbellatus* was reported to Stamets and Weil (1983) as being exceptionally effective against lung cancer. Aqueous extracts (tea) were given to patients directly after radiation therapy, with promising results. Polypores offer a tremendous resource for novel antiviral, antibacterial, tumorcidal, and immunopotentiating drugs.

The polypores covered in this book are Reishi, *Ganoderma lucidum*); Maitake, *Grifola frondosa* (=*Polyporus frondosus*); Zhu Ling, *Polyporus umbellatus* (=*Grifola umbellata*); and Turkey Tail, *Trametes versicolor*. Many other polypores, such as *Laetiporus sulphureus* (=*Polyporus sulphureus*), can be grown on stumps. Future editions of this book will expand on the number of polypore species that I have successfully cultivated. A short list of these candidates includes, but is not limited to.

> *Albatrellus* spp.
> *Daedalea quercina*
> *Fomes fomentarius*
> *Fomitopsis officinalis*
> *Ganoderma applanatum*
> *Ganoderma curtisii*
> *Ganoderma oregonense*
> *Ganoderma neo-japonicum*
> *Ganoderma sinense*
> *Ganoderma tsugae*
> *Inonotus obliquus*
> *Laetiporus sulfureus*
> *Oligoporus* spp.
> *Oxyporus nobilissimus*
> *Phellinus linteus*
> *Piptoporus betulinus*
> *Polyporus indigenus*
> *Polyporus tuberaster*
> *Polyporus saporema*
> *Trametes cinnabarinum* (=? *Pycnoporus* *cinnabarinus*)
> *Wolfiporia cocos*

Polypores are premier wood decomposers, producing annual or perennial fruitbodies. None are known to be poisonous, although some people have allergic reactions to certain species. Some people taking MAO-inhibitor antidepressant medication can have allergic reactions to the edible polypores containing tyramine. Chicken-of-the-Woods, *Laetiporus (Polyporus) sulphureus*, has been reported to contain alkaloids similar to those found in plants known to be psychoactive, like kava kava (Lincoff and Mitchel 1977).

The cultivation of these species can take several tracks. One track is to simply inoculate hardwood logs as with the cultivation of Shiitake. If you bury the inoculated logs in sawdust or soil, moisture is better preserved, and fruitings extend over several years. Stumps can also be inoculated, although if other fungi have already captured that niche, production is inhibited. In outdoor environments, the first flushes of mushrooms are often delayed, not showing for several years after inoculation. However, since polypores are naturally lower in moisture and require less water, outdoor patches require less maintenance than indoor methods.

By far the most dependable and rapid production system is the cultivation of polypores indoors under controlled environmental conditions. Several techniques lead to success. One of the main differences between the cultivation of polypores versus the fleshier, gilled mushrooms is that the polypores do not enjoy, nor require, the heavy watering schedules and high humidities of the gilled mushrooms. Like most mushrooms, the polypores are sensitive to carbon dioxide levels and light conditions. The development of the fruitbodies are extremely responsive to changes within the growing room environment. Many cultivators manipulate the environment to elicit substantial stem formation before cap development. Some polypore species produce better fruitings if the substrate block is compressed after colonization. Other differences, unique to each species, are outlined in the forthcoming growth parameters.

Reishi or Ling Chi of the Genus *Ganoderma*
Ganoderma lucidum (Wm. Curtis: Fries) Karsten

Introduction: A mushroom of many names, *Ganoderma lucidum* has been used medicinally by diverse peoples for centuries. The Japanese call this mushroom Reishi or Mannentake (10,000-Year Mushroom) whereas the Chinese and Koreans know it as Ling Chi, Ling Chih, or Ling Zhi (Mushroom or Herb of Immortality). Renowned for its health-stimulating properties, this mushroom is more often depicted in ancient Chinese, Korean, and Japanese art than any other. Ling Chi is traditionally associated with royalty, health and recuperation, longevity, sexual prowess, wisdom, and happiness. Ling Chi has been depicted in royal tapestries, often portrayed with renowned sages of the era. For a time, the Chinese even believed this mushroom could bring the dead to life when a tincture specifically made from it was laid upon the dead person's chest.

The use of *Ganoderma lucidum* spans more than two millennia. The earliest mention of Ling Chi was in the era of the first emperor of China, Shih-huang of the Ch'in Dynasty (221–207 B.C.). Henceforth, depictions of this fungus proliferated through Chinese literature and art. In the time of the Han Dynasty (206 B.C.–A.D. 220) while the imperial palace of Kan-ch'uan was being constructed, Ling Chi was found growing on timbers of the inner palace, producing nine "paired leaves." So striking was this good omen, that henceforth emissaries were sent far and wide in search of more collections of this unique fungus. Word of Ling Chi thus spread to Korea and Japan whereupon it was elevated to a status of near reverence.

A Tibetan Ling Chi "Tree" statuette made of wood, from pre-A.D.1600. Revered and protected in the Lama Temple, Beijing.

This mushroom is known by many in North America and Europe as one of the "Artist's Conk" fungi. (The true Artist Conk is *Ganoderma applanatum*.) As the fruitbody develops, the spore producing underlayer—the hymenium—is white and can be drawn upon. As the pores are crushed, a browning reaction occurs, thus allowing the artist to sketch an image.

An exquisite "antler" form of Ling Chi featured in a Chinese mushroom museum.

Small golden statue in the Rijks Museum, Amsterdam, of an antlered deer from the eighteenth century China ("Qing Period") in which the deer had within its mouth an ornate object, described as "a toadstool of everlasting life." In Eurasian art, the deer has been associated with both *Ganoderma lucidum* and *Amanita muscaria*.

Common Names: Reishi (Japanese for "Divine or Spiritual Mushroom")
Ling Chi, Ling Chih, Ling Zhi (Chinese for "Tree of Life Mushroom")
Mannentake (Japanese for "10,000-Year Mushroom," "Mushroom of Immortality")
Saiwai-take (Japanese for "Good-Fortune Mushroom")
Sarunouchitake (Japanese for "Monkey's Seat")
The Panacea Polypore

Taxonomic Synonyms and Considerations: *Ganoderma lucidum* is the type mushroom, the pivotal species around which the genus concept is centered. *Ganoderma lucidum* grows on oaks, and other hardwoods, whereas two close relatives, *G. tsugae* and *G. oregonense*, grow primarily on conifers. *G. tsugae* grows on hemlocks, as its name implies, while in the Southwest of North America, this species has been reported on white fir, *Abies concolor*. In culture, *G. lucidum* and *G. tsugae* develop long stems in response to manipulation of the environment. *G. oregonense* can be found on a variety of dead or dying conifers, including *Tsuga*. Given how mutable the formation of the stalk is under different cultural conditions, and that *Ganoderma lucidum* readily fruits on a variety of conifer and hardwood sawdust mixtures, delineation of these individuals based solely on habitat seems highly suspect.

G. oregonense is a much more massive mushroom than *G. lucidum* and is characterized by a thick pithy flesh in the cap. Also *G. oregonense* favors colder climates whereas *G. lucidum* is found in warmer regions. (*G. lucidum* has not been reported from the Rocky Mountain and Pacific Northwest regions.) *Ganoderma curtisii*, a species not recognized by Gilbertson and Ryvarden, but acknowledged by Zhao (1989) and Weber (1985),

G. *lucidum* (ATCC 52412) 4 and 10 days after inoculation onto malt extract agar. The zonations are a strain-specific feature.

G. *lucidum* (Forintek's 34-D) 7 days after inoculation onto malt extract medium.

grows in eastern North America, and is distinguished from others by the predominantly yellowish colored cap as it emerges. These North American "Reishis"— *Ganoderma lucidum, G. curtisii, G. oregonense,* and *G. tsugae*—represent a constellation of closely related individuals, probably stemming from a common ancestry. The argument for retaining them as separate species may be primarily ecological and host specific and not biological. One of the few cultural distinctions described by Adaskaveg and Gilbertson (1986) is that *G. lucidum* produces chlamydospores in culture whereas *G. tsugae* does not.

One historic and notable attempt to distinguish the North American from the Far Eastern taxa can be found in an article published by R. Imazeki (in Japanese) titled "Reishi and *Ganoderma lucidum* that grow in Europe and America: Their Differences," in 1937. Currently, the best treatises discussing the taxonomy of these polypores are

Gilbertson and Ryvarden's 1987 monograph, *North American Polypores: Vol. I and II* and Zhao's *The Ganodermataceae in China* of 1989. The spore size of *G. lucidum* is smaller than the inclusive range of 13–17 μ in length by 7.5–10 μ in width characteristic of *G. oregonense* and *G. tsugae*. Nevertheless, Gilbertson and Ryvarden did not consider this feature to be more significant than habitat when delineating these three taxa in their "Key to Species." Placing emphasis on habitat may also be a dubious distinction when considering these species produce fruitbodies on nonnative woods when cultivated. Features of higher taxonomic significance—such as interfertility studies and PCR-rDNA fingerprinting—are needed to support accurate and defensible species delineation. For instance, interfertility studies with some collections reveal that *G. curtisii* (Berk.) Murr. may merely be a yellow form of *G. lucidum* common to the southeastern United States. (See Adaskaveg and Gilbertson, 1986 and 1987, and Hseu and Wang, 1991.)

Buried log cultivation of Ling Chi in China.

In Asia, *Ganoderma lucidum* has a number of unique allies. Most notably, a black-stalked *Ganoderma* species, also considered to be a Reishi, is called *Ganoderma japonicum* Teng (=*Ganoderma sinense* Zhao, Xu et Zhang, colloquially known as Zi zhi.)

Shaded log cultivation of Reishi in Japan.

Description: Conk-like or kidney-like in shape, this woody-textured mushroom, 5–20 cm in diameter, has a shiny surface that appears lacquered when moist. The cap can be a dull red to reddish brown, and sometimes nearly black in color. Featuring pores on its underside, whitish, browning when touched. Areas of new growth whitish, darkening to yellow brown and eventually reddish brown at maturity, often with zonations of concentric growth patterns. Spores dispersed from the underside collect on the surface of the cap giving it a powdery brown appearance when dry. Stem white to yellow, eventually darkening to brown or black, eccentrically or laterally attached to the cap, usually sinuous, and up to 10 cm in length × .5–5.0 cm thick.

Distribution: This mushroom is widely distributed throughout the world, from the Amazon through the southern regions of North America and across much of Asia. This mushroom is less frequently found in temperate than in the subtropical regions.

Natural Habitat: An annual mushroom, growing on a wide variety of woods, typically on dead or dying trees, primarily on deciduous woods, especially oaks, maple, elm, willow, sweetgum, magnolia, locust, and in Asia, on plums. Found on stumps, especially near the soil interface, and occasionally on soils arising from buried roots. Occurring from May through November, and more common in warm temperate regions. In the southeastern and southwestern United States, *Ganoderma lucidum* is frequently found in oak forests. In the northeastern states, this species is most common in maple groves. This mushroom often rots the roots of aging or diseased trees, causing them to fall. (This is one of the "white rot" fungi that foresters know well.) From the darkened cavity of the upturned root wad, where carbon dioxide levels are naturally higher and light levels are low, long stalked mushrooms arise. These rare, multi-headed, antler-like forms are highly valued in Asia. (See page 353.)

Microscopic Features: Spores reddish brown, ellipsoid with a blunted end, roughened as in warty, 9–12 × 5.5–8 μ, two-walled, with spaced, internal "inter-wall pillars." Cystidia absent. Clamp connec-

Shaded log cultivation of Reishi in the United States (Louisiana).

tions present, otherwise hyphae aseptate. Hyphal system dimitic. Chlamydospores forming in cultured mycelium.

Available Strains: Yellow, red, purple, and black strains are widely available from most culture libraries. New strains are easily cloned from the wild, best taken from young fruitbodies from

After inoculation, the logs are placed into pot and covered with sand.

G. lucidum fruiting from pots containing inoculated oak logs topped with soil.

the central flesh leading to the disk or alternatively from the edge of the developing cap margin. Because of their woody texture, a sturdy and razor-sharp surgical scalpel is recommended. Recovery or "leap-off" may take 2 weeks. Once in culture, many strains grow rapidly and fruit on sawdust substrates. Each tissue culturist should note that those strains isolated from conifers might actually be *Ganoderma oregonense* or *Ganoderma tsugae* although they have the appearance of *Ganoderma lucidum*. Forintek's 34-D produces a reddish brown fruitbody and is popular among North American cultivators. American Type Culture Collection's #52412, which was used as the isotype for a taxonomic discussion by Wright and Bazzalo (*Mycotaxon* 16: 293–295, 1982), produces a multitude of rapidly grown antlers. (See page 353.)

Each strain is unique in its pattern of growth from agar media-to-grain-to-wood–based substrates. A notable difference between strains is that one group, when over-incubated, makes grain spawn nearly impossible to loosen into individual kernels upon shaking. The other, smaller group of strains allows easy separation, even when over-incubated.

Mycelial Characteristics: Longitudinally radial, nonaerial, initially white, rapid growing, becoming densely matted and appressed, yellow to golden brown, and often zonate with age. Some strains produce a brown hymenophore on MEA. A 1-cm square inoculum colonizes a 100 × 15-mm petri plate in 7 to 10 days at 75°F (24°C). Soon after a petri dish is colonized (2 weeks from inoculation), the mycelium becomes difficult to cut and typically tears during transfer. Culture slants can be stored for periods of 5 years at 35°F (1-2°C).

Fragrance Signature: Musty, mealy, not sweet, not pleasant.

Natural Method of Cultivation: *G. lucidum* can be grown via a wide variety of methods. In Southeast Asia, this mushroom is a common saprophyte on

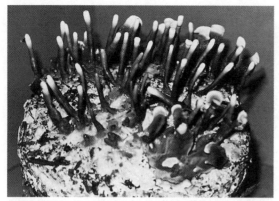 

Antler formation of *Ganoderma lucidum* is controlled by the carbon dioxide of the prevailing environment. (ATCC #52412)

diseased palm oil trees, giving plentiful "spontaneous" crops, which are now recognized as being a marketed commodity. Dead trees that support this fungus can be placed near newly-cut logs so that they can become infected. In China and Japan, the current method is to inoculate logs and lay them on the ground or shallowly bury them. (See pages 355–356.) The logs are placed in shady, naturally moist locations. By covering hoop-frames with shade cloth, light exposure and evaporation are reduced, creating an ambient environment conducive for fruitbody development. Typically 6 months to 2 years pass before substantial harvests begin, and continue for 4 to 5 years. (For more information, see Hengshan et al., 1991, and Thaithatgoon et al., 1993.) This method gives rise to natural-looking mushrooms.

Using Nature as an example, this mushroom grows prolifically on stumps. Hardwood stumps can be inoculated using any of the various methods described in this book. Cultivators living in high-humidity climates with prolonged growing seasons (such as Louisiana and elsewhere in the humid southeastern United States) have success growing *Ganoderma lucidum* on hardwood logs laid directly onto the ground. Individual preferences vary among cultivators who, by nature, tend be a secretive or reluctant-to-communicate breed.

A quasi-natural method is to inoculate short hardwood logs and place them into nursery-style pots. (See pages 356–357.) The pots are then filled with hardwood sawdust and topped with soil. Large greenhouses, covered with dense shade cloth, can house thousands of these individual containers that are simply laid out as a single layer over a gravel rock floor.

Rapid Cycle System for Indoor Cultivation: A unique combination of variables can be orchestrated to effect fruitbody formation. The indoor method I have developed calls for spawn inoculation onto a 50:50 (by volume) hardwood sawdust/wood chip mixture incubated in polyethylene space bags. For 3 to 4 days, the hardwood (alder/oak) wood chips are soaked/fermented in molasses-enriched water (50 ml molasses/20 liters water). After this soaking, the 17.50 × 8.25 × 4.75-inch bags are filled to 3 pounds wet weight. The bags are then sterilized for 2 hours at 15 psi. Upon cooling the bags are opened within a clean room. Grain or sawdust spawn is distributed equally into each bag. The bags are exposed to the airstream from a laminar flow bench, causing partial inflation. Directly thereafter, the bags are heat-sealed. The bags appear domed or inflated. In effect, an idealized, positive-pressurized, humidified environment is created. Slow gas

Some strains will not form a stem if carbon dioxide is maintained at atmospheric levels. White margin is new growth.

exchange occurs through the semipermeable microporous filter media patch.

Colonization is usually complete in 14 to 21 days at 75°F (24°C). Thirty to forty days after inoculation, the first mushrooms begin to emerge from the rough micro-topography of the sawdust/chip media. The emerging fruiting bodies are whitish to golden yellow in color and apically triangular in shape. Growth is slow, yet noticeable from day to day. The tongue-like formations yellow with age, and becoming progressively more reddish brown towards the base. By day 50 they have often achieved 4 inches in length, are branched, having arisen from multiple sites on the surface plane of the wood chip media. Frequently, these antler-stalks seek out the filter patch and become attached to it. Once the desired height of stalk formation has been achieved, the environment must be altered so as to proceed to the final stage. If the cultivator does not expose these

emerging antlers to near-natural atmospheric conditions, the opportunity for conk development will soon be lost.

For the previous two months, the entire growth cycle has occurred within the environment of the sealed plastic bag, wherein carbon dioxide and other gases accumulate to relatively high concentrations. Although the filter patch allows the slow diffusion of gases, it acts as a barrier to free air exchange. Often the interior plenum has carbon dioxide levels exceeding 20,000 ppm or 2%. Once the bag is opened, and a free rate of gas exchange prevails, carbon dioxide levels drop to near normal levels of 350 ppm or .035%. This sudden change in carbon dioxide levels is a clear signal to *Ganoderma lucidum* that cap development can begin.

In its natural habitat, this change is analogous to the stalk emerging from the rich carbon dioxide environment below ground level. Once the CO_2-sensitive stalk emerges into the open air, the photosensitive, spore-producing lateral cap develops. The caps form above the plateau of the ground and orient towards directional light. An indication of new growth is the depth and prominent appearance of a white band around the cap's edge. Under these conditions, cap formation is rapid and the time of harvest is usually indicated by the lack of new

When the plastic is removed, exposing the mycelium, *G. lucidum* will only form when a condensing fog environment is maintained for a prolonged period. If the bag is left on, less moisture is required.

DXN production crew filling bags with sawdust mixture for Reishi cultivation. (Malaysia)

margin growth and the production of rusty brown spores. The spores, although released from below, tend to accumulate on the upper plane of the cap.

The cultivator has two alternatives for eliciting conk development once antlers have begun to form. The plastic bag can be left on or stripped away from the mass of mycelium/wood chips/sawdust. If the protective plastic is removed and a fog-like environment does not prevail, massive evaporation will halt any fruitbody development. If the exposed block is maintained in a fog-like environment within the growing room, vertical stalk growth slows or abates entirely and the characteristic horizontal kidney-shaped cap begins to differentiate. Like the stalk, the margins of new growth are whitish while the aged areas take on a shiny burgundy brown appearance.

Cultivators in Asia inoculate 1- to 2-liter cylindrical bags or bottles, narrowly closed at one end and stopped with a cotton plug. Once inoculated, the bags or bottles are stacked horizontally in a wall-like fashion. After 30 to 60 days, depending upon the strain, inoculation rate, and growing conditions, the cotton filters are removed. The small opening channels CO_2, stimulating stem elongation. This same opening is also the only conduit for moisture loss. From this portal, finger-like primordial shoots emerge into the high-humidity environment of the growing room. With this method of cultivation, moisture is conserved and channeled to the developing mushrooms. Under low light conditions, stem elongation slows as the mycelium enters into the conk formation period. Cultivators of this method believe that the substantial substrate mass, protected from evaporation, produces better flushes than from a substrate exposed to the open atmosphere. With this second strategy, a condensing fog environment is not as critical as when the substrate is fully exposed to the air. Furthermore, contamination is less likely. Hence growers in Thailand are successful in growing *G. lucidum* in growing rooms featuring gravel floors and equipped with a minimum of environmental controls.

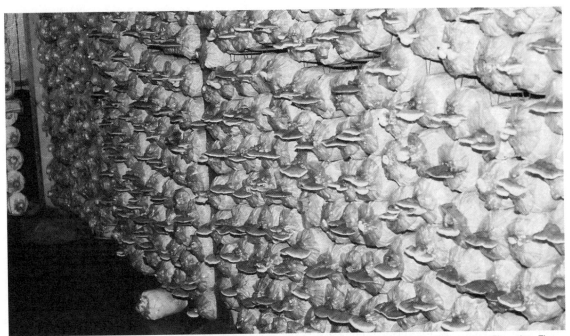

In Asia, many cultivators fruit *G. lucidum* from bags through the narrow opening that once hosted the cotton filter plug. The plastic is left intact to help retain the resident moisture.

Cylindrical bags are stacked upon one another to form a Reishi wall.

With either rapid cycle method, the cultivator can grow the archetypal form of *Ganoderma lucidum*, one long favored by the Chinese, Japanese, Koreans, and Thai people. From time of inoculation to time of harvest is less than 3 months. If only antler-shaped fruitbodies form, yields are approximately ¼ of the yield seen if caps are encouraged to form and mature. Should the cultivator's goals be solely that of yield and the development of the stalk is not desired, then yet another approach is recommended.

Five pounds of moistened hardwood sawdust/chips (60–70% moisture) are filled into the autoclavable spawn plastic bags. The bags with their tops folded over are loaded into the autoclave and sterilized. After sterilization, approximately 100–150 grams of rye grain spawn is inoculated into the bags subsequent to autoclaving. Each is heat-sealed. Colonization is characteristically rapid and complete in 10 to 20 days. After 30 days, the blocks are removed to the growing room whereupon the plastic is perforated. I use four-bladed stainless steel arrowheads mounted on a board. The bags are never opened. Each one is forcibly slammed downwards into the arrowheads. These "+" shaped slits become the sites for fruitbody formation. The bags are placed in the growing room and harvests of stemless conks usually begin within a month. (See page 210.)

Reishi are slow growing, taking weeks to mature.

Recommended Courses for the Exponential Expansion of Mycelial Mass to Achieve Fruiting: There are several courses for expanding the mycelium to the fruiting stage. Each step results in an exponential expansion of mycelial mass. The simplest method for most cultivators to follow is similar to the classic spawn expansion schedule employed by most laboratories. The first stage is to grow the mycelium on agar media in petri dishes. The next is transferring pure cultures onto sterilized grain, typically in jars (quart, half gallons, and gallons). At 75°F (24°C), 2 to 3 weeks pass before colonization is complete. Each of these grain masters can inoculate 10 gallon jars each containing 1,000–1,200 grams of sterilized rye grain. Once grown out, each gallon jar of spawn can readily inoculate ten 5-pound bags of hardwood sawdust (wet weight, 65–70% moisture). The resulting sawdust spawn is

the last step before inoculating a substrate capable of supporting fruitbodies.

At this juncture, the cultivator has several options: to grow fruitbodies on sawdust/chips; to grow fruitbodies on buried logs, to inoculate stumps; to fruit *Ganoderma lucidum* on vertically arranged columns of heat-treated sawdust. (This last method is currently under development by the author.) These four strategies all follow the same expansion schedule from agar-to-grain-to-sawdust-to-production block.

Suggested Agar Culture Media: MEA, OMYA, PDYA, and/or DFA. Uninhibited by gentamycin sulfate (¹⁄₁₅th gram/liter.)

1st, 2nd, and 3rd Generation Spawn Media: Rye grain, wheat grain, other cereal grains. Fruitbodies do not form on most grains except milo (a type of sorghum), whereupon fans of growth climb the inside surfaces of the spawn containers and fruit within. If mature grain spawn is not used directly after colonization, over-incubation results, making it difficult to disperse the grain kernels upon shaking. In this case, a transfer tool such as a sterilized spoon, knife, or similar tool is needed to break apart the spawn within the jar.

I prefer the liquid-inoculation method of generating spawn similar to that described in *The Mushroom Cultivator* by Stamets and Chilton (1983). Further, the mycelium of this species readily adapts to submerged fermentation. Submerged fermentation (liquid culture) of *Ganoderma lucidum* mycelium is now considered "traditional" in China.

Each unit of primary spawn can be expanded (3 cups grain in $^1/_2$ gallon jar) into 10 (5 to 20) units of 5 pounds sterilized, moist sawdust. Once inoculated, and incubated at 75°F (24°C) colonization is complete in 8 to 12 days. The 10 sawdust spawn blocks can be expanded into 100–200 3- to 5-pound sawdust/chip bags, which in turn, are colonized in a similar period of time. The need for additional supplements is strain-determined.

Substrates for Fruiting: Indoors on hardwood sawdust/chips. 5% supplementation of the sawdust with rice bran or sorghum enhances yields with most strains. From my experiences, I have found that over-supplementation with rice bran, beyond 15% of the dry mass of the substrate, inhibits fruitbody development.

Recommended Containers for Fruiting: Polypropylene bottles, bags, and/or similar containers.

Yield Potentials: From my experiments, yields on first flush average between 125–200 grams wet mass from 2,200–2,300 grams wet mass in 30 to 60 (90) days via the rapid cycle system. (Fruitbodies are 70–80% water, 10–20% less in moisture content than what is typical of the more fleshy fungi.) Second flushes 25–50% of first. Yields from log/stump culture are approximately 1–2 pounds per year.

Harvest Hints: If polypropylene bags are punctured, a stemless conk is produced under well-lit, low carbon dioxide conditions. With this strategy, a broad conk snaps off cleanly from a $^1/_4$ inch hole. If stems are encouraged to form by raising carbon

GROWTH PARAMETERS

Spawn Run:
Incubation Temperature: 70–80°F (21–27°C)
Relative Humidity: 95–100%
Duration: 10–20 days
CO_2: tolerated up to 50,000 ppm or 5%
Fresh Air Exchanges: 0–1 per hour
Light Requirements: n/a

Primordia ("Antler") Formation:
Initiation Temperature: 65–75°F (18–24°C)
Relative Humidity: 95–100%
Duration: 14–28 days
CO_2: 20,000–40,000 ppm
Fresh Air Exchanges: 0–1 per hour
Light Requirements: 4–8 hours at 200–500 lux

Primordia ("Young Conk") Formation:
Temperature: 70–80°F (21–27°C)
Relative Humidity: 95–100%
Duration: 14–28 days
CO_2: 5,000–2,000 ppm
Fresh Air Exchanges: as required for maintaining desired CO_2
Light Requirements: 12 hours on/off at 500–1,000 lux

Fruitbody Development
Temperature: 70–80°F (21–27°C)
Relative Humidity: 90–95%
Duration: 60 days
CO_2: <2,000 ppm
Fresh Air Exchanges: as required
Light Requirements: 12 hours on/off 750–1,500 lux

Cropping Cycle:
Two crops in 90–120 days

Brown spores, although released from the underside, tend to collect on the upper cap surfaces of *G. lucidum*.

dioxide or lowering ambient lighting, the mushrooms can be harvested by first twisting the stem base from the substrate and then trimming debris from the stem base. At 50% RH, mushrooms dry quickly in the open air at room temperature. After the mushrooms have dried, some cultivators short-cycle sterilize their mushrooms by placing them into the autoclave. This heat-treatment retards or prevents the birth of any insect larvae from eggs that may have been deposited during mushroom development.

Form of Product Sold to Market: Dried, whole mushrooms have been traditionally used in Oriental medicine. However, many other forms are marketed, including in pill, tea, and tincture forms. Ginseng and Ling Chi are often extracted and combined in liquid form by a number of Chinese pharmaceutical collectives. I have also seen cultured mycelium extracted for use in syrups. Antler forms are often preserved as works of art, portrayed in museums or temples, handed down through generations as family heirlooms, and even sold to tourists. (See page 210.) The use of Ling Chi in beers and wines as a medicinal/flavor additive is also popular in Japan and China.

Nutritional Content: Not known to this author.

Medicinal Properties: For centuries, the Chinese and Japanese literature has heralded this mushroom for its health-invigorating effects, especially attributing it with increasing longevity, treatment of cancer, and resistance to and recovery from diseases. Himalayan guides have used this mushroom to combat high-altitude sickness. Mayan Indians traditionally employed Ganoderma in teas to fight a variety of communicable diseases. Reishi has become the natural medicine of choice by many who are afflicted with immunological disorders.

A complex group of polysaccharides, especially the beta-glucans, have been isolated from this mushroom that reportedly stimulate the immune system. (One recent analysis shows our Reishi has 40.6% beta-glucans, an unusually high concentration.) One theory is that these polysaccharides stimulate the production of helper "T" cells, which attack infected cells. Ganoderic acids have also been isolated from Ling Chi and purportedly have anti-coagulating effects on the blood and lower cholesterol levels. (See Morigawa et al., 1986.) Some studies have been published showing its modulating effects on blood pressure and lipid levels (Kabir et al. 1989), influence on blood glucose levels (Kimura et al. 1989), and presence of immunomodulating proteins (Kino et al. 1989). T. Mizuno summarized the historical development of compounds extracted from this mushroom in *Chemical Times* (Volume 3: 50–60, 1989). Yun et al. (1995) found that in a new mouse model to study lung cancer, Reishi significantly reduced the incidence of tumors. Reishi has also been recommended as a chemopreventive agent against cancer, and markedly reduced the damaging effects of superoxide anions (Kim and Kim, 1999; Kim et al. 1999). Mizushina et al. (1998) reported two unique DNA polymerase alpha-type inhibitors from Reishi fruitbodies. Alcohol extracts of *Ganoderma lucidum* significantly reduced T-4 leukemia cells *in vitro* (Lovy et al. 1999). Curiously, these mushrooms produce compounds that are cytotoxic to the growth of cancer cells, while at the same time activating a host-mediated immune response. Studies have shown that, even at large doses, Reishi is non-toxic to healthy cells and is safe to consume (Kim et al. 1986). For more information on Reishi's medicinal properties, see Nishitoba et al. (1984, 1997), Jifeng et al. (1985), Hirotani et al. (1985), Sato et al. (1986), Tanaka (1989), and Mizuno (1995, 1996). For current updated information on research with medicinal mushrooms, see Stamets and Yao (1999).

Studies by Stavinoha (1990) on mice at the Texas Health Science Center in San Antonio showed that unextracted "gill powder" (sic) or ether extracts of the mushrooms showed significant anti-inflammatory activity, comparable to hydrocortisone. Weil (1993) noted that it is difficult to reconcile that a mushroom could be both an immune stimulator and an anti-inflammatory agent. His point is that anti-inflammatory agents generally suppress immune function, not enhance it. Many forms of rheumatoid arthritis are viewed as auto-immune disorders. If the mode of activity of *G. lucidum* is to act as an immunomodulator, not an immuno-stimulator, then this contradiction in viewpoints is reconcilable.

A novel antiviral compound has been isolated from these mushrooms and shown to reduce replication of HIV (Kim et al. 1994). More recently, an extract from this mushroom was found to protect DNA from damage from contact with free radicals and radiation (Kim and Kim 1999). Wang et al. (1997) found that polysaccharides isolated from Reishi potentiate macrophages and lymphocytes, drastically increasing interleukin-1 and interleukin-6. One of the most thorough descriptions of the active glucans, triterpenoids (ganoderic acids), can be found in Mizuno's publications (Mizuno 1988, 1995, and 1996). Nishitoba et al. (1987) found that, as the mycelium matured to form fruitbodies, the triterpene content correspondingly increased, and was more concentrated in older tissue.

I have been told by native Chinese that Ling Chi is traditionally given to men by women (or an intermediary messenger) to express sexual interest as Ling Chi purportedly stimulates sexual virility, especially in older men. The antler form is preferred. No medical evidence has been published, to my knowledge, to support this claim.

We allow Reishi grow to fully mature before harvesting.

Flavor, Preparation, and Cooking: Typically extracted in hot water for teas, tinctures, syrups, and soups. My family enjoys making a tea from fresh, living specimens, breaking them into pieces, boiling in water for 2 hours and then steeping for 30 minutes. The tea is reheated to a desired temperature, strained and served, without sweeteners. If a daily regimen of Ling Chi tea is followed, as little as 3–5 grams per person has been traditionally prescribed. The antler forms, with a reduced hymenial package, have a rich, mildly sweet, and soothing flavor reaction. The well-developed conk forms reveal a more bitter aftertaste. Yellow strains are more often bitter than the red and black strains.

Comments: A satisfying mushroom to grow and consume, *Ganoderma lucidum* is a mushroom whose transformations are mesmerizing. Responsive to the slightest changes in the environment, its unique growth habits have undoubtedly enchanted humans for centuries. The formation and development of the fruitbody is greatly affected by the surrounding gaseous environment. Stem growth is elongated under prolonged, elevated carbon dioxide levels (>20,000 ppm) whereas cell formation leading to cap development and hymenial development is activated when carbon dioxide levels fall below 2,000 ppm. *Ganoderma lucidum* can be easily grown in a variety of ways, indoors and outdoors. Yield may not be the only measure of this mushroom's value. Although more biomass is generated with a strategy promoting short stalks and large caps, the antler and capitate-antler form appeals to many as art.

As *Ganoderma lucidum* gains popularity with North Americans, feasibility studies on the wide-scale cultivation of Reishi on stumps are warranted. If markets could support the resulting yields, a whole new industry might emerge on lands currently providing little or no immediate economic return. This mushroom is a perfect fit in an agroforestry model, especially in the vast hardwood forest regions of the world.

For more information on the taxonomy of this group, see Zhao (1989) and Gilbertson and Ryvarden (1987). For general information on the historical and medicinal uses of this fungus, consult Wasson (1972), Willard (1990), Jones (1992), and Jong and Birmingham (1992). For more information on the cultivation of this mushrooms see Thaithatgoon et al. (1993). To stay updated on medicinal properties, consult the information booklet by Stamets and Yao (1999).

One year after the short logs were inoculated (see page 356), Reishi mushrooms form.

Elephant form of Reishi.

Maitake or Hen-of-the-Woods
Grifola frondosa (Dicks: Fr.) S.F.Gray

G. frondosa 4 and 10 days after inoculation onto malt extract agar.

Maitake is a delicious, soft-fleshed polypore with excellent nutritional and medicinal properties. Of the polypores currently being studied, *Grifola frondosa* continues to attract considerable attention from the nutriceutical and pharmaceutical industries, especially in Japan, Korea, and most recently, the United States. Several causal compounds appear to be at play, most notably the beta-glucans—especially the D-fraction constituents. As the density of the mycelium increases to create a mushroom, it is presumed that the available beta-glucans increase as well. (All Maitake mushrooms have within them the 1,3 and 1,6 D-fractions of beta-glucans.) The transformations Maitake undergoes as it grows—from gray mounds, labyrinthine folds, petals, to extended leaflets at full maturity—is one of nature's great displays of grace and being.

Wild fruiting of Maitake, *G. frondosa*, at base of oak tree in a colonial graveyard in upstate New York.

Common Names: Maitake ("Dancing
Mushroom")*
Kumotake ("Cloud
Mushroom")
Hen-of-the-Woods
The Dancing Butterfly
Mushroom

Taxonomic Synonyms and Considerations:
Synonymous with *Polyporus frondosus* Dick ex. Fr.
Closely allied to *Polyporus umbellatus* Fr. (also
known as *Grifola umbellata* Pers.: Fr.), which has
multiple caps arising from a common stem, a lighter

color, and a more fragile texture. The primordia of
G. frondosa are rich, dark gray brown to gray black
in color whereas the fruitbody initials of *G. umbellata* are light gray. Macroscopically, these two
mushrooms are easily distinguished by their
form. Microscopically, the spores *G. umbellata* are
substantially larger and more cylindrically shaped
than the spores of *G. frondosa*.

Description: A large, fleshy polypore, dark gray
brown when young, becoming lighter gray in age.
(Some varieties fade to a light yellow at maturity.)
Fruitbody is composed of multiple, overlapping

---

* The validity of the common name of Maitake or "Dancing Mushroom" naturally comes into question. According
to fable, the ingestion of Maitake caused a group of lost nuns to fall into an uncontrolled frenzied dance with a
band of woodcutters. The reaction attributed to the nuns finding this mushroom, their chance encounter with
the male woodcutters, and the party-like atmosphere that ensued, seems incongruous with the effects of this mushroom compared to, for instance, the effects of the well-known consciousness-raising psilocybian mushroom species
like *Gymnopilus spectabilis*, a far more likely candidate. (*G. spectabilis* is known as the Big Laughing Mushroom,
or O'warai-take.) The first report of Maitake was recorded in a tale from the eleventh century Japanese text *Konjaku
Monogatari*. Imazeki (1973) and Wasson (1973) first cast doubt about the authenticity of "Maitake" being *Grifola
frondosa*. Whatever species induced frenzied dance with the nuns and woodcutters, Maitake is today synonymous
with *G. frondosa*.

Mycologist Bill Chapman with a huge *Grifola frondosa*, known to Americans as Hen-of-the-Woods and to Japanese as Maitake. This specimen was 5 feet in circumference and estimated to weigh more than 40 pounds. Growing from a stately oak tree in a colonial graveyard in upstate New York, the location of this mushroom patch is a closely guarded secret. This patch consistently produces specimens of such magnitude and occasionally generates 100 pound clusters!

Natural Habitat: Found on stumps or at the base of dead or dying deciduous hardwoods, especially oaks, elms, maples, blackgum, beech, and occasionally on larch. According to Gilbertson and Ryvarden (1986), this mushroom has also been collected on pines (Douglas fir), although rarely so. *G. frondosa* is a "white rot" fungus. Although it is found at the bases of dying trees, most mycologists view this mushroom as a saprophyte, exploiting tree tissue dying from other causes.

Microscopic Features: Spores white, slightly elliptical (egg-shaped), smooth, hyaline, 6–7 × 3.5–5 μ. Hyphal system dimitic, clamp connections present in the generative hyphae, infrequently branching, with skeletal, non-septate hyphae.

Available Strains: Strains from the wild, unlike those of *Pleurotus*, rarely produce under artificial conditions. Of the strains I have tested that have been obtained from culture libraries, deposited there by taxonomically schooled, noncultivator mycologists, 90% of them do not fruit well on sterilized wood-based substrates. Therefore, screening and development of strains is necessary before commercial cultivation is feasible. Strains that produce fruitbody initials in 30 days are considered very fast. Most strains require 60 to 120 days of incubation before primordia formation begins.

Mycelial Characteristics: White, longitudinally linear, eventually thickly cottony on enriched agar media, non-rhizomorphic. The mycelium grows out unevenly, not forming the circular colonies typical of most mushrooms. Regions of the mycelium surge while other regions abate in their rate of growth. This pattern of growth seems characteristic of the species, as I have seen it in the majority of the twenty strains of *G. frondosa* that I have in my culture library. Often, the mycelium develops light tawny brown tones along the outside peripheral edges in aging. At maturity, the dense mycelial mat can be peeled directly off the agar media. Once on sawdust, many strains have mycelia that develop strong yellowish to orangish brown mottled zones, exuding a yellowish orange metabolite. Sawdust spawn, when

caps, 2–10 cm in diameter, arising from branching stems, eccentrically attached, and sharing a common base. Young fruitbodies are adorned with fine gray fibrils. The pores on the underside of the caps are white.

Distribution: Growing in northern temperate, deciduous forests. In North America, primarily found in eastern Canada and throughout the northeastern and mid-Atlantic states. Rarely found in the northwestern and in the southeastern United States. Also indigenous to the northeastern regions of Japan, and the temperate hardwood regions of China and Europe where it was first discovered.

Maitake usually fruits at the base of dead or dying trees. This is the same tree that yielded the 40-pound cluster on page 369. Some clusters have weighed in at 100 pounds apiece.

young is white. As the spawn matures, rust colors prevail.

Fragrance Signature: Richly fungoid and uniquely farinaceous, sometimes sweet. To me, rye grain spawn has a fragrance reminiscent of day-old fried corn tortillas. When mushrooms begin to rot, a strong fish-like odor develops.

Natural Method of Cultivation: The inoculation of hardwood stumps or buried logs is recommended. Given the size of the fruitbody and its gourmet and medicinal properties, this mushroom may well become the premier species for recycling stumps in hardwood forests. The occurrence of this mushroom on pines, Douglas fir, and larch is curious; confirming that some strains exist in nature that could help recycle the millions of stumps dotting the timberlands of North America. As forests decline from acid rain, future-oriented foresters would be wise to explore strategies whereby the dead trees could be inoculated and saprophytized by Maitake and similar immunopotentiating fungi.

Those experimenting with stump culture should allow 1 to 3 years before fruitings can be expected. High inoculation rates are recommended. Stumps do not necessarily have to be "virgin." Maitake is well known for attacking trees already being parasitized by other fungi, as does Zhu Ling (*Polyporus umbellatus*). However, it is not yet known under what conditions Maitake will dominate over other fungi in this situation. Therefore, for best results, the inoculation of recently cut stumps is recommended.

As the antiviral and immunopotentiating properties of Maitake become better understood, I envision the establishment of personal *Sacred Medicinal Mushroom Forests and Gardens* in the near future, using polypores like Maitake and Reishi. Those citizens living in harmony with the fungal kingdom will be gifted with their own mushroom medicines from the very landscape in which they live. This practice becomes a way of life: a spiritual experience of profound significance. Permaculturally oriented farms can benefit by inoculating hardwood stumps interspersed among multi-canopied shade trees. (See Chapter 5.) Clear economic, ecological, medical, moral, and spiritual incentives are in place for such models. Those with compromised immune systems would be wise to establish their own medicinal mushroom forests and gardens utilizing Maitake and other mushrooms. Maitake prefer midrange temperatures, fruiting in the fall whereas Reishi prefers the warmer months, fruiting in the summer. With proper foresight, the mycological landscape can provide medicinal mushrooms throughout several seasons.

Recommended Courses for Expansion of Mycelial Mass to Achieve Fruiting: Agar to cereal grain (rye, wheat, sorghum, milo) for the generation of spawn masters. These can be expanded by a factor of 10 to create second-generation grain spawn. This spawn can, in turn, be used to inoculate sawdust. The resulting sawdust spawn can either inoculate stumps outdoors, or for indoor cultivation, sterilized sawdust/chips/bran.

David Arora, author of *Mushrooms Demystified*, affected by wild Maitake.

Fully developed Maitake ready for harvest. Clusters average ½ to 1 pound from 5 pounds of sawdust/chips.

Most fruiting strains begin producing 6 to 8 weeks from inoculation onto sterilized, supplemented sawdust. Chung and Joo (1989) found that a mixture 15:5:2 of oak sawdust:poplar sawdust:corn waste generated the greatest yields. The Mori Mushroom Institute of Japan has successfully used larch sawdust, supplemented with rice bran, to grow this mushroom. As a starting formula, I recommend using the standard sawdust-chips-bran combination described for the cultivation of Shiitake and then amending this formula to optimize yields. If blocks of this substrate formula are mixed to a makeup weight of 4 pounds and then inoculated with $1/2$ pound of grain spawn, $1/2$- to 1-pound clusters of Maitake can be expected. However, by increasing the makeup weight to 7 pounds, Maitake clusters greater than 1 pound form. The only drawback is that through spawning is made more difficult using the standard bags available to the mushroom industry. With an increase in substrate mass, elevated thermogenesis is to be expected. Therefore, using sawdust spawn to inoculate bags of greater mass results in more thorough colonization with less evolution of heat. Variation between strain performance on various woods is considerable. Each cultivator must fine-tune his or her substrate formulas to each strain.

Suggested Agar Culture Media: MYA, MYPA, PDYA, or DFA.

1st, 2nd, and 3rd Generation Spawn Media: Grain-to-grain expansions. The final spawn media can be another generation of grain spawn, or alternatively sawdust. The sawdust spawn can be used to inoculate supplemented sawdust (i.e., sawdust/chips/bran) or plugs for outdoor cultivation on stumps or partially buried, vertically positioned, large-diameter hardwood logs. Considering the long incubation period, cultivators are well advised to weigh the advantages of creating another generation of sawdust spawn. The advantage of the grain spawn-inoculated bags is that they produce several weeks earlier than do the sawdust-inoculated blocks. However, the quality of colonization is better provided by spawn made of sawdust than from grain.

GROWTH PARAMETERS

Spawn Run:
Incubation Temperature: 70–75°F (21–24°C)
Relative Humidity: 95–100%
Duration: 14–30 days, then dormant for 30 days
CO_2: 20,000–40,000 ppm
Fresh Air Exchanges: 0–1 per hour
Light Requirements: n/a

Primordia ("mound") Formation:
Initiation Temperature: 50–60°F (10–15.6°C)
Relative Humidity: 95%
Duration: 5–10 days
CO_2: 2,000–5,000 ppm
Fresh Air Exchanges: 4–8 per hour
Light Requirements: 100–500 lux

Frond Development:
Temperature: 50–60°F (10–15.6°C)
Relative Humidity: 90–95%
Duration: 10–14 days
CO_2: 2,000–5,000 ppm
Fresh Air Exchanges: 4–8 per hour
Light Requirements: 100–500 lux

Fruitbody Development:
Temperature: 55–60° (65°)F [13–16° (18°)C]
Relative Humidity: 75–85%
Duration: 14–21 days
CO_2: <1,000 ppm
Fresh Air Exchanges: 4–8 per hour
Light Requirements: 500–1,000 lux

Cropping Cycle:
Every 3–4 weeks for a maximum of two flushes

Substrates for Fruiting: Hardwood sawdust, sawdust/chips/bran, particularly oak, poplar, cottonwood, elm, willow, and alder. Alder and poplar stumps are less likely to support outdoor fruitings, given the hold competitors like *Pleurotus ostreatus* and allies have on that niche. For indoor cultivation, yields substantially vary between various wood types. Oak is generally preferred, although strains growing on conifers are being developed.

Recommended Containers for Fruiting: Polypropylene bags with filter patches for air exchange. Polypropylene bottles and buckets have also been used.

Yield Potentials: $^1/_2$ to 1 pound mushrooms per 5–7 pounds of sterilized, enriched hardwood sawdust.

Harvest Hints: Relative humidity should be carefully lowered, as the fruitbody develops to prevent bacterial blotch. Overwatering can quickly cause the fruitbodies to abort. The thick base should be cut to remove substrate debris. Mushrooms wrapped in rice paper, and then refrigerated, have an extended shelf life up to 2 weeks at 35°F (1-2°C).

Form of Product Sold to Market: Fresh and dried mushrooms for the gourmet market. High-quality tablets and teas are being marketed in the United States by Maitake, Inc. and Fungi Perfecti. (For the addresses of these two companies, please refer to the Resource Directory on page 501.) Fresh Maitake has been relatively inexpensive and commonly available throughout the food markets of Japan. Availability of this mushroom is expected to fluctuate wildly as demand surges in response to its rapidly increasing medicinal reputation.

Nutritional Content: Approximately 27% protein (dry weight). Producing assorted vitamins—Vitamin B1: 1.5 mg%; Vitamin B2: 1.6 mg%; Niacin: 54 mg%; Vitamin C: 63 mg%; and Vitamin D: 410 IU. Saccharide content is nearly 50%. Assorted minerals/metals include magnesium at 67 mg%; iron at 0.5 mg%; calcium at 11.0 mg%; and phosphorus at 425 mg%. Moisture content of fresh specimens are approximately 80%, in contrast to more fleshy mushrooms which average 90% water. Our analysis show beta-glucan levels at 14.5% of dry mass.

Medicinal Properties: In vitro studies at the National Cancer Institute of the powdered fruitbodies (sulfated fraction) of *Grifola frondosa* show

Maitake is at the perfect stage for harvest when the leaflets fully extend and the edges darken. When Maitake begins to spoil, the tissue from which these leaflets arise becomes soft to the touch and emits a foul odor from the proliferation of bacteria.

significant activity against the HIV (AIDS) virus when tested through its Anti-HIV Drug Testing System under the Developmental Therapeutics Program. The National Institute of Health of Japan announced similar results in January of 1991. (The mycelium is not active.) Maitake extracts compared favorably with AZT but with no negative side effects. This is the first mushroom confirmed to have anti-HIV activity, in vitro, by both Japanese researchers and U.S. scientists. Studies with human subjects are currently ongoing in United States, Sweden, and England.

One of the polysaccharide fractions responsible for the immunostimulatory activity is three-branched ß 1,6 Glucan, known as grifolan. This polysaccharide was first characterized by Ohno et al. (1985), whose team found it in both the mycelium and the mushrooms. Their work showed strong antitumor activity in mice against murine solid tumor, Sarcoma 180 in only 35 days, causing complete tumor regression in one-third to one-half of the trials. Alkali extracts were found to be more effective than cold or hot water infusions. Human prostatic cells grown in vitro were effectively killed (90%+ in combination with carustine) through oxidative stress to the cell wall membrane when exposed to a purified fraction of ß 1,6 glucan (480 µ/ml.). In a similar study with cancer cells grown in vitro, water extracts of Maitake significantly slowed the reproduction of T-4 leukemia and Hela cervical cancer cells (Lovy et al. 1999). The molecular weight of the ß 1,6 polysaccharide constituent approaches nearly 1,000,000, and yet it acts as a cytotoxic agent to cancerous cells, causing apoptosis, in vitro. The mechanism of activity—their target specificities of cancer cells—is still being investigated (Konno et al., 2000; Borchers et al., 1999).

According to Chinese and Japanese reports, water extracts of this mushroom, when given to mice implanted with tumors, inhibited tumor growth by more than 86% (Nanba 1992; Ying et al. (1987). The dosage level was 1.0–10 milligrams of Maitake extract per kilogram of body weight over 10 days. No studies have yet been published ascertaining the extrapolation of this dosage level from a mouse to a human. Dr. Fukumi Morishige suggests that vitamin C be taken with Maitake (as well as with *Ganoderma lucidum*) as it helps reduce the polysaccharides into smaller, more usable chains of sugars, increasing their bioavailability. A recent analysis of organically-grown Maitake fruitbodies showed ß-glucan levels at 14.5%.

A protein-bound polysaccharide (D-fraction) is particularly effective via oral administration. How these polysaccharides activate the mammalian immune system is still being studied. However, an increase in helper "T" cells has been seen with some AIDS patients. With other AIDS patients, the decline in helper "T" cells is interrupted. Much more research will be published in the next few years.

Other medicinal claims for this mushroom include reduction of blood pressure, diabetes, cholesterol, chronic fatigue syndrome (CFS), and a wide variety of cancers.

Maitake fruiting in Japan from blocks of sterilized hardwood sawdust that were buried outside in gravelly soil months before.

For further information concerning the medicinal properties of this mushroom, refer to Adachi et al. (1987); Kabir and Yamaguchi (1987); Adachi et al. (1988); Hishida (1988); Yamada et al. (1990); and Nanba (1992). Also see *Chemical Pharmacological Bulletin* 35(1), 1987; *Japan Economic Journal* (May 1992); and Stamets and Yao (1999).

Flavor, Preparation, and Cooking: Towards the stem base, the flesh of this mushroom is thick and dense, and is better sliced. The upper petal-like caps are better chopped. This mushroom can be prepared in many ways, delighting the connoisseur mycophagist. Simply slicing and sautéing *à la* Shiitake is simple and straightforward. This mushroom can also be baked and stuffed with shrimp, sliced almonds, spices, and topped with melted cheese. The late Jim Roberts of Lambert Spawn once fed me this dish featuring a 1-pound specimen that I devoured at one sitting, making for a very satisfying meal and a sleepful night. Dried specimens can be powdered and used to make a refreshing tea. Refer to the recipes in Chapter 24 that describes several methods for preparing Maitake.

Comments: This species is delicious and much sought-after. Specimens weighing up to 100 pounds have been collected at the base of trees, snags, or stumps. Although primarily a saprophyte, *G. frondosa* behaves facultatively as a parasite, attacking trees dying from other causes, especially elms and oaks.

From my experiences, only a few strains isolated from the wild perform under artificial conditions. Those that do fruit mature best if the environment is held constant between 55–60°F (13–16°C). Substantial fluctuation beyond this temperature range arrests fruitbody development. The best fruitings are those that form slowly and are localized from one or two sites of primordia formation. When Maitake is incubated outside the ideal temperature range, the fruitbody initials fail to further differentiate. Should the entire surface of the block be encouraged to form primordia, an aborted plateau of short folds results.

Fruitbody development passes through four distinct phases. During initiation, the mycelium first undergoes a rapid discoloration from white undifferentiated mycelium to a dark gray amorphous mass on the exposed surface of the fully colonized block. During the second phase, the surface topography soon becomes contoured with dark gray black mounds that differentiate into a ball-like structure. The third phase begins when portions of this primordia ball shoot out multiple stems topped with globular structures. Each globular structure further differentiates with vertically oriented ridges or folds. The fourth and final phase begins when, from this primordial mass, a portion of the folds elongate into the petal-like sporulating fronds or "leaflets." (See page 373.) With some strains and under some conditions, the third phase is skipped.

The strategy for the successful cultivation of Maitake is in diametric opposition to the cultivation of Oyster mushrooms. If Maitake is exposed to substantial and prolonged light during the primordia formation period, the spore-producing hymenophore is triggered into production. This results in dome-shaped primordial masses, devoid of stems. If, however, minimal light is given, and carbon dioxide levels remain above 5,000 ppm, stem formation is encouraged. (Elongated stem formation with Oyster cultivation is generally considered undesirable.) Once the stems have branched and elongated to 2 or more inches, carbon dioxide levels are lowered and light levels are increased, signaling Maitake to produce the sporulating, petal-shaped caps. Humidity must be fluctuated between 80–95%. Maitake, being a polypore, enjoys less humid environments than the fleshier, non-polypore mushrooms.

If the mushrooms are growing in polypropylene bags, the bags should be opened narrowly at the top so that a forking bouquet is elicited. Stripping off all the plastic increases evaporation from the exposed surfaces of the block, jeopardizing the moisture bank needed for successful fruitbody development. In my growing rooms, I follow a compromise strategy. Given good environmental controls and management, I am successful at growing Maitake by fully exposing the upper surface of the mycelium once the gray primordial mounds have formed 45 to 60 days after inoculation. I leave the remainder of

Maitake mushrooms undergo distinct phases in the development of the fruitbodies.

the plastic around the block to ameliorate the loss of water. Holes are punched in the bottom of the bags for drainage.

As the mushrooms develop, less watering is needed in comparison to that needed by, for instance, Oyster mushrooms. Furthermore, cultivators should note that if too much base nutrition of the substrate is allocated to stem formation, the caps are retarded and often abort. And, if the sawdust is over-supplemented, bacteria blotch is triggered by the slightest exposure to excessive watering or humidity. Every strain behaves differently in this regard. Maitake cultivation requires greater attention to detail than most other mushrooms. Because of its unique environmental requirements, this mushroom cannot share the same growing room as many of the fleshier gourmet and medicinal mushrooms.

Once the production blocks cease producing, they can be buried outside in hardwood sawdust. In outdoor environments, the subterranean block becomes a platform for more fruitings, maximizing the use of every inoculated block of sterilized sawdust. Blocks planted in the spring often give rise to fruitings in the fall. The autoclavable plastic should be removed—unless made of cellulose or other biodegradable material. If you scratch the outer surfaces of the blocks, the internal mycelium comes into direct contact with the sawdust bedding, stimulating leap-off.

Zhu Ling or the Umbrella Polypore
Polyporus umbellatus Fries

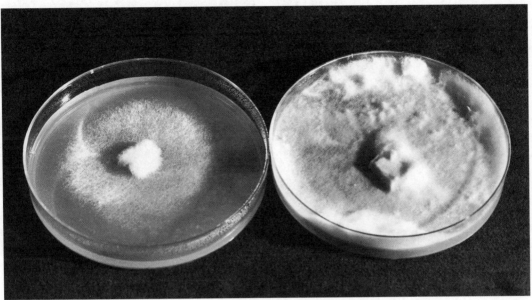

P. umbellatus mycelia 4 and 10 days after inoculation onto malt extract agar media.

Introduction: This fleshy polypore has long been a favorite edible. Of all the fruitbodies represented in the genus *Polyporus*, *P. umbellatus* is one of the most fragile and delicate. This mushroom forms an underground tuber-like structure, a sclerotium, which has figured prominently in Chinese pharmacopeia, especially in the treatment of lung cancer. The shape of the sclerotia resembles hog dung, except it is extremely hard and wood-like. After sitting dormant for months, the sclerotia soften as they swell with water, and sprout deliciously edible mushrooms.

Common Names: Zhu Ling (Chinese for "Hog Tuber")
Chorei-maitake (Japanese for "Wild Boar's Dung Maitake")
Tsuchi-maitake (Japanese for "Earth Maitake")
Umbrella Polypore
Chinese Sclerotium

Taxonomic Synonyms and Considerations: Gilbertson and Ryvarden (1987) follow tradition by keeping this mushroom within the genus *Polyporus*, i.e., *Polyporus umbellatus* Fr. This mushroom is commonly referred to as *Grifola umbellata* (Persoon: Fries) Donk and more infrequently called *Dendropolyporus umbellatus* (Pers.:Fr.) Julich.

Macroscopically, *Grifola frondosa*, Maitake, appears to be a close relative, but biologically the

Sclerotia of North American *P. umbellatus*. The tissue is as hard as wood.

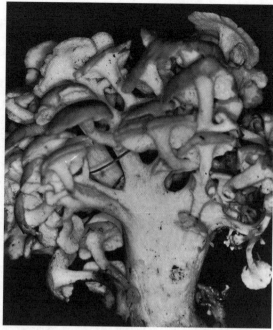

A soft, delicate, fleshy fruitbody arises from the sclerotia.

two have uniquely different life cycles. The close appearance of *P. umbellatus* and *G. frondosa* easily confuses amateur collectors (Lincoff, 1993). Microscopically, *G. frondosa* has smaller, non-cylindrical spores, lacks the sclerotial stage in its life cycle, and arises from a multiple forking base.

Description: The mushrooms arise from underground sclerotia. The gray to black sclerotium, resembling pigs' dung in form, but woody in texture, swells with water and generates multibranched, circular shaped mushrooms with umbellicate caps. These bouquets of mushrooms arise from a common stem base. The fruitbodies are whitish at first, becoming dingy brown with age, with an underside featuring circular to angular pores. They are often waterlogged due to their high carrying capacity for water.

Distribution: Infrequently occurring throughout the deciduous woodlands of north-central and northeastern North America, in the temperate regions of China, and in Europe where it was first described. Gilbertson and Ryvarden (1987) reported this mushroom has been found in Montana and in Washington State. If this mushroom indeed grows in the Pacific Northwest, it is exceedingly rare, as I have never found it and know no one who has.

Natural Habitat: Found on the ground, arising from dead roots or buried wood, on stumps, or in soils rich in lignicolous matter, preferring birches, maples, willows, and beeches. Predominantly growing in deciduous woodlands, this mushroom has been reported from coniferous forests, although rare. Weir (1917) reported this mushroom from Montana growing on spruce (*Picea* sp.) where, if it still occurs, is exceedingly difficult to find.

Microscopic Features: Spores 7–10 × 3–4 μ, white in deposit, smooth, cylindrical. Hyphal system dimitic, non-septate, clamp connections present on the generative hyphae.

Available Strains: Strains are available from most culture libraries. However, most of the strains that I have tested do not fruit in culture. Hence, strains that can produce under indoor, controlled conditions are needed.

Mycelial Characteristics: White, longitudinally linear, soon densely cottony, forming a thick, peelable mycelial mat on agar, in grain, and in sawdust media. On sterilized sawdust, the mycelium, as it ages, forms outer layers of yellowish, gelatinous exudate. This mushroom causes a white rot.

Fragrance Signature: Musty, sour, slightly bitter, and not pleasant.

Natural Method of Cultivation: The roots of stumps are inoculated by digging trenches into the root zones that have been already parasitized by, for instance, the Honey mushroom, *Armillaria mellea*. The trenches are partially filled with wood and burned. Logs of beech, birch, willow, maples, and/or oaks are given multiple cuts into which sawdust spawn or slices of fresh sclerotia are packed. The logs are reburied underneath a layer of sandy soil and covered with rich humus and deciduous leaves. After 3 years, the trenched logs are unburied and new sclerotia can be harvested. For sclerotia harvest, late spring is best. Fruitbodies are generated from the sclerotia in the late summer to early fall, when the ground temperature hovers between 50–60°F (10–15°C). For more information, consult *Fungi Sinica*, 1980.

Recommended Courses for Expansion of Mycelial Mass to Achieve Fruiting: Liquid culture to grain to sawdust to supplemented sawdust/chips. The cultivation of this mushroom initially parallels that of Maitake, *Grifola frondosa*, another fleshy polypore, according to Jong (1992). My own successes with this mushroom have been limited. *G. umbellata* can be grown via two methods: from a sclerotium or directly from the mycelial mat bypassing the sclerotial stage. Sclerotial production is stimulated by the microflora in soils, and by the absence of light. Therefore, formation of sclerotia under laboratory conditions is difficult. (Incubating microbially active soil substrates for prolonged periods of time

GROWTH PARAMETERS

Spawn Run:
Incubation Temperature: 60–75°F (15–24°C)
Relative Humidity: 90–100%
Duration: 21–30 days to 2 years
CO_2: >5,000 ppm
Fresh Air Exchanges: 1–4 per hour
Light Requirements: n/a

Sclerotia Formation:
Initiation Temperature: 50–60°F (10–16°C)
Relative Humidity: 90–100%
Duration: 60–90 days
CO_2: > 5,000 ppm
Fresh Air Exchanges: 1–4 per hour
Light Requirements: darkness required!

Primordia Formation:
Initiation Temperature: 40–50°F (4–10°C)
Relative Humidity: 90–100%
Duration: 30–60 days
CO_2: <5,000 ppm
Fresh Air Exchanges: 8 or more per hour
Light Requirements: modest to low light, no greater 200 lux

Fruitbody Development:
Temperature: 50–60°F (10–16°C)
Relative Humidity: 85–95%
Duration: 30–90 days
CO_2: <500 ppm or ambient natural
Fresh Air Exchanges: 4–8+ per hour
Light Requirements: 500–1,000 lux

Cropping Cycle:
Seasonal, typically occurring in the late summer and early fall

is fraught with failure.) However, there is an alternative strategy.

Zhu Ling behaves as a secondary saprophyte, depending upon the degradative abilities of other fungi to render a wood substrate into a usable platform for fruiting. In my experiments, fruitings did not result when the same formula was prepared from fresh starting material (i.e., sawdust/chips/bran). I have had limited success at fruiting this species on the remains of sterilized, recycled Shiitake and Reishi blocks. Nevertheless 45 to 60 days of incubation preceded any mushroom formation. This method has only resulted in short, lateral, hardened plateaus with pored surfaces, which achieved only a few centimeters in height. (The pores contained sporulating basidia.) This abbreviated fruitbody formation may be a function of an insufficiently developed sclerotial stage. (Sclerotia store nutrition prior to fruitbody initiation.) More research is required before indoor cultivation yields fruitings comparable to other polypores like Reishi or Maitake.

Suggested Agar Culture Media: MYPA, PDYA, OMYA, or DFA.

1st, 2nd, and 3rd Generation Spawn Media: Two expansions of grain spawn with the final spawn stage being sawdust.

Substrates for Fruiting: Outdoor plantings require the placement of either spawn or sclerotia into logs buried into beds of hardwood sawdust, around the root zones of beeches, birches, willows, or oaks. For indoor cultivation, I have had limited success using hardwood sawdust substrates rendered by other primary saprophytes. Expired Shiitake, Maitake, and Reishi blocks seem to work best. I recommend cultivators experiment by growing sclerotia on recycled substrates in darkness, removing the sclerotia when mature, and implanting them in lignicolous soils to stimulate fruitings. Currently the modified natural model is the main method of cultivation in China.

Recommended Containers for Fruiting: Polypropylene bags for indoor cultivation. Trench culture outdoors can be framed with nonchemically treated boards around the root zones of candidate trees and protected with shade cloth covered hoop-frames.

Yield Potentials: Not known. The natural model developed by the Chinese first yields sclerotia as the primary product, and secondarily mushrooms, which are more often used for food than for medicinal purposes. In liquid culture, I have achieved a 6% yield (dry weight/dry weight) of dry mycelium from a malt yeast slurry. Yield efficiencies of sclerotia and mushrooms from Chinese Zhu Ling tree nurseries (aka "Hog Tuber Farms") are not known to this author.

Form of Product Sold to Market: Sclerotia are exported from China, either whole or sliced, and in dried form. Packages of Zhu Ling often have a picture of Reishi on the label! The fresh fruitbodies are sold in markets in China. Currently, there are no producers of this mushroom in North America.

Nutritional Content: Protein: 8%; coarse fiber: 47%; carbohydrate: .5%; and ash: 7%. The sclerotia, with its woody texture, is likely to have substantially more polysaccharides but less protein than the fleshy fruitbody.

Medicinal Properties: This mushroom has been heralded to possess potent anticancer, immunopotentiating properties. However, few scientific studies have been conducted, and none by Western researchers. During a visit to the Institute of Materia Medica in October of 1983, Dr. Andrew Weil, Dr. Emanuel Salzman, Gary Lincoff, myself, and others were informed by excited researchers that water extracts (tea) of this mushroom, when given to lung cancer patients after radiation therapy, resulted in complete recovery in the majority of the patients. The patients continued a regimen of Zhu Ling tea for several years. Quality of life of the patients dramatically improved, characterized by increased appetites, absence of malaise, and normalization of immune function. The majority of those patients not given Zhu Ling died. I do not know if this study was ever published.

In 1983, Chang Jung-lieh presented results on the strong inhibitory effects Zhu Ling had on Sarcoma 180 tumors implanted in mice. Taken orally or intravenously, Zhu Ling is widely used as a traditional drug for preventing the spread of lung cancer. Modern day treatments using Zhu Ling often accompany radiation therapy. A juried paper, present by Han (1988) in the *Journal of Ethnopharmacology*, reported that mice, when implanted with Sarcoma 180 were given a dose of 1 mg of Zhu Ling per kilogram of body weight, tumors were reduced by 50% compared to the controls. Ying (1987) reported 70% reduction of tumor weight in mice. Miyaski (1983) also noted the anti-sarcoma properties of this fungus. According to Bo and Yun-sun (1980, p. 195), Chinese physicians are using extracts of the Zhu Ling sclerotia in the treatment of "lung cancer, cervix cancer, esophagus cancer, gastric cancer, liver cancer, intestine cancer, leukemia, mammary gland cancer, and lymphosarcoma."

Water extracts of Zhu Ling totally inhibited the growth of a pyrimethamine-resistant strain of malaria (*Plasmodium falciparum*) (Lovy et al. 1999). Many medical researchers have expressed extreme interest in all higher fungi that undergo a sclerotial stage for a potential source for novel antibiotics, antipathogen, anticancer, and antiviral drugs.

Flavor, Preparation, and Cooking: The fleshy, above-ground mushroom is easily broken or chopped, placed into a frying pan or wok, and sautéed as one would do with Shiitake.

The sclerotia are usually made into a tea by placing 20–25 grams in a liter of water and boiling for 20 minutes. Additional water is added to offset that lost to evaporation. Since the sclerotia are extremely tough, and are difficult to extract, I leave the sclerotia chips in the water for several more soakings before discarding. It should be noted that I have made tea of Zhu Ling and Reishi for more than 1,000 people at various mushroom conferences over the past 7 years without a single report of a negative reaction.

Comments: Zhu Ling sclerotia remain some of the least expensive of the imported medicinal fungi. Grown quasi-naturally in the temperate, mountainous Shansi Guu County of China, the sclerotia sell for about half the price of Reishi. Once the anticancer properties of this fungus become better understood, and if proven, the value of Zhu Ling sclerotia is likely to increase.

The sclerotia are used medicinally, whereas the fleshy and fragile fruitbodies, known as "Hog Tuber Flowers," are eaten as a gourmet mushroom. This is one of the softest polypores I have encountered. Its excellent flavor, unique life cycle, medicinal properties, and scarcity are all factors that should encourage the development of indoor, controlled-environment fruitings by cultivator entrepreneurs.

Zhu Ling, known to North Americans as the Umbrella Polypore (*Polyporus umbellatus*) growing from underground sclerotia.

Turkey Tail or Yun Zhi
Trametes versicolor (L:Fr.) Pilat

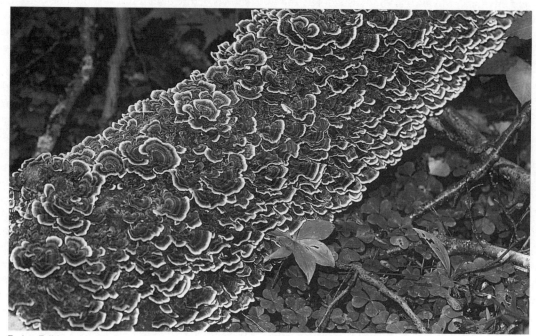

Turkey Tail is a ubiquitous woodland mushroom, growing on diverse hardwoods and conifers, and is found across much of the globe.

Introduction: One of the most easily identifiable polypores, and the most ubiquitous, this multi-colored mushroom is recognized throughout the world and is thought to have had a long history of use. Native to tropical, subtropical, and temperate zones, this species is highly adaptive, growing on the widest assortment of woods of any mushroom I have encountered. Not surprisingly, Turkey Tail is one of the most potent, and the best studied, of all medicinal mushrooms.

Common Names: Turkey Tail
Many-Colored Polypore
Kawaratake
Yun Zhi
Cloud Mushroom

Taxonomic Synonyms and Considerations: Better known as *Coriolus versicolor* (L.:Fr.) Quelet and formerly called *Polyporus versicolor* L.:Fr., this mushroom is probably a complex of many species with hundreds of varieties. Given their significance in nature, medicine, and bioremediation, Turkey Tails deserve more attention by taxonomists than they have previously received.

Description: A classic bracket fungus, with multi-colored, closely concentric zonations on fan-like shelves, often forming rosettes of overlapping fruit-bodies, with slightly hairy surfaces and usually

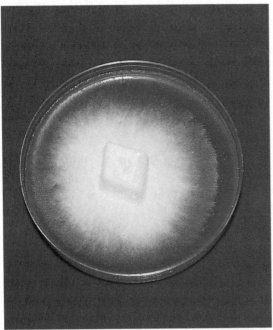 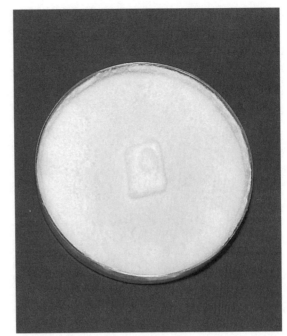

Trametes versicolor, Turkey Tail, mycelium at 10 and 20 days after inoculation.

with a wavy outer margin. Mushrooms are highly variable in color, rich with earth tones of grays and browns, often with bluish, greenish, reddish, or whitish zones. Underside covered with whitish pores. Although an annual mushroom, often surviving for several seasons, the fruitbodies become a dull grayish long after the life cycle has completed. Stem absent or if present, central and short. Often forming great colonies, especially on downed hardwood logs and stumps.

Distribution: Found in every state in the United States but not yet reported from Nevada (!). Apparently circumpolar, widely distributed throughout the boreal, temperate, subtropical, and tropical regions of the world. Few mushrooms can boast such adaptivity.

Natural Habitat: Turkey Tail is the most common polypore on dead hardwoods, more rarely on conifers. Most Shiitake cultivators have encountered this fungus as it is a common competitor on natural logs.

Microscopic Features: Spores 5–6 × 1.5–2 μ, cylindric, smooth, slightly curved, whitish in deposit. Hyphal system trimitic. Heterothallic mating strategy.

Available Strains: Wild strains are easy to come by. The mushrooms are very tough but pliable and hence only a super-sharp scalpel can cut into the fibrous tissue. Strains tend to be very aggressive in culture.

Mycelial Characteristics: Producing a whitish, tenacious, sheet-like, leathery mycelium, soon toughening in age, becoming aerial and exuding a yellowish metabolite. If the culture ages more than 2 weeks, the sharpest of scalpels only succeeds in tearing off skin-like sheaths of mycelium from the agar surface.

Fragrance Signature: Grain spawn sweet, pleasant, and distinctly "polypore-esque."

Natural Method of Cultivation: In logs, *à la* Shiitake, or in pots, *à la* Reishi. Plug, sawdust, or wedge spawn is an effective method of inoculation.

Turkey Tail grows prolifically on most sterilized hardwoods and is easy to cultivate.

Recommended Courses for Expansion of Mycelial Mass to Achieve Fruiting: Standard agar-to-grain, and then grain-to-grain transfers. Be forewarned that if the spawn is not used in a timely fashion, the permeation of the mycelium will be so dense as to make separation of the grain kernels, at transfer, very difficult. Grain spawn can inoculate unsupplemented hardwood sawdust and chips. The enzymes produced by this mushroom are so effective at releasing nutrients that I have seen little benefit from supplementation. On sawdust media, the mycelium eventually "pulps" the wood to the point of total decomposition—the wood totally loses its structure.

Suggested Agar Culture Media: MYPA, PDYA, OMYA, and/or DFA.

1st, 2nd, and 3rd Generation Spawn Media: Grain spawn throughout.

Substrates for Fruiting: This aggressive mushroom can decompose most hardwoods. I recommend oak, tan oak, alder, poplar, aspen, elm, ironwood, and eucalyptus. This is one of the few mushrooms that grow well on apple, cherry, and other fruit trees. Note that the softer hardwoods are quickly consumed by this species, leading to early, if not briefer flushes. The denser woods allow for fruitings over a longer period.

Recommended Containers for Fruiting: Plastic bags, bottles, and trays. This mushroom fruits equally well on horizontal and vertical surfaces.

Yield Potentials: A mushroom with much less moisture content than most (75–80%), a 4.5-pound (2-kg) block of sawdust yields 100–200 grams of mushrooms (dry).

Harvest Hints: Mushrooms will produce zonate "growth rings" that grow progressively over several weeks. The mushrooms are tough and strongly attached to the substrate. Using a serrated knife makes harvesting easier.

Form of Product Sold to Market: Powdered fruitbodies made into capsules and teas. Much of the medicinal products on the market are derived from fermented mushroom mycelia.

Nutritional Content: Not known to this author.

Medicinal Properties: *Trametes versicolor* is the source of PSK, commercially known as "Krestin," responsible for several hundred million dollars of sales as an approved anticancer drug in Asia. In clinical studies of 224 patients (Sugimachi et al. 1997) and 262 patients (Nakazato et al. 1994) afflicted with gastric cancer and treated with chemotherapy, patients showed a decrease in recurrence and an increase in the disease-free survival rate when treatment was combined with a regimen of using the protein-bound polysaccharide (PSK) from *Trametes versicolor*. By all measures, the treatment was clearly cost-effective. PSK reduces cancer metastasis (Kobayashi et al. 1995). PSK stimulated

GROWTH PARAMETERS

Spawn Run:
Incubation Temperature: 75–85°F (24–29°C)
Relative Humidity: 90–100%
Duration: 14–21 days
CO_2: >5,000 ppm
Fresh Air Exchanges: 1 per hour
Light Requirements: n/a

Primordia Formation:
Initiation Temperature: 50–75° (80°)F
  [10–24° (27°)C]
Relative Humidity: 95–100%
Duration: 7–14 days
CO_2: 400–800 ppm
Fresh Air Exchanges: 5–7 per hour
Light Requirements: 500–2,000 lux

Fruitbody Development:
Temperature: 65–75°F (18–24°C)
Relative Humidity: 85–90% (95%)
Duration: 45–70 days
CO_2: 500–1,000
Fresh Air Exchanges: 5–7 per hour
Light Requirements: 500–2,000 lux

Cropping Cycle:
Two to three flushes over three months

interleukin-1 production in human cells (Sakagami et al. 1993). PSK was also found to be a scavenger of free-radical oxidizing compounds (superoxide anions) through the production of manganese superoxide dismutases (Kobayashi et al. 1993; Kim et al. 1999). A highly water-soluble, low-cytotoxic polysaccharopeptide (PSP) isolated from this mushroom has been proposed as an antiviral agent inhibiting HIV replication based on an in vitro study (Collins and Ng 1997). PSP is a classic biological response modifier (BRM), inducing gamma interferon, interleukin-2, and T-cell proliferation, differing chemically from PSK in that it has rhamnose and arbinose while PSK does not but has fucose (Ng 1998). Dong et al. (1996, 1997) reported that a polysaccharide peptide (CVP) and its refined form (RPSP) has not only antitumor properties, but elicits an immunomodulating response by inhibiting the proliferation of human leukemia (HL-60) cells while not affecting the growth of normal human peripheral lymphocytes. Yang et al. (1992) also found that a smaller polypeptide (SPCV, 10,000 m.w.) significantly inhibited the growth of leukemia cells. Kariya et al. (1992) and Kobayashi et al. (1994) showed that the protein-bound polysaccharides of *Trametes versicolor* express superoxide dismutase (antioxidating) mimicking activity. Lin et al. (1996) showed that *Coriolus versicolor* polysaccharides (CVP) enhanced the recovery of spleen cells subsequent to gamma irradiation. Ghoneum et al. (1995, 1998) conducted two clinical studies in the United States using arabinoxylane, a product from fermenting *Trametes versicolor, Lentinula edodes* (Shiitake), and *Schizophyllum commune,* on rice that showed dramatic fivefold increase in NK activity within 2 months of treatment. PSK also has been found to be a strong antibiotic, effective against *Escherichia coli, Staphylococcus aureus, Pseudomonas aeruginosa, Candida albincans, Cryptococcus neoformans,* and other microbes pathogenic to humans (Sakagami and Takeda, 1993; Mayer and Drews, 1980). Both PSK and PSP are present in the mycelium and can be extracted from fermented cultures.

Flavor, Preparation, and Cooking: Tough like leather, the fruitbodies are extracted by boiling in water for use in soups or in teas. This mushroom has been used for centuries as a natural medicine.

Comments: Perhaps the most well studied of all polypore mushrooms, Yun Zhi has long been viewed as an ally to human health. Easy to identify in the wild, this zonate, multicolored mushroom is beautiful to behold, and once formed, resists rot. Traditionally, Turkey Tails have attracted artisans in the making of necklaces and other ornaments by native peoples worldwide. For our ancestors, I can think of few mushrooms that could serve so many useful purposes in treating disease.

Turkey Tail is not only remarkable in its medicines but also produces peroxidase enzymes, which are extremely effective in breaking down lignins, the structural fiber of wood. This mushroom has been the subject of studies for pulping wood (Katagiri et al. 1995) and for the degradation of polycyclic aromatic hydrocarbons (Field et al. 1992).

Although not yet studied in the scientific literature, I have heard from several patients fighting hepatitis C and benefiting from a regimen of drinking medicinal mushrooms in tea. One patient suffered from a swollen liver and spleen. He began a daily program of self-medication using a tea made from wild Turkey Tail and cultivated Reishi. After 2 weeks, the swelling was gone. Having a bad reaction from interferon treatments, he opted to continue on the tea, effectively reducing HVC from 1.3 million to 140,000 with liver enzymes normalizing. His health improved. His doctor was perplexed and confessed to being totally unaware of the beneficial properties of medicinal mushrooms. Since a substantial fraction of those ill with Hepatitis C do not have identifiable risk factors, and the vector through which the virus was communicated has not been identified, medicinal mushrooms hold promise for much needed, new antiviral medicines. Unique antiviral components have been isolated from other mushrooms, including *Lentinula edodes*, *Ganoderma lucidum* (Jong et al. 1992), and *Rozites caperata* (Piraino et al. 1999).

Trametes versicolor and *Ganoderma applanatum* co-inhabiting the same stump.

The Lion's Mane Mushroom

T he teethed fungi have many representatives, some of which are exquisitely edible. Species belonging to the genus *Hericium* are most notable. *Hericium erinaceus* and *H. coralloides* or *H. ramosum* produce prodigiously in culture and are the best flavored. *Hericium abietis*, a lover of conifers, is more difficult to cultivate. This group of mushrooms, with their distinctive snow-white icicle-like appearance has long been a favorite of woodspeople. Like Shiitake, the cultivation of this mushroom probably evolved from the astute observations of those who collected them from the wild.

Dusty Yao collecting one of the Lion's Mane species, *Hericium abietis*, in the old growth forest near Mt. Rainier in Washington state. This excellent edible mushroom is native to conifers, and can grow to a formidable size.

387

The Lion's Mane Mushroom of the Genus *Hericium*
Hericium erinaceus (Bulliard: Fries) Persoon

H. erinaceus mycelia 4, 7 and 14 days after inoculation onto malt extract agar media.

Introduction: *Hericium erinaceus* is one of the few mushrooms imparting the flavor of lobster when cooked. Producing a mane of cascading white spines, this mushroom can be grown on sterilized sawdust/bran or via the traditional log method first established for Shiitake. A favorite among woodspeople worldwide, this mushroom has gained renewed attention because of its medicinal properties.

Common Names: Lion's Mane
Monkey's Head
Sheep's Head
Bear's Head
Old Man's Beard
Hedgehog Mushroom
Satyr's Beard
Pom Pom
Yamabushi-take (Japanese for "Mountain-Priest Mushroom")
Houtou

Taxonomic Synonyms & Considerations: Formerly known as *Hydnum erinaceum* Fr. and sometimes cited as *Hericium erinaceum* (Fr.) Pers. *Hericium coralloides* and *Hericium abietis* are similar species, distinct in both their habitat preference and form. *H. coralloides* can also be cultivated on sawdust and differs from *H. erinaceus* in that its spines fork rather than emerge individually.

Description: Composed of downward, cascading, non-forking spines, up to 40 cm in diameter in the

Miniature *H. erinaceus* fruitbody forming on malt extract agar medium.

wild. Typically white until aged and then discoloring to brown or yellow brown, especially at the top.

Distribution: Reported from North America, Europe, China, and Japan. Of the *Hericium* species, this species is most abundant in the southern regions of United States.

Natural Habitat: On dying or dead oak, walnut, beech, maple, sycamore, and other broadleaf trees. Found most frequently on logs or stumps.

Microscopic Features: Spores white, 5.5-7.0 × 4.5-5.5 μ. Ellipsoid, smooth to slightly roughened. Clamp connections present, but infrequent.

Available Strains: ATCC #62771 is an excellent, high-yielding strain. Tissue cultures of wild collections vary significantly in the size of the fruitbody at maturity. I prefer to clone from the midsection of the stem or pseudo-stem of very young specimens.

Mycelial Characteristics: Whitish, forming triangular zones of collected rhizomorphs, radiating from the dense center section. (The mycelium can

Dr. Andew Weil in China with *H. erinaceus* forming on oak log.

GROWTH PARAMETERS

Spawn Run:
Incubation Temperature: 70–75°F (21-24°C)
Relative Humidity: 95–100%
Duration: 10–14 days
CO_2: >5,000–40,000 ppm
Fresh Air Exchanges: 0–1 per hour
Light Requirements: n/a

Primordia Formation:
Initiation Temperature: 50–60°F (10–15.6°C)
Relative Humidity: 95–100%
Duration: 3–5 days
CO_2: 500–700 ppm
Fresh Air Exchanges: 5–8 per hour
Light Requirements: 500–1,000 lux

Fruitbody Development:
Temperature: 65–75°F (18–24°C)
Relative Humidity: (85) 90–95%
Duration: 4–5 days
CO_2: 500–1,000 ppm
Fresh Air Exchanges: 5–8 per hour
Light Requirements: 500–1,000 lux

Cropping Cycle:
14 days apart

Fragrance Signature: Rich, sweet, and farinaceous.

Natural Method of Outdoor Cultivation: Inoculation of logs or stumps outdoors using sawdust or plug spawn *à la* the methods traditionally used for Shiitake. This is one of the few mushrooms that produces well on maple logs, oaks, beech, elm, and similar hardwoods. (The "paper"-barked hardwoods such as alder and birch are not recommended.) Once inoculated, the 3- to 4-foot-long logs should be buried to one-third of their length into the ground, in a naturally shady location. Oaks are comparatively slow to decompose due to its density, providing the outdoor cultivator with many years of fruitings. A heavy inoculation rate shortens the gestation period.

Recommended Courses for Expansion of Mycelial Mass to Achieve Fruiting: This mushroom, in my experience, requires greater attention to the details of mycelial development for the creation of spawn than most other species. The mycelium grows relatively slowly on nutrified agar media, with fruitbodies often forming before the mycelium has grown to a mere 25 mm in radius. Furthermore, the transferring of mycelium from agar to grain media using the traditional scalpel and wedge technique, results in comparatively slow growth, taking weeks to colonize unless a regimen of diligent and frequent shaking of the spawn jars is followed.

Hericium erinaceus is a classic example of a species that is stimulated by agitation in liquid culture. I wait until primordial colonies form upon the mycelial mat on agar media. At that time, cultures are cut into sections and placed into an Eberbach-like stirrer. Once blended, the myceliated fluid, now rich in the growth hormones associated with primordia formation, is expanded by transferring to sterilized grain-filled spawn jars. The result is an explosion of cellular activity.

Two colonies of mycelium grown out on standard 100 x 15-mm petri dishes are recommended for use with every 1,000 ml of sterile water. Once it is stirred, 20–50 ml of liquid-inoculum is poured into every liter to half-gallon jar. The jars are then thoroughly shaken to evenly distribute the liquid

resemble the structure of a glaciated mountain, i.e., Mt. Rainier, as seen from high overhead from an airplane.). If the top and bottom of the culture dishes are taped together, evaporation is lessened with an associated pooling of carbon dioxide. This stimulates the mycelium into aerial growth. As cultures age, the mycelia become yellow to distinctly pinkish. Islands of young fruitbodies form in petri dish cultures incubated at 75°F (24°C) in 2 to 3 weeks. Elongated, aerial spines ("spider-like"), which in age change from whitish to yellowish characterize the fruitbodies.

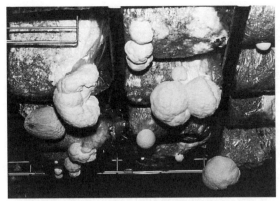

H. erinaceus fruiting from punctured bags in a growing room.

Substrates for Fruiting: Sterilized sawdust supplemented with rice bran for indoors. Hardwood and Douglas fir logs and stumps are recommended for outdoors. The pH range for fruiting falls between 5.0 and 6.5.

Yield Potentials: 550 grams fresh wet from 5 pounds hardwood (alder) sawdust, unenriched. 1-pound clusters are common using the above technique. With multiple sites forming exterior to the bag, maximum yield efficiency approaches 2 pounds

Harvest Hints: If inducing mushrooms to form through minute holes in plastic, the fruitbody snaps off with no need for further cleaning. Picking mushrooms in this fashion makes for some of the fastest harvesting I have seen. Mushrooms should be harvested when the spines dramatically elongate, but before the very top of the fruiting mass softens and becomes markedly yellowish or pink. The relative humidity in the growing room should be lowered

inoculum, placed on the spawn incubation rack, and left undisturbed for 7 to 12 days at 75°F (24°C). Soon thereafter evidence of mycelial recovery can be seen, often with numerous white dense spots. These dense white spots are sites of rapidly forming primordia, now numbering many times more than that which had formed in the original petri dishes. The spawn jars must be used immediately for further expansion either into more sterilized grain or sterilized wood, lest the primordia develop into sizable fruitbodies. Should the latter occur, further use of these spawn jars is not recommended since the developing mushrooms are damaged in the course of shaking, and then highly susceptible to contamination. The subsequent transfer of these smashed fruitbodies is often followed by a massive bacterial outbreak. If the spawn is pure, then each half-gallon (liter) of grain spawn is transferred into four 5-pound bags of sterilized sawdust/bran. Since this species adapts well to submerged fermentation, commercial cultivators might find the direct inoculation of sterilized grain with liquid mycelium is the most efficient path of spawn generation.

Suggested Agar Culture Media: MYPA, PDYA, or DFA.

1st, 2nd, and 3rd Generation Spawn Media: Grain: rye, wheat, milo, wheat, barley, corn, or millet for grain masters. Sawdust for subsequent generations.

The ³/₄-pound Lion's Mane pictured here arose from the minute hole in the plastic seen at the tip of the knife. With this method, mushrooms can be harvested and sold with no need for cleaning.

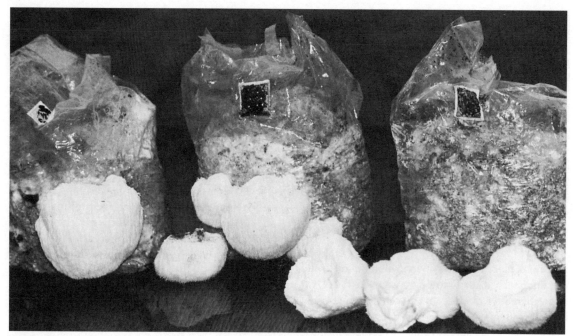

When properly initiated, mushrooms form and mature through the holes in the plastic. Once mushroom balls begin growing exterior to the bag, internal fruitings within the plenum of the bag are arrested or prevented.

to 80% for 4–8 hours prior to harvest to reduce surface moisture and prolong storability.

Form of Product Sold to Market: ¼ to ½-pound clusters carefully packaged in damage-proof containers are ideal. (The size of the clusters can be controlled by limiting the number of mushrooms that form on each flush. Gourmet Mushrooms, a specialty mushroom company in California, sells *Hericium*, marketing the mushrooms under the clever name "Pom Pom Blanc." The mushroom must be picked with great care to avoid creating a site for bacteria infestation that can quickly spread. If not used fresh, fruitbodies are dried, powdered, and presented in tablet form for oral ingestion. Extracts are also sold.

Nutritional Content: Protein 31.7 g/100 g; fat 4 g/100 g; ash 9.8 g/100 g; available carbohydrate 17.6 g/100 g; energy 233 kcal/100 g; dietary fiber 30 g/100 g; sodium 1.2 mg/100 g; phosphorus

1.22 g/100 g; iron 20.3 mg/100 g; calcium 1.3 mg/100 g; potassium 4.46 mg/100 g; mangesium 123 mg/100 g; thiamin 5.33 mg/100 g; riboflavin 3.91 mg/100 g; calciferol 240 IU/100 g; niacin 18.3 mg/100 g; ergosterol 381 mg/100 g.

Medicinal Properties: Chen (1992) reported that studies on subjects in the Third People's Hospital of Shanghai that *H. erinaceus*, in tablet form, proved to be effective on "ulcers, inflammations, and tumors of the alimentary canal." Ingestion of this mushroom is said to have a remarkable effect in extending the life of cancer-ridden patients. Ying (1987) reports that pills of this mushroom are used in the treatment of gastric and esophageal carcinoma. I know of no studies outside of China. A patent recently awarded in Japan (#05391544) showed that this mushroom produces Eninacines (sic) (= erinacines), which are strong stimulators to nerve growth factor synthesis (Kawagishi et al.

1991, 1994). These compounds stimulate neurons to regrow, potentially significant in the treatment of senility, Alzheimer's disease, repairing neurological trauma from strokes, improving muscle/motor response pathways, and cognitive function.

Flavor, Preparation, and Cooking: This is one of my family's favorite gourmet mushrooms. Its flavor is greatly affected by the maturity of the harvested mushrooms, their moisture content, the method of cooking, and particularly the other foods that are cooked with this mushroom. To some, this mushroom has a flavor similar to lobster; to others the flavor is reminiscent of eggplant.

We cut the mushrooms transverse to the spines into dials and cook them at high heat in canola (rapeseed) oil until the moisture has been reduced and the dials are light golden brown. (Garlic, onions, and almonds also go well with this mushroom.) The addition of a small amount of butter near the end of the cooking cycle brings out the lobster flavor. A combination of Shiitake and Lion's Mane, sautéed in this fashion, with a touch of soy or tamari, and added to white rice results in an extraordinarily culinary experience with complex, rich fungal tones.

Comments: This mushroom grows quickly and is acclaimed by most mycophagists. From a marketing point of view *H. erinaceus* has distinct advantages and few disadvantages. The snowball-like forms are most appealing. Picked individually and wrapped in rice paper or presented in a see-through container, this mushroom is best sold individually, regardless of weight. A major disadvantage is its high water content and white background, which makes bruising quite apparent, although the mushroom may be, as a whole, in fine shape. Once the brown bruises occur, the damaged tissue becomes a site for bacterial blotch, quickly spreading to the other mature parts of the mushroom. In short, the harvesters must handle this mushroom ever so carefully. By reducing humidity several hours before harvest to the 60–70% range, the mushroom loses sufficient water and tends not to bruise so readily.

A 1-pound specimen often develops when only one primordial site is allowed per 5-pound bag of supplemented sawdust/chips/bran.

Hericium erinaceus grows aggressively on hardwood sawdust enriched with bran. Incubation proceeds for 2 weeks, after which primordia occur spontaneously. Since fruitings off vertical faces of the plastic bags are more desirable than top fruitings, it is essential that holes be punched into the sides of the bags directly after colonization. Should primordia form unabated within the confines of the sealed bag, the number and quality of spines are adversely affected. Under these conditions, the spines elongate, are loosely arranged, and when they fully develop, the mass of the harvested mushroom is only a fraction of what it would have otherwise been.

Few studies on the cultivation of this mushroom have been published. Most are in Japanese or Chinese. For further information, please consult Huguang (1992).

Using the classic bottle method, a 2-kilometer tunnel proved ideal for growing Yamabushitake in Japan.

The Wood Ear Mushrooms

The Wood Ears are peculiar mushrooms that have captured the palate of Asian mycophagists for centuries. Extensive trading of Wood Ears persisted in the late 1800s from New Zealand to China and Hong Kong. Although not remarkably flavorful, these mushrooms rehydrate readily from a dried state, embellishing soups and sauces. Imparting a unique and pleasing texture to most meals, these mushrooms are a centerpiece of Asian cooking, and highly valued. In general, the cultivation methods for this species parallel that of Shiitake on logs or on sterilized sawdust.

The Wood Ears of the Genus *Auricularia*
Auricularia polytricha (Montagne) Saccardo

 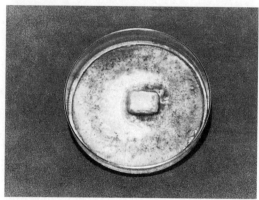

A. polytricha 6 and 12 days after inoculation onto malt agar media.

Introduction: According to records from China circa A.D. 600, this mushroom is heralded as the first species to be cultivated (Chang and Miles 1987 and 1989). This mushroom has an unusual, appealing texture when eaten but, by itself, is not remarkably flavorful. Nevertheless, *A. polytricha* is highly prized in Asia and is commonly used in soups. Upon drying, Wood Ears shrivel to a fraction of their original size, and upon contact with water, rehydrate and enlarge true to form.

Common Names: Wood Ear, Ear Fungus or Tree Ear
Yu er, Mu-er, or Mo-er (Chinese, often also used for *A. auricula*)
Maomuer, Yung Ngo, or Muk Ngo (Chinese)
Kikurage (Japanese for "Tree Jelly Fish")
Mokurage, Aragekikurage (Japanese)

Taxonomic Synonyms and Considerations: At least three very similar Wood Ear taxa occur in the Americas. *Auricularia polytricha* (Montagne) Saccardo has been reported in Louisiana, but is also common from the Americas through Mexico and south to Argentina. It is brownish and coarsely hairy on its outer surface with the hairs measuring up to 450 × 6 μ. The common northern temperate species of Wood Ear bearing strong resemblance to *A. polytricha* is *Auricularia auricula* (Hooker) Underwood. This species is brownish and finely hairy on its outer surface with hairs much shorter, measuring 100 × 6 μ. (See Lowy, 1971.) As a cultivator, I find that humidity has a great effect on this character. Hence, the delineation of these two taxa is often difficult, underscoring their close alliance. One other feature proposed for delineation is the presence or absence of clamp connections. *A. polytricha* has them; *A. auricula* does not. In Asia, *Auricularia auricula* is generally collected from the wild and not as widely cultivated. The Japanese call *A. auricula* "Senji," whereas *A. polytricha* is mostly referred by the name "Kikurage." Most cultivators not trained in the skills of taxonomy are likely to pass on the mistaken identifications of those before them. With the many "varieties" of Wood Ears circulating, the need for interfertility studies and DNA comparisons is evident.

The proper name for the cultivated *A. polytricha* should be *Auricularia cornea* (Ehr.:Fr.) Ehr. ex Endl. Until I can further study these taxa and their arguments, I am retaining the name *A. polytricha* for this mushroom.

Another closely related species is *Auricularia fuscosuccinea* (Montagne) Farlow, the most common Wood Ear in the southeastern United States. This mushroom is rosy to reddish brown and minutely hairy on its outer surface (with hairs measuring 80×5 µ). This species is distributed as far south as Argentina.

Although similar in shape to Ascomycetes, like *Peziza* species, this mushroom is a Basidiomycete as its spores are produced on a club-like structure, not within an ascus.

Description: A gelatinous cup fungus, ear-shaped, generally purplish grayish brown to dingy brown, 2–15 cm broad, sessile. Covered by a medulla of fine hairs. Surface smooth, wrinkled towards the center and upturned towards the outer edges. Firmly gelatinous in texture and readily rehydrating in water to natural form.

Distribution: Varieties of Wood Ears grow throughout the temperate hardwood forests of the world.

Natural Habitat: On conifer or hardwood logs or stumps, especially oaks, willows, locust, mulberry, locust-acacia, and other broadleaf trees. Commonly occurring in soils rich in wood debris, during the cool wet seasons throughout the temperate forests of the world. This mushroom generally favors cool weather and grows from sea level to tree line.

Microscopic Features: Spores white in deposit, otherwise hyaline, cylindrical to sausage shaped, 11–14 (17.5) × 4–6 µ, basidia 4-spored. Clamp connections present.

Available Strains: Strains are available from most culture libraries. Wild specimens abound and can be easily brought into culture. Strains of *A. auricula* and *A. polytricha* are often mislabeled due to the difficulty in separating these taxa from one another.

GROWTH PARAMETERS

Spawn Run:
Incubation Temperature: 75–85°F (24–30°C)
Relative Humidity: 90–95%
Duration: 25–40 days
CO_2: >5,000–20,000 ppm
Fresh Air Exchanges: 0–1 per hour
Light Requirements: not needed

Primordia Formation:
Initiation Temperature: 55–70°F (12–20°C)
Relative Humidity: 90–100%
Duration: 5–10 days
CO_2: 600–1,000 ppm
Fresh Air Exchanges: 5–8 per hour
Light Requirements: 500–1,000 lux

Fruitbody Development:
Temperature: 70–85°F (21–30°C)
Relative Humidity: 85–90%
Duration: 5–7 days
CO_2: 2,000–5,000 ppm
Fresh Air Exchanges: 4–5 per hour
Light Requirements: 500–1,000 lux

Cropping Cycle:
Every 2 to 3 weeks for three to five flushes

Mycelial Characteristics: Longitudinally linear, thickening with age to form a dense cottony white mycelial mat, becoming mottled with brown discoloration in age.

Fragrance Signature: Unpleasant, musty, and reminiscent of raw compost.

Natural Method of Cultivation: The most common technique used in Asia has been to cut logs approximately 1 yard in length, 5–12 inches in diameter, in the late fall to early spring. The logs are simply drilled with holes and spawn is pack tightly into the cavities. I prefer to inoculate logs with sawdust spawn that is packed into chain-saw cuts

A. polytricha in classic form.

Cylinders of sterilized sawdust are suspended in the growing room and punctured to elicit fruitings off the vertical faces.

a foot apart. The logs are kept moist in a shaded, well-ventilated forest. To initiate mushroom formation, submerge the logs in water for 24 hours.

Recommended Courses for Expansion of Mycelial Mass to Achieve Fruiting: From liquid-inoculated grain masters, a second generation of grain spawn in gallon jars is generated. Each gallon jar of spawn inoculates ten 5-pound bags of sawdust supplemented with rice bran. Chopped corn and rye flakes can also be used as supplements.

Suggested Agar Culture Media: MYA, MYPA, PDA, PDYA, or DFA.

1st, 2nd, and 3rd Generation Spawn Media: Millet, milo, rye, wheat, or sorghum all support the formation of a vigorous and luxuriant mycelial mat.

Substrates for Fruiting: Essentially the same hardwoods that are recommended for Shiitake support good fruitings of this species. In Asia, *Acacia* spp. are widely used. The ideal pH range is between 6.5

and 7.0. Wheat straw has proved successful as a medium (Jianjun 1991), especially when sawdust spawn is used.

Recommended Containers for Fruiting: Polypropylene bags and bottles. Each should be punctured with ten to twenty holes after full colonization (25 to 40 days after inoculation) to localize primordia formation. One method uses cylindrically shaped polypropylene bags, 6–8 inches in diameter, that are cut open at both ends and laid horizontally to build a wall of exposed surfaces. From these open ends the Wood Ear mushrooms emerge. Others use the type of polypropylene spawn bags fitted with a microporous filter patch. These bags are usually punctured to encourage mushroom formation around the bags, or are partially opened at the top to elicit a surface flush of mushrooms. I personally prefer the puncture hole technique.

Yield Potentials: $^1/_2$ to 1 pound of mushrooms per 5 pounds of supplemented sawdust. Logs produce for several years, yielding at best, 20% of their wet mass into fresh mushrooms over 3 to 5 years.

Harvest Hints: If mushrooms form through holes in the plastic, harvesting is fast and efficient. Clusters of ear-shaped mushrooms pop off without residual substrate debris.

Form of Product Sold to Market: Fresh and dried. The greatest volume of this mushroom is sold in dry form. Although dark when dried, these mushrooms lighten to brownish in color as they rehydrate, usually true to form. The rubbery and cartilaginous consistency is strangely appealing.

Nutritional Content: 8–10% protein, 0.8–1.2% fat, 84–87% carbohydrates, 9–14% fiber, and 4–7% ash (Ying 1987; Chang and Hayes 1978). Moisture content of fresh mushrooms is usually within a few percentage points of 90%.

Medicinal Properties: Ying (1987) reports that this mushroom is 80 and 90% effective against Ehrlich carcinoma and Sarcoma 180, respectively. The references are in Chinese. A hematologist at a medical school in Minnesota pricked his finger in a blood-clotting test, and when his blood failed to clot, the ensuing investigation traced the cause to the Wood Ear mushrooms he had eaten the night before at a Chinese Szechwan restaurant. In the United States during 1970s, when Chinese restaurants started serving Wood Ear mushrooms, some patrons noted the emergence of blotchy hemorrhages on their faces the day after consumption. Caucasian women were particularly susceptible. This phenomenon was later dubbed Szechwan Restaurant Syndrome, later becoming known as "Szechwan Purpura" (Hammerschmidt 1980; Benjamin 1995). This discovery lead to a new anticoagulant effective in breaking up blood clots.

Yuan et al. (1998) showed that the water-soluble polysaccharide fraction from Wood Ear had a hypoglycemic effect on genetically diabetic mice. A water-insoluble glucan similar to ß 1-3 D-glucans and ß 1-6 D-glucans (560,000–610,00 m.w.) were isolated from the hot-water extract of the fruiting bodies. In vitro tests showed potent anti-tumor

A. polytricha can be grown from a horizontal surface. However, most cultivators prefer to crop the mushrooms from vertical faces as featured on page 400.

activity against the solid form of Sarcoma 180 (Kio et al. 1991).

Flavor, Preparation, and Cooking: For many this mushrooms is not remarkably flavorful. Nevertheless, this mushroom adds another dimension to the culinary experience. *A. polytricha* has a most appealing brittle-gelatinous texture, potentiating the flavors of foods cooked with it. I have seen chefs embellish salads with this mushroom, uncooked, as a garnish.

Comments: This mushroom is extremely popular in Asia and to a much lesser extent, in Europe. In the United States, primarily those of Asian descent use this mushroom. Appealing for its ease of use, Wood Ear mushrooms dry and rehydrate quickly.

The method of cultivation closely parallels that of Shiitake. The punctured polypropylene bags should be placed in a 100% or condensing fog environment to encourage the emergence of mycelium. Once mushroom initials form, the atmosphere should clear of condensing fog but be held at 95–100% humidity. Watering two to four times a day brings on fruitbody formation within 5 to 10 days. If by third flush, substrate moisture has fallen below 50% and cannot be replenished through frequent watering, submerging the sawdust bags will induce one last substantial flush.

Imazeki et al. (1988) rates *A. auricula* as superior to *A. polytricha* in culinary terms. *A. auricula* can be grown in the same fashion as *A. polytricha* except that *A. auricula* thrives in the 50–60°F (10–15°C) range. (These differences may be varietal in nature—assuming that the taxa of *A. auricula* and *A. polytricha* are conspecific.)

Wood Ear fruiting from super-pasteurized sawdust.

The Jelly Mushrooms

Of the jelly fungi, the *Tremella* species have drawn the most attention. Existing in its simplest form as a yeast, this group of jelly-like mushrooms triggers into a mycelial stage when parasitizing other wood-loving fungi. When wet, they swell with water and are easy to recognize, trembling when you touch them. The bright yellow Witch's Butter, *Tremella mesenterica*, is probably the most common. Many are cold weather tolerant. They have unique basidia that are vertically partitioned. No poisonous species have been identified, at least so far, and several have had a long tradition of use.

White Jelly Mushroom of the Genus *Tremella*
Tremella fuciformis Berk

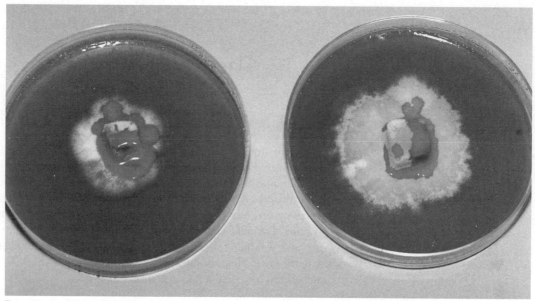

The cultures of *Tremella fuciformis* first grow as a slimy yeast that can spontaneously morph into whitish mycelium.

Introduction: One of the foremost medicinal and culinary mushrooms of China and Taiwan, this mushroom has one of the most unusual strategies for cultivation. *Tremella fuciformis* is a parasitic yeast, growing as a mucous-like substance until it encounters its preferred host, a common wood-decomposing Ascomycetous fungus, *Hypoxylon archeri*. This parasitic yeast is then triggered into its aggressive mycelial stage, and may ultimately form a translucent, whitish, jelly-like fruitbody utilizing the nutrients gained from the victimized host. A poor decomposer of cellulose and lignin, *Tremella fuciformis* is largely dependent upon the digestive abilities of a precursor fungus, which it parasitizes for essential nutrients leading to mushroom production.

Common Names: Yin Er
White Jelly Fungus
White Jelly Leaf
("Shirokikurage")
Silver Ear Mushroom
Snow Mushroom
Chrysanthemum Mushroom

Taxonomic Synonyms and Considerations: Although the yeast stage of *Tremella fuciformis* is well studied, its anamorphs are not. In its simplest, imperfect form—a yeast—this fungus simply buds as chains of individual cells, soon separating. The yeast stage can metamorphosize into a hyphal state, where whitish filamentous mycelia emerge.

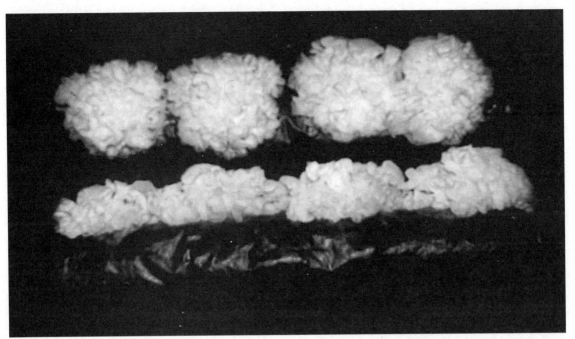

The Silver Ear or White Jelly Mushroom fruiting on sawdust.

(See photograph.) Once true mycelium develops, mushrooms can form, provided sufficient nutrients are available. In its perfect (sexual) stage, this species forms basidia upon a gelatinous, jelly-like fruitbody similar to *T. mesenterica*, the common Witch's Butter mushroom of North America. Because of its complex life cycle, and its ability to generate progeny through a variety of mechanisms, the complexity of the life cycle of *Tremella fuciformis* is still not fully understood. *Tremella aurantia*, the Golden Ear mushroom, is similar except for color, and can be cultivated in a similar manner using *Stereum hirsutum*, a basidiomycete, as the parasitic host.

Description: When wet, young fruitbodies swell to resemble blobs of translucent to whitish jelly. When dried the fruitbodies collapse but upon rehydration, the mushrooms reform and regrow, eventually forming ear-shaped leaflets, and as maturing, resembling a chrysanthemum flower, up to 25 cm in diameter.

Distribution: Indigenous to the southern United States, Mexico, and the subtropics of Asia, this species is probably more widely distributed than the literature presently indicates, and has, on occasion, been found in temperate regions of the world.

Natural Habitat: An inhabitant of woodland environments, *Tremella* species are found worldwide, and often coinhabit with *Hypoxylon* species, especially *H. archeri*. *Tremella* species primarily grow attached to the barks of deciduous trees.

Microscopic Features: Spores ovoid to egg-shaped, 4-7.5 × 7-12 μ, produced by a tetra-longitudinally septate basidium, whitish in mass.

Available Strains: For the purposes of fruiting, *Tremella fuciformis*, a parasitic yeast, must be matched with a susceptible host, in this case the common wood saprophyte, an Ascomycete, belonging to the genus *Hypoxylon*. When cloning

GROWTH PARAMETERS

Spawn Run:

Incubation Temperature: 73–77°F (23–25°C)
Relative Humidity: 90–100%
Duration: 7–15 days
CO_2: >5,000 ppm
Fresh Air Exchanges: 1 per hour
Light Requirements: n/a

Primordia Formation:

Initiation Temperature: 68–77°F (20–25°C)
Relative Humidity: 90–95%
Duration: 5–7 days
CO_2: 400–800 ppm
Fresh Air Exchanges: 5–7 per hour
Light Requirements: 500–3,000 lux

Fruitbody Development:

Temperature: 73–77°F (23–25°C)
Relative Humidity: 88–90% (92%)
Duration: 7–12 days
CO_2: Less than 400–2,000 ppm
Fresh Air Exchanges: 5–7 per hour
Light Requirements: 500–3,000 lux

Cropping Cycle:

Every 2 to 3 weeks for two to three flushes

---

specimens from the wild, mycologists have noted that two strains of fungi are commonly isolated. The tandem cultivation of these native fungi from the same location assures compatibility. The paired cultures that I have received from China (*T. fuciformis* and Xing Hui, "Friendly Fungus" or "Feather Fungus," probably *Hypoxylon archeri*) perform better. These cultures are first grown in isolation from each other. Isolating this parasitic yeast and its host from the same natural habitat allows for the creation of a mixed colony more suitable for commercial cultivation than two isolates from foreign environments. Although the American Type Culture Collection has cultures of both, their terri-

torial origins are not always identified, and hence compatibility is not assured. Please note that mixing these two fungi together and trying to grow them as a combined culture has limited success over the long term. They are best grown separately and then joined together during fermentation.

Mycelial Characteristics: This parasitic species is a classic dimorphic fungus, whose primary life cycle is that of a slow-growing yeast, gelatinized in appearance, and becoming filamentous spontaneously, especially on the outer margins of the culture, but also in reaction to the encounter with a host, such as its preferred victim, the Ascomycetous wood-lover *Hypoxylon archeri* and allies. Some authors have noted the formation of oidia, chlamydospores, and conidiophores from *Tremella fuciformis*, but it is not clear that the cultures studied originated from single spore isolates or from "mixed" cultures. The repertoire of alternative means for spore dispersal in *Tremella fuciformis* is impressively complex, and underscores that the simplest of fungi—the yeasts—have much to be understood.

Fragrance Signature: Grain spawn sweet at first, musty in age.

Natural Method of Cultivation: The traditional method for the cultivation of this fungus has been on logs, as pioneered in China and Taiwan, and is modeled after Shiitake. The first success at cultivation inoculated logs by inserting broken fruitbodies directly into drilled holes, but as with Shiitake, the use of pure culture sawdust spawn dramatically improved yields. Once sawdust spawn is packed into the holes, molten wax is used to cover the holes that are then capped with pieces of bark. The logs produce in the late summer to early fall. This mushroom, although more than 90% water, has the uncanny ability to dry and rehydrate, especially when young, and still survive for full maturation.

Recommended Courses for Expansion of Mycelial Mass to Achieve Fruiting: Several strategies have been recommended, through trial and error, for the expansion of *Tremella fuciformis*. The uniqueness of this species has caused various researchers to pursue different paths. Since the *Tremella* culture is dependent upon the parasitization of another fungus to achieve fruiting, the proper matching of cultures is critical for success. One approach has been to use a "mixed" culture, which can be successful for a few generations, but from my experience, fails in the long run, due to a progressive dominance of *Hypoxylon* over *Tremella*. The method I am describing here works well, and has not been described in any of the references I have seen. Every *Tremella* cultivator will undoubtedly develop their own preferred method given their unique sets of experiences.

After the yeast-like culture has colonized 50% of a 100 mm PDYA petri culture dish, I recommend the liquid fermentation of the culture for a period of 4 to 5 days. Two culture dishes are placed into 500 ml of water in a high-speed blender (the Eberbach brand is one I recommend). After a few seconds of high-speed stirring, 500 ml of this fragmented culture is transferred into a 2,000 ml Erlenmeyer flask containing 1,000 ml of a sterile nutrient broth (30 g malt, 3 g yeast, 3 g powdered grain). The fermentation continues at 300 rpm for 4 to 5 days. Upon close examination, small stellar-like colonies of wispy white mycelia can be seen. After the fermentation cycle has been completed, a nearly fully colonized 100 mm petri dish of *Hypoxylon archeri* is placed into 1,000 ml of water and fragmented at high speed. (I use a 1,000 ml Eberbach stirrer for this also.) Only 1–2 ml of this *Hypoxylon* culture (equivalent to 30 to 50 drops!) is added to the 1,500 ml of fermented *Tremella* culture. Although this represents much less than a 1:1,000 rate of inoculation (given that the *Tremella* culture has been growing three-dimensionally over the past 4 days), the *Tremella* is "activated" into explosive hyphal growth. This newly combined broth is then diluted by a factor of 10 (= 15 liters), and about 50 ml are injected or poured onto 300 2- to 3-kilo containers (bags or jars) of sterilized grain. Colonization is complete in 4 to 5 days and the spawn should be used promptly before conidiophores form and release spores. (The need for immediate use is primarily so those spores do not cross-infect other cultures in the laboratory.) The above-mentioned spawn strategy is sufficient for inoculating 10,000+ fruiting containers. Of course, if this is too much, the expansion schedules can be assuaged to meet the needs of the cultivator.

Suggested Agar Culture Media: MYPA, PDYA, OMYA, and/or DFA, balanced to a pH within 5 to 6.

1st, 2nd, and 3rd Generation Spawn Media: Grain spawn for the first two generations, and then sawdust spawn for the final stage. With each stage, spawn should be used within 3 to 4 weeks of inoculation and not older as the *Hypoxylon* may increasingly dominate, with the cultures becoming more and more speckled with yellowish green to blackish green conidia. Should these conidia become airborne in a clean room, the laboratory integrity can be jeopardized. Some cultivators rely on laboratories for the initial master spawn, and then generate spawn by pasteurizing sawdust in as clean an environment as they can provide.

Substrates for Fruiting: Broadleaf hardwood sawdust (oak, tan oak, alder, beech, birch, elm, melaleuca, etc.) supplemented to 20% with rice, oat, and/or wheat bran, 1% gypsum, and 1% sugar. Water is added to bring up the moisture content to 65–70%. An alternative formula recommended by Chang and Miles (1989) utilizes sugarcane bagasse as the base material in lieu of sawdust with the addition 20% bran, 1% calcium carbonate, and 1% soybean powder. Other formulas have been tested successfully in China. One uses 75% cottonseed hulls, 20% bran, 2.5% gypsum, and .4% magnesium sulfate. Another uses 38% cottonseed hulls, 38% chopped corncobs, 20% bran, 3% gypsum, and 1% urea. This mushroom is adaptive to a wide variety of formulations making it one of the easier mushrooms to grow. Traditionally, in China, the substrate material is then super-pasteurized at 210°F (<100°C) for 10 to 12 hours at atmospheric pressure. An alternative method, available for those have autoclaves, is to sterilize the substrate at 252°F

(120°C) at 15 psi (1 kg/sq cm) for 4 to 6 hours depending upon density of fill and spacing of bags.

Recommended Containers for Fruiting: Perforated plastic bags (often black) and/or columns made of reinforced plastic fabric, laid horizontally or vertically.

Yield Potentials: S. T. Chang (in Chang and Miles, 1989) reports that yield in China average between 160–180 kg of fresh mushrooms for every 160–180 kg of cottonseed hulls (not including other amendments), a remarkably high biological efficiency. However, such consistently high yields are only representative of the most skilled cultivators. Note that fresh mushrooms are 92–94% water, exceptionally high for any mushroom. Actual dried mushrooms will lessen the biological efficiency in comparison to most mushrooms using the same fresh-weight-of-yield to dry-weight-of-substrate formulas.

Harvest Hints: Mushrooms are best harvested when the fruitbodies lose their translucence and become white, at a time when the "flower petals" have fully extended, typically when the mushrooms are between 3 and 5 inches (8 and 12 cm). (See photograph.) When the mushrooms overmature, they darken with age, and the leaflets soften and wilt. Care should be taken to cut the mushrooms closest to the point of origin, and to discard any yellowing tissue. Solar drying is a common practice in the tropics. An ideal temperature for rapid drying is 113°F (45°C).

Form of Product Sold to Market: Mushrooms are most often sold whole, dried, in Chinese food and herbal stores. Most often added to soups, they are commonly sugared and sold as candies. A number of medicinal mushroom formulas use *Tremella fuciformis* as an ingredient. Of interest to me is the ambiguity associated with these medicinal mushroom formulations sold in the marketplace. Most formulas do clearly state whether they are made from only the fruitbodies, from only the yeast stage through fermentation, or from the co-culture with *Hypoxylon archeri* on carrier materials such as grains. If the product is speckled with greenish black nodules, they are likely the dark brown spores from

Hypoxylon perithecia. Obviously, those products derived directly from fruitbodies will be more potent than from mixed cultures. Studies on the medicinal properties of this mushroom should properly delineate materials derived from fermented products, with or without *Hypoxylon* versus pure fruitbodies. This is yet another reason for the need for standardization in the medicinal mushroom industry. Most fermented products probably have *Hypoxylon* spores and mycelia as well, which have not yet been medically analyzed.

Nutritional Content: When dried, mushrooms have 4.6% protein, 0.2% fat, 1.4% fiber, and 0.4% ash.

Medicinal Properties: *Tremella fuciformis* produces highly water-soluble acidic hetero-polysaccharides, containing xylose, glucuronic acid, mannose, and glucose (Misaki and Kakuta 1995). *Tremella* polysaccharides have been studied for their antitumor properties (Ukai et al. 1972; Zhou et al. 1987; Xia et al. 1989; Ukai et al. 1992; Ma et al. 1992). Gao et al. (1996) found that three heteroglycans isolated from the fruitbodies induced human monocytes to produce interleukin-1 and interleukin-6. Extracts from this mushroom are reported to protect canine liver cells from radiation damage (Zhao et al. 1982), and for the treatment of liver diseases. One of the more complete descriptions of its medicinal properties can be found in Christopher Hobb's book *Medicinal Mushrooms*, Botanica Press. Many of the references supporting *T. fuciformis* originate from China, where there has been the longest history of use.

Flavor, Preparation, and Cooking: Mushrooms are usually dried, sealed in plastic, and then sold for addition to soups. They are also made into candies by immersing in water, boiling down, and adding copious amounts of sugar. This mushroom is highly regarded for its gelatinous texture and its primary use as a culinary/medicinal mushroom is in soups.

Comments: The preferred method of inoculation from sawdust spawn into the supplemented substrate varies depending upon nuances of each cultivator. For instance, the traditional method has been to super-pasteurize the supplemented substrate, allow it to cool, and then aseptically punch holes into the bags, stuffing the holes with peanut-sized chunks of spawn using forceps or one of the sawdust-plug spawn inoculators developed for the Shiitake industry. Once inoculated, the holes should be covered with a paper-like "masking" tape, which is torn off after 2 weeks, when primordia first become visible on a minority of the bags. (If a callus has formed on the spawn containers, then the stroma-like layer should be torn off and removed, a practice also common to users of aged Oyster and Reishi spawn.) For me, an easier method is to use the four-bladed "arrowhead" technique I describe in this book for promoting primordia in the cultivation of Oyster and Lion's Mane mushrooms. The punching of the bags with sharp arrowheads create four flaps of plastic that open as the primordia enlarge. What is best for you depends upon many factors. No one situation is exactly alike.

For further reading, I recommend the 1989 article by S. T. Chang and P. G. Miles, "*Tremella*— Increased Production by a Mixed Culture Technique: Chapter XV" in *Edible Mushrooms and their Cultivation* (CRC Press, Boca Raton, Florida).

The Cauliflower Mushrooms

A group unto itself, these unique wood lovers have been the favorite of many mushroom hunters who could return to the same location year after year, sometimes finding specimens so huge that they have been difficult to carry! Cauliflower mushrooms belong to the genus *Sparassis* with five species thus far published: *S. crispa, S. laminosa, S. radicata, S. herbstii,* and *S. spathulata.* An eastern form, the same as is found in Europe, with a short stem has been called *Sparassis crispa* while the western form, sporting a long taproot, has been identified as *Sparassis radicata.* They may be two forms of the same species.

The Cauliflower Mushroom of the Genus *Sparassis*

Sparassis crispa Wulf ex Fries
= *Sparassis radicata* Weir
= ? *Sparassis herbstii* Peck

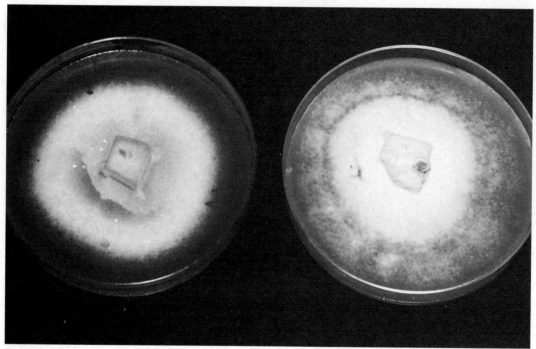

Sparassis crispa mycelia at 10 and 20 days.

Introduction: Popular in northern Europe and United States, *Sparassis crispa* can grow to enormous sizes (up to 50 lbs!), especially those native to the old-growth fir forests of northwestern North America. One of the most popular of edible mushrooms, and easiest to identify, the cultivation of this mushroom has only recently been developed, largely through the isolation of productive strains.

Common Names: The Cauliflower Mushroom
Ruffles
Clavaire Cre'pue

Taxonomic Synonyms and Considerations: A constellation of species, *S. crispa*, *S. radicata*, *S. laminosa*, and *S. herbstii* look very similar to one another. The classic radicate stem ("tapering root") of *S. radicata* is typical of those feeding off the root zone of mature conifer trees of the West and is so named. Those forms found on oaks in the eastern United States and in Europe typically lack the long, tapered root-stem, tend to have thicker folds, and are generally smaller. The hardwood-degrading

Azureus Stamets holding 20-pound cluster of *Sparassis radicata* collected from the old-growth forest near Bagby Hotsprings, Oregon.

GROWTH PARAMETERS

Spawn Run:
Incubation Temperature: 70–75°F (21–25°C)
Relative Humidity: 90–100%
Duration: 20–25 days
CO_2: >5,000 ppm
Fresh Air Exchanges: 1 per hour
Light Requirements: n/a

Primordia Formation:
Initiation Temperature: 50–60°F (10–16)°C
Relative Humidity: 95–100%
Duration: 7–14 days
CO_2: 500–1,000 ppm
Fresh Air Exchanges: 5–7 per hour
Light Requirements: 1,000–1,500 (2,000) lux

Fruitbody Development:
Temperature: 55–65°F (13–18°C)
Relative Humidity: 85–90% (95%)
Duration: 7–14 days
CO_2: 500–1,000 ppm
Fresh Air Exchanges: 5–7 per hour
Light Requirements: 1,000–1,500 lux

Cropping Cycle:
One flush indoors on sawdust—once cropped, the blocks can be buried outside for subsequent flushes in the fall

species have been suggested to be a different species, properly called *Sparassis herbstii* or *S. spathulata*.

Description: When young, whitish to yellowish white, looking like a cauliflower, then a brain, tightly-covered egg noodles, soon with ridges elongating into flattened, wavy leaflet-like structures, diverging from the center, smooth, whitish to creamy yellow, with the margins darkening during age or drying. Growing up to 2 ft. (60 cm) across with a central stem, often tapering into a long taproot, darkening towards the base. Cultivated specimens are much smaller than their wild cousins, with mushrooms from the Pacific Northwest old-growth forests providing the largest specimens.

Distribution: Found throughout the temperate regions of Europe, northeastern and western North America.

Natural Habitat: Growing at the base of coniferous trees (pines, Douglas firs) and in the eastern North America and Europe on oaks from late July through November, preferring older forests.

Microscopic Features: Spores smooth, ellipsoid, 4-7 × 3-4 μ, whitish in deposit.

Available Strains: The strains from Europe appear to be more adaptive to cultivation than the isolates I have obtained from the Pacific Northwest of North America. The difference may be simply that of

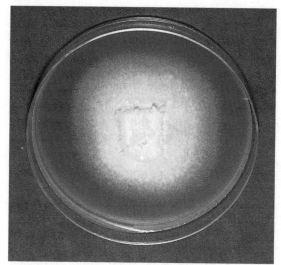

Culture of *Sparassis laminosa,* 10 days old.

habitat—*Sparassis crispa* isolated from firs are more difficult to grow than those growing on hardwoods. If you are cloning, the best region for taking tissue cultures is from the upper stem base regions giving rise to the diverging leaflets. The flesh in this region can be very wet, and tissue cultures should be carefully excised, paying attention to avoid the excess moisture that flows from the cut cells, or else bacteria will proliferate.

Mycelial Characteristics: Longitudinally radial, diffuse to grayish white, forming random, cottony pseudo-fruitbodies on enriched MEA media, not rhizomorphic. Although several authors report this mushroom causes a brown rot (Arora 1986; Lincoff 1981), the mycelium is white, and when rotting wood, the pulp is whitish in color, not dark brown.

Fragrance Signature: Grain spawns musty smelling, not pleasant.

Cauliflower mushrooms fruiting on sterilized hardwood sawdust.

Natural Method of Cultivation: In the hopes of establishing a perennial patch, burying blocks of mycelium adjacent to oak and fir trees, is an option, although it is not clear whether *Sparassis crispa* is an opportunistic parasite or a benign saprophyte. Some mycologists believe it to parasitize the root zones of fir trees, causing a brown rot (Arora 1985). However, in culture, the mycelium on sawdust is white. If the tree accepts this fungal neighbor, a perennial mushroom patch can be established lasting for many years. Another method is to bury an intact mass of mycelium into a shady location in the spring for a flush the following fall.

Recommended Courses for Expansion of Mycelial Mass to Achieve Fruiting: Standard agar-to-grain, grain-to-grain transfers or via liquid fermentation for the beginning of the grain sequence.

Suggested Agar Culture Media: MYPA, PDYA, OMYA, and/or DFA.

1st, 2nd, and 3rd Generation Spawn Media: Grain spawn throughout.

Substrates for Fruiting: Sawdust from oak, alder, cottonwoods/poplars, aspen, and elm support mature fruitbody development. Supplementation rates, from my experience, should not exceed 15% of the sawdust mass.

Recommended Containers for Fruiting: From bags or trays; this mushroom is naturally a ground dweller fruiting best from horizontal surfaces. If fruiting in bags, leave a 4-inch (10-cm) edge of plastic to help keep in humidity and promote primordia development.

Yield Potentials: Fairly good, 1 pound (½ kg) per 4-pound (2-kg) block of sterilized, supplemented sawdust.

Harvest Hints: Best harvested when the leaflets have fully extended, but before the edges brown from age. As a cultivator, I prefer the smaller fruitbodies, with tight folds.

Form of Product Sold to Market: Fresh and dried. Common in market places in Switzerland, and becoming more available across Europe.

Nutritional Content: Not known to this author.

Medicinal Properties: Not known. I would not be surprised if this mushroom contained potent polysaccharides and novel antibiotics. Its ability to grow to a prodigious size and its residence in old-growth forests make it a prime candidate for novel constituents.

Flavor, Preparation, and Cooking: A "leafy" fruitbody, best torn into pieces and then cooked in mass. *Sparassis* species are crunchy, with a nutty, mild, but enjoyable flavor. I prefer tearing the leaflets apart, cooking them in a wok, with lots of vegetables, and using the excess water from the mushrooms to provide the steam. The Cauliflower mushroom is exceptionally good when dipped in beer-based batter and deep-fried. Also excellent when added to casseroles, soups, and stews.

Comments: Strains of this mushroom native to hardwoods are easier to grow than those endemic to conifers. This mushroom shares many similarities in its growth parameters to Maitake (*Grifola frondosa*) and can be co-cultivated in the same growing rooms. One advantage of the cultivated forms, besides year-round availability, is that the mushrooms are free of the soil debris in the recesses of the folds typical of wild collected specimens. First pioneered by Swiss mycologists, the races of *Sparassis crispa* in Europe may be quite distinct from North American forms.

The Morels (Land-Fish* Mushrooms)

Morels embody the mystique of mushrooms. They are elusive, highly camouflaged, and appear for just a few days in any one place. Finding Morels in the wild tests the skills of even the most experienced mushroom hunter. Once found, Morel patches are guarded in secrecy and treasured like family heirlooms. Revealing the location of a Morel patch is an admirable expression of friendship and trust. And it may be the most foolish thing any Morel hunter can do!

In northern latitudes, from April to June, Morels grow in several specialized habitats. They thrive in abandoned apple orchards; at the bases of dying and dead elms; around living cottonwoods, oaks, and poplars; in sandy gravel soils along rivers and streams; in "beauty bark" used for landscaping; at the bases of young firs; in the tracks left by bulldozers punching new roads through forests; in limed soils; and especially in the wastelands left by fires.

What in the world could be the common denominator shared by all these habitats? Who knows? The only habitats I can easily recreate are the aftermath of the fire, the beauty bark bed, and the gravel roadbed. Of these, the fire-treated habitat has proved the most reliable and reproducible. Fire destroys competitors and reduces nutrient levels of carbon and nitrogen while proportionately increasing levels of calcium, potassium, and mineral salts. This wasteland supports little life, save for the treasured Morel.

When nature suffers a catastrophe, the Morel life cycle lies ready in the waiting. How Morels thrive in the charred desolation left by a forest fire is, in itself, mystifying. In the spring of 1989, after the massive Yellowstone fires of the previous summer, huge fruitings of Morels emerged from the ashen landscape. Some of the largest Morel finds in history were discovered one year after the Mount St. Helen's eruption. So abundant were the fruitings that excited collectors filled their pickup

---

* "Land-Fish" is a Native American name, possibly Mohawk, aptly given to Morels.

415

The aftermath of the Yellowstone fire created an ideal habitat that yielded huge quantities of edible and choice Morels.

trucks to the brim with hundreds of pounds of Morels. To their dismay and disbelief, the gritty ash made the mushrooms entirely inedible!

Stories of unusual Morel fruitings are as enticing as they are bewildering, and in certain cases approach legendary status. Here are a few examples.

- A massive Morel fruiting occurred several weeks after sludge from a Washington pulp company was flooded into a tree nursery. Thousands of Morels sprang up.

- After a flood in eastern Oregon, where a family's backyard was under a foot of water for more than a week, Morels weighing several pounds soon followed. (See page 417.)

- A rain-soaked and decomposing straw bale in the middle of a wheat field yielded an enormous Morel weighing several pounds.

- From the ruins of a house destroyed by fire in Idaho, huge Morels were found in a basement coal bin. The strain I cloned from these mushrooms is known as "M-11" and is featured in this book.

- A local nursery was selling phlox plants and from every pot Morels were sprouting. (See page 417.)

- An old timer recently told me that, after shooting a chicken-killing dog one autumn, he was shocked to find Morels fruiting in a circle around the decomposing carcass the following spring. (I forgot to ask him whether or not he ate the Morels....)

- An excited woman once called me from Napa Valley, California. To her family's utter disbelief they found a Morel fruiting from the ashes of their indoor fireplace. The fireplace

This 4-pound Morel was discovered in a backyard after floodwaters receded.

The freak occurrence of Morels popping up from nursery-grown, potted phlox plants still mystifies me. This discovery steered me on a fruitless path of experimentation.

had not been used for 6 months. They have no idea how it got there.

- Near Vancouver, Washington, one of my students planted Morel spawn into the ashes of his barbecue grill that was located on a verandah of his condominium. He was amazed to find Morels popping up from his hibachi several months later.

- I discovered the only Morel patch naturally growing on our 17-acre farm through a pre-cognitive dream. Upon telling my dinner guests of the dream I had had that morning, we went directly to the location. In exactly the 20 × 20-ft. plot of ground seen by me in dream-travel, Morels were popping up. No where else on our property has a native Morel patch been discovered in the years hence. . . .

Baffling and beguiling, Morels continue to tease us with their peculiar sense of humor. If any readers know of similarly unusual encounters of the Morel kind, I would like to know. Please write me c/o Fungi Perfecti, P.O. Box 7634, Olympia, WA 98507 USA.

The Morel Life Cycle

Morel spores germinate quickly. Their mycelia race through the environment—up to 4 inches per day. Morel mycelium can colonize a vast territory in a relatively short time. But when they encounter a physical boundary, a non-nutritional zone, or competitors, the mycelium stops expanding. After experiencing environmental shock (such as a fire), the mycelium collapses and forms a subterranean structure called a *sclerotium*. Understanding

Sclerotia of the Giant Morel, *Morchella crassipes*, forming in jars. Soil is placed onto colonized grain. The mycelium grows into the soil and, after several weeks, forms sclerotia. As with most sclerotia-forming mushroom species, this phenomenon is encouraged by darkness during incubation.

The harvested Morel sclerotia can grow to several inches in diameter.

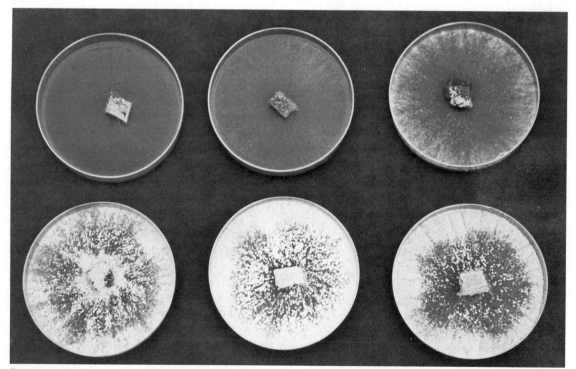

Six-day sequence of the growth of *Morchella angusticeps* (Stamets strain #M-11). Note rapid rate of growth and the formation of "micro-sclerotia." This strain loses the ability to form micro-sclerotia when propagated more than five petri dishes from the original culture. Downstream inoculations into all bulk substrates are similarly affected.

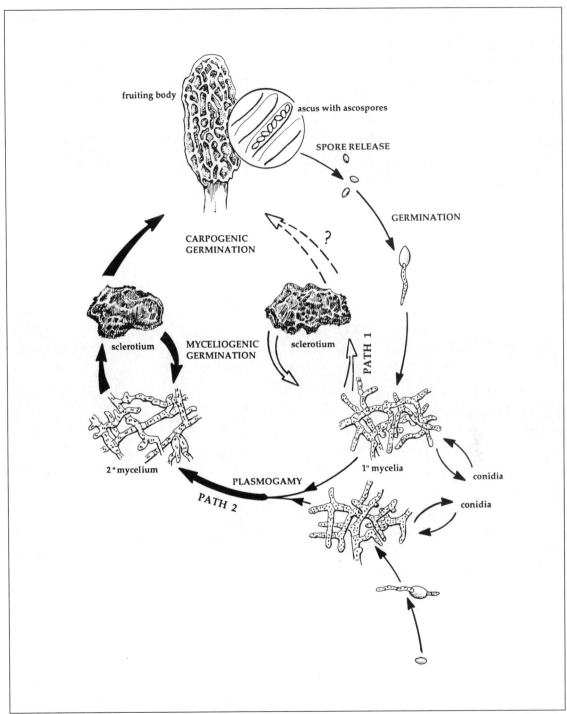

Life Cycle of the Morel by Kandis Elliot, copyrighted 1990 by Thomas J. Volk. Reprinted with permission from *Frontiers in Industrial Mycology*, ed. by Gary Leatham.

A Morel primordium forms (just right of center).

Morchella esculenta fruiting in trays at Morel Mountain. The technique they have developed is patented.

sclerotia (plural of sclerotium) is the key to Morel cultivation.

A sclerotium is a hardened, asexual mass of cells, which roughly resembles a gold nugget or walnut in form and size. Sclerotia are produced by many mushroom species, including *Agrocybe arvalis*, *Collybia tuberosa*, *Conocybe cyanopus*, *Hypholoma tuberosum*, *Polyporus tuberaster*, *Polyporus umbellatus*, *Pleurotus tuberregium*, *Poria cocos*, *Psilocybe mexicana*, *Psilocybe tampanensis*, and dozens of others. The mushrooms that form sclerotia tend to be soil-dwellers. The sclerotia represent a nutrient storage and resting stage, allowing the mushroom species to survive inclement weather. They can be dried to the point of flammability, and upon rehydration, the cells swell with moisture and the sclerotium springs to life and either transforms into a mushroom or reemerges as a new mycelial network. Morel mycelium produces sclerotia naturally, without the interference of humans, on many habitats from peat moss to sand to straw.

The sclerotia of the Black Morel are uniquely different than all the other Morel species. The sclerotia of the Yellow Morel, *Morchella esculenta,* and the so-called Giant Morel, *Morchella crassipes,* are dense, slippery, and heavy when fresh, and dark and walnut-like. The sclerotia of the Black Morel, *Morchella angusticeps,* are abrasive, golden yellow to orange, light in weight when fresh, and pumice-like. Although studies on the sclerotia formation of *Morchella esculenta* and *Morchella crassipes* have been published (Ower 1982; Ower et al. 1986, 1988; and Volk and Leonard 1989, 1990), no studies have been published on *Morchella angusticeps* until now. One other difference: The sclerotia of the Black Morels form by the thousands per cubic foot whereas the sclerotia of the Yellow Morels are comparatively fewer in number, but larger in size. The differences between these two groups of Morels are soon seen after the clones or spores are put into culture.

Although Ron Ower (1986) was the first to note that Morels arise from sclerotia, the first to propose a complete Morel life cycle was Thomas Volk (1990), a rendition of which is on page 419.

Absent from this life cycle is an asexual phase wherein sterile cells are borne on short hyphal branches, similar to oidia. An abundance of these asexual spores form a powdery mildew, and has been called *Costantinella cristata* Matr. (Costantin 1936). I only see it in outdoor-inoculated sawdust substrates.

The Development of Indoor Morel Cultivation

All attempts at controlled, indoor Morel cultivation failed, until Ron Ower succeeded in 1982. By his own admission, the discovery of Morel cultivation was more by accident than design. Ron Ower told me that his experiences growing *Psilocybe* mushrooms combined with an "accident in the laboratory" led to his success. (He revealed he had used the casing formula outlined in my first book, *Psilocybe Mushrooms and Their Allies* [Stamets 1978] as the sclerotia-forming formula.) Under pressure from venture capitalists, Ron Ower applied for and was awarded two patents, along with G. Mills and J. Malachowski, for the cultivation of *Morchella esculenta,* the Yellow or White Morel (Ower et al. 1986 and 1988). The patent describes a technique whereby the Morel mycelium is grown from a nutritious food source into a non-nutritious medium. The mycelium forms sclerotia in the nutritionally impoverished substrate. After a period of dormancy the sclerotia are saturated with water, swell, and when subject to *specific environmental conditions,* metamorphosize into Morel mushrooms.

I have some insights that may be useful in the cultivation of a related species, the famous Black Morel, *Morchella angusticeps.* My research has concentrated on the cultivation of the Black Morel because this species is the most common Morel endemic to western Washington State, where I live. All the readers of this book are encouraged to experiment with Morel cultivation, and develop their own, unique techniques that do not come into legal conflict with the patented processes. To date, all cultivated mushrooms can be grown by more than one method. For instance, Oyster mushrooms can be grown on straw or sawdust or coffee plants. Over the years many techniques can be used for culturing Shiitake, Nameko, Reishi, and Shaggy Manes. With more experimentation, Morels are no exception.

The Morels of the Genus *Morchella*
The Black Morels
Morchella angusticeps Peck complex

Morchella angusticeps mycelia 3, 5, and 7 days after inoculation onto malt agar media.

Introduction: Few mushrooms have gained such an avid following as this mushroom. Morels are one of the most enigmatic of all fungi, appearing only in the spring in the weirdest assortment of places—abandoned apple orchards, burned grounds, flood plains, and underneath cottonwoods, elms, oaks, and firs. Morels are one of the few mushrooms that love construction sites and devastated habitats, appearing well before other forms of life. In fact, they may be keystone species—the first in a sequence to rehabilitate catastrophically damaged environments, such as those seen from forest fires. Humans, deer, bears, small varmints, and insects all compete for this delicious mushroom, and in doing so, spread its spores.

Common Names: The Black Morel
The Conic Morel
Peck's Morel

Taxonomic Synonyms and Considerations: Morel taxonomy, to put it politely, is horribly confused. From the same culture, I have grown Morels totally dissimilar in appearance, bolstering my suspicions about the divisions between "species." My experiences reveal that the growth environment has a radical effect on morphology. However, from a cultivator's point of view, I see some natural groupings.

The Morel taxa, which include all the white, yellow, and black forms, are far too numerous to list here in their entirety. However, the Black Morels are a naturally definable cluster, including *Morchella angusticeps, M. conica*, and *M. elata*. In culture, they behave similarly. I would not be surprised if they are

Portrayal of the archetypal Black Morel, *Morchella angusticeps complex*.

The unique *Morchella atretomentosa*. These two specimens were collected from a burn site in Kamilche Pt., Washington, in the spring of 1988. Note the coat of fine hairs.

all found to be the same species in the broadest sense. A new, totally Black Morel, covered with a fine fuzzy coat, is called *Morchella atretomentosa* (Moser) Bride, a mushroom that was uncommon in North America until the year after the Yellowstone fires. This Morel is so unique in its appearance that I would be surprised if it shared synonymy with any other. The Yellow or White Morels include *Morchella esculenta*, *M. deliciosa*, and *M. crassipes*. These morels are extremely similar and probably cross over taxonomically. The Half-free Morel, *Morchella semilibera*, which has a short cap overhanging the stem, stands apart from these other Morels. New DNA studies are soon to be published that should shed light onto the abyss of Morel taxonomy.

Description: A honeycombed, ribbed species with black edges, the cap is typically conical shaped, measuring 2–6 cm wide × 2–8 cm high. Stem white,

hollow, with a granular texture, measuring 5–12 cm long by 2–4 cm thick. White mycelium is attached to the base of the stem.

Distribution: Widely distributed throughout the temperate regions of the world.

Natural Habitat: Common in the spring in a variety of habitats, particularly in sandy soils of mixed woods along rivers, in burned areas (1 to 2 years after burning), and less frequently in conifer forests. On the West Coast of North America, this mushroom is commonly found in newly laid wood chips ("beauty bark"). The Black Morels can be found in burnt grounds the first 2 years after a fire. In the Pacific Northwest, Black Morels are also found directly underneath cottonwoods and neglected apple trees. David Arora notes that along coastal California, Morels can be found throughout the year, although they are more frequent in the spring (Arora 1986). In general, fall fruitings are rare and

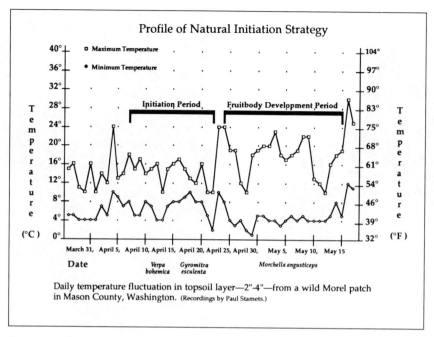

Profile of Natural Initiation Strategy

Daily temperature fluctuation in topsoil layer—2"-4"—from a wild Morel patch in Mason County, Washington. (Recordings by Paul Stamets.)

Temperature fluctuation for a period prior to and during outdoor fruitings of Morels (*Morchella angusticeps*). Note that the temperature fluctuation for primordia formation occurred within 40-60°F (5-15°C).

unpredictable. In Colorado's Front Range, just outside of Denver, Morels are a spring mushroom, but in the mountains around Telluride, Black Morels can be occasionally found in August, under spruce.

Microscopic Features: Spores light creamy brown, 24–28 × 12–14 µ, ellipsoid and smooth, forming 8 at a time in sac-like cells called asci. Mycelia typically multinucleate, with frequent side branching at maturity, clustering and swelling into micro-sclerotia that conglomerate into larger forms.

Available Strains: Strains are readily obtained from wild clones or spore germinations. The mycelium out-races most competitors and can easily be isolated from contaminants. Success in cultivation is highly strain dependent. I believe the primary reason most mycologists have not been able to reproduce the techniques described in the patent is because only a small subset of wild strains can be cultivated with this method.

Mycelial Characteristics: Mycelium at first fine, divergent, fast-running, nonaerial and initially gray, soon thickening becoming gray-brown, with young clones developing numerous brown with orangish to golden nodules, which I call "micro-sclerotia." (See page 418.) As the mycelium matures, the nutrified agar media becomes stained dark brown. (By viewing the petri dish cultures from underneath, the staining of the medium is clearly seen.) As cultures overmature, the mycelium resembles a squirrel's fur. When the mycelium is implanted into unsterilized wood chips, powdery gray mildew forms on the surface. This asexual stage, resembling oidia, has been been classified as *Costantinella cristata*. I do not see this form on agar or grain media.

Fragrance Signature: Mycelium pleasant, smelling liked crushed, fresh Morel mushrooms. After transferring each jar of grain spawn, I am compelled to deeply inhale the residual gases still within each container (a sure symptom of a Morel addict).

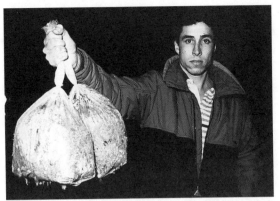

Two bags of sawdust spawn (Stamets strain #M-11) were inoculated into a burn site December 27th. Morels appeared May 4th.

The patch was located in a field (poor location) where no Morels would likely grow naturally. A shade cloth protected the patch from sun and deer. Morels formed in the spring. (See page 426.)

Natural Method of Cultivation: Although my attempts to grow the White Morel (*Morchella esculenta*) indoors have only produced stunted mushrooms, I have had fairly consistent success at growing the Black Morels (*Morchella angusticeps*) outdoors in burned areas topped with peat moss or hardwood sawdust (oak or alder) supplemented with calcium sulfate. When Black Morel mycelium is dispersed into an outdoor burn site, the cultivator relinquishes control to the natural weather conditions. In effect, *nature becomes the cultivator*. In tracking temperatures, I believe the temperature fluctuation from day to night, a circadian cycle, is critical for the formation and development of Morels. From my charts, I found that the limited temperature changes from 40–60°F (4–16°C) for several weeks during early spring triggers mushroom formation. Once the seasonal temperatures move beyond this critical period of fluctuation, no new Morels form. Temperatures as high as 60°F (15.6°C) cause the young mushrooms that have formed to rapidly develop to maturity.

Morels are easy to grow by buying spawn and implanting it into your backyard. Fall plantings have the highest success rates that I estimate at 50%. However, it takes only a few minutes to install a Morel bed, especially if you have ready access to wood ash or a burn site. Once a fruiting Morel patch

is established, the success rate jumps to about 90% for subsequent years, provided a new habitat is created and additional spawn is introduced at the same time each year. In Nature, every ecological niche is unique. The trick is not only finding the right location for your Morel patch, but having conducive weather conditions in the early spring. Protecting the Morel patch from the sun and marauding animals is absolutely essential.

First, acquire sawdust spawn of a proven, fruiting strain of Morchella. Five pounds of Morel spawn is roughly equivalent to a gallon. This amount of spawn is good for inoculating a 4 × 4-ft. to 10 × 10-ft. bed. Inoculations are best conducted in the summer to early fall. Black Morel spawn is available from a number of spawn companies whose addresses are listed in the Resource Directory section in Appendix 4.

You have two options for creating a Morel patch. The first is the simplest. Select a burn site (campfire, fire pit, bonfire site, etc.) that has had a fire in the past month and is well shaded. (Avoid sites where chemically treated wood, paper, or plastics have been used.) Spade and mix the Morel spawn deep into the ashes and burnt wood. Heavily water the site and *ignore it* until early next April. If rain does not fall for more than 2 to 3 days in April and May, a few minutes of watering in the morning and

From the spawn featured on page 425, several forms of Morels showed. The large specimen with the pitted stem and conic cap weighed $1/4$ pound The round-headed one weighed approximately half as much. A taxonomist would be hard pressed to call these the same species, but a cultivator sees such differences in form frequently from the same culture.

evening can only help. Well water is fine. Do not use chlorinated water.

If you do not have a burn site, but have a wood-stove or like to barbecue in the summer, then those ashes can be mixed with other ingredients to create a Morel patch. Mix equal portions of the following ingredients.

10 gallons peat moss
5 gallons ash
1 gallon gypsum (calcium sulfate)

Mix the ingredients in dry form. Find a shady, well-drained location and remove all topsoil until "mineral-earth" is exposed. Lay down the mixture to a depth of 4 inches to cover as broad an area as this volume makes. Water until saturated. Using a shovel or spade, mix in spawn throughout the depth of the substrate. Heavily water the site, provide shade and then institute the strategy of benign neglect—ignore it until early spring.

When Morel spawn is planted late in the year, between the months of October and December, the mushroom patch remains localized to the inoculation site. In contrast, Morel spawn that is planted in the spring often gives rise to fruitings remote from

Another burn site inoculated with "M-11" Morel spawn produced these succulent fruitbodies.

the inoculation site. A professor from the Portland State University planted her spawn in the early summer. The next spring she found a convoy of Morels fruiting from the site of inoculation extending several hundred feet along a walkway. This also illustrates that, by locating your Morel patch in an area generally conducive to Morel growth, a substantially larger patch than one just 10 ft. × 10 ft. in size can be created. From my experience, the best sites, always shaded, are around freshly laid wood chips of elm, oak, poplar, cottonwood, and/or Douglas fir; in apple orchards; along gravel driveways; in washes from overflowing streams; and of course, in soils where a fire has swept through. The greater the access to these types of favorable habitats that you give the Morel mycelium, the larger your potential Morel patch. The larger the mycelial mat, the more opportunities for widespread, underground sclerotia formation. Once the spawn is in place, you relinquish control to natural forces. In effect, you allow Nature to do what it does best.

I have always envisioned, being the mad scientist at heart, of aerially bombarding prospective habitats with Morel sclerotia. Every time I see a television report of airplanes using a fire retardant to quell a forest fire, I imagine their returning a week or two later and bombing the same sites with a sclerotial slurry of Morel spawn. I happily volunteer to be the spawn maker and the bombardier!

Outdoor Morel beds often support other mushrooms, some of which I view as "indicator" species. Their presence is a sure sign that the habitat is suitable for Morels. The most common and welcome indicator species are the brown cup fungi species belonging to the genus *Rhizinia (R. undulata) Discinia (D. perlata)*, and the genus *Peziza, P. phyllogena (= Peziza badio-confusa)*. Since I find Morels fruiting abundantly in among these cup fungi, I do not view them as true competitors. Furthermore, the False Morel, *Gyromitria esculenta*, and the Early Morel, *Verpa bohemica*, precede Morels by 2 or more weeks. (See page 424.)

In the Pacific Northwest, Morels are found directly at the base of cottonwood trees. A new hybrid strain of cottonwoods, a cross between an eastern and western variety, is being planted *en masse* for pulp production. The mating of these two strains has yielded a "super strain" of cottonwoods, which grow up to an inch per day. These cottonwoods, with their accelerated life cycles, seem like ideal candidates for the companion cultivation of Morels outdoors.

A similar approach might work with apple trees. By locating a Morel bed directly underneath apple trees, the cultivator may create a perennial Morel patch. Orchards, both small and large scale, could provide a bumper crop each spring. Once established, the Black Morel is well known to frequent the same apple orchard for decades. Most other habitats provide only a temporary home for Morels.

Since cottonwoods enjoy especially wet conditions, often unsuitable for pines, their soils are characterized as having naturally higher moisture content. This environment is ideal for the natural cultivation of a number of mushrooms outdoors, including those on logs and on chip/sawdust mounds. Mature cottonwoods can be harvested and inoculated with a wide variety of gourmet and medicinal mushrooms. Logs can be impregnated with Reishi (*Ganoderma lucidum*), Maitake (*Grifola frondosa*), Shiitake (*Lentinula edodes*), or Lion's Mane (*Hericium erinaceus*) mycelium. Stumps, branches or log sections can be used for outdoor cultivation. Once mushrooms have fully produced on these wood substrates, the remaining material can be recycled for species sequencing, as fodder for animal consumption, as a base for mycofiltration, as a supplement for soil enhancement, or even as pulp for paper manufacturing. Mushroom mycelium naturally pulps the wood on which it grows.

I believe the judicious combination of factors—mushroom strain, tree type, and site location—can be juxtaposed to create a synergistic model for myco-forestry. Bear in mind that the ways of Nature are enigmatic. Some of these interactions may be far more complex than science can currently comprehend. I encourage readers of this book to explore these concepts and develop them further. This idea fits neatly into Chapter 5: Permaculture with a Mycological Twist.

Leonard and Volk (1990) reported the co-occurrence of Morels (in the *M. esculenta* complex) fruiting with begonia plants (*Begonia tuberhybrida*). I have seen Morels growing with a variety of potted ornamentals, and in the wild, with leeks. We know of no direct relationship between Morels and these lower plants—their co-occurrence may merely be coincidental. The direct association of Morels and certain trees like cottonwoods, elms, oaks, firs, and apples is consistent and long-term. Dahlstrom et al. (2000) noted that some *Morchella* species appear to form a mycorrhizal-life relationship with some Pinaceae trees (*Pinus contorta, Pinus ponderosa, Pseudotsuga menziesii,* and *Larix occidentalis*).

Recommended Courses for Expansion of Mycelial Mass to Achieve Fruiting: Currently the only successful method for indoor cultivation for any Morels is the one developed by Ower et al. (1986, 1988). However, I know of no skilled cultivators or professional mycologists who have been able to grow Morel mushrooms by precisely following the patented techniques. I am currently developing a method for *Morchella angusticeps* on cased rye grass seed but have only been successful to the stage where white "fuzz balls" emerge from a sea of brown mycelium. From the center of these fuzzy formations, finger-like Morel primordia form but abort due to some unknown environmental or genetic shortcoming.

Morel mycelium can grow across nutrient agar media in a 100 × 15-mm petri dishes in 3 to 4 days, and is clearly the fastest growing of all mushrooms. A conidial, or asexual stage, also develops from the mycelium wherein a spore is generated from the hyphal network and once germinated, produces more mycelia and more sclerotia. Soon after the Morel mycelium colonizes the surface of the nutrified agar media, sclerotia form. The Black Morel produces a unique form of sclerotia, whose sheer numbers run into the thousands per cubic foot. These sclerotia resemble small golden nodules that eventually amass together. When they are broken during their early stage of development, many more sclerotia grow. Micro-sclerotia form abundantly on nutrified, sterilized media, especially MYA,

This mycological experimenter started a patch of Morels by simply broadcasting "M-11" spawn into a rocky debris field resplendent with burnt wood and upturned earth. Inoculated in November, a dozen or more Morels appeared in late March. Square pieces of paper indicate locations of nearly invisible Morel colonies.

OMYA, and the many variations of these formulas. (See page 418.)

When sawdust is inoculated with grain spawn, the sclerotia form most abundantly against the glass or plastic container, or along other interfaces. With most mushroom species producing these unique formations, sclerotia grow better in darkness than in light (Heim et al. 1967; Stamets and Chilton 1983; Volk and Leonard 1990). Although light can totally inhibit the formation of sclerotia in the *Morchella esculenta* group, strains of *Morchella angusticeps* are less affected. Buller (1934) first noted that the spore-generating cells, the asci, orient towards light when spores are ejected.

Clones of *M. angusticeps* have a soon-lost ability to form these golden micro-sclerotia. Black Morel sclerotia amass as hundreds of small, hardened, pumice-like structures that can become golf-ball sized. The sclerotia-forming ability is lost with continuously expanded mycelium. Only by using cultures close to their genetic origins is this sclerotia-forming ability preserved. When the mycelium declines in vigor, sclerotia are not only absent, but the mycelium changes into a mulatto form—cottony white mixed through golden brown, aerial mycelium. When cultures are transferred for more than ten generations of petri dishes (100 × 15-mm dishes filled with MYA), the mycelium ceases to form micro-sclerotia. Clearly, maintaining a stock culture closest to their wild origins is critical for success in sclerotia production, and by inference, mushroom production.

Morel mycelium adapts to the liquid fermentation (Gilbert 1960) and injection techniques described in this book and in *The Mushroom Cultivator* (1983). Sclerotia formation is substantially greater on annual rye grass seed than on coarser grains like rye, wheat, or sorghum. Sclerotia from rye grass seed can be harvested after several weeks of incubation in low light or darkness.

Suggested Agar Culture Media: MYA, MYPA, PDYA, and OMYA. Volk and Leonard (1990) noted that Morels of the *Morchella esculenta*, when grown on standard media preparations, notably Difco's PDA and Difco's Mycological Agar, sclerotia failed to form spontaneously.

1st, 2nd, and 3rd Generation Spawn Media: Rye grain for the first two generations. I prefer using soaked annual rye grass seed for the third or final generation, buffered with 5% calcium sulfate (by dry weight). Mix the rye grass seed and calcium sulfate together in dry form, submerge in water at 2 times its makeup volume, and allow it to sit overnight. Add 5 pounds of moist hardwood sawdust into polypropylene bags. Place 1 cup (.24 liters) of moistened rye grass seed as an even layer to the top surface of each sawdust-filled bag. Fold bags closed and sterilize for 2 to 3 hours at 15 psi. Inoculate the top layer with rye grain or liquid spawn. *Do not through-mix.* Incubate in low light and/or darkness for 2 weeks.

GROWTH PARAMETERS*

Spawn Run:
Incubation Temperature: 70–75°F (21–24°C)
Relative Humidity: 100%
Duration: 10–14 days
CO_2: >5,000 ppm
Fresh Air Exchanges: 0–1 per hour
Light Requirements: n/a

Sclerotia Formation:
Incubation Temperature: 60–70°F (16–21°C)
Relative Humidity: 90–100%
Duration: 20–30 days
CO_2: >5,000 ppm
Fresh Air Exchanges: 0–1 per hour
Light Requirements: darkness
Separate sclerotia from mycelium and plant into moist blend of sand, calcium carbonate, and peat moss blend.

Primordia Formation:
Initiation Temperature: 36–42°F (2–5°C)
Relative Humidity: 85–95%
Duration: 10–12 days
CO_2: < 5,000 ppm
Fresh Air Exchanges: 2–4 per hour
Light Requirements: 200–800 lux

Fruitbody Development:
Temperature: 40–50°F (4.4–10°C)
Relative Humidity: 85–95%
Duration: 10–20 days
CO_2: <5,000 ppm
Fresh Air Exchanges: 2–4 per hour
Light Requirements: 200–800 lux

Cropping Cycle:
One crop indoors-spawn can be placed outdoors for creating natural patches

---

* This is a proposed initiation strategy based on my observations of natural fruitings of the Morel and laboratory research with *Morchella angusticeps*, the Black Morel. This strategy will be amended and improved over time.

Substrates for Fruiting: When fully colonized and resplendent with sclerotia, take the above and invert the bags so that the seed layer is on the bottom. Place 4–6 inches of a moist layer of peat moss that has been buffered with 10% calcium sulfate (by volume) over the morel seed/sawdust spawn. This concoction can be placed outside to benefit from natural spring initiation strategies or used in the attempt to grow Morels indoors. Note that separation of the seed from the substrate is not critical. The 10- to 12-inch depth from the bottom seed layer to the top surface of the buffered peat moss is sufficient to encourage Black Morel fruitbody formation. A fine layer (1-inch) of moistened vermiculite aids aeration. Since the physical separation of the nutritional seed layer from the nutritionally poor peat moss is not necessary, this technique is uniquely different from those which have been patented by Ower et al. The pH optimum for fruiting falls between 6.5 and 8.0.

Recommended Containers for Fruiting: Tray of sufficient depth (10–12 inches) to accommodate the above with holes for drainage.

Yield Potentials: Estimating from the photographs of the indoor method used by Morel Mountain, yields appear to be in the 1-pound-per-square-foot range. Since Morels are hollow, the number of mushrooms per square foot weighs less than the same number of similarly sized solid mushrooms, for instance, Shiitake.

Harvest Hints: Black Morels are best picked before the ridge edges become darkened and spores are released. When the mushrooms cease developing, the margins of the folds darken, wilt, and dry.

Form of Product Sold to Market: Fresh, dried, and powdered. Morel Mountain has fresh Morels year-round. Fungi Perfecti sells Black Morel spawn for outdoor inoculation. Their addresses are listed in the Resource Directory section in the Appendix.

Nutritional Content: 20% protein (N × 4.38), 4.8% fat, 8.7% fiber, 64.4% carbohydrates.*

Medicinal Properties: Not known to this author.

Flavor, Preparation, and Cooking: A superb edible, this mushroom should be well cooked as many individuals are sensitive to it in the raw state. Morels work well in stir-fries and in a wide range of preparations. Please refer to the recipes in Chapter 24.

Comments: Morels cannot be easily co-cultivated with other gourmet and medicinal mushrooms in the same growing room. Once a cultivator succeeds in getting to the stage of "white tuft" formation, humidity and temperature are critical for fruitbody development. At this stage, a different set of environmental stimuli is introduced. I am proposing such a strategy in the above Growth Parameter section. Humidity levels should be lowered below the range specified for most mushrooms. One nice overview of cultivation techniques can be found at *www.gorsky.com/~pdilley*.

For more information, consult Ower (1982), Ower et al. (1986, 1988), Leonard and Volk (1992), Volk and Leonard (1989), Volk (1990), and Sanderson (1969).

Morels are omni-directional when they form, for reasons unknown.

---

* This analysis based on *Morchella esculenta.* Taken from *The Biology and Cultivation of Edible Mushrooms*, ed. by Chang and Hayes, 1978.

CHAPTER 22

Maximizing the Substrate's Potential through Species Sequencing

At our farm, I have found that the spent substrate generated in the course of Shiitake cultivation is in itself a valuable by-product. More mushrooms can be grown upon it! The mushroom cultivator can implement a circuit of recycling by sequencing species on the same substrate, resulting in the maximum yield of mushrooms imaginable. Each decomposer produces its own unique set of enzymes that can only partially break down a wood-based substrate. Once the life cycle of one mushroom has been completed, the life cycle of another species can begin on the same substrate utilizing its own unique set of enzymes, taking advantage of the remaining undecomposed wood fiber and the dead mycelium of the predecessor mushroom. After this second decomposer exploits the remaining lignin-cellulose to its fullest ability, a third species can be introduced. And so on.... I have been able to grow four species in sequence with this method. After several generations of mushroom species, the mass of final substrate is a mere fraction of the original formula. The end material is reduced to a soft loam and is best used for compost or soil enhancement.

After running several species through the same substrate, Chang and Miles (1989, p. 332) found that the net available nitrogen in the waste substrate actually increased, proportionately. Using cotton waste, the total nitrogen of the fresh compost waste was 0.63%. After the Paddy Straw mushroom produced on it, the residual nitrogen became 1.54%. After taking this same waste substrate and inoculating it with Oyster mycelium (*P. ostreatus* var. *florida*), the nitrogen increased to 1.99%. (The effect that spawn had on the substrate was not described. A 10% spawning rate with rye could substantially affect these figures. If "substrate spawn" were used, the net effect would be much less.) The end result of species sequencing is the production of rich humus, ideal for gardening. This concept is further incorporated into a permaculture model described in Chapter 5.

The sequence of species introduction, however, is most important. The Shiitake model is the easiest to understand. After Shiitake mushrooms stop producing on supplemented sawdust/chips, the now-blackened blocks are broken apart until they resemble sawdust in texture. Calcium sulfate and/or carbonate enhance particle separation, drainage, and adjust the pH to the 6.5 to 7.5 range. (Try 1 cup of gypsum/chalk for every twenty blocks and adjust accordingly.) The type of wood initially used becomes the overriding factor affecting proper formulation. Water is slowly added until good moisture is achieved. (I prefer moisture content of 60–65%, less than the ideal 75%. Higher moisture contents often result in a higher percentage of bags spoiling due to fermentation.) The moistened sawdust mixture is filled into polypropylene bags or other suitable containers, and sterilized. If water collects at the bottoms of the bags, then the substrate has been over-moistened.

After sterilizing, the bags are inoculated according to the procedures in this book. I have found that Oyster mushrooms grow profusely on the waste Shiitake substrate and needs little or no amendment. King Oyster and Maitake also fruit, although 10% supplementation with rice bran or corn substantially improves yields. After the second species in sequence has run its course, the waste substrate is collected, remixed, sterilized, and finally inoculated with King Stropharia (*Stropharia rugosoannulata*) or Shaggy Mane (*Coprinus comatus*). However, if the spent substrate is understerilized and/or too much water is added at makeup, contamination during incubation is likely. Keep in mind that waste substrates host far more microorganisms than fresh sawdust. Hence, sterilization may have to be prolonged to insure killing all the resident contaminants.

Each time one of the above species (except *Stropharia rugosoannulata*) is grown through the sterilized sawdust-based substrate, approximately 10% of the dry mass (= 25% wet weight) of the substrate yields fresh mushrooms. Depending upon the species and many other variables, between 20 and 40% of the dry mass evolves into gases, mostly carbon dioxide, nitrogen, and ethylene. The first species, in this case Shiitake, easily produces 1.5 pounds of mushrooms from the original 6-pound

substrate (75% moisture). At least 1.5 pounds is lost through carbon dioxide evolution and evaporation. At the end of the Shiitake fruiting cycles, a 3-pound waste substrate remains with moisture content approaching 50%. After Oyster mushroom mycelium has taken its turn, the reduction in mass undergoes another 50% reduction in mass. Now, our sample has been reduced from an original 6 pounds to 1.5 pounds. At this stage, the remaining material, without supplementation, supports vigorous growth of the King Stropharia (*Stropharia rugosoannulata*) or the Shaggy Mane (*Coprinus comatus*). Once colonized, the mycelia of these species are best used as spawn to inoculate outdoor substrates. At this final stage, the nutritional base of the substrate is largely expired, and the fruitings are anemic.

In all, more than 20% of the substrate (dry weight to dry weight) is converted into edible mushrooms. At least that amount is liberated as gases. The remaining material can be added to garden composts as a supplement. The process of reduction/conversion is substantially prolonged if the cultivator utilizes large-particle wood chips in the original substrate formulas. When denser hardwoods are used, the cycle is prolonged. If the waste wood substrate is further supplemented, the cycle can be extended.

This is but one path of species sequencing. Many others naturally come to mind. For instance, when production blocks of recycled Oyster, Maitake, or Reishi (or others) have stopped producing indoors, they can be implanted outdoors into beds of sawdust. Additional fruitings arise from the buried blocks in 3 to 6 months, depending, of course, upon the weather. I am always fascinated by the fact that these outdoor fruitings are often better than those indoors. Mushrooms seem to always benefit when nature is used as an ally. The implanted blocks of mycelium have the ability to draw additional nutrients from the surrounding habitat. By launching the expired blocks from the growing rooms into supportive outdoor habitats, the cultivator maximizes the potential of the mycelial mass. One of my natural culture beds supports a succession of three species—first Morels in the spring, then King Stropharia in the summer, and an assortment of *Hypholoma* and allied species in the fall. I see this approach as the "Zen" of mushroom growing.

Whatever path is chosen, the implications are profound. These courses of decomposition are occurring daily in our forests' ecosystems. Ecologists should also find this model especially fascinating in understanding the concurrence of many species living in the same habitat. This model may also be useful for those living in desert, island, or other environments where substrate materials for wood decomposers are costly and hard to acquire. I encourage all readers of this book to push these concepts forward with new innovations and applications, incorporating more sets of organisms. By understanding the nuances within the mycosphere, I envision the creation of complex biospheres wherein fungi play determinant roles in supporting other life cycles. And I am not alone in believing that mushrooms can be instrumental in generating food for humans in the exploration of space.

CHAPTER 23

Harvesting, Storing, and Packaging the Crop for Market

Mushrooms can be compared to fish in their perishability. Once harvested, they are quick to spoil unless properly cared for. One advantage of growing gourmet and medicinal mushrooms is that, historically, they have been used in dried form for centuries. In Asia, more Shiitake is sold dried than fresh. Asians long ago discovered that the flavor of Shiitake is actually enhanced in drying. Further, having a readily available supply of dried mushrooms, which can be stored for months at room temperature in airtight containers with no special care, is very convenient for consumers. Because the storage problem is compounded by the lack of refrigeration in many developing countries, dried mushrooms make good sense for both producers and consumers. In the United States, Canada, and Europe, more mushrooms are sold fresh than dried. In these markets, cultivators first supply the needs of the fresh market and then dry the surplus. Dried mushrooms can be sold as is or powdered for soup mixes, spices, or teas.

Harvesting the Crop

Simple guidelines prevail in the proper harvesting and storing of mushrooms. First, young mushrooms last much longer after harvest than aged mushrooms. Once spores have developed on the face of the gills, perishability is accelerated. If mushrooms have partial veils, like the Button mushroom (*Agaricus brunnescens*) or the Black Poplar mushroom (*Agrocybe aegerita*), they are best picked while the partial veils are intact, in other words when the mushrooms are still young. Partial veils protect the gills, limiting moisture loss, preventing spore release, and rupture only as the caps expand.

The cultivator must constantly counterbalance maximum yield with marketability. The comment I most often hear, after presenting my Oyster mushrooms to a distributor who has been purchasing them from afar, is "I didn't know Oyster mushrooms could look like this!" Because

Oyster mushrooms readily suffer from shipping and handling, local producers can easily usurp the markets of distant growers. Oyster mushrooms have a functional life span of only 5 days, after which marketability drastically declines.

Another rule is that clusters yield, pound for pound, higher-quality mushrooms than mushrooms grown individually. Bouquets of mushrooms have obvious advantages, both from the point of view of harvesting as well as marketing. They can be picked with ease, needing minimum handling and trimming. Once harvested, the mushrooms protect one another by being bunched together, and as with many Oyster species, frequently are joined together from a common stem base. Harvesting mushrooms in clusters limits the damage caused by individual loose mushrooms jostling against one another. Most importantly, bouquets, at the ideal stage for harvest, are composed of younger mushrooms. Mushroom bouquets can then be sold much like broccoli. All these features combined extend shelf life far beyond that of individual mushrooms and make clusters highly desirable for most cultivators. All the gilled mushrooms described in this book can be harvested as bouquets.

Each species passes through an ideal stage as the mushrooms mature. For most species, the ideal stage is when the caps are still convex, and before flattening out. After adolescence, the mushroom changes in its form while not appreciably increasing in its mass. Cultivators have long noted that the flesh of

Harvesting Paddy Straw mushrooms in Thailand.

a mushroom at the "drumstick" stage is much thicker than when the mushroom is fully mature. The loss of flesh, directly above the gills to the top of the cap, appears to be reproportioned to the leading edge of the expanding cap, and perhaps more significantly, to the extension of the spore-producing gills. No real advantage, in terms of weight, is

Before and after cleaning the stem butts of Himematsutake, the Almond Portobello, *Agaricus blazei.*

Bouquets of *Pleurotus ostreatus* facilitate harvesting, extend shelf life, and increase marketability. The bouquets usually snap off, originating from a small, localized point of formation.

realized in picking a fully mature mushroom versus one that is a mature adolescent. In fact, mature adolescents store longer and taste better.

The ideal stage for harvest of each species is described in Chapter 21 under each species' growth parameters. Please refer to that chapter for helpful hints in harvesting. Since cropping is labor-intensive, more efficient harvesting methods are always being explored. Cropping mushrooms in the most cost-effective manner largely depends on the structure of the fruiting frames. In Holland, mechanical harvesters have been devised to harvest mushrooms from horizontal beds, eliminating the largest labor contingent in a mushroom farm—the pickers.

Packaging and Storing the Crop for Market

Once mushrooms have been harvested, they must be quickly chilled. Most pickers at mushroom farms place mushrooms directly into open-grate plastic baskets that are frequently ferried to the cold room. The larger farms utilize blast chillers, which precipitously drop the temperature of the mushroom from room temperature to near freezing. The common mistake many growers make is to place their fresh mushrooms directly into cardboard boxes after picking. Cardboard boxes insulate the mushrooms after harvest; essentially preventing them from being rapidly cooled. During or after cooling, mushrooms are sorted and packaged. Once cooled, the mushrooms must not be rewarmed until delivery. The ideal temperature for storage is 34°F (1–2°C). (See Lomax, 1990 and Hardenburg, 1986.)

Mushrooms are sorted according to markets to which they are destined. The Japanese are by far the

Four strains of *Pleurotus ostreatus*, in 5-pound boxes, ready for delivery to restaurants.

connoisseurs of the world in terms of quality standards for marketing. So strict are their standards that many North American growers have been unable to penetrate into the Japanese market. The Japanese also have the advantage of having a large pool of specialty growers who can coordinate their product lines to best fill their complex market requirements. Mushrooms are carefully graded according to type, size, and form. In North America, the markets are relatively unsophisticated and the primary concern is for freshness. Currently, in Seattle, Shiitake mushrooms are selling for $11.99 per pound at the local grocery store. With stores making a 40% markup, the grower would then be paid $7.20 per pound. In the United States, some growers, buyers, and sellers follow a loosely adhered-to grading system. Number 1 Shiitakes are usually 3–5 inches across, dark brown in color, with incurved margins, usually adorned with veil remnants. Number 2 are basically number ones that have more or less fully expanded. Number 2s are often lighter in color and exceed 4–5 inches in diameter. Number 3s show some damage, either to the gills or cap margin, and are often deformed. Number 3s vary in size from tiny to excessive large mushrooms. I find it interesting that Americans, as a culture, have historically favored large mushrooms. Currently, in markets in San Francisco, large Shiitake are selling for several dollars per pound more than small ones.

Once mushrooms are sorted to grade, they are packaged for market. Restaurants generally prefer 5- to 7-pound boxes. Packages for consumers typically weigh 3, 5, or 7 ounces, a trick employed by many marketers to disguise the actual price per pound. (It's not easy for the consumer to divide 16 ounces (1 pound) by 3, 5, or 7 to determine the actual price per pound.) In the United States, packages of fresh mushrooms should be small enough so that they can be grasped by one hand, and ideally retail at or below $2.00. Once the sale price to the consumer exceeds the $2.00 threshold, a precipitous decline in sales is seen. If every 3-ounce package sold for $2.00, the retail price is $10.66 per pound. Most retailers consider a 40% markup fair. This gives the growers $6.40 per pound at the wholesale level.

Another tactic commonly used with Button mushrooms is to sell the mushrooms loose in a tray, and have the consumers fill small paper bags imprinted with information on handling, cooking, etc. The consumer can be more selective in picking the mushrooms most desirable. However, every time the mushrooms are rummaged through, they suffer in quality. Although Button mushrooms are often sold loose, the gourmet mushrooms, being more fragile, are best sold packaged.

Covered with clear, anticondensate, breathable plastic, mushrooms can be preserved for extended periods of time. A patent was awarded to Asahi-Dow Ltd. for a vapor-permeable film specifically designed for extending the shelf of Shiitake. [See Japanese Patent #57,163,414 (82,163,414).] The rate of diffusion of carbon dioxide giving the best results was within 5,000 and 40,000 milliliters per square meter at atmospheric pressure over 24 hours. The optimal range of oxygen diffusion was 2,000–20,000 milliliters per square meter at atmospheric pressure in 24 hours. This new generation of anticondensate, gas-permeable films must be carefully matched with a cardboard base or strawberry-like basket. Even with shelf life being extended, mushrooms should be rotated through stores at least twice weekly to ensure the highest-quality product. Oyster mushrooms in particular are quick to spoil.

The greatest insult to marketing gourmet mushrooms can be seen by vendors who buy large quantities of Oyster mushrooms from production factories whose main concern is yield, not quality. Oyster, Enoki, and other mushrooms, when they spoil, cause severe abdominal cramping, nausea, and gastrointestinal upset. (See page 439.) Once customers have experienced these "gourmet" mushrooms, they are unlikely to ever buy them again. Remember, mushrooms are the first suspected and first blamed for any type of food poisoning, whether they are at fault or not. (For more information on the proper handling of mushrooms after harvesting, please consult Murr and Morris, 1975.)

An example of poor packaging. Note mushrooms lie on Styrofoam base. They were covered with plastic. This package was photographed directly after purchase. This is the "sajor-caju" strain of *Pleurotus pulmonarius*. Thousands of primordia are forming on the adult mushrooms as they rot. Mushrooms in this condition, if eaten, cause severe cramping, diarrhea, and painful gastrointestinal discord.

Drying Mushrooms

By drying mushrooms, cultivators recapture much of the revenue that would otherwise be lost due to overproduction. Most mushrooms are approximately 90% water. Reishi mushrooms, being woody in texture, are usually between 70 and 80% water. When Shiitake are grown outside, especially in the Donko (cracked cap) form, moisture content is often only 80%. When mushrooms are young, moisture contents are usually higher than when they are mature. Mature mushrooms, with their gills exposed, dry faster than young, closed mushrooms.

Shiitake, Oyster, Morel, Reishi, and many other mushrooms dry readily and can be stored for many months. Mushrooms can be sold in their natural form or powdered for soups, spice mixtures, teas, etc. Some cultivators actually sterilize their dried mushrooms, without harm, to prolong storage. Sterilization assures no bacteria, insect eggs, or other microorganisms consume the crop when stored for prolonged periods of time. Once dried, the mushrooms should be hermetically sealed, and ideally *frozen* until needed.

Many types of dehydrators can be used for drying mushrooms. The smallest are those also marketed for home use in the drying of fruits, meat, and fish. For most growers, home dehydrators have insufficient capacity, so many fabricate their own dehydrators. Window screens can be stacked within a vertical framework, 3–4 inches apart. At the bottom a heat source, often heat lamps or an electric coil, is positioned. Ample air inlets are located near ground level. The vertical framework is solid save for a hinged door on one face that allows easy insertion and retrieval of trays. A fan is located at the top, drawing air out of the dehydrator. This arrangement ensures a chimney effect whereby heated air is drawn through the bottom and exhausted out the top. The humidity of the incoming air greatly affects the efficiency of this type of dryer. Some growers locate their dryers in hot rooms, typically low-humidity greenhouse-like environments, which helps the drying process.

The best commercial dryer I have seen is also the simplest. Mushrooms are placed onto trays and stacked into vertical racks equipped with wheels. The wheeled racks are inserted into a large plastic wind tunnel. (See page 440.) The plastic wind tunnel can be kept inflated by hoops of plastic pipe and through the force of a large blower located at one end. Trays with fresh mushrooms are moved into the wind tunnel furthest downstream. The fully dried mushrooms are retrieved through an overlapping "flap-door" nearest the fan. For most cultivators, this type of commercial dehydrator does not require a heat source. The huge volume of air removes the moisture through evaporation.

Depending upon the species and the final product desired, mushrooms can be placed gills down or gills up. If you place Shiitake mushrooms with

A commercial dehydrator utilizing a large volume of air to remove moisture from mushrooms. Fresh mushrooms are placed onto screened shelves on wheeled racks, entering the drier downstream from the drying mushrooms. Hundreds of pounds of mushrooms can be dried at one time, inexpensively.

their gills down, the mushrooms remain flatter in drying and take on a more brittle texture. Most experienced Shiitake growers find that by drying mushrooms, gills facing up, that the cap curls inwards, giving the mushroom an overall tighter and more resilient texture. This form is the one most recognized by Asians.

Dried mushrooms are then packaged, sometimes shrink-wrapped into plastic bags, and usually sold in 3- to 5-ounce packages. In most cases, the shelf life of dried mushrooms is about a year. If there is any danger of fly larvae or insect infestation, low-pressure steam sterilization is recommended.

Marketing the Product

In the United States, markets for fresh mushrooms have surged over the past 30 years, from a total market value of $68 million in 1969 to $665 million in 1992. Fresh gourmet mushrooms were virtually unavailable in 1980. In 1992, gourmet mushrooms represented $17 million of total fresh mushroom sales, a 22% increase over the same period from the previous year. The per capita consumption of mushrooms continues to rise in the United States, going from 3.70 pounds per capita in 1990 to 4.10 pounds in 1999. Although Canada consumes 50% more mushrooms per capita than in the United States, the rate of increase in consumption in the U.S. is more than double Canada's. In 1999 specialty mushroom consumption in the United States (excluding Portobellos) increased 35% from the previous year, with Shiitake representing 66% of that market. For updated information please consult United States Dept. of Agriculture Market Reports at *usda.mannlib.cornell.edu/reports/nassr/other/zmu-bb/* and the United Nations Food and Agricultural Organization at *usda.mannlib.cornell.edu/data-sets/specialty/94003*. (Approximately four times more Shiitake are sold in this country than Oyster mushrooms.) The upward trend in terms of price, production, diversity, and markets is expected well into the future.*

Before producing mushrooms on a commercial level, the cultivator is advised to conduct mini-trials. With a little experimentation, the cultivator can refine their techniques. Each failure and success is useful in determining the proper mushroom strain, substrate formula, temperature tolerance, lighting level, harvesting methods, and marketing strategies. Note that yields from mini-culture experiments often exceed average values from commercial scale operations. I strongly encourage cultivators to increase their production levels slowly, and according to their skills in both mushroom technology and business management.

A number of organizations help growers find markets for their mushrooms. Some co-operative marketing organizations coordinate production and sales. Co-op marketing becomes a necessity when multiple growers overwhelm local markets. (Refer to the Resource Directory, Appendix 4 for a list of some marketing organizations.) The United States Department of Agriculture can sometimes assist growers in contacting marketing outlets.

I am a strong believer in growing mushrooms organically. Once certified organic, local producers can sell mushrooms to natural food co-ops and some upscale grocery chains for a premium. The restaurant trade, from my experience, seems little impressed whether or not the mushrooms are organically grown. In either case, the key to the financial success of a mushroom farm centers on its ability to market mushrooms successfully. The person in charge of marketing must foster a close, professional, and personal relationship with the buyers.

In Asia, marketing gourmet mushrooms has benefited from a long tradition while in North America, gourmet mushrooms are a relatively new phenomena, having been available for less than

10-kilogram bags of dried Shiitake being sold in a market in China.

thirty years. (See Farr, 1983.) With more growers coming into production, the markets are likely to fluctuate in response to the sudden increase in the availability of mushrooms. Cycles of over- and underproduction are typical in any new, expanding marketplace and should be expected. Growers must adapt their production schedules and product lines so they do not become overextended. As the public becomes increasingly aware of the health stimulating properties of mushrooms, markets should expand enormously. Those growers who are able to survive this early period of market development will become key players in an industry that is destined to become a centerpiece of the new environmental economy.

---

* Data derived from *Mushrooms*, Agricultural Statistics Board, National Agricultural Statistics Service, United States Department of Agriculture, Washington, D.C. See Resource Directory section in Appendix 4.

Mushroom Recipes: Enjoying the Fruits of Your Labors

Versatile, tasteful, and nutritious, gourmet mushrooms enhance any meal. Multidimensional in their flavor qualities, mushrooms appeal to both vegetarians and meat-eaters. The following recipes can be used with almost any of the mushrooms described in this book. The simplest way to prepare gourmet mushrooms is in a stir-fry (*à la wok*) at medium-high heat with light oil and frequent stirring. Other condiments (onions, garlic, tofu, nuts, etc.) can be added after the mushrooms have been well cooked.

Here are some of the favorite recipes from notable mushroom dignitaries. I hope you enjoy these recipes as much as I have. For more recipes, please consult the references listed at the end of this chapter. Bon appetit!!!

Our family prefers to tear Shiitake rather than cutting them. This preserves flavor and minimizes damage to the cells.

David Arora's Mediterranean Mushroom Recipe

"Nothing could be simpler or more delicious," according to David.

King Stropharia, Oyster,
Hon-Shimeji, or Shiitake
mushrooms

Salt

Olive oil

Shallots or garlic

Choose the meatiest caps of King Stropharia (young), Oyster, Hon-Shimeji, or Shiitake mushrooms, lightly salt the gills, and dab them with olive oil. Stuff slivers of shallots or garlic between a few of the gills. Broil or grill the mushrooms over hot coals for a few minutes on each side, until tender.

Arleen and Alan Bessette's Dragon's Mist Soup

1 cup thinly sliced fresh Shiitake mushrooms

1 tablespoon vegetable oil

1 (14 ½-ounce) can chicken broth

1½ cups water

4 cloves garlic, minced

2 scallions with tops, minced

½ cup drained, rinsed, 2- to 4-inch long bamboo shoots

2 tablespoons soy sauce

¼ teaspoon white pepper

5 ounces tofu (cut into ½-1 inch cubes)

1 teaspoon salt

1 teaspoon sesame oil

In a saucepan, sauté mushrooms in vegetable oil over medium-low heat for approximately 5 minutes. Add broth, water, and garlic; bring to a boil, reduce heat and simmer 10 minutes. Add all remaining ingredients except sesame oil. Return to boil, reduce heat and simmer 5 minutes. Just before serving, stir in sesame oil. Serves 4.

Jack Czarnecki's Shiitake in Burgundy Butter Sauce*

½ cup chopped onions

3 tablespoons melted butter

1 cup water

½ teaspoon ground chili powder

1 teaspoon lemon juice

¼ cup red wine

1 teaspoon sugar

½ teaspoon ground coriander

1 teaspoon crushed fresh garlic

¼ teaspoon ground black pepper

1 teaspoon salt

1 tablespoon soy sauce

1 pound fresh Shiitake caps

1½ tablespoons cornstarch mixed
with ⅓ cup water

In a skillet, sauté the onions in the butter until transparent, then add water. Add the other ingredients, except the mushrooms and cornstarch, and stir for 1 minute. Add the mushrooms, and turn the heat to low. Cover the skillet with a tight-fitting lid, and let simmer for 30 minutes. Thicken the mixture with the cornstarch-and-water mixture, and serve alone or over rice. According to Jack, the sauce actually enhances the flavor of the Shiitake. Serves 4.

Jack Czarnecki's Chicken with Oyster Mushrooms*

1½ cups heavy cream

1 teaspoon crushed fresh garlic

1 tablespoon finely chopped onion

½ pound Oyster mushrooms,
cut into 2-inch strips

Salt and pepper to taste

1 tablespoon cream sherry

1 tablespoon prosciutto

12 ounces cooked chicken or turkey meat
from breast, sliced into 2-inch strips

½ cup chicken stock

2 tablespoons cornstarch mixed with
⅓ cup water

Combine all the ingredients except the cornstarch mixture in a heavy skillet. You may also want to save the salting until the dish is slightly heated. Heat until simmering over a medium flame, then continue simmering over a low flame for 5 minutes. Thicken with the cornstarch-and-water mixture, and adjust for salt as necessary. Serves 2.

---

* Gratefully reprinted, with permission of the author, from *Joe's Book of Mushroom Cookery*, 1988, Atheneum/Macmillan Publishing Co., New York, NY.

Larry Lonik's Morel Quiche*

¼ cup bacon (or "bacon bits")	1½ cups milk
1 pound Morels	¾ cup Bisquick
½ cup chopped onion	3 eggs
½ cup chopped green pepper	1 teaspoon salt
1½ cups shredded baby Swiss cheese	¼ teaspoon pepper

Preheat oven to 400°F (200°C). In a 10-inch, lightly greased pie pan, mix bacon, mushrooms, onion, green pepper, and cheese. In medium-sized bowl, add milk, Bisquick, eggs, salt, and pepper. Beat until smooth. Pour into pie pan. Bake 35–40 minutes or until inserted toothpick comes out clean. Serves 2 to 4.

---

* Gratefully reprinted, with permission of the author, from *The Curious Morel: Mushroom Hunter's Recipes, Lore and Advice.*

Irene and Gary Lincoff's Pickled Maitake (Hen-of-the-Woods)

1 pound Maitake mushrooms

Marinade:

1 cup white wine (dry)	A few slices of onion
⅓ cup olive oil	A few sprigs of parsley
Juice from 1 lemon	A few cloves of garlic
	Salt, peppercorns, or other spices

Clean 1 pound of mushrooms and separate the "leaves" of Maitake into bite-sized pieces. Throw them in boiling water for 2 minutes, drain, and let cool on paper towels.

Combine marinade ingredients and season with a sprinkle of salt, peppercorns, bouquet garni, or other preferred spices. Bring marinade to a boil and cook the mushrooms in the marinade until done to taste (crunchy to soft).

Remove mushrooms. Discard solids from marinade. Store mushrooms covered with marinade in glass jars, with a thin layer of olive oil on top to help preserve. Store refrigerated. 4 to 8 servings.

(The Lincoffs wish to thank Jean-Paul and Jacqueline Latil for this recipe.)

Hope and Orson Miller's Hot Mushroom Dip Especial

1 pound fresh mushrooms*

6 tablespoons butter

1 tablespoon lemon juice

2 tablespoons minced onion

1 pint carton low-fat sour cream

2 vegetable or chicken bouillon cubes (or 2 teaspoons granules)

Salt and pepper to taste

2 tablespoons soft butter or margarine

2 tablespoons flour

Chop mushrooms quite fine and sauté in pan with butter and lemon juice. Let simmer 5–10 minutes. Add onions, sour cream, bouillon, salt, and pepper. Simmer 5–10 minutes more. Make a paste of the remaining butter and flour. Add to hot mixture and stir until thickened. Serve hot, in fondue pot or chafing dish, with chips, crackers, or fresh vegetables. (Note: If thickened with seasoned breadcrumbs, fresh dill may be added as filling for Mushroom Squares. Use crescent roll dough. Pat the dough into a small 9-inch-square baking pan, spread filling and cover with more dough. Bake 20–30 minutes at 375°F. Cut into squares and serve hot.) Hope has prepared this dish to the kudos of mycologists throughout the world. Serves 2 to 4.

---

*Morels, Lion's Mane, Shiitake, or Shimeji. Gleaned, with permission of the author, from *Hope's Mushroom Cookbook,* by Hope Miller (Mad River Press, 1993).

Scott and Alinde Moore's Cheese-Mushroom Quiche

1 ready-made pie crust

1½ cups grated Swiss cheese

1 tablespoon butter

1 medium onion, chopped

¼ pound mushrooms, chopped

Dash of salt, pepper, and thyme

¼ teaspoon salt

1 ½ cups milk

¼ teaspoon dry mustard

4 eggs

3 tablespoons flour

Paprika

Cover bottom of pie crust with grated Swiss cheese. Meanwhile, sauté butter, onions, mushrooms, salt, pepper, and thyme. Add the sautéed ingredients to the pie crust. Beat together the remaining five ingredients and pour over mushroom layer in piecrust. Sprinkle with paprika. Bake at 375°F (190°C) for 40–45 minutes, or until center is firm. Serves 2 to 4.

Maggie Roger's Oyster Mushrooms
with Basmati Rice and Wild Nettles

Serve over basmati rice, fluffy white rice, or lightly toasted sourdough bread cubes. Serve with steamed wild nettles or freshly steamed asparagus on the side and a clear light wine. "Friends will chirp with satisfaction once they get over the wildness of it all." Serves 2 to 3.

6 dozen nettles or 1 loose, full grocery bag

Salt and pepper to taste

1½ cups chopped Oyster (or Shimeji) mushrooms

1 tablespoon butter or olive oil

1 thinly sliced scallion

¼ cup thinly sliced celery

¼ cup chopped carrot

¼ teaspoon basil

¼ teaspoon thyme

¼ cup very dry sherry

1 cup chopped cooked turkey breast

Bouillon granules

½ cup cold water

2 tablespoon flour, or 1 tablespoon cornstarch

1 cup half-and-half cream

From Maggie's notes: At the time you're finding spring Oyster mushrooms, the wild nettles are about to flower and ready for picking. Use leather gloves or native wisdom to keep from getting stung. Break off just the top three leaf levels of each stalk. Prepare them by rinsing in cold water, steaming over boiling water for 7 minutes and lightly salting and peppering them. (You can pick nettles in the spring, blanch for 6 minutes, and freeze several packages for future use.) If served as below, no melted butter is needed; they have their own flavor.

Sauté mushrooms in butter or olive oil with scallion, celery, and carrots. Let simmer in its juices for 10 minutes or until slightly reduced. Add basil, thyme, sherry, turkey bits, bouillon granules, and simmer for another 10 minutes. Taste, add salt and freshly ground pepper as needed.

Make a thickening of the water and flour (or cornstarch). Stir into mushroom mixture slowly and let simmer for another 5 minutes or more, or until thickened, stirring occasionally.

Just before serving, add the half-and-half, simmer for 5 minutes or so, then pour into a warmed bowl.

Robert Rosellini's Broiled Rockfish in an Oyster or Shimeji Mushroom and Ginger Sauce

The following recipe uses a yellow-eye rockfish from the Pacific Northwest, which is firm and delicately textured with a low fat content. The absence of fat in this particular fish provides its "clean" and pure quality. The following recipe is quick, simple, and easy for home preparation.

6 ounces fillet of rockfish

Olive oil

1 ounce white wine

¼ pound fresh, thinly sliced Oyster or Shimeji mushrooms

½ ounce preserved ginger

1 ounce sweet butter (unsalted)

Brush the fillet of fish with oil and broil (or bake at 450°F) until the translucent flesh turns opaque, 7–10 minutes. Remove fish from broiling pan, place the same pan on oven burner at high heat, and deglaze pan with white wine. When wine has been reduced by two-thirds, add sliced mushrooms and preserved ginger. Sauté for about 1 minute, add butter, and swirl ingredients together until butter forms a smooth texture, about 16–20 seconds. Remove immediately and pour over fish fillet.

Note: Achieving the smooth texture with butter requires practice to avoid breaking the butter. Once this simple technique is accomplished, there are many variations of this recipe procedure. Butter contains half the fat content of most oils, and thus these recipes should be consistent with low-fat dining strategies. Serves 2.

Paul and Dusty's Killer Shiitake Recipe

Once in a while, you come across a simple recipe that elicits enthusiastic exclamations of joy. Recently we came up with one. We get inundated with requests for it. So here it is.

1 tablespoon olive oil

2 tablespoons sesame oil

1 tablespoon tamari or soy sauce

2–3 tablespoons white wine

Pinch of black pepper

1–2 cloves of crushed garlic

1 pound whole, fresh Shiitake mushrooms

Mix the oils, tamari, wine, and spices in a small bowl. Stir vigorously as the ingredients tend to separate. Set aside.

Cut the mushroom stems from the caps. Place gills facing up. Do not slice mushrooms! (The stems can be dried and used for a soup base or discarded.) Baste the sauce onto the gills of the mushrooms, making sure the gills become saturated with the sauce.

In a 350°F oven, bake mushroooms uncovered for 30–40 minutes. Or you can barbecue on an open grill. The smoky flavor makes it even better. Serve hot with seafood, rice, pasta, or whatever. Unbelievably good. Yum! Serves 2–4.

Fresh Shiitake Omelet

8 eggs	½ small onion, chopped
¼ cup water	2 cloves garlic, chopped
2 tablespoons tamari	½ cup cashews
½ pound fresh sliced Shiitake	1 cup grated cheddar cheese
2 tablespoons canola oil	Salt and pepper to taste

Mix the eggs and water in a large bowl and beat thoroughly. Add tamari to mushrooms and sauté in a frying pan with oil until water is cooked out. Add onions and garlic, cook for 1 minute, then add cashews. In a medium-sized skillet, spray a thin coat of oil or butter, add egg mixture and cook for 1 minute. Add layer of cheese. Pour mushroom mixture over the cheese. Fold over and cook for 2–3 minutes. Add salt and pepper to taste. Serves 4.

Larry Stickney's Morel Crème Superieur

Larry's recipe is quick and easy, ideal for tasting Morels at primitive campsites during remote Morel forays in the mountains. "Only the medically forbidden should entertain thoughts of dietary restraint during the brief Morel season when long, hard hours in the forest set up a healthy appetite in the wake of massive calorie burning and likely liquid deprivation.... Delectation is soon at hand." He makes no note for those athletically inclined Morel cultivators who must walk from their house to their backyard to harvest Morels. Presumably, the same advice holds.

1–2 pints heavy cream	½ pound (or more) fresh Morels
2–3 tablespoons butter	(cut into ¼-1 inch cartwheel sections)

Warm cream and butter over a camp stove or wood fire in a small pot. Do not let cream burn onto the pan's bottom. When the cream simmers, place the sliced Morels into pot, and cook until cream returns to a simmer for at least 5 minutes. Morels should not become limp before tasting, using toothpicks or forks to retrieve them. After the last Morel is removed, Larry notes that lots must be drawn to determine the lucky soul who has the privilege of downing the heavenly, rich, spore-darkened cream soup. Serves 2 to 4.

Andrew Weil's Shiitake Teriyaki

1 cup dried Shiitake

Hot water

¼ cup sake

¼ cup soy sauce

2 tablespoons light brown sugar

2 chopped scallions

A few drops (roasted) sesame oil

Reconstitute 1 cup dried Shiitake by covering with hot water and let stand till caps are completely soft. (Or cover with cold water, microwave on high for 2 minutes, and let stand.) Cut off and discard stems. Squeeze excess liquid from caps and slice into ¼-inch pieces. Place pieces in saucepan with sake, soy sauce, and brown sugar. Bring to boil and simmer, uncovered, till liquid is almost evaporated, tossing mushrooms occasionally. Remove from heat, cool, and chill. Sprinkle with finely chopped scallions and a few drops of dark (roasted) sesame oil. Serve as appetizer, side dish, or over rice.

Shiitake or Maitake Clear Soup

½ ounce dried mushrooms
   (or ¼ pound fresh mushrooms)

2 cups water

1 teaspoon soy sauce

2 tablespoons miso

1 pound tofu, coarsely chopped

¼–½ cup chopped onions

Soak cut or broken dried mushrooms in 2 cups cold water for 15 minutes. Drain off broth and save. Cover mushrooms with more cold water and soak for another 20 minutes. Add saved mushroom broth back into preparation and boil it for a few minutes. Season soup with soy sauce and miso. Add tofu and onions. (If needed, add 1–2 cups more water.) Bring back to a boil for 1–2 minutes and turn down heat. Allow to simmer for 5 minutes. Serves 4.

Shiitake Hazelnut Vegetarian Pâté*

4 ounces Shiitake mushrooms

3 tablespoons butter

1 clove garlic, minced

⅛ teaspoon thyme

¼ teaspoon salt

⅛ teaspoon pepper

1 teaspoon fresh parsley leaves

¼ cup toasted hazelnuts

3 ounces Neufchâtel cheese

2 teaspoons dry sherry

Trim and discard woody ends from mushrooms. In a food processor, finely chop mushroom caps and stems. In medium skillet, melt butter. Add mushrooms and garlic and sauté for at least 5 minutes. Stir in thyme, salt, and pepper. In food processor, chop parsley. Add hazelnuts and process. Add Neufchâtel cheese and process until smooth. Add sherry and mushroom mixture. Process until well mixed. Spread or mold in serving dish. Cover. Chill at least 1 hour. Serve with crackers. Yield: 1 cup. (Other mushrooms can be substituted for or combined with Shiitake.)

---

* Credit for this recipe is gratefully given to Timmer, Pershern and Miller's "Cooking American with an Oriental Favorite: Recipe Development with Shiitake Mushrooms" from Shiitake Mushrooms: A National Symposium Trade Show, University of Minnesota Center for Alternative Plant and Animal Products College of Natural Resources, College of Agriculture, St. Paul, Minnesota, May 3–5, 1989.

Maitake "Zen" Tempura

This recipe can also be used as a base to make tempura shrimp, whitefish, and/or assorted vegetables (zucchini, potatoes, onions, etc.).

1 ounce dried or
⅓ pound fresh Maitake

2 cups flour (⅓ pound)

2 eggs

2 cups plus ½ cup cold water

Vegetable or canola (rapeseed) oil
for frying

Tempura sauce

If using dried Maitake, soak cut or broken mushrooms in 2 cups cold water for 15 minutes. Discard water. Cover mushrooms with more cold water and soak for another 20 minutes. Drain and discard water. In a separate bowl, mix flour with eggs and cold water. Dip and roll the mushrooms into the flour/egg mixture. Deep-fry the Maitake mushrooms in hot oil (356°F / 180°C) for 1 minute. Remove any damp excess oil with paper towel. Serve with tempura sauce. Serves 2.

Stuffed Portobello Mushrooms

8 large portobello mushrooms

1 tablespoon olive oil

2 red bell peppers, seeded and finely chopped

2 green bell peppers, seeded and finely chopped

1 large onion, finely chopped

3 scallions (spring onions), green and white parts, thinly sliced

5 to 8 cloves garlic, finely chopped, to taste

½ teaspoon oregano

½ teaspoon dried basil

½ teaspoon thyme

Salt and freshly ground pepper to taste

6 ounces goat's cheese (optional)

Additional sliced scallions, for garnish

Remove the stems from the portobellos, chop and reserve. Place the whole mushroom caps smooth side down on a lightly greased baking sheet and bake in a preheated 425°F (220°C) oven for 15 minutes. Meanwhile, heat the oil in a skillet over moderate heat and sauté the mushroom stems, bell peppers, onion, scallions, and garlic until tender, 8 to 10 minutes. Add the herbs and cook an additional 2 minutes. Spoon the vegetable mixture into the mushroom caps and top with the cheese. Bake an additional 10 minutes, or until the mushrooms are tender and the cheese has melted. Sprinkle with sliced scallions and serve immediately. Serves 4 (2 mushrooms per person).

Recommended Mushroom Cookbooks

Cooking with Mushrooms: A Culinary Guide to Chef Pisto's Favorite Fungi by John Pisto, 1997. Pisto's Kitchen, Pacific Grove, California.

A Cook's Book of Mushrooms by Jack Czarnecki, 1995. Artisan Books, Workman Publishing, New York.

Edible Wild Mushrooms of North America: A Field-to-Kitchen Guide by David W. Fischer and Alan E. Bessette, 1992. University of Texas Press, Austin, Texas.

Hope's Mushroom Cookbook by Hope Miller, 1993. Mad River Press, Eureka, California.

Joe's Book of Mushroom Cookery by Jack Czarnecki, 1988. Atheneum, New York.

Mushroom Cookery by Rosetta Reitz, 1945. Gramercy Publishing Co., New York.

The Mushroom Feast by Jane Grigson, 1975. Alfred Knopf, New York.

Mushrooms Wild and Tamed by Rita Rosenberg, 1995. Fisher Books, Tucson, Arizona.

Wild about Mushrooms for Foresters and Feasters by Louise and Bill Freedman, 1987. Addison-Wesley, Reading, Massachusetts.

Wild Mushroom Cookery edited. Mike Wells and Maggie Rogers, 1988. The Oregon Mycological Society, Portland, Oregon.

CHAPTER 25

Cultivation Problems and Their Solutions: A Troubleshooting Guide

T his troubleshooting guide should be used in conjunction with the Growth Parameters for Gourmet and Medicinal Mushroom Species (Chapter 21) and the Six Vectors of Contamination described in Chapter 10. Individual contaminants are not specifically characterized in this book. If the vector introducing contamination is blocked using the techniques described here, the competitor organisms are effectively stopped. For extensive descriptions on contaminants, please refer to *The Mushroom Cultivator* by Stamets and Chilton (1983) and *The Pathology of Cultivated Mushrooms* by Houdeau and Olivier (1992).

The following guide lists the most frequently encountered problems, their probable causes, and effective solutions. A combination of solutions can often solve problems whose causes cannot be easily diagnosed. Most can be prevented through process refinement, structural redesign, improvement in hygiene maintenance, and/or replacement of personnel. Most importantly, the manner of the cultivator has the overriding influence on success or failure. I strongly encourage that, at every stage in the cultivation process, the cultivator leaves one petri dish, spawn jar, sawdust bag, etc. *uninoculated* to help determine whether or not ensuing contaminants are unique to the media preparation process versus the inoculation method. These "blanks" are extremely helpful in diagnosing the probable vector of contamination.

Cultivators should note that when one error in the process occurs, many symptoms could be expressed. For instance, diseases attacking mature mushrooms are to be expected if the humidity is maintained at too high levels during cropping. If the growing room is kept at 100% RH, the surfaces of the mushrooms remain wet and become perfect environments for parasitic fungi and bacteria. Bacterial blotch attacks developing mushrooms. Green molds proliferate. Mites eat mold spores. Flies carry mites and spores. If these organisms spread to developing primordia, massive deformation and contamination ensues. Those mushrooms that do survive have exaggeratedly short shelf lives after harvest. So, in this instance, one problem—humidity being too high—results in multiple symptoms. The lesson here: What is good for one contaminant is good for many! Controlling the vector of contamination must be coupled with creating an environment more conducive to the growth of mushrooms than competitors.

Population explosions of Sciarid and Phorid flies defeat Oyster mushroom cultivators more than any other competitor. Fly control measures have ranged from simple sticky pads to the use of chemical pesticides, a recourse I abhor. The use of chemical pesticides, although rampant among many "old school" cultivators, is totally unnecessary for gourmet and medicinal mushroom cultivation given a balance of preventative measures. Several biopesticides have been successfully used for prevention of fly maturity. Two are notable: a parasitic nematode, *Steinernema feltia,* which attacks the fungus fly larvae, and the well known bacterium, *Bacillus thuringiensis,* aka "BT-14." Applications are typically weekly, and are especially useful for those using manure-based substrates, or whenever a casing layer is applied.

Bug lights should be positioned at the entrance of every door. The bug traps I find that work the best are those that feature a circular black light and centrally located fan that creates a negative pressure vortex, features that greatly extend their effective range. These bug lights should also have sticky pads affixed below them that trap "fly-bys" or "near-misses." Coupled with the frequent washing down of the growing room, at least twice a day, population explosions can be forestalled or precluded.

There is one final control measure I recommend highly and that occurred naturally in our growing rooms. For the past 5 years, our growing rooms have sustained a population of small tree frogs. The only food source for the frogs, which have ranged in number from 2 to 8 in a 1,000-square-foot growing room, are flies. Each frog consumes between 20 to 100 flies per day. The growing rooms—with their resident flies and frogs—are in a state of constant biological flux, readjusting to maintain a delicate equilibrium. When the fly population declines, so do

A highly effective bug trapper. The circular light attracts flies to the vacuum-vortex that throws the flies into a clear plastic bag. By attaching "sticky paper" underneath the light, hovering flies are also captured. The clear bag allows the easy, daily counting of flies and helps predict impending outbreaks.

the frogs, and vice versa. An unexpected and purely aesthetic benefit: At night, the frogs' chorus while sitting on the mushrooms has a mesmerizing resonance that is joyful to hear. I consider these frogs to be faithful guardians, deserving respect for their valiant efforts at fly feasting, an activity that has an immediate beneficial impact in protecting the mushroom crop and limiting the spread of disease.*

My growing rooms have harbored a resident population of tree frogs for more than 10 years. Each frog consumes dozens of fungus gnats each day. I highly recommend frogs as an effective and natural method for limiting fly infestation. In this case, a frog lies in wait, perched upon Reishi mushrooms fruiting from a decomposing copy of *The Mushroom Cultivator*.

---

*The historical symbology here, between mushroom and frogs (or toads), should not go unnoticed. The word toadstool may well have originated from this association.

Problem	Cause	Solution
Agar Culture		
Media will not solidify	Not enough agar	Add more agar >20 grams/liter
	Insufficiently mixed	Agitate before pouring
	Bacteria	Increase sterilization time to at least 30 minutes at 15 psi
Spores will not germinate	Inviable spores	Obtain fresh spores
		Soak in 5 cc of sterilized water for 48 hours; place one drop of spore solution per petri dish
	Improperly formulated media	See pages 85–87
Spores germinate with mold contaminants ("powdery mildew")	Contaminated spores	Isolate and transfer "white spots" from one another to new media dishes; always move mycelium *away* from competitors
Spores germinate with bacterial contaminants ("slime")	Contaminated spores	Use antibiotic media ($^1/_{15}$–$^1/_{20}$) g/liter of gentamycin sulfate)
		"Sandwich" spores between two layers of antibiotic media; isolate mycelium when it appears on top surface
Mycelium grows, then dies back	Poor media formulation	See pages 85–87
	Over-sterilized media	Sterilize for less than 1 hour at 15 psi
	Poor strain	Acquire new strain
Contaminants appear along inside of petri dish	Contamination airborne in laboratory	Filter lab air, clean lab Look for source
		Wrap edges of petri dishes wit tape of elastic film after pouring and after inolculation
Contaminants localized to point of transfer	Culture contaminated	Isolate new culture
	Scalpel contaminated	Flame-sterilize scalpel longer

Problem	Cause	Solution
Agar Culture, continued		
Contaminants appear equally over the surface of agar media	Airborne contamination	Filter air and use good sterile technique
	Contaminated media— insufficient sterilization	Increase sterilization time
	Hands upstream of cultures during inoculation in airstream of laminar flow hood	Keep hands downstream of inoculation site
Media evaporates, cracks, before colonization is compelte	Humidity too low	Increase lab RH to 50% or wrap petri dishes with tape or elastic film
	Culture in airstream	Place cultures outside airstream
Grain Culture		
Grain spawn contaminates before opening, before inoculation	Bacteria endemic to grain	Soak grain overnight to trigger endospores into germination, making them susceptible to heat sterilization
	Contaminants enter during cooldown	Filter air during cooldown or open pressure vessel at 1 psi in clean room
Mycelium does not grow	Too dry	Over-sterilized; cook 1–2 hours at 15 psi
		Reduce surface area or porosity of filter media
	Bacterial contamination	Soak grain spawn overnight
		Boil grain in a cauldron before sterilization
	Over-sterilized	Reduce sterilization time
	Culture not receptive to media formula	Alter media formula

Problem	Cause	Solution
Grain Culture, continued		
Mycelium grows but spottily/incompletely	Insufficient distribution of mycelium through grain	Shake jars with greater frequency
	Insufficient inoculation rate	Increase amount of mycelium placed into each grain jar
	Bacterial contamination	Clean up strain
		Increase sterilization time of grain
Grain spawn difficult to break up	Over-incubation	Use spawn sooner
	Excessive water	Reduce water in formula by 10–20%
		Use different type of grain; use rye or wheat, not millet
		Add gypsum
Grain spawn appears pure, but contaminates with mold or bacteria after inoculation	Mycelium endemically contaminated—coexisting with other organism(s)	Return to stock cultures or clean strain
	Over-incubation	Use spawn sooner; normal for old spawn to eventually support other microorganisms
	Underside of filter laden with organic debris, providing a medium for contamination to grow through	Soak filter disks in bleach solution in between spawn runs
	Contamination airborne or from lab personnel	Install micron filters; observe good sterile technique
Jars crack/Bags break	Radical fluctuation in temperature and/or pressure	Reduce temperature and pressure flux
	Inadequate quality	Acquire higher-quality heat-tolerant jars and bags

Problem	Cause	Solution
Straw Culture		
Mushrooms fail to form	Improper initiation strategy	Consult chapter 21; alter moisture, temperature, light, CO_2, etc.
	Bad strain	Obtain younger strain of known vitality and history
	Chlorinated or contaminated water	Use activated charcoal water filters to eliminate chemical contaminants
Mycelium produces aborted mushrooms	Poor fruitbody development strategy	Consult chapter 21
	Bad strain	Obtain younger strain of known vitality and history
	Mite contamination	Discard, disinfect, and begin anew
	Nematode contamination	Minimize contact with soils and increase pasteurization time
	Insect damage from developing larvae	Shut down growing room, "bleach bomb" for 24 hours, install bug lights and/or frog population, and refill with new crop
Second and third crops fail to produce substantially or at all	Anaerobic contamination in core of substrate	Increase aeration or decrease depth of substrate mass
	Growing room mismanagement	Better management
	Bad strain	Acquire better strain
	Insufficient spawning rate	Increase spawning rate
Green and black molds appearing on straw	Insufficient pasteurization	Increase pasteurization time
	Prolonged exposure to elevated carbon dioxide levels	Lower CO_2 levels—increase air exchange
	Incubation at too high a temperature	Lower incubation temperature

Problem	Cause	Solution
Supplemented Sawdust Culture		
Mycelium fails to grow through within 2 weeks	Bags inoculated too hot	Allow to cool before substrate is inoculated
	Insufficient spawning rate	Increase spawning rate
	Inadequately mixing of spawn through sawdust	Increase mixing and/or spawn rate
	Mismatch of mycelium and wood type	Use woods native to mushrooms
	Sawdust too dry	Increase moisture
	Sawdust over-sterilized	Reduce sterilization time
Mycelium grows and then stops. Often accompanied by foul odors, slimy fluids. Yellow, green or black molds	Presence of contaminants—bacteria or molds	Consult chapter 10: The Six Vectors of Contamination
	Inadequate sterilization	Increase sterilization time
	Sterilization sufficient but contaminated during cooldown	Filter air during cooldown, check autoclave seals, drains, etc.
	Grain or sawdust spawn infected	Use pure spawn
	Person inoculating introduced contaminants	Adhere to good sterile technique
	Bags not separated to allow heat loss during incubation	Space bags 2 inches apart, maintain air temperature at 75°F (24°C)
	Airborne contamination	Use HEPA filters and good sterile technique
	Excessive carbon dioxide levels during incubation (>25%)	Increase surface area, or transpiration rate of filter media
	One of the above	Incubate the bacterially contaminated bags at 40–50°F (4–10°C) for month (40–100% are often salvageable)

Problem	Cause	Solution
Pre-Harvest Period		
Mycelium grows but fails to produce mushrooms	Monokaryotic strain—absence of clamp connections	Mate with compatible monokaryons—check for clamp connections
	Bad strain	Acquire new strain
	Mismatch of strain with substrate formula	Redevise substrate formula
	Virus/bacteria/parasitic fungi/nematodes	Regenerate spawn from clean stock cultures
	Inhibited by environmental toxins	Remove source of toxins
Harvest Stage		
Mushrooms form, but abort	Bad strain	Acquire new strain
	Poor environmental conditions	Consult chapter 21 for species
	Competitors: molds (*Mygogone, Verticillium, Trichoderma*) and bacteria	Consult chapter 10: The Six Vectors of Contamination
		Imbalanced growing room environment; CO_2 and RH too high; consult chapter 21: Growth Parameters for Gourmet and Medicinal Mushroom Species
		Mist mushrooms with water containing $1/2$ teaspoon of elemental sulfur per gallon; equivalent to 1 lb. sulfur per 100 gal. of water
	Chemical contamination from solvents, gases, chlorine, etc.	Remove toxins
Mushrooms form, but stems are long; caps underdeveloped	Inadequate light	Increase or adjust light to correct wavelength
	Excessive CO_2	Increase air exchanges

Problem	Cause	Solution
Harvest Stage, continued		
Massive numbers of mushrooms form; few develop	Poor strain	Obtain better strain
	"Over-pinning"	Shorten primordia formation period
	Lack of oxygen, inadequate light	Adjust according to the species' growth parameters; see chapter 21
	Inadequate substrate nutrition	Reformulate
	Dilate spawning rate	Increase spawning rate
Mushrooms deformed	Competitor organisms (*Mycogone, Verticillium,* bacteria, etc.)	Rebalance growing room environment to favor mushrooms and disfavor competitors
	Inadequate air circulation	Increase air circulation
	Excessive humidity or watering	Reduce humidity to prescribed levels; surface water must evaporate from mushrooms several times per day
	Bad strain	Acquire better strain
	Chemical contamination	Remove toxins
Mushrooms produced on first flush, fail to produce subsequent flushes	Inadequate substrate nutrition	Reformulate
	Competitors	Consult chapter 10: The Six Vectors of Contamination
	Window of opportunity missed	Reinitiate
	Mycelium "panned"	Disturb substrate: break apart and repack, allow to recover and reinitiate
	Poor growing room management	Improve management
	Bad strain	Acquire new strain

Problem	Cause	Solution
Harvest Stage, continued		
Flies endemic to growing room	Inadequate pasteurization	Extend pasteurization period
	Open doors, vents, etc.	Improve seals at doors and windows
	Poor growing room maintenance	Wash down growing room 2–3 times per day; place 1 cup bleach in drain to kill flies, once per day
	Slow cycling of crops	Increase crop rotation
	Inadequate cleanup of growing rooms between "runs"	Remove debris, wash ceilings, walls, and floors using bleach solution
		"Bleach bomb" for 24 hours
		Bug traps
		Frogs
Post-Harvest		
Mushrooms quick to spoil	Mushrooms too mature when harvested	Harvest when younger
	Mushrooms too warm before packaging	Chill mushrooms before placing in marketing containers
	Mushrooms too wet when harvested	Reduce humidity several hours before harvesting
	Mushrooms improperly packaged	Mushrooms need to breathe; cellophane or anticondensate, gas-permeable wrapping films recommended
	Mushrooms stored beyond shelf life	Sell mushrooms sooner

Ceramic sculpture depicting mycophiles dancing around a sacred mushroom from Colima, western Mexico, circa 100 B.C.

Description of Environments for a Mushroom Farm

For you to best understand the individual components making up a mushroom farm, a comparative description of the environments and activities occurring in each room is helpful. Some rooms can accommodate more than one activity, as long as schedules do not conflict. This list is especially useful for designers, architects, and engineers who are employed to design a complete facility. Since I do not employ caustic chemicals, pesticides, and other toxins, these descriptions may not fully address the needs of farms that handle manure-based substrates and use toxic remedies.

The Laboratory Complex

Ideally a specially constructed building or a space within an existing building is retrofitted with the following parameters in mind. Please refer to Appendix 2 for more information on the necessary equipment and rules of behavior within the laboratory environment.

Purposes: To isolate and develop mushroom cultures.
    To generate pure culture spawn.

Facility: A building well separated from the growing room complex

Maximum Temperature: 80°F (26–27°C)

Minimum Temperature: 70°F (21–22°C)

Humidity: 35–50%

Light: 500–1,000 lux

Insulation: R16–R32

Positive Pressurization: Yes, through HEPA filters

Additional Comments: Ideally, the laboratory should be uphill from the growing rooms so that passage of spawn is aided by gravity as it is

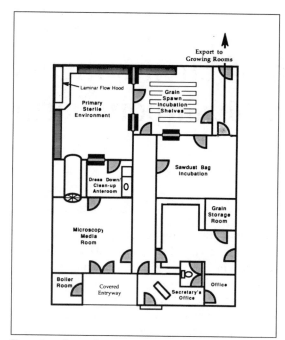

Floor plan of a mushroom spawn laboratory. Most of the substrate enters the clean room through the autoclave (midway left). Spawn is exported from laboratory to the growing rooms. Spawn is rotated frequently out of the laboratory.

transported. The laboratory is a relatively dry environment, encouraging the growth of mycelium only in protected containers (petri dishes, jars, and bags). Condensation surfaces must be minimized. After construction, every seam should be sealed with silicone caulking.

The Growing Room Complex

The growing room complex can house all nonlaboratory activities within one building. Each room has different requirements according to their functions. These recommendations should be used as general guidelines, subject to amendment. Growers in humid tropical requirements face a set of problems uniquely different from growers in cold, temperate climates. Ancillary storage and shop maintenance buildings are not listed.

Environment 1: The Growing Rooms

Purpose: To grow as many mushrooms as possible

Facility: Rooms vary in size from 10 × 20 feet to 40 × 100 feet with 10- to 20-foot ceilings and are usually rectangular in shape with large doors at both ends. Growing rooms should have cement floors with large drains and be equipped with water and/or steam lines. Electrical boxes and lights must be waterproofed. Internal walls should be constructed of non-degradable materials. Use of wood should be minimized. Entries into each room should be 8–16 feet wide to allow for easy access using forklifts or other equipment. Inside surfaces should be made with exterior, weather-resistant materials.

Maximum Temperature: 80°F (26–27°C)

Minimum Temperature: 45°F (7–9°C)

Humidity: 50–100%

Light: 50–1,000 lux

Insulation: R8–R16 or as needed

Positive Pressurization: Yes, through electrostatic filters

Additional Comments: Ideally, a flow-through design is followed, both in consideration of fresh mushrooms as well as the entry and exit of substrate mass. Removing contaminated substrates into the same corridor through which freshly spawned substrate is being transferred causes cross-contamination. Spent substrate should be exited out of the opposite end of the growing rooms. Many farms bring their fresh mushrooms into the main hallway, en route to the sorting and cold storage rooms.

Environment 2: The Spawning Room

Purpose: A room adjacent to the pasteurization chamber wherein inoculations into bulk substrates are conducted.

Facility: The room must be constructed of materials that will not harbor mold colonies and can be washed down with ease. The height should be

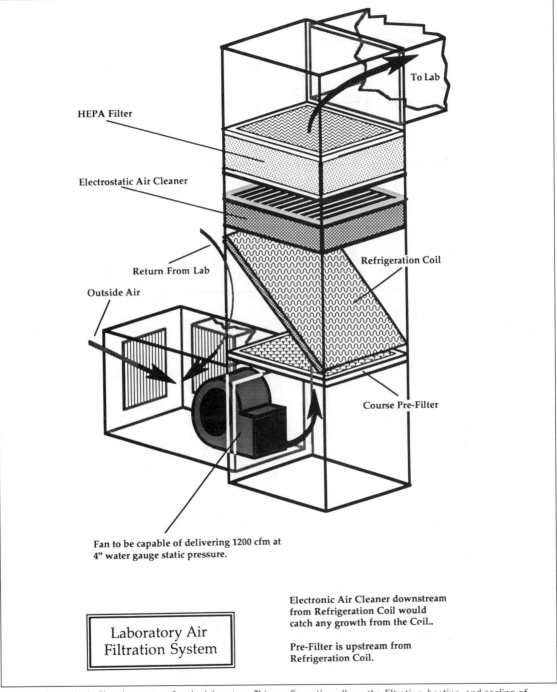

HEPA Filter

Electrostatic Air Cleaner

Return From Lab

Outside Air

To Lab

Refrigeration Coil

Course Pre-Filter

Fan to be capable of delivering 1200 cfm at
4" water gauge static pressure.

Laboratory Air
Filtration System

Electronic Air Cleaner downstream
from Refrigeration Coil would
catch any growth from the Coil..

Pre-Filter is upstream from
Refrigeration Coil.

Exteriorly located, air filtration system for the laboratory. This configuration allows the filtration, heating, and cooling of incoming and recirculated air. The filters must be removed periodically for cleaning and/or replacement.

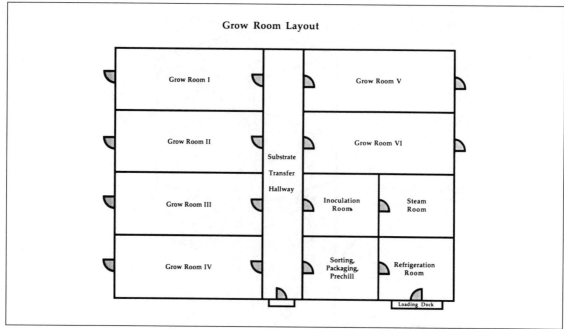

Grow Room Layout

Floor plan of standard growing-room complex. This configuration allows for 6 growing rooms and processing areas to be housed under one roof.

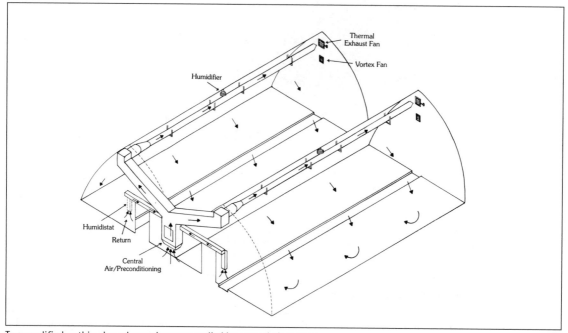

Two modified gothic-shaped greenhouses supplied by central air system.

sufficient to accommodate the unloading of the steam box, conveyors (if used), elevated platforms and funnels for filling columns (12–16 feet). Cement floors and moisture-proof electrical fixtures are essential for safety.

Maximum Temperature: 90°F (32–33°C)

Minimum Temperature: Ambient

Humidity: Fluctuating from ambient to 100%

Light: 200–500 lux. Needed only for ease of personnel—skylights or moisture-proof fluorescents

Insulation: None needed

Positive Pressurization: Yes, through HEPA filters

Additional Comments: Once this room is thoroughly washed down with a dilute bleach solution prior to spawning, the fan/filter system is activated for positive pressurization. The filtration system is ideally located overhead. Air is passively or actively exhausted near the floor. During inoculation, this room becomes very messy, with spawn and substrate debris accumulating on the floor. Since the spawning room is only used directly after each run through the pasteurization chamber, it can serve more than one purpose during non-spawning days. The spawning room should be adjacent to the main corridor leading to the growing rooms to facilitate substrate handling.

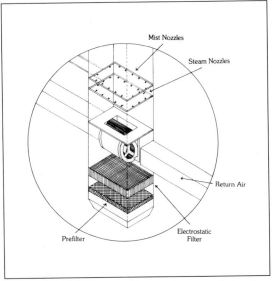

Close-up of air supply system for growing rooms featured. Air enters from below, passing through a coarse and an electrostatic air filter. A large squirrel cage blower pushes air into main duct system where two "rain trees" allow the introduction of cold or hot water into the airstream. In-house thermostats coupled to solenoid valves regulate the cold or hot (steam) water supply. Recirculation chutes enter the plenum from the side. The degree of recirculation can be controlled within the growing rooms. Preconditioning the air quality to 70–80% of desired levels is recommended before entry into the growing rooms. Provisions for excess condensation must be engineered into system.

Environment 3: The Pasteurization Chamber or Phase II Room

Purpose: To pasteurize bulk materials (straw, bagasse, etc.) by subjecting the substrate to steam for a prolonged period of time (2 to 24 hours)

Facility: Usually rectangular, the pasteurization chamber is a highly insulated room with a false floor, usually screened or grated, under which steam is injected. Two drains are recommended per pasteurization box. The drains should have a screened basket over them to prevent clogging. Ideally the drain line should have a check or gate valve to prevent contaminants being drawn into the pasteurization box during cooldown. The walls and floors must be constructed in such a manner to

withstand radically fluctuating temperatures and humidities. Rooms are often constructed of cinder block, cement formed, or temperature-tolerant fiberglass-reinforced plastic (FRP). Wood construction is strongly discouraged. The pasteurization chamber should have ample head space. Large farms use "walking floors" or a net pulled by a winch, facilitating off-loading. (See page 172.)

Maximum Temperature: 210°F (99–100°C)

Minimum Temperature: Ambient

Humidity: 10–100%

Light: Minimal or none

Insulation: R30+

Positive Pressurization: Yes, through HEPA filters

Additional Comments: Pasteurization tunnels take lot of abuse from the loading and unloading of substrate. The screened floors should be removable so waste debris can be gathered after each run. Once emptied, the rooms should be doused with a bleach solution to limit the growth of any mold colonies. I have seen pasteurization tunnels made from reconverted old saunas, silage vessels, grain silos, semi-trucks, ocean cargo containers, beer fermentation vats, etc. Any doors or openings must be tightly gasketed. Provisions for the ease of filling and unloading aid production efficiency.

Environment 4: The Main Corridor—a Highway for Substrate and Product Flow

Purpose: A central hallway connecting the essential environments to one another, facilitating movement of materials, products, and personnel

Facility: Often, a metal-framed, open-truss building, is centrally located, broadside to the growing rooms. Ceilings should be at least 12 feet, ideally 20 feet high. Here again, wood surfaces should be minimized. Floors should be cement with drains and have access to waterlines for maintenance and cleanup.

Maximum Temperature: Ambient

Minimum Temperature: Ambient

Humidity: Ambient: 20–80%

Light: Skylights or moisture-proof fluorescents

Insulation: None needed

Positive Pressurization: Not needed

Additional Comments: Since this is a high-traffic area, two-way passing is essential. Those using forklifts must have ample turnaround space for maximum mobility. Many cultivators have charts and/or remote temperature sensors constructed into the walls by each growing room. Bug traps are placed at several locations to intercept winged intruders before possible entry into the mushroom growing rooms.

Environment 5: Sorting, Grading, and Packing Room

Purpose: To sort, grade, and package mushrooms into their end-user containers

Facility: A well-lit room with gravity conveyors, sorting tables, and often a blast-chiller that quickly cools the mushrooms prior to packaging and storage in the refrigeration room. This room is usually located at the end of the main corridor and is immediately adjacent to the refrigeration and shipping rooms.

Maximum Temperature: 50°F (10°C)

Minimum Temperature: 35°F (1–2°C)

Humidity: 50–75%

Light: 500–1,000 lux

Insulation: R30

Positive Pressurization: No

Additional Comments: I have seen a number of configurations of this type of processing room. Mushrooms usually arrive in open-grate plastic carrier baskets. The baskets are placed into the airstream from the blast-chiller (or in the cold room) prior to sorting into cardboard end-user boxes. Although this room is kept cool, the packaging personnel are comfortable at this temperature, busily sorting, weighing, labeling, and arranging boxes for distribution.

Environment 6: The Refrigeration Room

Purpose: To chill mushrooms to 35°F (1–2°C) so they can be maximally preserved

Facility: A standard refrigeration room

Maximum Temperature: 38°F (3°C)

Minimum Temperature: 32°F (0°C)

Humidity: 60–80%

Light: Only as needed for personnel

Insulation: R30–R60

Positive Pressurization: No

Additional Comments: Installers should engineer systems with mushroom preservation in mind, paying particular attention to humidity concerns. Standard refrigeration systems usually suffice, except that humidity must be kept between 60 and 85% RH to prevent sudden dehydration of the product. Humidity in excess of 90% often causes mushrooms to "re-vegetate," causing a grayish fuzz and accelerating spoilage. As of 1995, Freon has been banned as a refrigerant and non-ozone-destroying substitutes will be used. New refrigeration systems, including nonmechanical, carbon dioxide-based designs, are being developed. I have no information concerning their applicability to chilling and preserving mushrooms. Sufficient airflow is essential to effect slow evaporation off the cap surfaces. Still-air refrigeration systems cause mushrooms to quickly rot unless the evaporation rate is increased to compensate.

Environment 7: Shipping and Receiving Room

Purpose: To transfer mushrooms from the cold storage to the shipping/receiving area. Many farms have loading docks at the same elevation as the beds of the produce trucks picking up and delivering the product. Most facilities are equipped with over-hanging doors.

Facility: An open, high-ceiling room with direct access to the main corridor, the sorting room, and/or the refrigeration room

Maximum Temperature: Ambient

Minimum Temperature: Ambient

Humidity: Ambient

Light: As needed for personnel. Fluorescents do have lenses.

Insulation: Minimal

Positive Pressurization: None

Additional Comments: A high-traffic area, the shipping and receiving room, besides its obvious purposes, is also used as a buffer, limiting the impact of environmental fluctuations from the outside.

Mushrooms are usually weighed and then 10% is added to offset weight loss due to evaporation during shipment.

Environment 8: Production/Recapture Open-Air Growing Room

Purpose: To recapture as many mushrooms possible which cannot be realized in controlled-environment growing rooms. This building can solve a dilemma constantly confronting the growing room personnel: to maximize mushroom yield while not jeopardizing future crops as contaminants become more common with the cycle coming to completion.

By the third or fourth flush, yields are in a state of precipitous decline. Rather than discarding this mycelium, the cultivator can realize additional harvests, with minimum effort, if the substrate is placed outside during conducive weather conditions. In the temperate regions of the world, these favorable weather conditions span several months. During these moist months, Oyster and Shiitake mushrooms produce prolifically outdoors. I am continually amazed at the size of mushrooms that can be harvested outside from "spent" straw or sawdust that has been exported from the indoor growing rooms. Two types of buildings serve this purpose well.

Facilities: Either a hoop-frame structure covered with "bug-out" or shade cloth or a covered building with walls constructed of the same, draping from the outer roof joists

Maximum Temperature: Ambient

Minimum Temperature: Ambient

Humidity: Ambient, augmented to 85–100% by overhead sprinklers

Light: Ambient. Indirect natural light coming from the sides is best

Insulation: None needed

Positive Pressurization: n/a

Additional Comments: Two structures meet these needs well. The first is the simplest. By constructing a hoop-type greenhouse and covering it with 70–80% shade or "bug-out" cloth, moisture can penetrate through to interior and airflow is naturally high. If the pore spacing is fine enough, as in the commercially available antibug screens ("bug-out"), then flies will be hindered from entry. If a metal-roofed, open-sided haybarn is used, then draping this fabric from the outer frame to create fabric walls will accomplish a similar function. In either environment, a simple overhead misting system, activated by a timer or hand controls, will promote additional mushroom crops. Compared to the details needed for the controlled environment, high-efficiency growing rooms, the construction of these types of rooms is self-explanatory and open to modification.

Designing and Building a Spawn Laboratory

Mushroom cultivation is affected as much by psychological attitude as it is by scientific method. The synergistic relationship between the cultivator and his/her cultures becomes the overwhelming governing factor in determining laboratory integrity. Since contamination is often not evident for days, even weeks, after the mistake in technique has occurred, the cultivator must develop a super-sensitive, prescient awareness. Practically speaking, this means that every time I enter the laboratory, I do so with a precautionary state of mind.

The laboratory is designed and built for the benefit of the mushroom mycelium. The role of the cultivator is to launch the mycelium onto appropriate sterile substrates in the laboratory and then leave. The less nonessential time spent by humans in the laboratory the better, since humans are often the greatest threat to the viability of the mushroom cultures. Growing mushrooms successfully is not just a random sequence of events scattered throughout the week. One's path through the facilities, growing rooms, and laboratory can have profound implications on the integrity of the entire operation.

The growing of mushroom mycelium in absence of competitors is in total contradiction to nature. In other words, the laboratory is an artificial environment; one designed to forestall the tide of contaminants seeking to colonize the same nutritious media that has been set out for the mushroom mycelium. This could be, and is, a frightening state of affairs for most would-be cultivators until books like *The Mushroom Cultivator* by Stamets and Chilton (1983) and this one offered simple techniques for making sterile culture practical for mushroom cultivators. These volumes represent, historically, a critical step in the passing of the power of sterile tissue culture to the masses at large.

Before the advent of HEPA filters*, sterile culture work succeeded only by constantly battling legions of contaminants with toxic disinfectants, presenting real health hazards to the laboratory personnel. Now, the use of disinfectants is minimized because the air is constantly being refiltered and cleaned. Once airborne contamination is eliminated, the other vectors of contamination become much easier to control. (Please consult Chapter 10 for the six vectors of contamination.)

Most people reading this book will retrofit a bedroom or walk-in pantry in their home. Large-scale, commercial operations will require a separate building. In either case, this chapter will describe the parameters necessary for designing and building a laboratory. If you are building a laboratory and pay strict attention to the concepts outlined herein, contamination will be minimized. Like a musical instrument, the laboratory must be fine-tuned for best results. Once the lab is up and running, a sterile state of equilibrium will preside for a short time. Without proper maintenance, the lab, as we say, "crashes." The laboratory personnel must constantly clean and stay clean while they work. In effect, the laboratory personnel become the greatest threat to the lab's sterility. Ultimately, they must shoulder the responsibility for every failure.

The laboratory should be far removed from the growing rooms, preferably in a separate building. The air of the laboratory is always kept free of airborne particulates while the growing rooms' air becomes saturated with mushroom spores. *The growing rooms are destined to contaminate.* Even the spores of mushrooms should be viewed as potential contaminants threatening the laboratory. If both the laboratory and growing rooms are housed in the same building, the incidence of contamination is much more likely. Since the activities within the laboratory and growing rooms are distinctly different, separate buildings are preferred. I know of several large mushroom farms that built their spawn laboratory amid their growing rooms. Their ability to generate pure culture spawn is constantly being jeopardized by the contaminants coming from the growing rooms. Every day, the laboratory manager faces a nightmarish barrage of contaminants.

A good flow pattern of raw materials through the laboratory and of mature cultures out of the laboratory is essential. Farms with bad flow patterns must constantly wage war against seas of contaminants. The concepts are obvious. The positioning of the growing room exhaust fans should be oriented to avoid directing a "spore stream of contaminants" into the laboratory filtration system. Furthermore, the design of a mushroom farm's buildings should take into account prevailing wind direction, sunlight exposure, shade, the positioning of the wetting or compost slabs, the location of the bulk substrate storage, and the overall flow patterns of raw materials and finished goods.

The major problem with having a laboratory within a home is the kitchen—a primary breeding ground for contamination. Rotting fruits, foods spoiling in the refrigerator, and garbage containers represent a triple-barreled threat to the laboratory's integrity with the air and the cultivator as carrier vectors. However, good sterile technique coupled with the use of HEPA filters can make a home laboratory quite functional for the small and midsized cultivator. Most importantly, the cultivator must have a heightened awareness of his/her path through the sources of contamination before attempting sterile tissue culture. I prefer to do my laboratory work in the mornings after showering and putting on newly washed clothes. Once the lab work is completed, the laboratory personnel can enter the packaging and growing rooms. Otherwise, these areas should be strictly off-limits. In a mushroom facility, duties must be clearly allocated to each person. If you are working alone, extra attention to detail is critical to prevent cross-contamination.

---

* HEPA, or High Efficiency Particulate Air Filters, eliminate particulates down to .3 microns with an efficiency rating of 99.99%. ULPA (Ultra-Particulate Air) filters screen out particles down to .1 μ with 99.999% efficiency.

Design Criteria for a Spawn Laboratory

The design criteria for constructing a spawn laboratory are not complicated. A short description of one of my labs might help the reader understand why it works so well. My first laboratory is housed in a 1,440-square-foot building. A 15 horsepower boiler is located in its own room and generates steam for the 54-inch diameter, 10-foot-long, double-door retort. The walls and ceilings are covered with FRP (fiber reinforced plastic). The lights are covered with waterproof, dustproof lenses. Plug-ins for the remote vacuum system are handy and well used. To enter, you must pass through three doors before reaching the clean room. In the clean room, a 2-foot high by 12-foot long homemade laminar flow bench gives me ample freedom of hand movement and surface area. Fresh outside air is serially filtered and positive-pressurizes the lab from overhead through a coarse prefilter, an electrostatic filter, and finally a 24 × 24-inch HEPA filter. A return duct, recycling the room's air should be on the floor but, I admit, is not. If the return duct is located low in the laboratory, contaminants are constantly being pushed to and skimmed off the floor. My laminar flow bench—with its massive surface area—recirculates the air in the entire room once every 1.3 minutes. This is far more than what is minimally necessary.

Here are a few key concepts in designing laboratory, whether the clean room is in the home or in its own building. If incorporated into the design of your facility, contamination vectors will be minimized. Following this list are helpful suggestions of behavior, which, in combination, will give rise to an efficient, steady-state clean room.

- **Positive-pressurized laboratory.** The laboratory should be continuously positive-pressurized with fresh air. The fresh, outside air is serially filtered, first through a coarse prefilter (30% efficient at 1 μ), then an electrostatic filter (95% efficient at 1μ), and finally a HEPA filter (99.99% efficient at .3 μ). Blowers must be properly sized to overcome the cumulative static pressures of all the filters.

In most cases, the combined static pressure approaches 1.25 inches. (Static pressure of 1 inch is the measure of resistance represented by the movement of water 1 inch, in a 1-inch diameter pipe.) Air velocity off the face of the final filter should be at least 200 feet per minute. For a 400-square-foot clean room, 2 foot × 2 foot × 6-inch filters should be employed. The construction of the intake air system should allow easy access to the filters so they can be periodically removed, cleaned, and replaced if necessary. (See page 469.) Fresh air exchange is essential to displace the carbon dioxide and other gases being generated by the mushroom mycelium during incubation. Should carbon dioxide levels escalate, the growth of contaminants becomes more likely. Sensitive cultivators can determine the quality of the laboratory immediately upon entering with their sense of smell.

- **Stand-alone laminar flow bench.** A laminar flow bench constantly recirculates the air within the laboratory. The air entering the laboratory has been already filtered from the positive pressurization system described in the previous paragraphs. By having two independent systems, the life span of the filter in the laminar flow bench is greatly extended. And, the clean room becomes easy to maintain. Furthermore, I am a strong believer in creating a laboratory that is characterized by turbulent air streams, with a high rate of impact through micron filters. Turbulent, filtered air in the laboratory is far more desirable than a still air environment. The key idea: *If airborne particles are introduced, they are kept airborne with turbulence. If kept airborne, particles are soon impacted into the micron filters.* This reduces the stratification of contaminant populations in the laboratory, and of course, temperature variations. It is important to note that this concept is diametrically opposite to the "old-school" concept that still air in the laboratory is the ideal environment for handling pure cultures.

- **Double- to triple-door entries.** There should be at least two doors, preferably three doors, separating the clean room from the outside environment. Double-door entries are a standard in the industry. Doors with windows have obvious advantages in preventing accidents. Furthermore, the doors should be gasketed with dirt skirts. When the doors swing outwards as you exit the innermost clean room, the export of mature spawn or blocks is made easier. (I prefer to kick the doors open upon exiting as, often, my hands are full.) As workers travel towards the clean room, they enter rooms hygienically cleaner than the previous, and into increasingly higher pressure zones. Floor decontamination pads, otherwise known as "sticky mats," are usually placed before each doorway to remove debris from the feet.

  With ultra-modern clean rooms, a double-door anteroom, called a "decontamination chamber," utilizes the downflow of HEPA-filtered air over the worker who stands on a metal grate. The air is pushed from above and actively exhausted through the floor grate to the outside. The principle concept here is valid: *The constant descension of airborne particulates improves laboratory integrity.* Another variation of this concept is the replacement of the solid inner doors with downflowing air curtains. However, decontamination chambers and air curtains should be the last projects on a long list of other priorities for the financially conservative investor.

- **Interior surfaces not biodegradable.** Interior surfaces such as the walls, countertops, shelving, etc. should not be able to support mold growth. Wood and Sheetrock should be avoided. The floors should be painted several times or overlaid with a chemically resistant, cleanable mat. When using paint, use a non-mildewing enamel. (*Caution:* Do not use paint containing fungicides, particular any containing tributyl tin oxide, an extremely dangerous toxin to both humans and mushrooms.) Countertops can be made of stainless steel or a hardened laboratory-grade Formica. The shelves storing the incubating bags should be wire meshed, and not solid, so that the heat generated from incubation is dissipated. Petri dish cultures can be stored on solid shelves. (See page 78.)

- **Walls and ceiling well insulated.** Ambience of temperature is critical for maintaining a laboratory. Temperature fluctuation causes two problems. When temperatures within the lab radically change from day to night, condensation forms within the spawn containers and on the upper inside of the petri dishes. These water droplets will carry otherwise dormant contaminants down to the rich media. Bacteria particularly love condensation surfaces. When a laboratory is run at 50% relative humidity and 75°F (24°C) condensation should dissipate within 24 hours after autoclaving. The other problem caused by temperature fluctuation is that the outer walls of the laboratory, especially those made of concrete, sweat. I had one small home laboratory that grew an enormous colony of white mold on a painted, white cinder-block wall. The whole laboratory contaminated despite my best efforts. Only when I washed the wall did I find the source. If your only option is a laboratory with an outside facing cinder block wall, make sure the pores have been sealed with a thick coat of paint and permanently place an electric baseboard heating unit facing the wall to eliminate the possibility of condensation.

- **Lights covered with dustproof covers.** Fluorescent lights should be covered with lens coverings. Uncovered lights ionize particulates that will collect as dust layers on oppositely charged surfaces. Over time, a habitat for contaminants builds. Ionizers are similar in their effect and are greatly overrated. (See *Consumer Reports*, October 1992.)

- **Remote vacuum cleaning system.** Since constant cleaning must occur throughout the inoculation process, having the ability to quickly pick up spilled grain and sawdust greatly enhances the ease of inoculation. When inoculations are done quickly, the likelihood of airborne fall-out (primarily from the cultivator) is minimized. Brooms should *never* be used in the laboratory. Wet/dry remote vacuums run the risk of clogging and then breeding contaminants. Therefore, a "dry" remote vacuum system is recommended.

Good Clean Room Habits: Helpful Suggestions for Minimizing Contamination in the Laboratory

- **No shoes are allowed in the laboratory.** The lab is strictly a "shoes-off" environment. Disposable booties are often used over socks. No outer clothing that has been exposed to the outside air should be worn into the laboratory.

- **Wear newly laundered clothes and/or a laboratory coat.** Once your clothes have come into contact with contaminants, these contaminants will become airborne within the laboratory. Two primary sources of contamination are people's pets and their car seats. Once the laboratory personnel come in contact with contamination sources, their usefulness in the laboratory has been jeopardized.

- **Wash hands frequently with antibacterial soap and isopropanol.** Personnel should thoroughly wash their hands before entering the laboratory and, with frequency, every half-hour during the course of inoculations. Isopropanol ("rubbing" alcohol) is used for wiping countertops and hands, and topically sterilizing tools. Other disinfectants are available from the hospital supply industry.

- **Frequently mop floors with a 10% bleach solution.** The lab floors should be mopped at least once a week, and directly after each major run. Two buckets are used: one for bleach and one for rinsing the dirt-ladened mop. Mop heads should be frequently replaced.

- **Do not conduct inoculations when you are sick with a cold, the flu, or other contagious illnesses.** I know of cases where cultivators have inadvertently cultivated Staphylococcus bacteria and reinfected themselves and others. Face masks should be worn if you have no option but to work when you are sick.

- **Do not speak, exhale, or sing while conducting inoculations.** Your breath is laden with bacteria that thrive in the same media designed for the mushroom mycelium. If you have a telephone in your laboratory, be aware that it often becomes a redistribution point for contamination.

- **Remove trash, and contaminated and over-incubated cultures daily.** I do not have wastebaskets in my laboratory, forcing me to remove trash constantly and preventing a site for contamination. Over-incubated cultures are likely to contaminate so they are best rotated out of the laboratory with frequency.

- **If cloning a specimen, never bring sporulating mushrooms into the laboratory.** Ideally, have a second small portable laminar flow hood specifically used for cloning. I use this same laminar flow hood as a "Micron Maid" to help keep airborne particulates at reduced levels in downstream environments. New petri dish cultures from clones should be wrapped with elastic film or tape to prevent the escape of molds, bacteria, and mites into the laboratory. If sporulating molds are visible, isolate in a still-air environment.

- **Isolate cultures by placing petri dishes on "sticky mats."** I came up with this innovation when fighting mites and trying to prevent cross-contamination. Sticky mats are also known as decontamination floor pads. See page 78.

- **Establish a daily and weekly regimen of activity.** Daily and weekly calendar schedules for managing the laboratory will help give continuity to the production stream. Since so many variables affect the outcome of mushroom cultivation, try to establish as many constants as possible.

- **Rotate spawn frequently.** Do not let cultures and spawn over-incubate. Over-incubated Oyster cultures are especially a hazard to the lab's integrity. After 3 to 4 weeks, Oyster mushrooms will fruit within their containers, often forcing a path through the closures. If unnoticed, mushrooms will sporulate directly in the laboratory, threatening all the other cultures.

The laboratory's health can be measured by the collective vitality of hundreds of cultures, the lack of diseases, and the diversity of strains. Once filled to capacity with the mycelia of various species, the lab can be viewed as one thermodynamically active, biological engine. The cultivator orchestrates the development of all these individuals, striving to synchronize development, *en masse,* to meet the needs of the growing rooms.

Success in mushroom cultivation is tantamount to not cultivating contaminants. Confounding success is you, the caretaker cultivator, resplendent with legions of microflora. Individuals vary substantially in their microbial fall-out. Smokers, pet owners, and even some persons are endemically more contaminated than others. Once contamination is released into the laboratory, spores soon find suitable niches, from which a hundredfold more contaminants will spring forth at the earliest opportunity. As this cycle starts, all means must be enacted to prevent outright mayhem. Contamination outbreaks resemble dominoes falling, and are soon overwhelming to all but the best prepared. The only recourse is the mandatory shutting down of the entire laboratory—the removal of all incubating cultures and the necessary return to stock cultures. After purging the lab of virtually everything, a strong solution of bleach is used for repetitive cleaning in short sequence. Three days in row of repetitive cleaning is usually sufficient. Clearly, prevention is a far better policy than dealing with contaminants after the fact.

No matter how well the laboratory is designed, the cultivator and his/her helpers ultimately hold the key to success or failure. Each individual can differ substantially in their potential threat to a clean room. Here's a poignant example. At one time when contamination was on an upward spiral, I had eliminated all the vectors of contamination except one: the MCUs, mobile contamination units—which includes people and other mobile organisms. Determined to track down the source, I brought in an expensive airborne particulate meter, used commonly by the computer industry to judge the quality of clean rooms. This unit measured airborne contamination per cubic meter through a range of particle sizes, from .10 microns to >10 microns.

Several fascinating results were observed. One obvious measurement was that, in a calm air room, 100 times more particulates were within 1 foot of the floor than were within a foot of the ceiling. *Truly, the air exists as an invisible sea of contaminants.* What was most surprising was the contamination fall-out from each employee. Standing each employee in the airstream coming from the laminar flow bench, I recorded downwind particle counts. The contamination source was immediately identified. An employee was generating nearly 20 times the contamination fall-out than anyone else. The dirty employee was summarily banned from the laboratory. Soon thereafter, the integrity of the laboratory was restored. The lesson learned—that humans carry their own universe of contaminants—and are the greatest threat to clean room integrity.

The Growing Room: An Environment for Mushroom Formation and Development

The first attempts at growing mushrooms indoors were in caves in France late in the eighteenth century. They provided an ideal environment for the Button mushroom (probably *Agaricus brunnescens*): constant, cool temperature and high humidity. To this day, cave culture for the Button mushroom is still widely practiced. One of the largest mushroom farms in the world utilizes an extensive network of caves in Butler County, Pennsylvania. Cave culture has one major drawback for gourmet mushroom production: darkness.

The Button mushroom does not require nor is sensitive to light. All the other mushrooms described in this book are phototropic. This major difference—the need for light—presents a financial obstacle to the retrofitting of Button mushroom farms into gourmet mushroom production facilities. Many gourmet mushroom farms must build customized growing rooms. But, in many cases, other types of structures can be retrofitted for commercial production. Here are some examples:

airplane hangars	airplane shells
ammunition bunkers	army barracks
barns	basements
bomb shelters	cargo containers
car washes	caves
dairies	greenhouses
hog farms	mines
missile silos	poultry sheds
Quonset huts	ship hulls
slaughterhouses	train cars
train and highway tunnels	volcano (lava) tubes
warehouses	potato bunkers

In general, custom-designed growing rooms will perform better than structures that have been engineered for other purposes. However, with wise modifications, any of the above structures can be made into intensive growing chambers.

MOUTH OF MUSHROOM-CAVE NEAR PARIS

BOTTOM OF SHAFT OF MUSHROOM-CAVE

Caves were used as growing chambers in France circa 1868. (Reprinted from Robinson's *Mushroom Culture*, 1885.)

Whereas the laboratory is maintained at a constant temperature and humidity, the growing room's environment is fluctuated during the development of the mushroom crop. These changes in environment are specific, and sometimes must be radical, to trigger the switchover to mushroom formation and development. Whole new sets of skills are demanded of the growing room manager, which are distinctly not needed by the laboratory technician. The ability of the manager to implement these changes is directly affected by the design of the growing rooms. Here are some of the design criteria that must be satisfied for creating a functioning growing room.

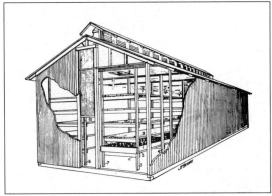

The first growing rooms resembled chicken houses. Reprinted from a 1929 USDA circular: *The Mushroom Growing House.*

Design Criteria for the Growing Rooms

Shape. The general shape of a growing room should be rectangular. I have never seen a square or circular growing room function well. The growing room should be at least twice as long as it is wide. This configuration allows for air to be distributed down a central ductwork. The rectangular shape is naturally process oriented: permitting the flow-through of substrate materials and fresh mushrooms. Rectangular rooms are simply easier to manage.

Interior walls. The inside skin of a growing room must be built of water- and mold-resistant materials. Fiberglass, polycarbonate, acrylic, glass, and galvanized metals can be used for the interior of a growing room. The material of choice, by most professional cultivators, is called FRP (for fiberglass reinforced plastic). This high-temperature-extruded fiberglass material has a smoothed finish, and its pliability makes installation simple. Furthermore, FRP will not be degraded by mold fungi, does not out-gas toxic fumes, and is tolerant to most cleaning agents. Wood or metal surfaces can be painted with a mold-/rust-resistant glazing. Cultivators should check with local ordinances so that the materials used in their growing rooms fully comply with food production and building code standards.

For those with limited budgets, the cheapest material is polyethylene plastic sheeting used by the greenhouse industry. This material usually survives no more than 2 or 3 years under the conditions used for growing mushrooms. I have attached greenhouse sheeting using galvanized staples over lengths of thick plastic tape.

Doors. As with the laboratory, the growing rooms should be protected from the outside by at least two doors. The first door from the outside leads into an operations room or hallway where the second door opens into the growing room. Doors should be at least 4 × 8 feet. Some farms have two 5 × 10-foot double-opening bay doors, or a 10 × 10-foot sliding door. These large doors allow the easy filling and emptying of the growing rooms. A small door is sometimes inner-framed within one of the larger doors so the growing room environment is not jeopardized when only personnel need to enter. In any event, the doors should accommodate small forklifts or similar machinery that need access to the growing rooms. The doors should be made of a material that does not support the growth of molds. The bottom of the door is often fitted with a brush-skirt that discourages insects from entering. The doorjambs are usually gasketed to assure a tight seal when closed.

At the opposite end of the growing room, a similarly sized exit door should be installed. This door

Modified gothic-framed buildings have several advantages. Simple and quick to build, these buildings have walls with an inside curvature that keeps condensation from dripping onto the crop. The condensation adheres to the walls and streams to the floor where it is re-evaporated or drained off. The peak roof allows for the removal of excess heat and/or the mixing of humidity and air prior to contact with the mushroom crop.

facilitates the emptying of the growing room after the cropping cycle has been completed. To bring aged substrate, which is often contaminated after the fourth or fifth flush, into the same corridor that leads to other growing rooms presents a significant cross-contamination vector.

Many farms employ pass-through curtains, usually made of clear, thick, overlapping strips of plastic. These are especially useful in the sorting, packaging, and refrigeration rooms. However, pass-through curtains do not afford a tight enough seal for protection of the growing rooms from other environments.

Structures insulated. Environmental control systems will function far better in growing rooms that are insulated than in those that are not. Some strains of mushrooms are far more sensitive to temperature fluctuation than others. In localities where temperature fluctuation is extreme, insulation is essential. Some cultivators partially earth-berm portions of their growing rooms for this purpose. When using insulation, make sure it is water-repellent and will not become a food source for molds.

Inside roof. The inside roof should be curved or peaked for heat redistribution by the air circulation system. Furthermore, the slope of the inside roof should be angled so that condensation adheres to the sloped roof surface and is carried to the walls, and eventually spills onto the floor. This allows for the re-evaporation from the floor back into air. The height of a growing room should be at least 10 feet, preferably 12–16. At least 4–6 feet of free air space should be above the uppermost plateau of mushrooms.

Flat roofs encourage condensation, and a microclimate for contamination growth. If the cultivator has no choice but to work with a flat roof, I recommend the installment of lengths of 6- to 12-inch diameter drain-field pipe, perforated every 2 feet with 1-inch holes, down the length of each growing room at the junction of the wall and the ceiling. By installing a "T" midspan, and locating a downward flowing duct fan at the base of the "T," you can draw air into the holes. This scheme will eliminate the dead-air pockets that form along the corner of the wall and ceiling. With fine-tuning, entrainment of the air can be greatly improved. (Entrainment can be measured by a "smoke or steam" test, observing the swirling air patterns.)

Floors. Floors should be cement, painted with a USDA-approved, dairy-grade cement, and sloped to a central drain. Channel-like drains used in dairies work well, although they need not be wider than 6 inches. Before the entrance to the drain-field, a

A Texas Reishi production facility. With modified hoop-framed greenhouses covered with an open-sided, metal roofed super-structure, growing rooms could be constructed at low cost.

An insulated growing room constructed on the interior and exterior with weatherproof metal siding.

screened basket fashioned out of metal mesh prevents clogging. This basket should be easily removed for daily cleaning. Once every several days, a cup of bleach is washed into the drain to discourage flies from breeding.

Many cultivators install a footbath prior to each growing room to disinfect footwear. (Shoes are a major vector of contamination of soil-borne diseases into the growing rooms.) These footbaths are built into the cement slab before pouring so that a drain can be installed. A 2 foot × 3 foot × 2-inch recessed footbath is ideal. The drain is capped and filled water. Bleach (chlorine) is added as a disinfectant. A plastic, metal, or sponge-like grate helps remove debris from the feet and enhances the effectiveness of the footbath.

Rate of air exchange. Air exchanges control the availability of fresh oxygen and the purging of carbon dioxide from the respiring mushroom mycelium. High air exchange rates may adversely affect humidity, especially prior to and at primordia formation when aerial mycelium abounds. Should aerial mycelial die back, or "pan," potential yields are substantially depressed. On the other hand, to prevent malformation of the fruitbody, bacterial blotch, and mold infestation, the movement of air—*turbidity*—is a substantial factor in preventing disease vectors.

The need for adequate air exchange is a direct reflection of the species being grown, its rate of metabolism (as measured by CO_2 generation), and the "density of fill." Fast-growing, tropical strains generate more CO_2 than cold-weather strains due to their higher rate of metabolism. The density of fill is the fraction of space occupied by substrate versus the total volume of the growing room. Button mushrooms growers often fill up to $1/4$ of the growing room space with substrate. This high rate of fill is impractical with most gourmet mushrooms. I recommend filling the growing rooms to no more than $1/6$th, and preferably $1/8$th, of capacity.

At 1,000 cubic feet per minute (cfm) of free air delivery, the air in an *empty* 10,000-cubic-foot room will be exchanged every 10 minutes, equivalent to 6 air exchanges per hour. This rate of air exchange is near the minimum required for gourmet mushroom cultivation. At 2,000 cfm, this same room will be exchanged every 5 minutes, or 12 air exchanges per hour. *Another way to calculate the needed CFM is to assume 2 CFM/square foot in each growing room.* I recommend designing growing rooms that can operate within this rate of air exchange, i.e., between 6 and 12 air exchanges per hour. The actual rate of air exchange will be lower, affected by the rate of fill and limited by the avenues of exhaust. The growing rooms should always remain positive-pressurized to limit contamination vectors from the outside. A strip of cloth or plastic above the doorjamb works well as a simple visual indicator of positive pressurization.

A 400–600 cfm thermal exhaust fan is recommended for a growing room of the above-described dimensions. This fan is typically located at the apex of the growing room, opposite the incoming air. A thermostat, preset by the cultivator to skim off excess heat, activates this fan. Another fan, which I call a "vortex fan" having a 200–400 cfm capacity, is located below the thermal exhaust fan, usually at head level, above the exit door. The vortex fan helps enhance the cyclonic entrainment of the air as it moves down the growing room. Both fans should be covered, *from the inside,* with a bugproof, non-mildewing cloth. This cloth will prevent the entry of insects when the fans are not in operation. Furthermore, the thermal exhaust and vortex fan should have louver shutters that close when not in use.

This is but one configuration of a growing room. Ideally, the growing room environment rotates as a giant wind tunnel, providing a homogeneously mixed atmosphere. Simplicity of design makes operation easy. Each growing room should be independently controlled so crops can be cycled and managed according to their stage of development.

Filtration of fresh air supply. Fresh air is brought in from the outside and passed through a series of filters. The growing rooms do not require the degree of filtration that is necessary for the laboratory. Since the growing room will, at times, have full air exchanges of 4 to 10 times per hour, the filters must have sufficient carrying capacity. Air is first filtered through a standard Class 2 prefilter. These filters are relatively coarse, filtering particles down to 10 microns with 30% efficiency. Prefilters are disposable and should be replaced regularly, in most cases every 1 to 3 months. The next filter is usually electrostatic. Electrostatic filters vary substantially in their operating capacities and airflow parameters. Typically, particulates are filtered down to 1 micron with 95% efficiency. Electrostatic filters can be

removed, periodically, for cleaning with a soapy solution. Their performance declines as dust load increases. For most growing rooms of 10,000–20,000 cubic feet, a 25 × 20 × 6-inch electrostatic filter suffices when combined with a fan sending 1,000–2,000 cubic feet per minute airstream into each growing room.

Filtration of recirculated air. Recirculated air from a growing room is relatively free of airborne particulates during the colonization phase. When the cropping cycle begins, the air becomes thick with mushroom spores. (This is especially the case with Oyster mushrooms and much less so with Button mushroom cultivation.) I have seen the spore load of Oyster mushrooms become so thick as to literally stop the rotation of high-volume cfm fans!

The design of an air system should allow partial to full recirculation of the air within the growing room. Usually, a recirculation duct is centrally located directly below the incoming air. A damper door controls the degree of recirculation. If the recirculated air is passed through filters, these filters will quickly clog with mushroom spores, and airflow will radically decline. If electing to use filters, they should be changed every day during the cropping cycle. A simple way of cleaning the recirculated air is to position mist nozzles in the recirculation ductwork. The air will be largely rinsed clean of their spore load from the spray of water. Ideally, these downward flowing nozzles are located directly above a drain. Having spray nozzles located below the inside roofline and misting downwards will also facilitate the downward flow of spores to the floor. This concept can have many permutations.

Humidification. With the 6 to 12 air exchanges per hour required during the primordia formation period, full humidification within the growing room is difficult if drawing in dry outside air. This problem is solved through the conditioning of outside air in an intermediate chamber called a *preconditioning plenum.* The preconditioning plenum is usually located outside of the growing room. Its purpose to elevate humidity and alter temperature to levels adjustable by the in-room environmental systems.

One preconditioning plenum can supply several growing rooms, if properly designed.

In cold climates or during the cold winter months, the preconditioning plenum can be largely humidified with steam. Steam provides both moisture and heat. Thermostats located in the preconditioning plenum and/or growing room activate solenoid valves on live steam lines coming from an on-duty boiler. Steam is sent downstream to a square-shaped grid of interconnected pipes. Holes have been drilled to orient the flow of steam towards the center. The main air system blower pushes the steam from the preconditioning box into the growing rooms. Each growing room has its own high-volume axial fan that inflates the ducting and distributes the humidified and heated air. Since moisture will collect in the polyethylene ducting, provisions must be made for removing this condensate. The simplest solution is to slant the duct at a slight angle. The condensation can then drip directly into the channel drain running lengthwise down the center of the room.

When the temperature in the growing room exceeds prescribed levels, the thermostat will close the in-line solenoid valve. Humidity will fall. A humidistat sensing the humidity in the preconditioning plenum takes control, opening a separate solenoid valve, sending cold water downline, activating the mist nozzles. The cold water lines should be positioned in a fashion so that they are not damaged when the steam lines are in operation. The goal here is to elevate humidity to 75–80% RH. Independent humidifiers or mist nozzles located high in the growing rooms control the remaining 25% of the humidity required for primordia formation. Some companies offer systems that utilize compressed-air (100 to 400 psi) misting systems that emanate a fog-like cloud of humidity. With proper filtration and maintenance these systems work equally as well although typically are much more expensive to install.

Swamp coolers generate humidity *and* lower temperature of the incoming air. They work especially well when the outside air temperature is high and the humidity is low. These types of evaporative coolers can also be incorporated into the design of a preconditioning box. Heat exchangers must be carefully engineered for maximum effect. The cultivator is encouraged to consult a reputable HVAC (heating, ventilation, and air conditioning) specialist before installation of any of the above-mentioned systems. Often, independent systems with manual control overrides perform better, in the long run, than all-in-one packages. The cultivator must have alternatives for humidity should equipment fail for any reason. The value of one saved crop can easily offset the expense of a simple backup system.

Insect control. Flies are the bane of the mushroom cultivator. Once introduced, a single pregnant fly can give rise to hundreds of voracious offspring in a few weeks. (For a complete description of the flies and their life cycles, please consult *The Mushroom Cultivator,* 1983, by Stamets and Chilton.) Many models of insect traps are available. "Bug zappers" electrocute flies when they come in between two opposite-charged metal screens. Many of the flies endemic to mushroom culture are minute and pass, unaffected, between the electrified panels. I prefer circular bug lights that use an interior fan. A cone-shaped vortex is formed well beyond the light. The bugs are thrown into a plastic bag that allows for easy counting by the growing room manager. By attaching yellow sticky pads to the lights, the number of flies that can be caught greatly increases. Astute managers rely on the increasing numbers of flies as an indication of impending disaster. These lights should be positioned on or beside every door within the growing room, and especially in the rooms prior to the growing rooms. Many of the fly problems can be circumvented through some of the good practices described below.

Managing the Growing Rooms: Good Habits for the Personnel

Integral to the success of a mushroom farm is the daily management of the growing rooms. The environment within the growing room is constantly in a dynamic state of change. Events can quickly cascade, drastically, and sometimes inalterably, affecting the outcome of the crop. The daily activities of the personnel especially impact the quality of the crop. The head cultivator can perfectly execute his or her duties only to have employees unwittingly sabotage the crop. In many cases, a simple redirection of the sequence of activities can correct the problem. With consistently followed well-defined rules of conduct, the growing rooms can function to their fullest potential. I recommend establishing a daily and weekly schedule that defines the activities of the personnel. By following a calendar, the crew becomes self-organizing according to the days of the week.

Maintain personal hygiene. Shower every day. Wear newly laundered clothes. Avoid contact with molds, soils, pets, etc. Body odors are a result of growing colonies of bacteria. Those with poor personal hygiene threaten the stability of a farm. If employees cannot maintain good personal hygiene, then they must be fired.

Keep the property clean. Particular emphasis should be on removing any organic debris that accumulates directly after a production run or harvest. All excess materials should be placed in a specifically allocated, dry storage location. Waste materials should be composted a safe distance away from the production facility.

Isolate or minimize contact between growing room managers, pickers, and other workers. Growing room managers present the same problem doctors do in hospitals: They spread disease. All personnel should wash their hands several times a day and/or wear gloves. The growing room manager should point out contamination for workers to remove. The growing room manager, unless alone, should not handle contaminants. Furthermore, laboratory personnel and growing room personnel should not share the same lunchroom, a common site for the redistribution of contaminants.

Use footbaths. Step in disinfectant footbaths prior to going into each growing room. Change baths daily.

Frequently wash down growing rooms. The activity most affecting the maintenance of a growing room is its thorough washing down once in the morning and once in the evening. After spraying the ceilings, walls, and floors with water (untreated), direct any debris to the central drain channel, and collect and remove it. After a washing, the room takes on a freshness that is immediately apparent. *The growing room feels fresh.*

Record maximum/minimum/ambient temperature at the same time each day. Temperature profiling at different elevations is deemed necessary only in poorly insulated rooms. With turbulent air circulation, in a highly insulated environment, temperature stratification is minimized. Manual temperature readings should occur at the same time every day. If constant-duty chart recorders or computer sensing systems are in place, they should be reviewed daily and compared. Ambience of temperature within 3–5°F (2–3°C) is preferred.

Chart relative humidity. Relative humidity requirements change with each species, each day. Charting humidity can be a valuable tool in training the growing room manager to delicately balance the environment in favor of mushroom formation and development, but limiting the growth of molds. Wood-based substrates are particularly susceptible to green mold growth if humidity is too high directly after removal of the blocks from the soaking tank. For instance, Shiitake blocks benefit from a fluctuating humidity environment whereas Oyster mushrooms require sustained 90%+ humidity.

Dispose of contamination once a day. Once you have handled contamination, consider yourself contaminated. If removing *Trichoderma* (green mold) contaminated blocks of Shiitake, that person is then unworthy of any activity where susceptible substrates are handled. For many growers, this mandates that contaminated cultures be disposed at the end of the day. If contamination must be dealt with early in the workday, a person nonessential to the production stream should be the designated disposer of contaminated cultures.

APPENDIX 4

Resource Directory

The following resource directory is not an endorsement of any organization, company, individual, or author. The sole intent is to give the reader the broadest range of resources so that mushrooms can be grown successfully. An omission of any group is probably not intentional. If you know of any resource that should be listed, feel free to write to the author so that future editions of *Growing Gourmet and Medicinal Mushrooms* will include them.

Recommended Mushroom Field Guides

Cultivators must constantly return to nature for new strains. If mushrooms are not properly identified, the cultures will be inaccurately labeled. Once these cultures are passed on to other individuals, the misidentification may not be discovered for years, if ever. An excellent example of this is what happened with cultures of "*Pleurotus sajor-caju,*" most of which are in fact *P. pulmonarius*. (See the discussion of the taxonomy of *P. pulmonarius* on page 316.) Cultivators are encouraged to photograph the wild mushroom in fresh condition, retain a dried specimen, and make notes about its location and habitat. Without this data, accurately identifying the mushroom (and therefore the culture) will be difficult.

The following field guides are designed primarily to help amateurs in the field. Professional mycologists (although some are loathe to admit it) refer to these manuals also, particularly when they need a quick overview of the species complexes. Scientific monographs are ultimately used for confirming the identity of a mushroom to species. Amateurs should be fully aware that even professional mycologists make mistakes.

All that the Rain Promises and More by David Arora, 1991. Ten Speed Press, Berkeley, California.

The Audubon Society Field Guide to North American Mushrooms by Gary Lincoff, 1991. A. A. Knopf, New York.

A Field Guide to Southern Mushrooms by Nancy Smith Weber and Alexander H. Smith, 1985. University of Michigan Press, Ann Arbor, Michigan.

A Field Guide to Western Mushrooms by Alexander Smith, 1958. University of Michigan Press, Ann Arbor, Michigan.

Fungi of Japan by R. Imazeki et al. 1988. Yama-Kei Publishers, Tokyo, Japan.*

The Mushroom Book by Laessoe, DelConte, and Lincoff, 1996. DK Publishers, New York, New York.

The Mushroom Hunter's Field Guide by Alexander Smith and Nancy Smith Weber, 1980. University of Michigan Press, Ann Arbor, Michigan.

Mushrooms Demystified by David Arora, 1986. Ten Speed Press, Berkeley, California.

Mushroooms of Northeastern North America by Alan E. Bessette, Arleen R. Bessette, and David W. Fisher, 1997. Syracuse University Press, Syracuse, New York.

Mushrooms of the Adirondacks: A Field Guide by Alan E. Bessette, 1988. North Country Books, Utica, New York.

The New Savory Wild Mushroom by Margaret McKenny, Daniel E. Stuntz, ed. by Joseph Ammirati, 1987. University of Washington Press, Seattle, Washington.

---

*This field guide, although in Japanese, is probably one of the best, if not *the* best field guide published to date. The photography is *exceptional*. I highly recommend and admire this volume.

Mushroom Book Suppliers

Fungi Perfecti, LLC
(Field Guides, Cultivation Books, General Treatises)
P.O. Box 7634
Olympia, WA 98507
(360) 426-9292 FAX: (360) 426-9377
www.fungi.com

FS Book Company
(General Books)
P.O. Box 417457
Sacramento, CA 95841-7457

Lubrecht & Cramer, Ltd
(General and Erudite Monographs)
38 County Rte. 48
Forestburgh, NY 12777
(914) 794-8539

Mad River Press
(General, Publisher of Mycological Texts)
141 Carter Lane
Eureka, CA 95501
(707) 443-2947

Mushroom People
P.O. Box 220
Summertown, TN 38483-0220
(615) 964-2200

Mycologue Publications
331 Daleview Place
Waterloo, ON N2L 5M5
Canada

Raymond M. Sutton, Jr.
(Supplier of old and rare mushroom books)
Mycology Books
430 Main St.
Williamsburg, KY 40769

Annual Mushroom Festivals and Events

Many of the mycological societies stage annual mushroom exhibits. Here is a short list of some of the more notable annual mushroom exhibitions. Most are held in September or October unless otherwise indicated. Please contact them for more specific information.

Asheville Annual Labor Day Mushroom Foray

(Early September)
Nature Center, Gashes Cr. Rd.
Asheville, NC

Boyne City's Mushroom Festival

(Early May)
Boyne City Chamber of Commerce
Boyne City, MI
(616) 582-6222

Breitenbush Mushroom Weekend

(Late October)
Breitenbush Hot Springs Resort
Detroit, OR 97342
(541) 854-3314

Colorado Mycological Society Mushroom Show

(Mid-August)
P.O. Box 9621
Denver, CO 80209-0621

Gulf States Mycological Society Foray

(Mid-July)
University of Mississippi, Gulfport
c/o Dr. Bill Cibula and Anna Pleasanton
1,000 Adams St.
New Orleans, LA 70118

Lincoln County Mushroom Society

(Mid-October)
207 Hudson Loop
Toledo, OR 97391

Mount Pisgah Arboretum

Fall Festival and Annual Mushroom Show
(Mid-October)
Eugene, OR

Mycological Society of San Francisco

Annual Mushroom Show
P.O. Box 882163
San Francisco, CA 94188-2163

Oregon Mycological Society

Annual Fall Mushroom Show
(Mid-October)
c/o Maggie Rogers
1943 S.E. Locust
Portland, OR 97214

Puget Sound Mycological Society Annual Mushroom Exhibit

(Mid-October)
University of Washington
Urban Horticulture
GF-15
Seattle, WA 98195-0001

Richmond's Annual Mushroom Festival

(1st weekend of May)
c/o Richmond Chamber of Commerce
108 West Main
Richmond, MO 64085

The Santa Cruz Fungus Fair

Fungus Federation of Santa Cruz
1305 E. Cliff St.
Santa Cruz, CA 95062-3722
+27-21-809 3483 FAX: +27-21-809 3491
adriaan@infruit.agric.za

Annual Mushroom
Festivals and Events (continued)

South Africa Wild Mushroom Conference and Festival

c/o Dr. W. Adriaan Smit
Mushroom Research Centre
ARC Infruitec-Nietvoorbij
Private Bag X5013
Stellenbosch 7599
South Africa

Spokane Mycological Society's Priest Lake Foray

(Late September)
P.O. 2791
Spokane, WA 99220-2791

Vancouver Mycological Society Mushroom Show

(Mid-October)
c/o Van Deusen Botanical Gardens
Paul Kroeger
395 E. 40th Ave.
Vancouver, BC V5W 1M1
Canada
(604) 322-0074

Wild Mushrooms: Telluride

(Late August)
Fungophile, Inc.
P.O. Box 480503
Denver, CO 80248-0503
(303) 296-9359

Mushroom Cultivation Seminars and Training Centers

Arunyik Mushroom Center

Mr. Satit Thaithatgoon
P.O. Box 1
Bangkok 10162
Thailand
FAX: 662-441-9246

Mushroom Experimental Station

P.O. Box 6042
5960 AA Horst
The Netherlands

Forest Resource Center

1991 Brightsdale Rd.
Rte. 2 Box 156 A
Lanesboro, MN 55949
(507) 467-2437

International Mushroom School

Mushroom Information
OK Press S.R.L.
Via Poggiorenatico, 2
40016 San Giorgio di Piano (Bo)
Italy
051/893-768

Mushroom Society of India

National Centre for Mushroom Research and Training
Chambaghat, Solan 173213 (H.P.)
India

Pennsylvania State College Short Course

Dept. of Plant Pathology
College of Agriculture
The Pennsylvania State University
210 Buckhout Laboratory
University Park, PA 16802-4507

The Stamets Seminars

c/o Fungi Perfecti, LLC
P.O. Box 7634
Olympia, WA 98507
(360) 426-9292 FAX: (360) 426-9377
www.fungi.com

Mushroom Study Tours and Adventures

David Arora's International Mushroom Adventures

c/o David Arora
343 Pacheco Ave.
Santa Cruz, CA 95062
(831) 425-0188

Adirondack Mushroom Weekends

c/o Alan Bessette
Utica College of Syracruse University
1600 Burrstone Rd.
Utica, NY 13502
(315) 792-3132

Gerry Miller's Amazon Forays

Box 126
East Haddam, CT 06423
(203) 873-8286

Fungophile's Mushroom Study Tours

w/ Gary Lincoff and Dr. Emanuel Salzman
P.O. Box 480503
Denver, CO 80248

North American Mycological Association

10 Lynne Brook Pl.
Charleston, WV 25312-9521
www.namyco.org

International Mushroom Associations

Associazione Micologica Bresadola

Via S. Croce 6
C.P. 396, 38100
Trento, Italy

International Mycological Association

CAB International Mycological Institute
Ferry Lane, Kew
Surrey TW9 3AF
United Kingdom
www.lsb380.plbio.lsu.edu/ima/index.html

International Society for Mushroom Science (ISMS)

Executive Secretary, ISMS
196 Rugby Road
Leamington Spa
Warwickshire CV32 6DU
United Kingdom
www.hri.ac.uk/isms/

International Network of Organic Medicinal Mushroom Growers

www.fungi.net

North American Mushroom Societies and Associations

Arkansas Mycological Society

4715 W. Hensley Rd.
Hensley, AR 72065

Alaska Mycological Society

P.O. Box 2526
Homer, AK 99603-2526

Albion Mushroom Club

Whitehouse Nature Center
Albion, MI 49224

Arkansas Mycological Society

5115 S. Main St.
Pine Bluff, AR 71601-7452

Asheville Mushroom Club

Nature Center
Gashes Cr. Rd.
Asheville, NC 28805

Berkshire Mycological Society

Pleasant Valley Sanctuary
Lenox, MA 01240

North American Mushroom Societies and Associations (continued)

Blue Ridge Mushroom Club
P.O. Box 2032
North Wilkesboro, NC 28659-2032

Boston Mycological Club
855 Commonwealth Ave.
Newton Centre, MA 02159

Central New York Mycological Society
343 Randolph St.
Syracruse, NY 13205-2357

Cercle des Mycologues de Montreal
4101 E. Rue Sherbrooke, #125
Montréal, P.Q.
Canada

Cercle des Mycologues de Rimouski
University of Quebec
Rimouski, P.Q.
Canada

Cercle Mycologues de Quebec
Pav. Comtois
Univ. Laval
Ste. Foy, P.Q.
Canada

Cercle des Mycologues du Saguenay
438 Rue Perrault
Chicoutimi, P.Q.
Canada

Chibougamau Mycological Club
804 5e Rue
Chibougamau, P.Q.
Canada

Colorado Mycological Society
P.O. Box 9621
Denver, CO 80209

COMA
Connecticut-Westchester Mycological
Association
c/o Sandy Sheine
93 Old Mill River Road
Pound Ridge, NY 10576-1833

Connecticut Valley Mycological Association
c/o Marteka
Jobs Pond Rd.
Portland, CT 06480

Fungus Federation of Santa Cruz
1305 E. Cliff Dr.
Santa Cruz, CA 95062-3722

Gulf States Mycological Society
1000 Adams St.
New Orleans, LA 70118-3540

Humboldt Bay Mycological Society
P.O. Box 4419
Arcata, CA 95521-1419

Illinois Mycological Association
4020 Amelia Ave.
Lysons, IL 60534

Kaw Valley Mycological Society
601 Mississippi St.
Lawrence, KS 66044

Kitsap Peninsula Mycological Society
P.O. Box 265
Bremerton, WA 98310-0054

Lewis County Mycological Society
196 Taylor Rd. S.
Chehalis, WA 98532

North American Mushroom Societies and Associations (continued)

Lincoln County Mycological Society
207 Hudson Loop
Toledo, OR 97391

Los Angeles Mycological Society
Biology Dept.
5151 State University Drive
Los Angeles, CA 90003

Lower East Shore Mushroom Club
R.R. 1 Box 94B
Princess Anne, MD 21853-9801

Maine Mycological Society
c/o Sam Ristich
81 Sligo Rd.
N. Yarmouth, ME 04021

Michigan Mushroom Hunters Association
15223 Marl Dr.
Linden, MI 48451

Mid Hudson Mycological Association
43 South St.
Highland, NY 12528

Minnesota Mycological Society
7637 East River Rd.
Fridley, MN 55432

Missouri Mycological Society
2888 Ossenfort Rd.
Glencoe, MO 63038

Montshire Mycological Club
c/o Carlson
P.O. Box
Sunapee, NH 03782

Mount Shasta Mycological Society
623 Pony Trail Rd.
Mount Shasta, CA 96067

Mycological Association of Washington
12200 Remington Drive
Silver Springs, MD 20902

Mycological Society of America
Linda Harwick
P.O. Box 1897
Lawrence, KS 66044-8897
www.erin.utoronto.ca/~w3msa/

Mycological Society of San Francisco
P.O. Box 882163
San Francisco, CA 94188-2163

Mycological Society of Toronto
4 Swallow Court
New York, ON
Canada

Mycological Society of Vancouver
c/o Tamblin
403 3rd St.
New Westminster, BC
Canada

New Jersey Mycological Society
1187 Millstone River Rd.
Somerville, NJ 08876

New Hampshire Mycological Society
84 Cannongate III
Nashua, NH 03063

New Mexico Mycological Society
1511 Marble Ave. NW
Albuquerque, NM 87104

New York Mycological Society
c/o Gary Lincoff
157 W. 95th St.
New York, NY 10025

North American Mushroom Societies and Associations (continued)

North American Mycological Association
10 Lynne Brook Pl.
Charleston, WV 25312-9521

North American Truffling Society
P.O. Box 296
Corvallis, OR 97339

North Idaho Mycological Association
E. 2830 Marine Drive
Post Falls, ID 83854

Northwestern Mushroomers Association
c/o Marlowe
831 Mason St.
Bellingham, WA 98225-5715

Northwestern Wisconsin Mycological Society
311 Ash St.
Frederic, WI 54837

Nutmeg Mycological Society
c/o Kovak
191 Mile Creek Rd.
Old Lyme, CT 06371

Ohio Mycological Society
10489 Barchester Dr.
Concord, OH 44077

Olympic Mountain Mycological Society
P.O. Box 720
Forks, WA 98331-0720

Oregon Coast Mycological Society
P.O. Box 1590
Florence, OR 97439

Oregon Mycological Society
14605 S.W. 92nd Ave.
Tigard, OR 97224

Pacific Northwest Key Council
1943 S.E. Locust Ave.
Portland, OR 97201-2250

Parkside Mycological Club
5219 85th St.
Kenosha, WI 53142-2209

Prairie States Mushroom Club
310 Central Drive
Pella, IA 50219-1901

Puget Sound Mycological Society
Center for Urban Horticulture
GF-15 University of Washington
Seattle, WA 98195

Rochester Area Mycological Association
54 Roosevelt Rd.
Rochester, NY 14618-2933

Snohomish Mycological Society
Box 2822, Claremont Station
Everett, WA 98203

Sonoma County Mycological Association
1218 Bennett Lane
Calistoga, CA 94515

South Idaho Mycological Association
P.O. Box 843
Boise, ID 83701

South Sound Mushroom Club
c/o Ralph and Bonnie Hayford
6439 32nd Ave.
Olympia, WA 98507

Spokane Mushroom Club
N. 2601 Barder Rd. Sp. 5
Otis Orchards, WA 99027

North American Mushroom Societies and Associations (continued)

Tacoma Mushroom Club
P.O. Box 99577
Tacoma, WA 98499-0577

Texas Mycological Society
7445 Dillon
Houston, TX 77061

Triangle Area Mushroom Club
P.O. Box 61061
Durham, NC 27705

Vancouver Mycological Society
403 Third St.
New Westminster, BC
Canada

Wenatchee Valley Mushroom Society
287 N. Iowa Ave.
East Wenatchee, WA 98802

West Michigan Mycological Society
923 E. Ludington Ave.
Ludington, MI 49431

Willamette Valley Mushroom Society
2610 East Nob Hill S.E.
Salem, OR 97303

Wisconsin Mycological Society
800 W. Wells St. Rm. 614
Milwaukee, WI 53233

For a continually updated list of mycological societies, go to:
*www.//clubs.yahoo.com/clubs/
mushroomhuntersabide*

Mushroom Growers Associations in the United States

Many of the previously listed mycological societies have established "Cultivation Committees," which organize members interested in mushroom cultivation. The following organizations are independent of any mycological society.

Alabama Shiitake Growers Association
c/o Hosea Nall
Cooperative Extension Service
Alabama A & M University
819 Cooke Ave.
Normal, AL 35762

American Mushroom Institute
One Massachusetts Ave N.W., Suite 800
Washington, DC 20001-1401

Appalachian Mushroom Growers Associations
c/o Margey Cook
Rt. 1, Box BYY
Haywood, VA 22722

Canadian Mushroom Growers Association
310-1101
Prince of Wales Drive
Ottawa, ON K2C 3W7
Canada

Carolina Exotic Mushroom Association
c/o Ellie Litts
P.O. Box 356
Hodges, SC 29653

Florida Mushroom Growers Association
c/o Charlie Tarjan
3426 S.W. 75th St.
Gainesville, FL 32607

The Mushroom Council
11875 Dublin Blvd.
Suite D-21
Dublin, CA 94568

Sources for Mushroom Cultures

American Type Culture Collection

10801 University Blvd.
Manassus, VA 20110-2209

Arunyik Mushroom Center

P.O. Box 1 Nongkhaem
Bangkok 10162
Thailand

Centre for Land and Biological Research

Research Branch, Agriculture Canada
Ottawa, ON K1A OC6
Canada

Collection of Higher Basidiomycetes Cultures

Komarov Botanical Institute
Academy of Sciences
Leningrad, Russia

Culture Collection of Fungi

Commonwealth Mycological Institute
Kew, United Kingdom

Culture Collection of Basidiomycetes

Department of Experimental Mycology
Videnska 1083, 142 20 Praha 4 Krc.
Czech Republic

Forest Research Institute

Private Bag
Rotorua, New Zealand

Fungi Perfecti, LLC

Stamets Culture Collection
P.O. Box 7634
Olympia, WA 98507
USA
(360) 426-9292
www.fungi.com

Instituto de Ecologia, A.C.

AP 63 K.M. 2.5 Antigua Carretera
A Coatepec
Xalapa, Veracruz
Mexico

INRA

Station de Recherches sur les Champignons
Villenave D'Ornon
France

Institute for Fermentation

17-85 Juso-honmachi 2-chome
Yodagawa-ku
Osaka 532
Japan

Japanese Federation of Culture Collections

Tokyo Agriculture College
Sakuragaoka 1-1-1
Setagaya-ku, Tokyo 156
Japan

Mushroom Research Centre

ARC Infruitec-Nietvoorbij
Privage Bag X5013
Stellenbosch 7599
South Africa
+27-21-809 3483 FAX: +27-21-809 3491
adriaan@infruit.agric.za

Mushroom Research Centre

Huettenallee 235
4150 Krefeld
Germany

National Type Culture Collection

Forest Pathology Branch
Forest Research Institute and Colleges
Dehra Dun
India

Sources for Mushroom Cultures
(continued)

Pennsylvania State University
Mushroom Research Station
Buckhout Laboratory
University Park, PA 16802
USA

Somycel
Centre de Recherches sur les Champignons
Langeais
France

Tottori Mycological Institute
Furukage 211
Tottori, 689-11
Japan

Sources for Mushroom Spawn

Alpha Spawn
R.D. #1, Box 34
Avondale, PA 19311

Amycel, Inc.
553 Mission Vineyard Rd.
P.O. Box 1260
San Juan Bautista, CA 95045

Field and Forest Products
N 3296 Kozuzek Rd.
Peshtigo, WI 54157

Fungi Perfecti, LLC
P.O. Box 7634
Olympia, WA 98507
(360) 426-9292 FAX (360) 426-9377
www.fungi.com

Gourmet Mushrooms
P.O. Box 515
Graton, CA 95444

International Spawn Laboratory
Beechmount, Navan
Meath, Ireland

Lambert Spawn
P.O. Box 407
Coatesville, PA 19320
(610) 384-5031

MushroomPeople
(Shiitake Spawn and Inoculation Equipment, Books)
P.O. Box 220
Summertown, TN 38483-0220
(615) 964-2200

Northwest Mycological Consultants
702 N.W. 4th St.
Corvallis, OR 97330
(541) 753-8198

Rainforest Mushroom Spawn
P.O. Box 1793
Gibsons, BC V0N 1V0
Canada
(604) 886-7799

J.B. Swayne Spawn
P.O. Box 618
Kennett Square, PA 19348
(610) 444-0888

Sun Shiitake Brazilian Spawn Company
www.sunshiitake.com.br

Sylvan Spawn
87 Lakes Blvd.
Dayton, NV 89403

International Mushroom Growers Associations and Sources for Marketing Information

American Mushroom Institute

One Massachusetts Ave. N.W., Suite 800
Washington, DC 20001
(215) 388-7806

Canadian Mushroom Growers Association

310-1101 Prince of Wales Drive
Ottawa, ON K2C 3W7
Canada

Hungarian Mushroom Growers Association

Baross Gabor utca 15
H-1165 Budapest
Hungary

International Mushroom Growers Association

1st Khoroshevski Pr., 3A Room 417
Moscow 125284
Russia

Mushroom Growers Association of Great Britain and N. Ireland

Agriculture House
Knightsbridge, London SWIX 7NJ
United Kingdom

National Agricultural Statistics Service

United States Dept. of Agriculture
ERS/NASS "Mushrooms" (National marketing trends)
P.O. Box 1608
Rockville, MD 20849-1608
(800) 999-6779

National Medicinal Mushroom Growers Organization

2-33-5 Nihonbashi Hama-cho, Chuo-ku
Tokyo 103-0007
Japan

Mushroom Newsletters and Journals

Fungal Diversity

An International Journal of Mycology
The Centre for Research in Fungal Diversity
Department of Ecology and Biodiversity
Kadoorie Biological Sciences Building
The University of Hong Kong
Pokfulam Road, Hong Kong S.A.R.
China

Fungiflora A/S

P.O. Box 95
Blindern
N-0314 Oslo 3
Norway

International Journal of Medicinal Mushrooms

Solomon P. Wasser
Begell House
New York, NY

Mushroom Farming and Life Magazine

Han-young Lee
4th Floor, Haeng-oon Building
150-5 Pyungchang-dong Chongno-gu
Seoul, Korea 110-012
www.mushroom98.co.kr

McIlvainea, Journal of Amateur Mycology and The Mycophile

North American Mycological Association
10 Lynne Brook Pl.
Charleston, WV 25312-9521

The Mushroom Growers' Newsletter

c/o The Mushroom Company
P.O. Box 5065
Klamath Falls, OR 97601

Mushroom Information

La Rivista del Fungicoltore Moderno
40016 S. Giorgio di Plano (B0)
Postale Grupo III/70
Bologna, Italy

Mushroom Newsletters and Journals
(continued)

Mushroom, The Journal
Box 3156
Moscow, ID 83843

Mushroom Journal of the Tropics
The International Mushroom Society for the
Tropics
c/o Department of Botany
Chinese University of Hong Kong
Shatin, New Territories, Hong Kong

Mushroom News
American Mushroom Institute
907 East Baltimore Pike
Kennett Square, PA 193587
(215) 388-7806

Mushroom Research
International Journal of Research and
Development
National Centre for Mushroom Research and
Training
Chambaghat, Solan 173 213 (HP)
India

Mycologia, Official Publication of the
Mycological Society of America
The New York Botanical Garden
Bronx, NY 10458

Mycological Research
Cambridge University Press
North American Branch
40 West 20th St.
New York, NY 10011-4211

Mycotaxon
P.O. Box 264
Ithaca, NY 14850

hiitake News
Forest Resource Center
Rt. 2 Box 156 A
Lanesboro, MN 55949

Mushroom Museums

The Mori Mushroom Institute
Kiryu-shi Gumma 376
Kiryu, Japan
0277-22-0591

Mushroom Museum of Holland
S.O.M./Tweede Walstrat 99
6511 Lr Nijmegen
Nijemgen, 6511 Holland

The Mushroom Museum at Phillips Place
909 E. Baltimore Pike
Kennett Square, PA 19348 USA

Sources for Medicinal Mushroom Products

Fungi Perfecti, LLC
P.O. Box 7634
Olympia, WA 98507
(360) 426-9292; (800) 780-9126
www.fungi.com

Maitake, Inc.
P.O. Box 1354
6 Aster Court
Paramus, NJ 07653
(800) 747-7418

Sources for Medicinal Mushroom Products (continued)

Maizteca Gourmet Mushrooms

(Cultured Cuitlacoche)
160 Amboy Court
Athens, GA 30605
(706) 543-1629

North American Reishi

P.O. Box 1780
Gibsons, BC VON 1VO
Canada
(604) 886-7799

Organotech

7960 Cagnon Road
San Antonio, TX 78252-2202

Mycological Resources on Internet

Cornell WWW Virtual Library for Mycologists

www.muse.bio.cornell.edu/taxonomy/ fungi.htlml

Econet

www.myco.html

The Edible Mycorrhizal Mushroom Research Group (The Chantarelle Project)

www.mykopat.slu.se/mycorrhiza/ kantarellfiler/texter/home.htm

Enviromarket Watch—Markets for Mushrooms in Europe

www.agrar.com

Fungus

www.mtjeff.com/fungi

International Mycological Network

www.fungi.net

The Loss of Mushroom Diversity and Species Extinction in Europe

www.people.cornell.edu/pages/p118/end.html

The Mushroom Genome and Mycodiversity Preservation Project

www.mycodiversity.org

The Mushroom Growers Newsletter

www.mushroomcompany.com

Mycelium

www.econet.apc.org/mushroom/welco.html

MycoInfo

www.mycoinfo.com/

Mycological Society of America: Inoculum Newsletter

www.erin.utoronto.ca/~w3msa

Mycology Resources-Cornell University

www.keil.ukans.edu/~fungi/MycoPage
www.inf.unitn.it/~mflorian/mycopage.html

Paul Stamets's Home Pages:

www.fungi.com
www.fungi.net

Sun Shiitake Brazilian Spawn Company

www.sunshiitake.com.br

United Nations Food and Agricultural Organization

www.usda.mannlib.cornell.edu/data-sets/ specialty/94003

USDA National Agriculture Statistic Service

www.usda.gov/nass/

United States Dept. of Agriculture Market Reports

www.usda.mannlib.cornell.edu/reports/nassr/ other/zmu-bb/

Analysis of Basic Materials Used in Substrate Preparation

DRY ROUGHAGES OF FIBROUS MATERIALS

Material	Total dry matter Per ct.	Protein Per ct.	Fat Per ct.	Fiber Per ct.	N-free extract Per ct.	Total minerals Per ct.	Calcium Per ct.	Phos-phorus Per ct.	Nitro-gen Per ct.	Potas-sium Per ct.
Alfalfa hay, all analyses	90.5	14.8	2.0	28.9	36.6	8.2	1.47	0.24	2.37	2.05
Alfalfa hay, very leafy (less than 25% fiber)	90.5	17.2	2.6	22.6	39.4	8.7	1.73	0.25	2.75	2.01
Alfalfa hay, leafy (25-28% fiber)	90.5	15.8	2.2	27.4	36.6	8.5	1.50	0.24	2.53	2.01
Alfalfa hay, stemmy (over 34% fiber)	90.5	12.1	1.4	36.0	33.4	7.6	1.10	0.18	1.94	1.68
Alfalfa hay, before bloom	90.5	19.0	2.7	22.6	36.7	9.5	2.22	0.33	3.04	2.14
Alfalfa hay, past bloom	90.5	12.8	2.1	31.9	36.2	7.5	—	—	2.05	—
Alfalfa hay, brown	87.9	17.3	1.6	24.5	35.1	9.4	1.37	0.26	2.77	—
Alfalfa hay, black	83.1	17.5	1.5	29.1	25.3	9.7	—	—	2.80	—
Alfalfa leaf meal	92.3	21.2	2.8	16.6	39.7	12.0	1.69	0.25	3.39	—
Alfalfa leaves	90.5	22.3	3.0	14.2	40.5	10.5	2.22	0.24	3.57	2.06
Alfalfa meal	92.7	16.1	2.2	27.1	38.2	9.1	1.32	0.19	2.58	1.91
Alfalfa stem meal	91.0	11.5	1.3	36.3	34.8	7.1	—	—	1.84	—
Alfalfa straw	92.6	8.8	1.5	40.4	35.1	6.8	—	0.13	1.41	—
Alfalfa and bromegrass hay	89.3	12.4	2.0	28.6	38.1	8.2	0.74	0.24	1.98	2.18
Alfalfa and timothy hay	89.8	11.1	2.2	29.5	40.3	6.7	0.81	0.21	1.78	1.78
Alfilaria, dry (Erodium cicutarium)	89.2	10.9	2.9	23.4	40.2	11.8	1.57	0.41	1.74	—
Alfilaria, dry, mature	89.0	3.5	1.5	31.4	44.1	8.5	—	—	0.56	—
Atlas sorghum stover	85.0	4.0	2.0	27.9	44.2	6.9	0.34	0.09	0.64	—
Barley hay	90.8	7.3	2.0	25.4	49.3	6.8	0.26	0.23	1.17	1.35
Barley straw	90.0	3.7	1.6	37.7	41.0	6.0	0.32	0.11	0.59	1.33
Bean hay, mung	90.3	9.8	2.2	24.0	46.6	7.7	—	—	1.57	—
Bean hay, tepary	90.0	17.1	2.9	24.8	34.7	10.5	—	—	2.74	—
Bean pods, field, dry	91.8	7.1	1.0	34.8	45.0	3.9	0.78	0.10	1.14	2.02
Bean straw, field	89.1	6.1	1.4	40.1	34.1	7.4	1.67	0.13	0.98	1.02
Beggarweed hay	90.9	15.2	2.3	28.4	37.2	7.8	1.05	0.27	2.43	2.32
Bent grass hay, Colonial	88.5	6.6	3.0	29.5	42.8	6.6	—	0.18	1.06	1.42
Bermuda grass hay	90.6	7.2	1.8	25.9	48.7	7.0	0.37	0.19	1.15	1.42
Bermuda grass hay, poor	90.0	5.8	0.9	38.8	37.7	6.8	—	—	0.93	—
Berseem hay, or Egyptian clover	91.7	13.4	2.7	21.0	42.7	11.9	3.27	0.28	2.14	2.05
Birdsfoot trefoil hay	90.5	13.8	2.1	27.5	41.2	5.9	1.13	0.22	2.35	1.52
Black grass hay (Juncus Gerardi)	89.7	7.5	2.5	25.1	47.3	7.3	—	0.09	1.20	1.56
Bluegrass hay, Canada	89.3	6.6	2.3	28.2	46.4	5.8	—	0.20	1.06	1.94

Material	Total dry matter Per ct.	Protein Per ct.	Fat Per ct.	Fiber Per ct.	N-free extract Per ct.	Total minerals Per ct.	Calcium Per ct.	Phos- phorus Per ct.	Nitro- gen Per ct.	Potas- sium Per ct.
Bluegrass hay, Kentucky, all analyses	89.4	8.2	2.8	29.8	42.1	6.5	0.46	0.32	1.31	1.73
Bluegrass hay, Kentucky, in seed	87.3	5.5	2.5	31.0	41.9	6.4	0.23	0.20	0.88	1.48
Bluegrass hay, native western	91.9	11.2	3.0	29.8	39.9	8.0	—	—	1.79	—
Bluejoint hay *(Calamagrostis Canadensis)*	88.5	7.2	2.3	32.9	39.6	6.5	—	—	1.15	—
Bluestem hay *(Andropogon,* spp.)	86.6	5.4	2.2	30.2	43.4	5.4	—	—	0.86	—
Bromegrass hay, all analyses	88.1	9.9	2.1	28.4	39.5	8.2	0.20	0.28	1.58	2.35
Bromegrass hay, before bloom	89.0	14.5	2.3	24.6	37.9	9.7	—	—	2.32	—
Broom corn stover	90.6	3.9	1.8	36.8	42.4	5.7	—	—	0.62	—
Buckwheat hulls	88.6	3.0	1.0	42.9	40.1	1.6	0.26	0.02	0.48	0.27
Buckwheat straw	88.6	4.3	1.0	36.2	38.8	8.3	1.24	0.04	0.69	2.00
Buffalo grass hay *(Bulbilis dactyloides)*	88.7	6.8	1.8	23.8	46.2	10.1	0.70	0.13	1.09	1.36
Bunchgrass hay, misc. varieties	91.7	5.8	2.0	30.4	44.1	9.4	—	—	0.93	—
Carpet grass hay	92.1	7.0	2.2	31.8	40.9	10.2	—	—	1.12	—
Cat-tail, or tule hay *(Typha angustifolia)*	90.8	5.8	1.7	30.8	44.3	8.2	—	—	0.93	—
Cereals, young, dehydrated	92.8	24.5	4.7	16.1	33.1	14.4	0.66	0.46	3.92	—
Chess, or cheat hay *(Bromus,* spp.)	91.7	6.9	2.1	29.2	46.1	7.4	0.29	0.25	1.10	1.47
Clover hay, alsike, all analyses	88.9	12.1	2.1	27.0	39.9	7.8	1.15	0.23	1.94	2.44
Clover hay, alsike, in bloom	89.0	13.4	3.2	26.9	37.7	7.8	1.32	0.25	2.14	2.27
Clover hay, Alyce	89.0	10.9	1.6	35.4	35.5	5.6	—	—	1.74	—
Clover hay, bur	92.1	18.4	2.9	22.9	37.8	10.1	1.32	0.45	2.94	2.96
Clover hay, crimson	89.5	14.2	2.2	27.4	37.0	8.7	1.23	0.24	2.27	2.79
Clover hay, Ladino	88.0	19.4	3.2	20.7	34.9	9.8	1.32	0.29	3.10	2.78
Clover, Ladino, and grass hay	88.0	16.3	2.2	20.7	41.7	7.1	1.05	0.26	2.61	1.97
Clover hay, mammoth red	88.0	11.7	3.4	29.2	37.0	6.7	—	0.24	1.87	—
Clover hay, red, all analyses	88.1	11.8	2.6	27.2	40.1	6.4	1.35	0.19	1.89	1.43
Clover hay, red, leafy (less than 25% fiber)	88.1	13.4	3.1	23.6	40.8	7.2	—	—	2.14	—
Clover hay, red, stemmy (over 31% fiber)	88.2	10.1	2.1	34.1	36.0	5.9	0.99	0.15	1.62	1.77
Clover hay, red, before bloom	88.1	18.3	3.6	18.0	41.1	7.1	1.69	0.28	2.93	2.26
Clover hay, red, early to full bloom	88.1	12.5	3.5	26.1	39.7	6.3	1.47	0.22	2.00	1.73
Clover hay, red, second cutting	88.1	13.4	2.9	24.5	40.4	6.9	—	—	2.14	—
Clover hay, sweet, first year	91.8	16.5	2.5	24.6	39.7	8.5	1.37	0.26	2.64	1.57
Clover hay, sweet, second year	90.7	13.5	1.9	30.2	37.6	7.5	1.25	0.23	2.16	1.78
Clover hay, white	88.0	14.4	2.4	22.5	40.9	7.8	1.16	0.24	2.30	1.66
Clover leaves, sweet	92.2	26.6	3.2	9.5	41.9	11.0	—	—	4.26	—
Clover stems, sweet	92.7	10.6	1.1	38.0	35.6	7.4	—	—	1.70	—
Clover straw, crimson	87.7	7.5	1.5	38.8	32.9	7.0	—	—	1.20	—
Clover and mixed grassy, high in clover	89.7	9.6	2.7	28.8	42.2	6.2	0.90	0.19	1.54	1.46
Clover and timothy hay, 30 to 50% clover	88.1	8.6	2.2	30.3	41.2	5.8	0.68	0.20	1.38	1.47
Corn cobs, ground	90.4	2.3	0.4	32.1	54.0	1.6	—	0.02	0.37	0.37
Corn fodder, well-eared, very dry (from barn or in arid districts)	91.1	7.8	2.2	27.1	47.6	6.4	0.24	0.16	1.25	0.82
Corn fodder, high in water	60.7	4.8	1.4	16.7	34.2	3.6	0.16	0.11	0.77	0.55
Corn fodder, sweet corn	87.7	9.2	1.8	26.4	41.3	9.0	—	0.17	1.47	0.98
Corn husks, dried	85.0	3.4	0.9	28.2	49.6	2.9	0.15	0.12	0.54	0.55
Corn leaves, dried	82.8	7.7	1.9	23.9	42.6	6.7	0.29	0.10	1.23	0.36
Corn stalks, dried	82.8	4.7	1.5	28.0	43.3	5.3	0.25	0.09	0.75	0.50

Material	Total dry matter Per ct.	Protein Per ct.	Fat Per ct.	Fiber Per ct.	N-free extract Per ct.	Total minerals Per ct.	Calcium Per ct.	Phos-phorus Per ct.	Nitro-gen Per ct.	Potas-sium Per ct.
Corn stover (ears removed), very dry	90.6	5.9	1.6	30.8	4.65	5.8	0.29	0.05	0.94	0.67
Corn stover, high in water	59.0	3.9	1.0	20.1	30.2	3.8	0.19	0.04	0.62	0.44
Corn tops, dried	82.1	5.6	1.5	27.4	42.0	5.6	—	—	0.90	—
Cotton bolls, dried	90.8	8.7	2.4	30.8	42.0	6.9	0.61	0.09	1.39	3.18
Cotton leaves, dried	91.7	15.3	6.8	10.3	43.5	15.8	4.58	0.18	2.45	1.36
Cotton stems, dried	92.4	5.8	0.9	44.0	37.5	4.2	—	—	0.93	—
Cottonseed hulls	90.7	3.9	0.9	46.1	37.2	2.6	0.14	0.07	0.62	0.87
Cottonseed hull bran	91.0	3.4	0.9	37.2	46.7	2.8	—	—	0.54	—
Cowpea hay, all analyses	90.4	18.6	2.6	23.3	34.6	11.3	1.37	0.29	2.98	1.51
Cowpea hay, in bloom to early pod	89.9	18.1	3.2	21.8	36.7	10.1	—	—	2.90	—
Cowpea hay, ripe	90.0	10.1	2.5	29.2	41.8	6.4	—	—	1.62	—
Cowpea straw	91.5	6.8	1.2	44.5	33.6	5.4	—	—	1.09	—
Crabgrass hay	90.5	8.0	2.4	28.7	42.9	8.5	—	—	1.28	—
Durra fodder	89.9	6.4	2.8	24.1	51.4	5.2	—	—	1.02	—
Emmer hay	90.0	.97	2.0	32.8	36.4	9.1	—	—	1.55	—
Fescue hay, meadow	89.2	7.0	1.9	30.3	43.2	6.8	—	0.20	1.12	1.43
Fescue hay, native western *(Festuca,* spp.)	90.0	8.5	2.0	31.0	42.8	5.7	—	—	1.36	—
Feterita fodder, very dry	88.0	8.0	2.1	18.7	51.5	7.7	0.30	0.21	1.28	—
Feterita stover	86.3	5.2	1.7	29.2	41.9	8.3	—	—	0.83	—
Flat pea hay	92.3	22.7	3.2	27.7	32.0	6.7	—	0.30	3.63	2.02
Flax plant by product	91.9	6.4	2.1	44.4	33.1	5.9	—	—	1.02	—
Flax straw	92.8	7.2	3.2	42.5	32.9	7.0	0.48	0.07	1.15	0.73
Fowl meadow grass hay	87.4	8.7	2.3	29.7	39.5	7.2	—	—	1.39	—
Furze, dried	94.5	11.6	2.0	38.5	35.5	7.0	—	—	1.86	—
Gama grass hay *(Tripsacum dactyloides)*	88.2	6.7	1.8	30.4	43.1	6.2	—	—	1.07	—
Grama grass hay *(Bouteloua,* spp.)	89.8	5.8	1.6	28.9	45.6	7.9	0.34	0.18	0.93	—
Grass hay, mixed, eastern states, good quality	89.0	7.0	2.5	30.9	43.1	5.5	0.48	0.21	1.12	1.20
Grass hay, mixed, second cutting	89.0	12.3	3.3	24.8	41.7	6.9	0.79	0.31	1.97	1.15
Grass straw	85.0	4.5	2.0	35.0	37.8	5.7	—	—	0.72	—
Guar hay *(Cyamposis psoraloides)*	90.7	16.5	1.3	19.3	41.2	12.4	—	2.64	—	
Hegari fodder	86.0	6.2	1.7	18.1	52.5	7.5	0.27	0.16	0.99	—
Hegari stover	87.0	5.6	1.8	28.0	41.7	9.9	0.33	0.08	0.90	—
Hops, spent, dried	93.8	23.0	3.6	24.5	37.4	5.3	—	—	3.68	—
Horse bean hay	91.5	13.4	0.8	22.0	49.8	5.5	—	—	2.14	—
Horse bean straw	87.9	8.6	1.4	36.4	33.1	8.4	—	—	1.38	—
Hyacinth bean hay *(Dilichos lablab)*	90.2	14.8	1.4	33.6	33.6	6.8	—	—	2.37	—
Johnson grass hay	90.1	6.5	2.1	30.4	43.7	7.4	0.87	0.26	1.04	1.22
June grass hay, western *(Koeleria cristata)*	88.3	8.1	2.5	30.4	40.5	6.8	—	—	1.30	—
Kafir fodder, very dry	90.0	8.7	2.6	25.5	44.2	9.0	0.35	0.18	1.39	1.53
Kafir fodder, high in water	71.7	6.5	2.7	21.6	37.6	3.3	0.28	0.14	1.04	1.23
Kafir stover, very dry	90.0	5.5	1.8	29.5	44.3	8.9	0.54	0.09	0.88	—
Kafir stover, high in water	72.7	3.8	1.3	23.7	36.6	7.3	0.44	0.07	0.61	—
Koahaole forage, dried	88.7	12.7	1.9	29.8	39.2	5.1	—	—	2.03	—
Kochia scoparia hay	90.0	11.4	1.5	23.6	40.7	12.8	—	—	1.82	—
Kudzu hay	89.0	15.9	2.5	28.6	35.1	6.9	2.78	0.21	2.54	—
Lespedeza hay, annual, all analyses	89.2	12.7	2.4	26.7	42.2	5.2	0.98	0.18	2.03	0.91

Material	Total dry matter Per ct.	Protein Per ct.	Fat Per ct.	Fiber Per ct.	N-free extract Per ct.	Total minerals Per ct.	Calcium Per ct.	Phosphorus Per ct.	Nitrogen Per ct.	Potassium Per ct.
Lespedeza hay, annual, before bloom	89.1	14.3	2.7	22.7	43.0	6.4	1.04	0.19	2.29	1.06
Lespedeza hay, annual, in bloom	89.1	13.0	1.8	26.5	42.7	5.1	1.02	0.18	2.08	0.94
Lespedeza hay, annual, after bloom	89.1	11.5	1.9	32.6	38.6	4.5	0.90	0.15	1.84	0.82
Lespedeza hay, perennial	89.0	13.2	1.7	26.5	42.7	4.9	0.92	0.22	2.11	0.98
Lespedeza leaves, annual	89.2	17.1	2.9	19.7	43.1	6.4	1.30	0.20	2.74	0.92
Lespedeza stems, annual	89.2	8.3	1.0	38.5	37.7	3.7	0.64	0.13	1.33	0.89
Lespedeza straw	90.0	6.8	2.3	29.2	47.1	4.6	—	—	1.09	—
Lovegrass hay, weeping	91.2	9.2	2.8	30.9	43.4	4.9	—	—	1.47	—
Marsh or swamp hay, good quality	90.2	7.7	2.3	28.2	44.3	7.7	—	—	1.23	—
Millet hay, foxtail varieties	87.6	8.2	2.7	25.3	44.7	6.7	0.29	0.16	1.31	1.70
Millet hay, hog millet, or proso	90.3	9.3	2.2	23.9	47.6	7.3	—	—	1.49	—
Millet hay, Japanese	86.8	8.3	1.6	27.7	40.8	8.4	0.20	—	1.33	2.10
Millet hay, pearl, or cat-tail	87.2	6.7	1.7	33.0	36.8	9.0	—	—	1.07	—
Millet straw	90.0	3.8	1.6	37.5	41.6	5.5	0.08	—	0.61	1.44
Milo fodder	88.5	8.0	3.3	21.9	48.4	6.9	0.35	0.18	1.28	—
Milo stover	91.0	3.2	1.1	29.1	48.1	9.5	0.58	0.11	0.51	—
Mint hay	88.3	12.7	2.1	20.3	45.6	7.6	1.51	0.19	2.03	—
Mixed hay, good, less than 30% legumes	88.0	8.3	1.8	30.7	41.8	5.4	0.61	0.18	1.33	1.47
Mixed hay, good, more than 30% legumes	88.0	9.2	1.9	28.1	42.8	6.0	0.90	0.19	1.47	1.46
Mixed hay, cut very early	90.0	13.3	2.7	25.3	39.4	9.3	—	—	2.13	—
Napier grass hay	89.1	8.2	1.8	34.0	34.6	10.5	—	—	1.31	—
Natal grass hay	90.2	7.4	1.8	36.8	39.2	5.0	0.45	0.29	1.18	—
Native hay, western mt. states, good quality	90.0	8.1	2.1	29.8	43.2	6.8	0.39	0.12	1.30	—
Native hay, western mt. states, mature and weathered	90.0	3.9	1.4	33.6	43.6	7.5	—	—	0.62	—
Needle grass hay (*Stipa,* spp.)	88.1	7.2	2.0	30.8	41.9	6.2	—	—	1.15	—
Oak leaves, live oak, dried	93.8	9.3	2.7	29.9	45.3	6.6	—	—	1.49	—
Oat chaff	91.8	5.9	2.4	25.7	46.3	11.5	0.80	0.30	0.94	0.86
Oat hay	88.1	8.2	2.7	28.1	42.2	6.9	0.21	0.19	1.31	0.83
Oat hay, wild (*Avena fatua*)	92.5	6.6	2.6	32.5	44.0	6.8	0.22	0.25	1.06	—
Oat hulls	92.8	4.5	1.3	29.7	50.8	6.5	0.20	0.10	0.78	0.48
Oat straw	89.7	4.1	2.2	36.1	41.0	6.3	0.19	0.10	0.66	1.35
Oat grass hay, tall	88.7	7.5	2.4	30.1	42.7	6.0	—	0.14	1.20	1.36
Orchard grass hay, early-cut	88.6	7.7	2.9	30.5	40.7	6.8	0.19	0.17	1.23	1.61
Picnic grass hay (*Panicum,* spp.)	92.1	8.3	2.3	29.5	44.9	7.1	—	—	1.33	—
Para grass hay	90.2	4.6	0.9	33.6	44.5	6.6	0.35	0.35	0.74	1.44
Pasture grasses and clovers, mixed, from closely grazed, fertile pasture, dried (northern states)	90.0	20.3	3.6	19.7	38.7	7.7	0.58	0.32	3.25	2.18
Pasture grasses, mixed, from poor to fair pasture, before heading out, dried	90.0	14.1	2.3	19.4	43.2	11.0	0.41	0.12	2.26	0.74
Pasture grass, western plains, growing, dried	90.0	11.6	2.5	28.0	40.2	7.7	0.37	0.24	1.86	—
Pasture grass, western plains, mature, dried	90.0	4.6	2.3	31.9	45.3	5.9	0.34	0.14	0.74	—
Pasture grass, western plains, mature and weathered	90.0	3.3	1.8	34.1	44.5	6.3	0.33	0.09	0.53	—

Material	Total dry matter Per ct.	Protein Per ct.	Fat Per ct.	Fiber Per ct.	N-free extract Per ct.	Total minerals Per ct.	Calcium Per ct.	Phosphorus Per ct.	Nitrogen Per ct.	Potassium Per ct.
Pasture grass and other forage on western mt. ranges, spring, dried	90.0	17.0	3.1	14.0	49.1	6.8	1.21	0.38	2.72	—
Pasture grass and other forage on western mt. ranges, autumn, dried	90.0	8.8	4.3	17.4	51.4	8.1	—	—	1.41	—
Pea hay, field	89.3	14.9	3.3	24.3	39.1	7.7	1.22	0.25	2.38	1.25
Pea straw, field	90.2	6.1	1.6	33.1	44.0	5.4	—	0.10	0.98	1.08
Pea-and-oat hay	89.1	12.1	2.9	27.2	39.1	7.8	0.72	0.22	1.94	1.04
Peanut hay, without nuts	90.7	10.1	3.3	23.4	44.2	9.7	1.12	0.13	1.62	1.25
Peanut hay, with nuts	92.0	13.4	12.6	23.0	34.9	8.1	1.13	0.15	2.14	0.85
Peanut hay, mowed	91.4	10.6	5.1	23.8	42.2	9.7	—	—	1.70	—
Peanut hulls, with a few nuts	92.3	6.7	1.2	60.3	19.7	4.4	0.30	0.07	1.07	0.82
Peavine hay, from pea-cannery vines, sun-cured	86.3	11.9	2.4	23.0	42.2	6.8	1.48	0.16	1.90	—
Prairie hay, western, good quality	90.7	.57	2.3	30.4	44.9	7.4	0.36	0.18	0.91	—
Prairie hay, western, mature	91.7	3.8	2.4	31.9	47.1	6.5	0.28	0.09	0.61	0.49
Quack grass hay	89.0	6.9	1.9	34.5	38.8	6.9	—	—	1.10	—
Ramie meal	92.2	19.2	3.8	20.1	35.9	13.2	4.32	0.22	3.07	—
Red top hay	91.0	7.2	2.3	29.3	45.3	6.9	0.33	0.23	1.15	1.93
Reed canary grass hay	91.1	7.7	2.3	29.2	44.3	7.6	0.33	0.16	1.23	—
Rescue grass hay	90.2	9.8	3.2	24.6	44.5	8.1	—	—	1.57	—
Rhodes grass hay	89.0	5.7	1.3	31.7	41.8	8.5	0.35	0.27	0.91	1.18
Rice hulls	92.0	3.0	0.8	40.7	28.4	19.1	0.08	0.08	0.48	0.31
Rice straw	92.5	3.9	1.4	33.5	39.2	14.5	0.19	0.07	0.62	1.22
Rush hay, western *(Juncus,* spp.)	90.0	9.4	1.8	29.2	44.2	5.4	—	—	1.50	—
Russian thistle hay	87.5	8.9	1.6	26.9	37.4	12.7	—	—	1.42	—
Rye grass hay, Italian	88.6	8.1	1.9	27.8	43.3	7.5	—	0.24	1.30	1.00
Rye grass hay, perennial	88.0	9.2	3.1	24.2	43.4	8.1	—	0.24	1.47	1.25
Rye grass hay, native western	87.4	7.8	2.1	33.5	37.6	6.4	—	—	1.25	—
Rye hay	91.3	6.7	2.1	36.5	41.0	5.0	—	0.18	1.07	1.05
Rye straw	92.8	3.5	1.2	38.7	45.9	3.5	0.26	0.09	0.56	0.90
Salt bushes, dried	93.5	13.8	1.6	22.1	38.8	17.2	1.88	0.11	2.21	4.69
Salt grass hay, misc. var.	90.0	8.1	1.8	28.8	39.5	11.8	—	—	1.30	—
Sanfoin hay *(Onobrychis viciaefolia)*	84.1	10.5	2.6	19.7	44.2	7.1	—	—	1.68	—
Seaweed, dried *(Fucus,* spp.)	88.7	5.2	4.2	9.4	53.6	16.3	—	—	0.83	—
Seaweed, dried *(Laminaria,* spp.)	83.7	11.4	1.1	8.6	45.8	16.8	—	—	1.82	—
Sedge hay, eastern *(Carex,* spp.)	90.7	6.1	1.7	29.2	46.3	7.4	—	—	0.98	—
Sedge hay, western *(Carex,* spp.)	90.6	10.1	2.4	27.3	44.0	6.8	0.60	0.24	1.62	—
Seradella hay	89.0	16.4	3.2	29.8	32.0	7.6	—	0.33	2.62	1.25
Sorghum bagasse, dried	89.3	3.1	1.4	31.3	50.0	3.5	—	—	0.50	—
Sorghum fodder, sweet, dry	88.8	6.2	2.4	25.0	48.1	7.1	0.34	0.12	0.99	1.29
Sorghum fodder, sweet, high in water	65.7	4.5	2.4	16.6	37.6	4.6	0.25	0.09	0.72	0.96
Soybean hay, good, all analyses	88.0	14.4	3.3	27.5	35.8	7.0	0.94	0.24	2.30	0.82
Soybean hay, in bloom or before	88.0	16.7	3.3	20.6	37.8	9.6	1.53	0.27	2.67	0.86
Soybean hay, seed developing	88.0	14.6	2.4	27.2	36.5	7.3	1.35	0.25	2.34	0.78
Soybean hay, seed nearly ripe	88.0	15.2	6.6	24.0	38.2	4.0	0.86	0.32	2.43	0.81
Soybean hay, poor quality, weathered	89.0	9.2	1.2	41.0	30.4	7.2	0.94	—	1.47	—

Material	Total dry matter Per ct.	Protein Per ct.	Fat Per ct.	Fiber Per ct.	N-free extract Per ct.	Total minerals Per ct.	Calcium Per ct.	Phos- phorus Per ct.	Nitro- gen Per ct.	Potas- sium Per ct.
Soybean straw	88.8	4.0	1.1	41.1	37.5	5.1	—	0.13	0.64	0.62
Soybean and Sudan grass hay, chiefly Sudan	89.0	7.4	2.2	31.1	43.4	4.9	—	—	1.18	—
Spanish moss, dried	89.2	5.0	2.4	26.6	47.7	7.5	—	0.04	0.80	0.46
Sudan grass hay, all analyses	89.3	8.8	1.6	27.9	42.9	8.1	0.36	0.26	1.41	1.30
Sudan grass hay, before bloom	89.6	11.2	1.5	26.1	41.3	9.5	0.41	0.26	1.79	—
Sudan grass hay, in bloom	89.2	8.4	1.5	30.7	41.8	6.8	—	—	1.34	—
Sudan grass hay, in seed	89.5	6.8	1.6	29.9	44.4	6.8	0.27	0.19	1.09	—
Sudan grass, young, dehydrated	88.0	14.5	2.5	20.4	41.2	9.4	0.52	0.39	2.32	—
Sudan grass straw	90.4	7.1	1.5	33.0	42.3	6.5	—	—	1.14	—
Sugar cane fodder, Japanese, dried	89.0	1.3	1.8	19.7	64.3	1.9	0.32	0.14	0.21	0.58
Sugar cane bagasse, dried	95.5	1.1	0.4	49.6	42.0	2.4	—	—	0.18	—
Sugar cane pulp, dried	93.8	1.7	0.6	45.6	42.2	3.7	—	—	0.27	—
Sweet potato vine, dried	90.7	12.6	3.3	19.1	45.5	10.2	—	—	2.02	—
Teosinte fodder, dried	89.4	9.1	1.9	26.5	41.7	10.3	—	0.17	1.46	0.88
Timothy hay, all analyses	89.0	6.5	2.4	30.2	45.0	4.9	0.23	0.20	1.04	1.50
Timothy hay, before bloom	89.0	9.7	2.7	27.4	42.7	6.5	—	—	1.55	—
Timothy, full bloom	89.0	6.4	2.5	30.4	44.8	4.9	0.23	0.20	1.02	1.50
Timothy hay, in bloom, nitrogen fertilized	89.0	9.7	2.1	31.6	42.6	3.9	0.40	0.21	1.41	1.41
Timothy hay, late seed	89.0	5.3	2.3	31.0	45.9	4.5	0.14	0.15	0.85	1.41
Timothy hay, in bloom, dehydrated	89.0	7.7	2.3	28.3	45.5	5.2	—	—	1.23	—
Timothy hay, second cutting	88.7	15.0	4.6	25.4	36.5	7.2	—	—	2.40	—
Timothy and clover hay, one-fourth clover	88.8	7.8	2.4	29.5	43.8	5.3	0.51	0.20	1.25	1.48
Velvet bean hay	92.8	16.4	3.1	27.5	38.4	7.4	—	0.24	2.62	2.20
Vetch hay, common	89.0	13.3	1.1	25.2	32.2	6.2	1.18	0.32	2.13	2.22
Vetch hay, hairy	88.0	19.3	2.6	24.5	33.1	8.5	1.13	0.32	3.09	1.96
Vetch-and-oat hay, over half vetch	87.6	11.9	2.7	27.3	37.5	8.2	0.76	0.27	1.90	1.51
Vetch-and-wheat hay, cut early	90.0	15.4	2.2	28.8	36.4	7.2	—	—	2.46	—
Wheat chaff	90.0	4.4	1.5	29.4	47.1	7.6	0.21	0.14	0.70	0.50
Wheat hay	90.4	6.1	1.8	26.1	50.0	6.4	0.14	0.18	0.98	1.47
Wheat straw	92.5	3.9	1.5	36.9	41.9	8.3	0.21	0.07	0.62	0.79
Wheat grass hay, crested, cut early	90.0	9.2	2.0	32.2	40.2	6.4	—	—	1.47	—
Wheat grass hay, slender	90.0	8.0	2.1	32.2	41.0	6.7	0.30	0.24	1.28	2.41
Winter fat, or white sage, dried (Eurotia lanata)	92.6	12.9	1.9	27.4	40.8	9.6	—	—	2.06	—
Wire grass hay, southern (Aristida, spp.)	90.0	5.5	1.4	31.8	47.9	3.4	0.15	0.14	0.88	—
Wire grass hay, western (Aristida, spp.)	90.0	6.4	1.3	34.1	41.0	7.2	—	—	1.02	—
Yucca, or beargrass, dried	92.6	6.6	2.2	38.6	38.3	6.9	—	—	1.06	—

CONCENTRATES

Material	Total dry matter Per ct.	Protein Per ct.	Fat Per ct.	Fiber Per ct.	N-free extract Per ct.	Total minerals Per ct.	Calcium Per ct.	Phos- phorus Per ct.	Nitro- gen Per ct.	Potas- sium Per ct.
Acorns, whole (red oak)	50.0	3.2	10.7	9.9	25.0	1.2	—	—	0.51	—
Acorns, whole (white and post oaks)	50.0	2.7	3.0	9.3	33.7	1.3	—	—	0.43	—
Alfalfa-molasses feed	86.0	11.4	1.2	18.5	46.2	8.7	—	—	1.82	—
Alfalfa seed	88.3	33.2	10.6	8.1	32.0	4.4	—	—	5.31	—

Material	Total dry matter Per ct.	Protein Per ct.	Fat Per ct.	Fiber Per ct.	N-free extract Per ct.	Total minerals Per ct.	Calcium Per ct.	Phosphorus Per ct.	Nitrogen Per ct.	Potassium Per ct.
Alfalfa seed screenings	90.3	31.1	9.9	11.1	33.1	5.1	—	—	4.98	—
Apple-pectin pulp, dried	91.2	7.0	7.3	24.2	49.4	3.3	—	—	1.12	—
Apple-pectin pulp, wet	16.7	1.5	0.9	5.8	7.9	0.6	—	—	0.24	—
Apple pomace, dried	89.4	4.5	5.0	15.6	62.1	2.2	0.10	0.09	0.72	0.43
Apple pomace, wet	21.1	1.3	1.3	3.7	13.9	0.9	0.02	0.02	0.21	0.10
Atlas sorghum grain	89.1	11.3	3.3	2.0	70.6	1.9	—	—	1.81	—
Atlas sorghum head chops	88.0	9.5	2.8	10.7	60.2	4.8	—	—	1.52	—
Avocado oil meal	91.4	18.6	1.1	17.6	36.0	18.1	—	—	2.98	—
Babassu oil meal	92.8	24.2	6.8	12.0	44.6	5.2	0.13	0.71	3.87	—
Bakery waste, dried (high in fat)	91.6	10.9	13.7	0.7	64.7	1.6	—	—	1.74	—
Barley, common, not including Pacific Coast states	89.4	12.7	1.9	5.4	66.6	2.8	0.06	0.37	2.03	0.49
Barley, Pacific Coast states	89.8	8.7	1.9	5.7	70.9	2.6	—	—	1.39	—
Barley, light weight	89.1	12.1	2.1	7.4	64.3	3.2	—	—	1.94	—
Barley, hull-less, or bald	90.2	11.6	2.0	2.4	72.1	2.1	—	—	1.86	—
Barley feed, high grade	90.3	13.5	3.5	8.7	60.5	4.1	0.03	0.40	2.16	0.60
Barley feed, low grade	92.0	12.3	3.45	14.7	56.2	5.3	—	—	1.97	—
Barley, malted	93.4	12.7	2.1	5.4	70.9	2.3	0.06	0.42	2.03	0.37
Barley screenings	88.6	11.6	2.7	9.1	61.3	3.9	—	—	1.86	—
Beans, field, or navy	90.0	22.9	1.4	4.2	57.3	4.2	0.15	0.57	3.66	1.27
Beans, kidney	89.0	23.0	1.2	4.1	56.8	3.9	—	—	3.68	—
Beans, lima	89.7	21.2	1.1	4.7	58.2	4.5	0.09	0.37	3.39	1.70
Beans, mung	90.2	23.3	1.0	3.5	58.5	3.9	—	—	3.73	—
Beans, pinto	89.9	22.5	1.2	4.1	57.7	4.4	—	—	3.60	—
Beans, tepary	90.5	22.2	1.4	3.4	59.3	4.2	—	—	3.56	—
Beechnuts	91.4	15.0	30.6	15.0	27.5	3.3	0.58	0.30	2.40	0.62
Beef scraps	94.5	55.6	10.9	1.2	0.5	26.3	—	—	8.90	—
Beet pulp, dried	90.1	9.2	0.5	19.8	57.2	3.4	0.67	0.08	1.47	0.18
Beet pulp, molasses, dried	91.9	10.7	0.7	16.0	59.4	5.1	0.62	0.09	1.71	1.63
Beet pulp, wet	11.6	1.5	0.3	4.0	5.3	0.5	0.09	0.01	0.24	0.02
Beet pulp, wet, pressed	14.2	1.4	0.4	4.6	7.1	0.7	—	—	0.22	—
Blood flour, or soluble blood meal	92.2	84.7	1.0	1.1	0.7	4.7	0.68	0.50	13.55	—
Blood meal	91.8	84.5	1.1	1.0	0.7	4.5	0.33	0.25	13.52	0.09
Bone meal, raw	93.6	26.0	5.0	1.0	2.5	59.1	23.05	10.22	4.16	—
Bone meal, raw, solvent process	93.1	25.7	1.0	1.0	1.9	63.5	24.02	10.65	4.11	—
Bone meal, steamed	96.3	7.1	3.3	0.8	3.8	81.3	31.74	15.00	1.14	0.18
Bone meal, steamed, solvent process	96.8	7.2	0.4	1.5	3.7	84.0	—	—	1.15	—
Bone meal, steamed, special	97.7	13.5	7.9	1.0	5.1	70.2	31.88	13.48	2.16	—
Bone meal, 10 to 20% protein	97.2	14.6	6.5	1.5	3.6	71.0	26.00	12.66	2.34	—
Bread, white, enriched	64.1	8.5	2.0	0.3	52.0	1.3	0.06	0.10	1.36	0.10
Brewers' grains, dried, 25% protein or over	92.9	27.6	6.5	14.3	40.9	3.6	0.29	0.48	4.42	0.10
Brewers' grains, dried, below 25% protein	92.3	23.4	6.4	16.1	42.5	3.9	—	—	3.74	—
Brewers' grains, dried, from California barley	91.1	20.0	5.7	18.1	43.6	3.7	—	—	3.20	—
Brewers' grains, wet	23.7	5.7	1.6	3.6	11.8	1.0	0.07	0.12	0.91	0.02
Broom corn seed	89.7	9.2	3.7	5.1	69.1	2.6	—	—	1.47	—
Buckwheat, ordinary varieties	88.0	10.3	2.3	10.7	62.8	1.9	0.09	0.31	1.64	0.45

Material	Total dry matter Per ct.	Protein Per ct.	Fat Per ct.	Fiber Per ct.	N-free extract Per ct.	Total minerals Per ct.	Calcium Per ct.	Phos-phorus Per ct.	Nitro-gen Per ct.	Potas-sium Per ct.
Buckwheat, Tartary	88.1	10.1	2.4	12.7	60.9	2.0	0.13	0.31	1.62	0.44
Buckwheat feed, good grade	89.3	18.5	4.9	18.2	43.5	4.2	—	0.48	2.96	0.66
Buckwheat feed, low grade	88.3	13.3	3.4	28.6	39.8	3.2	—	0.37	2.13	0.68
Buckwheat flour	88.1	10.2	2.1	0.9	73.4	1.5	0.01	0.09	1.63	0.16
Buckwheat kernels, without hulls	88.0	14.1	3.4	1.8	66.5	2.2	0.05	0.45	2.26	0.49
Buckwheat middlings	88.7	29.7	7.3	7.4	39.4	4.9	—	1.02	4.76	0.98
Buttermilk	9.4	3.5	0.6	0	4.5	0.8	0.14	0.08	0.56	0.07
Buttermilk, condensed	29.7	10.9	2.2	0	12.6	4.0	0.44	0.26	1.74	0.23
Buttermilk, dried	92.4	32.4	6.4	0.3	43.3	10.0	1.36	0.82	5.18	0.71
Carob bean and pods	87.8	5.5	2.6	8.7	68.5	2.5	—	—	0.88	—
Carob bean pods	89.5	4.7	2.5	8.7	70.9	2.7	—	—	0.75	—
Carob bean seeds	88.5	16.7	2.6	7.6	58.4	3.2	—	—	2.67	—
Cassava roots, dried	94.4	2.8	0.5	5.0	84.1	2.0	—	—	0.45	—
Cassava meal (starch waste)	86.8	0.9	0.7	4.6	78.8	1.8	—	0.03	0.14	0.23
Cheese rind, or cheese meal	91.0	59.5	8.9	0.4	10.7	11.5	—	—	9.52	—
Chess, or cheat, seed	89.6	9.7	1.7	8.2	66.4	3.6	—	—	1.56	—
Chick peas	90.0	20.3	4.3	8.5	54.0	2.9	—	—	3.24	—
Citrus pulp, dried	90.1	5.9	3.1	11.5	62.7	6.9	2.07	0.15	0.94	—
Citrus pulp and molasses, dried	92.0	5.3	2.8	9.3	66.6	8.0	—	—	0.84	—
Citrus pulp, wet	18.3	1.2	0.6	2.3	12.8	1.4	—	—	0.19	—
Clover seed, red	87.5	32.6	7.8	9.2	31.2	6.7	—	—	5.22	—
Clover seed screenings, red	90.5	28.2	5.9	10.2	40.3	5.9	—	—	4.51	—
Clover seed screenings, sweet	90.1	21.7	3.7	14.7	41.1	8.9	—	—	3.47	—
Cocoa meal	96.0	24.3	17.1	5.1	43.7	5.8	—	—	3.89	—
Cocoa shells	95.1	15.4	3.0	16.5	49.9	10.3	—	0.59	2.46	2.16
Coconut oil meal, hydr. or exp. process	93.2	21.3	6.7	10.7	48.3	6.2	0.21	0.64	3.41	1.95
Coconut oil meal, high in fat	93.7	21.0	10.6	11.3	44.4	6.4	—	—	3.36	—
Coconut oil meal, solvent process	91.1	21.4	2.4	13.3	47.4	6.6	—	—	3.42	—
Cod-liver oil meal	92.5	50.4	28.9	0.7	9.6	2.9	0.18	0.61	8.06	—
Corn, dent, Grade No. 1	87.0	8.8	4.0	2.1	70.9	1.2	0.02	0.28	1.41	0.28
Corn, dent, Grade No. 2	85.0	8.6	3.9	2.0	69.3	1.2	0.02	0.27	1.38	0.27
Corn, dent, Grade No. 3	83.5	8.4	3.8	2.0	68.1	1.2	0.02	0.27	1.34	0.27
Corn, dent, Grade No. 4	81.1	8.2	3.7	1.9	66.2	1.1	0.02	0.26	1.31	0.26
Corn, dent, Grade No. 5	78.5	7.9	3.6	1.9	64.0	1.1	0.02	0.25	1.26	0.25
Corn, dent, soft or immature	70.0	7.2	2.3	2.5	56.5	1.5	—	0.24	1.16	0.26
Corn, flint	88.5	9.8	4.3	1.9	71.0	1.5	—	0.33	1.57	0.32
Corn, pop	90.0	11.5	5.0	1.9	70.1	1.5	—	0.29	1.84	—
Corn ears, including kernels and cobs (corn-and-cob meal)	86.1	7.3	3.2	8.0	66.3	1.3	—	0.22	1.17	0.29
Corn ears, soft or immature	64.3	5.8	1.9	7.8	47.7	1.1	—	—	0.93	—
Corn, snapped, or ear-corn chops with husks	88.8	8.0	3.0	10.6	64.8	2.4	—	—	1.28	—
Corn, snapped, very soft or immature	60.0	5.3	1.8	8.2	42.7	2.0	—	—	0.85	—
Corn bran	90.6	9.7	7.3	9.2	62.0	2.4	0.03	0.27	1.56	0.56
Corn feed meal	88.6	9.8	4.7	2.9	69.2	2.0	0.03	0.34	1.57	0.28
Corn germ meal	93.0	19.8	7.8	8.9	53.2	3.3	—	0.58	3.17	0.21
Corn gluten feed, all analyses	90.9	25.5	2.7	7.6	48.8	6.3	0.48	0.82	4.08	0.54

Material	Total dry matter Per ct.	Protein Per ct.	Fat Per ct.	Fiber Per ct.	N-free extract Per ct.	Total minerals Per ct.	Calcium Per ct.	Phos-phorus Per ct.	Nitro-gen Per ct.	Potas-sium Per ct.
Corn gluten feed, 25% protein guarantee	91.1	26.6	3.0	7.2	48.2	6.1	—	—	4.26	—
Corn gluten feed, 23% protein guarantee	91.4	24.8	2.6	7.8	49.8	6.4	—	—	3.97	—
Corn gluten feed with molasses	88.8	22.6	2.1	6.8	50.9	6.4	—	—	3.62	—
Corn gluten meal, all analyses	91.4	43.1	2.0	4.0	39.8	2.5	0.13	0.38	6.90	0.02
Corn gluten meal, 41% protein guarantee	91.4	42.9	2.0	3.9	40.1	2.5	—	—	6.86	—
Corn grits	88.4	8.5	0.5	0.6	78.4	0.4	—	—	1.36	—
Corn meal, degerminated, yellow	88.7	8.7	1.2	0.6	77.1	1.1	0.01	0.14	1.39	—
Corn meal, degerminated, white	88.4	8.6	1.2	0.7	76.1	1.8	0.01	0.14	1.38	—
Corn oil meal, old process	91.7	22.3	7.8	10.3	49.0	2.3	0.06	0.56	3.57	—
Corn oil meal, solvent process	91.7	23.0	1.5	10.4	54.6	2.2	0.03	0.50	3.68	—
Corn-starch	88.6	11.6	0.6	0.1	0.2	87.6	0.1	—	—	0.10
Corn-and-oat feed, good grade	89.6	11.9	4.0	5.4	65.9	2.4	0.05	0.30	1.90	0.34
Corn-and-oat feed, low grade	89.6	9.1	2.9	13.4	59.0	5.2	—	—	1.46	—
Cottonseed, whole	92.7	23.1	22.9	16.9	26.3	3.5	0.14	0.70	3.70	1.11
Cottonseed, immature, dried	93.2	20.5	15.9	24.1	29.0	3.7	—	—	3.28	—
Cottonseed, whole pressed, 28% protein guarantee	93.5	28.2	5.8	22.6	32.2	4.7	—	—	4.51	—
Cottonseed, whole pressed, below 28% protein	93.5	26.9	6.5	24.7	30.8	4.6	0.17	0.64	4.30	1.25
Cottonseed feed, below 36% protein	92.4	34.6	6.3	14.1	31.5	5.9	0.26	0.83	5.54	1.22
Cottonseed flour	94.4	57.0	7.2	2.1	21.6	6.5	—	—	9.12	—
Cottonseed kernels, without hulls	93.6	38.4	33.3	2.3	15.1	4.5	—	—	6.14	—
Cottonseed meal, 45% protein and over	93.5	46.2	7.7	8.6	24.9	6.1	0.22	1.13	7.39	—
Cottonseed meal, 43% protein grade, not including Texas analyses	92.7	43.9	7.1	9.0	26.3	6.4	0.23	1.12	7.02	1.45
Cottonseed meal, 43% protein grade, Texas analyses	92.5	42.7	6.4	10.6	27.0	5.8	0.19	0.96	6.83	1.34
Cottonseed meal, 41% protein grade, not including Texas analyses	92.8	41.5	6.3	10.4	28.1	6.5	0.20	1.22	6.64	1.48
Cottonseed meal, 41% protein grade, Texas analyses	92.1	41.0	6.0	11.6	27.6	5.9	—	—	6.56	—
Cottonseed meal, below 41% protein grade	92.4	38.2	6.2	12.3	29.4	6.3	0.23	1.29	6.11	1.57
Cottonseed meal, solvent process	90.8	44.4	2.6	12.7	24.3	6.8	—	—	7.10	—
Cowpea seed	89.0	23.4	1.4	4.0	56.7	3.5	0.11	0.46	3.74	1.30
Crab meal	92.4	31.5	2.0	10.7	5.0	43.2	15.15	1.63	5.04	0.45
Darso grain	90.0	10.1	3.1	1.9	73.5	1.4	0.02	0.32	1.62	—
Distillers' dried corn grains, without solubles	92.9	28.3	8.8	11.4	41.9	2.5	0.11	0.47	4.53	0.24
Distillers' dried corn grains, with solubles	93.1	28.8	8.9	9.0	41.7	4.7	0.16	0.74	4.61	—
Distillers' dried corn grains, solvent extracted	93.7	33.4	1.4	8.6	46.4	3.9	—	—	5.34	—
Distillers' dried rye grains	93.9	18.5	6.4	15.6	51.0	2.4	0.13	0.43	2.96	0.04
Distillers' rye grains, wet	22.4	4.4	1.5	2.5	13.3	0.7	—	—	0.70	—
Distillers' dried wheat grains	93.7	28.7	6.1	13.0	42.2	3.7	—	—	4.59	—
Distillers' dried wheat grains, high protein	94.7	46.2	5.7	10.9	30.0	1.9	—	—	7.39	—
Distillers' solubles, dried, corn	93.0	26.7	7.9	2.6	48.4	7.4	0.30	1.41	4.27	1.75
Distillers' solubles, dried, wheat	94.0	28.2	1.5	2.8	58.9	2.6	—	—	4.51	—
Distillery stillage, corn, whole	7.9	2.3	0.6	0.7	4.0	0.3	0.006	0.05	0.37	—
Distillery stillage, rye, whole	5.9	1.9	0.3	0.5	2.9	0.3	—	—	0.30	—

Material	Total dry matter Per ct.	Protein Per ct.	Fat Per ct.	Fiber Per ct.	N-free extract Per ct.	Total minerals Per ct.	Calcium Per ct.	Phos-phorus Per ct.	Nitro-gen Per ct.	Potas-sium Per ct.
Distillery stillage, strained	3.8	1.1	0.4	0.2	1.8	0.3	0.004	0.05	0.18	—
Durra grain	89.8	10.3	3.5	1.6	72.4	2.0	—	—	1.64	—
Emmer grain	91.1	12.1	1.9	9.8	63.6	3.7	—	0.33	1.94	0.47
Feterita grain	89.4	12.2	3.2	2.2	70.1	1.7	0.02	0.33	1.96	—
Feterita head chops	89.6	10.7	2.6	7.4	65.7	3.2	—	—	1.71	—
Fish-liver oil meal	92.8	62.8	17.3	1.2	5.4	6.1	—	—	10.04	—
Fish meal, all analyses	92.9	63.9	6.8	0.6	4.0	17.6	4.14	2.67	10.22	0.40
Fish meal, over 63% protein	92.7	66.8	5.3	0.5	4.5	15.6	—	—	10.69	—
Fish meal, 58-63% protein	93.1	60.9	8.1	0.8	3.5	19.8	—	—	9.74	—
Fish meal, below 58% protein	93.2	56.2	11.0	0.7	2.9	22.4	—	—	8.99	—
Fish meal, herring	93.5	72.5	7.3	0.7	1.5	11.5	2.97	2.08	11.60	—
Fish meal, menhaden	93.6	62.2	8.5	0.7	4.2	18.0	5.30	3.38	9.96	—
Fish meal, redfish	94.2	56.7	11.4	0.9	0.9	24.3	4.01	2.44	9.07	—
Fish meal, salmon	92.8	59.4	9.8	0.3	4.3	19.0	5.49	3.65	9.50	—
Fish meal, sardine	93.1	67.2	5.0	0.6	5.4	14.9	4.21	2.54	10.76	0.33
Fish meal, tuna	90.1	58.2	7.9	0.7	3.4	19.9	4.80	3.10	9.31	—
Fish meal, whitefish	90.4	63.0	6.7	0.1	0.1	20.5	—	—	10.08	—
Fish solubles, condensed	49.5	29.3	8.4	—	2.2	9.6	—	—	4.69	—
Flaxseed	93.8	24.0	35.9	6.3	24.0	3.6	0.26	0.55	3.84	0.59
Flaxseed screenings	91.1	16.4	9.4	12.7	45.8	6.8	0.37	0.43	2.62	—
Flaxseed screenings oil feed	91.9	25.0	7.1	11.7	40.3	7.8	—	—	4.00	—
Garbage	39.3	6.0	7.2	1.1	22.2	2.8	—	—	0.96	—
Garbage, processed, high in fat	95.9	17.5	23.7	20.0	21.8	12.9	—	0.33	2.80	0.62
Garbage, processed, low in fat	92.3	23.1	3.5	13.5	38.1	14.1	—	—	3.70	—
Grapefruit pulp, dried	91.7	4.9	1.1	11.9	69.6	4.2	—	—	0.78	—
Grape pomace, dried	91.0	12.2	6.9	30.2	36.7	5.0	—	—	1.96	—
Hegari grain	89.7	9.6	2.6	2.0	73.9	1.6	0.18	0.30	1.54	—
Hegari head chops	89.6	10.0	2.1	11.9	60.6	5.0	—	—	1.60	—
Hempseed oil meal	92.0	31.0	6.2	23.8	22.0	9.0	0.25	0.43	4.96	—
Hominy feed, 5% fat or more	90.4	11.2	6.9	5.2	64.2	2.9	0.22	0.71	1.79	0.61
Hominy feed, low in fat	89.7	10.6	4.3	5.0	67.4	2.4	—	—	1.70	—
Horse beans	87.5	25.7	1.4	8.2	48.8	3.4	0.13	0.54	4.11	1.16
Ivory nut meal, vegetable	89.4	4.7	0.9	7.2	75.5	1.1	—	—	0.76	—
Jack beans	89.3	24.7	3.2	8.2	50.4	2.8	—	—	3.96	—
Kafir grain	89.8	10.9	2.9	1.7	72.7	1.6	0.02	0.31	1.74	0.34
Kafir head chops	89.2	10.0	2.6	6.9	66.4	3.3	0.08	0.27	1.60	—
Kalo sorghum grain	89.2	11.8	3.2	1.6	70.9	1.7	—	—	1.89	—
Kaoliang grain	89.9	10.5	4.1	1.6	71.8	1.9	—	—	1.68	—
Kelp, dried	91.3	6.5	0.5	6.5	42.6	35.2	2.48	0.28	1.04	—
Lamb's-quarters seed	90.0	20.6	4.5	15.1	40.2	9.6	—	—	3.30	—
Lespedeza seed, annual	91.7	36.6	7.6	9.6	32.8	5.1	—	—	5.86	—
Lespedeza seed, sericea	92.3	33.5	4.2	13.5	37.3	3.8	—	—	5.36	—
Lemon pulp, dried	92.8	6.4	1.2	15.0	65.2	5.0	—	—	1.02	—
Linseed meal, old process, all analyses	91.0	35.4	5.8	8.2	36.0	5.6	0.39	0.87	5.66	1.24
Linseed meal, o.p., 37% protein or more	90.9	38.0	5.9	7.7	33.7	5.6	0.39	0.86	6.08	1.10
Linseed meal, o.p., 33-37% protein	91.0	35.0	5.7	8.3	36.4	5.6	0.41	0.86	5.60	1.14

Material	Total dry matter Per ct.	Protein Per ct.	Fat Per ct.	Fiber Per ct.	N-free extract Per ct.	Total minerals Per ct.	Calcium Per ct.	Phos- phorus Per ct.	Nitro- gen Per ct.	Potas- sium Per ct.
Linseed meal, o.p., 31-33% protein	91.0	32.4	5.9	8.3	38.7	5.7	0.36	0.90	5.18	1.40
Linseed meal, solvent process, older analyses	90.4	36.9	2.9	8.7	36.3	5.6	—	—	5.90	—
Linseed meal and screenings oil feed (linseed feed)	90.5	31.2	5.4	10.1	37.0	6.8	0.43	0.65	4.99	—
Liver meal, animal	92.3	66.2	16.4	1.4	1.9	6.4	0.62	1.27	10.59	—
Locust beans and pods, honey	88.4	9.3	2.4	16.1	57.1	3.5	—	—	1.49	—
Lupine seed, sweet, yellow	88.9	39.8	4.9	14.0	25.7	4.5	0.23	0.39	6.37	0.81
Malt, barley	90.6	14.3	1.6	1.8	70.6	2.3	0.08	0.47	2.29	—
Malt sprouts	92.6	26.8	1.3	14.2	44.3	6.0	—	—	4.29	—
Meat scraps, or dry-rendered tankage, 60% protein grade	93.8	60.9	8.8	2.4	1.1	20.6	6.09	3.49	9.74	—
Meat scraps, or dry-rendered tankage, 55% protein grade	93.9	55.8	9.3	2.1	1.3	25.4	8.33	4.04	8.93	—
Meat scraps, or dry-rendered tankage, 55% protein grade, low fat	93.0	56.0	3.5	2.6	1.5	29.4	—	—	8.96	—
Meat scraps, or dry-rendered tankage, 52% protein grade	93.1	52.9	7.3	2.2	4.3	26.4	—	—	8.46	—
Meat and bone scraps, or dry-rendered tankage with bone, 50% protein grade	93.9	51.0	10.1	2.1	1.6	29.1	9.71	4.81	8.16	—
Meat and bone scraps, or dry-rendered tankage with bone, 45% protein grade	94.5	46.3	12.0	2.0	2.3	31.9	11.21	4.88	7.41	—
Mesquite beans and pods	94.0	13.0	2.8	26.3	47.4	4.5	—	—	2.08	—
Milk, cow's	12.8	3.5	3.7	0	4.9	0.7	0.12	0.09	0.56	0.14
Milk, ewe's	19.2	6.5	6.9	0	4.9	0.9	0.21	0.12	1.04	0.19
Milk, goat's	12.8	3.7	4.1	0	4.2	0.8	0.13	0.10	0.59	0.15
Milk, mare's	9.4	2.0	1.1	0	5.9	0.4	0.08	0.05	0.32	0.08
Milk, sow's	19.0	5.9	6.7	0	5.4	1.0	—	—	0.94	—
Milk albumin, or lactalbumin, commercial	92.0	49.5	0.9	1.0	12.8	27.8	—	—	7.92	—
Milk, whole, dried	96.8	24.8	26.2	0.2	40.2	5.4	—	—	3.97	—
Millet seed, foxtail varieties	89.1	12.1	4.1	8.6	60.7	3.6	—	0.20	1.94	0.31
Millet seed, hog, or proso	90.4	11.9	3.4	8.1	63.7	3.3	0.05	0.30	1.90	0.43
Millet seed, Japanese	89.8	10.6	4.9	14.6	54.7	5.0	—	0.44	1.70	0.33
Milo grain	89.4	11.3	2.9	2.2	71.3	1.7	0.03	0.30	1.81	0.36
Milo head chops	90.1	10.2	2.5	6.9	66.2	4.3	0.14	0.26	1.63	—
Molasses, beet	80.5	8.4	0	0	62.0	10.1	0.08	0.02	1.34	4.77
Molasses, beet, Steffen's process	78.7	7.8	0	0	62.1	8.8	0.11	0.02	1.25	4.66
Molasses, cane, or blackstrap	74.0	2.9	0	0	62.1	9.0	0.74	0.08	0.46	3.67
Molasses, cane, high in sugar	79.7	1.3	0	0	74.9	3.5	—	—	0.21	—
Molasses, citrus	69.9	4.0	0.2	0	61.3	4.4	—	—	0.64	—
Molasses, corn sugar, or hydrol	80.5	0.2	0	0	77.8	2.5	—	—	0.03	—
Mustard seed, wild yellow	95.9	23.0	38.8	5.0	23.6	5.5	—	—	3.68	—
Oat clippings, or clipped-oat by-product	92.2	8.8	2.3	25.3	44.9	10.9	—	—	1.41	—
Oat kernels, without hulls (oat groats)	90.4	16.3	6.1	2.1	63.7	2.2	0.08	0.46	2.61	0.39
Oat meal, feeding, or rolled oats without hulls	90.8	16.0	5.5	2.7	64.2	2.4	0.07	0.46	2.56	0.37
Oat middlings	91.4	15.9	5.2	3.3	64.6	2.4	0.08	0.45	2.54	0.57
Oat mill feed	92.4	5.6	1.8	27.9	50.8	6.3	0.13	0.16	0.90	0.60

Material	Total dry matter Per ct.	Protein Per ct.	Fat Per ct.	Fiber Per ct.	N-free extract Per ct.	Total minerals Per ct.	Calcium Per ct.	Phos-phorus Per ct.	Nitro-gen Per ct.	Potas-sium Per ct.
Oat mill feed, poor grade	92.4	4.3	1.8	30.5	50.2	5.6	—	—	0.69	—
Oat mill feed, with molasses	92.4	5.5	1.4	24.1	55.0	6.4	—	—	0.88	—
Oats, not including Pacific Coast states	90.2	12.0	4.6	11.0	58.6	4.0	0.09	0.34	1.92	0.43
Oats, Pacific Coast states	91.2	9.0	5.4	11.0	62.1	3.7	—	—	1.44	—
Oats, hull-less	90.0	15.4	4.2	2.6	65.7	2.1	—	—	2.46	—
Oats, light weight	91.3	12.3	4.7	15.4	54.4	4.5	—	—	1.97	—
Oats, wild	89.0	12.7	5.5	15.2	50.9	4.7	—	—	2.03	—
Olive pulp, dried, pits removed	95.1	14.0	27.4	19.3	31.0	3.4	—	—	2.24	—
Olive pulp, dried, with pits	92.0	5.9	15.6	36.5	31.5	2.5	—	—	0.94	—
Orange pulp, dried	87.9	7.7	1.5	8.0	67.3	3.4	—	—	1.23	—
Palm-kernel oil meal	91.4	19.2	6.7	11.9	49.7	3.9	—	0.69	3.07	0.42
Palm seed, Royal	86.5	6.1	8.3	22.8	43.8	5.5	—	—	0.98	—
Palmo middlings	94.1	16.1	9.7	6.7	56.3	5.3	—	—	2.58	—
Pea feed, or pea meal	90.0	17.7	1.4	23.7	43.7	3.5	—	—	2.83	—
Pea hulls of seeds, or bran	91.5	4.8	0.4	48.5	34.3	3.5	—	—	0.77	—
Pea seed, field	90.7	23.4	1.2	6.1	57.0	3.0	0.17	0.51	3.74	1.03
Pea seed, field, cull	89.7	24.8	2.5	7.1	52.0	3.3	—	—	3.97	—
Pea seed, garden	89.2	25.3	1.7	5.7	53.6	2.9	0.08	0.40	4.04	0.90
Peanut kernels, without hulls	94.6	30.4	47.7	2.5	11.7	2.3	0.06	0.44	4.86	—
Peanut oil feed	94.5	37.8	9.6	14.3	26.2	6.6	—	6.04	—	—
Peanut oil feed, unhulled, or whole pressed peanuts	93.1	35.0	9.2	22.5	21.4	5.0	—	—	5.60	—
Peanut oil meal, old process, all analyses	93.0	43.5	7.6	13.3	23.4	5.2	0.16	0.54	6.96	1.15
Peanut oil meal, o.p., 45% protein and over	93.4	45.2	7.4	12.1	23.7	5.0	—	—	7.23	—
Peanut oil meal, o.p., 43% protein grade	92.8	43.1	7.6	13.9	23.0	5.2	—	—	6.90	—
Peanut oil meal, o.p., 41% protein grade	93.8	41.8	7.8	12.7	25.9	5.6	—	—	6.69	—
Peanut oil meal, solvent process	91.6	51.5	1.4	5.7	27.2	5.8	—	—	8.24	—
Peanut skins	93.8	16.3	23.9	11.8	39.1	2.7	—	—	2.61	—
Peanut screenings	93.6	23.8	11.5	18.9	33.0	6.4	—	—	3.81	—
Peanuts, with hulls	94.1	24.9	36.2	17.5	12.6	2.9	—	0.33	3.98	0.53
Perilla oil meal	91.9	38.4	8.4	20.9	16.0	8.2	0.56	0.47	6.14	—
Pigeon-grass seed	89.8	14.4	6.0	17.3	45.8	6.3	—	—	2.30	—
Pigweed seed	90.0	16.8	6.2	15.9	47.8	3.3	—	—	2.69	—
Pineapple bran, or pulp, dried	85.3	4.0	1.9	19.4	57.2	2.8	0.20	0.10	0.64	—
Pineapple bran, or pulp, and molasses, dried	87.4	3.9	1.0	15.9	63.4	3.2	—	—	0.62	—
Poppy-seed oil meal	89.2	36.6	7.9	11.6	20.7	12.4	—	—	5.86	—
Potato meal, or dried potatoes	92.8	10.4	0.3	2.0	75.8	4.3	0.08	0.22	1.66	1.97
Potato pomace, dried	89.1	6.6	0.5	10.3	69.0	2.7	—	—	1.06	—
Pumpkin seed, not dried	55.0	17.6	20.6	10.8	4.1	1.9	—	—	2.82	—
Raisin pulp, dried	89.4	9.6	7.8	16.1	50.6	5.3	—	—	1.54	—
Raisins, cull	84.8	3.4	0.9	4.4	73.1	3.0	—	—	0.54	—
Rape seed	90.5	20.4	43.6	6.6	15.7	4.2	—	—	3.26	—
Rape-seed oil meal	89.5	33.5	8.1	10.8	30.2	6.9	—	—	5.36	—
Rice, brewers'	88.3	7.5	0.6	0.6	78.8	0.8	0.04	0.10	1.20	—
Rice, brown	87.8	9.1	2.0	1.1	74.5	1.1	0.04	0.25	1.46	—
Rice, polished	87.8	7.4	0.4	0.4	79.1	0.5	0.01	0.09	1.18	0.04

Material	Total dry matter Per ct.	Protein Per ct.	Fat Per ct.	Fiber Per ct.	N-free extract Per ct.	Total minerals Per ct.	Calcium Per ct.	Phos-phorus Per ct.	Nitro-gen Per ct.	Potas-sium Per ct.
Rice bran	90.9	12.5	13.5	12.0	39.4	13.5	0.08	1.36	2.00	1.08
Rice grain, or rough rice	88.8	7.9	1.8	9.0	64.9	5.2	0.08	0.32	1.26	0.34
Rice polishings, or rice polish	89.8	12.8	13.2	2.8	51.4	9.6	0.04	1.10	2.04	1.17
Rubber seed oil meal	91.1	28.8	9.2	10.0	37.6	5.5	—	—	4.61	—
Rye grain	89.5	12.6	1.7	2.4	70.9	1.9	0.10	0.33	2.02	0.47
Rye feed	90.4	16.1	3.3	4.6	62.7	3.7	0.08	0.69	2.58	0.83
Rye flour	88.6	11.2	1.3	0.6	74.6	0.9	0.02	0.28	1½79	0.46
Rye flour middlings	90.6	16.5	3.5	4.2	63.1	3.3	—	—	2.64	—
Rye middlings	90.2	16.6	3.4	5.2	61.2	3.8	—	0.44	2.66	0.63
Rye middlings and screenings	90.4	16.7	3.8	6.1	59.5	4.3	—	—	2.67	—
Safflower seed	93.1	16.3	29.8	26.6	17.5	2.9	—	—	2.61	—
Safflower seed oil meal, from hulled seed	91.0	38.0	6.8	21.0	17.0	8.2	—	—	6.08	—
Safflower-seed oil meal from unhulled seed	91.0	18.2	5.5	40.4	24.1	2.8	—	—	2.91	—
Sagrain sorghum grain	90.0	9.5	3.5	2.1	73.4	1.5	0.43	0.39	1.52	—
Screenings, grain, good grade	90.0	15.8	5.2	9.2	54.3	5.5	—	—	2.53	—
Screenings, grain, chaffy	91.5	14.3	4.4	18.3	46.1	8.4	—	—	2.29	—
Schrock sorghum grain	89.1	10.2	3.0	3.4	70.8	1.7	—	—	1.63	—
Sesame oil meal	93.7	42.8	9.4	6.2	22.8	12.5	2.02	1.61	6.84	1.35
Sesbania seed	90.8	31.7	4.3	13.5	38.0	3.3	—	—	5.07	—
Shallu grain	89.8	13.4	3.7	1.9	68.9	1.9	—	—	2.14	—
Shallu head chops	90.5	12.7	3.5	9.2	61.9	3.2	—	—	2.03	—
Shark meal	91.2	74.5	2.7	0.5	0	13.5	3.48	1.92	12.69	—
Shrimp meal	89.7	46.7	2.8	11.1	1.3	27.8	—	—	7.47	—
Skimmilk, centrifugal	9.5	3.6	0.1	0	5.1	0.7	0.13	0.10	0.58	0.15
Skimmilk, gravity	10.1	3.6	0.8	0	5.0	0.7	0.13	0.10	0.58	0.15
Skimmilk, dried	94.2	34.7	1.2	0.2	50.3	7.8	1.30	1.03	5.56	1.46
Sorghum seed, sweet	89.2	9.5	3.3	2.0	72.8	1.6	0.02	0.28	1.52	0.37
Soybean seed	90.0	37.9	18.0	5.0	24.5	4.6	0.25	0.59	6.06	1.50
Soybean flour, medium in fat	92.9	47.9	6.7	2.4	29.9	6.0	—	—	7.66	—
Soybean flour, solvent extracted	91.5	48.5	0.8	2.6	33.0	6.6	—	—	7.76	—
Soybean mill feed, chiefly hulls	90.8	11.8	2.7	34.0	38.1	4.2	—	—	1.89	—
Soybean oil meal, expeller or hydraulic process, all analyses	90.0	44.3	5.3	5.7	29.6	6.0	0.29	0.66	7.09	1.77
Soybean oil meal, exp. or hydr. process, 44-45% protein guarantee	91.3	45.4	5.3	5.4	29.3	5.9	0.31	0.68	7.26	1.92
Soybean oil meal, exp. or hydr. process, 43% protein guarantee	91.2	44.6	5.3	5.8	29.4	6.1	0.30	0.67	7.14	—
Soybean oil meal, exp. or hydr. process, 41% protein guarantee	90.9	44.2	5.3	5.7	29.7	6.0	0.26	0.59	7.07	—
Soybean oil meal, solvent process	90.6	46.1	1.0	5.9	31.8	5.8	0.30	0.66	7.38	1.92
Starfish meal	96.5	30.6	5.8	1.9	14.3	43.9	—	—	4.90	—
Sudan-grass seed	92.4	14.2	2.4	25.4	38.4	12.0	—	—	2.27	—
Sunflower seed	93.6	16.8	25.9	29.0	18.8	3.1	—	0.55	2.69	0.66
Sunflower seed, hulled	95.5	27.7	41.4	6.3	16.3	3.8	0.20	0.96	4.43	0.92
Sunflower-seed oil cake, from unhulled seed, solvent process	89.2	19.6	1.1	35.9	27.0	5.6	—	—	3.14	—

Material	Total dry matter Per ct.	Protein Per ct.	Fat Per ct.	Fiber Per ct.	N-free extract Per ct.	Total minerals Per ct.	Calcium Per ct.	Phosphorus Per ct.	Nitrogen Per ct.	Potassium Per ct.
Sunflower-seed oil cake, from hulled seed, hydr. process	90.6	36.3	13.5	14.2	20.2	6.4	0.43	1.04	5.81	1.08
Sweet clover seed	92.2	37.4	4.2	11.3	35.8	3.5	—	—	5.98	—
Sweet potatoes, dried	90.3	4.9	0.9	3.3	77.1	4.1	0.21	0.18	0.78	—
Tankage or meat meal, digester process, 60% protein grade	93.1	60.6	8.5	2.0	1.8	20.2	6.37	3.23	9.70	0.46
Tankage with bone, or meat and bone meal, digester process, 50% protein grade	93.5	51.3	11.5	2.3	2.3	26.1	10.97	5.14	8.21	—
Tankage with bone, or meat and bone meal, digester process, 40% protein grade	94.7	42.9	14.1	2.2	4.1	31.4	13.49	5.18	6.86	—
Tomato pomace, dried	94.6	22.9	15.0	30.2	23.4	3.1	—	—	3.66	—
Velvet bean seeds and pods (velvet bean feed)	90.0	18.1	4.4	13.0	50.3	4.2	0.24	0.38	2.90	1.20
Velvet beans, seeds only	90.0	23.4	5.7	6.4	51.5	3.0	—	—	3.74	—
Vetch seed	90.7	29.6	0.8	5.7	51.5	3.1	—	—	4.74	—
Whale meal	91.8	78.5	6.7	0	3.1	3.5	0.56	0.57	12.56	—
Wheat, average of all types	89.5	13.2	1.9	2.6	69.9	1.9	0.04	0.39	2.11	0.42
Wheat, hard spring, chiefly northern plains states	90.1	15.8	2.2	2.5	67.8	1.8	—	—	2.53	—
Wheat, hard winter, chiefly southern plains states	89.4	13.5	1.8	2.8	69.2	2.1	—	—	2.16	—
Wheat, soft winter, Miss. valley and eastward	89.2	10.2	1.9	2.1	73.2	1.8	—	—	1.63	—
Wheat, soft, Pacific Coast states	89.1	9.9	2.0	2.7	72.6	1.9	—	—	1.58	—
Wheat bran, all analyses	90.1	16.9	4.6	9.6	52.9	6.1	0.14	1.29	2.70	1.23
Wheat bran, chiefly hard spring wheat	91.1	17.9	4.9	10.1	52.2	6.1	0.13	1.35	2.86	—
Wheat bran, soft wheat	90.5	16.1	4.3	8.7	55.7	5.7	—	—	2.58	—
Wheat bran, winter wheat	89.9	15.5	4.2	8.9	55.1	6.2	—	—	2.48	—
Wheat bran and screenings, all analyses	90.0	16.8	4.5	9.6	53.0	6.1	0.14	1.21	2.69	—
Wheat brown shorts	88.7	16.9	4.2	7.1	56.0	4.5	—	—	2.70	—
Wheat brown shorts and screenings	88.7	17.0	4.1	7.0	56.0	4.6	—	—	2.72	—
Wheat flour, graham	88.1	12.5	1.9	1.8	70.4	1.5	0.04	0.36	2.00	0.46
Wheat flour, low grade	88.4	15.4	1.9	0.5	69.7	0.9	—	—	2.46	—
Wheat flour, white	88.0	10.8	0.9	0.3	75.6	0.4	0.02	0.09	1.73	0.05
Wheat flour middlings	89.2	18.3	4.2	3.8	59.8	3.1	0.09	0.71	2.93	0.89
Wheat flour middlings and screenings	89.6	18.2	4.5	5.2	57.8	3.9	0.14	0.68	2.91	—
Wheat germ meal, commercial	90.8	31.1	9.7	2.6	42.2	5.2	0.08	1.11	4.98	0.29
Wheat germ oil meal	89.1	30.4	4.9	2.6	46.4	4.8	—	—	4.86	—
Wheat gray shorts	88.9	17.9	4.2	5.7	56.9	4.2	0.13	0.84	2.86	—
Wheat gray shorts and screenings	88.6	17.6	4.0	5.8	57.0	4.2	—	—	2.82	—
Wheat mixed feed, all analyses	89.7	17.2	4.5	7.2	56.1	4.7	0.11	1.09	2.76	—
Wheat mixed feed, hard wheat	89.8	18.7	4.8	7.7	53.6	5.0	0.11	1.09	2.99	—
Wheat mixed feed and screenings	89.3	17.5	4.3	7.1	55.7	4.7	0.11	0.96	2.80	—
Wheat red dog	89.0	18.2	3.6	2.6	61.9	2.7	0.07	0.51	2.91	0.60
Wheat red dog, low grade	89.2	17.9	4.8	4.9	57.9	3.7	—	—	2.86	—
Wheat screenings, good grade	90.4	13.9	4.7	9.0	58.2	4.6	0.44	0.39	2.22	—
Wheat standard middlings, all analyses	89.6	18.1	4.8	6.5	55.8	4.4	0.09	0.93	2.90	1.04

Material	Total dry matter Per ct.	Protein Per ct.	Fat Per ct.	Fiber Per ct.	N-free extract Per ct.	Total minerals Per ct.	Calcium Per ct.	Phos-phorus Per ct.	Nitro-gen Per ct.	Potas-sium Per ct.
Wheat standard middlings and screenings, all analyses	89.7	18.0	4.7	7.4	55.1	4.5	0.15	0.88	2.88	—
Wheat white shorts	89.7	16.1	3.1	2.9	65.0	2.6	—	—	2.58	—
Whey, from cheddar cheese	6.9	0.9	0.3	0	5.0	0.7	0.05	0.04	0.14	0.19
Whey, skimmed	6.6	0.9	0.03	0	5.0	0.7	—	—	0.14	—
Whey, condensed	57.3	8.8	0.6	0	42.0	5.9	—	—	1.41	—
Whey, dried	93.5	12.2	0.8	0.2	70.4	9.9	0.86	0.72	1.96	—
Whey solubles, dried	96.3	17.5	2.0	0	62.8	14.0	—	—	2.80	—
Yeast, brewers', dried	93.8	49.3	1.0	3.7	31.9	7.9	0.13	1.56	7.89	—
Yeast, irradiated, dried	93.9	48.7	1.1	5.5	32.2	6.4	0.07	1.55	7.79	2.14
Yeast, dried, with added cereal	90.2	12.3	3.7	3.2	68.5	2.5	0.09	0.45	1.97	—
Yeast, molasses distillers', dried	91.0	38.8	1.9	6.1	30.2	14.0	—	—	6.21	—

(These tables have been adapted from "U.S.-Canadian Tables of Feed Composition"; Publication 1684; Committee on Animal Nutrition and National Committee on Animal Nutrition, Canada, National Academy of Sciences—National Research Council, Washington, D.C. 1969)

Data Conversion Tables

Weights and Volumes

1 liter water = 1,000 ml = 1,000 grams = 1 kilogram = 2.205 lbs.
1 gallon water = 3,785 ml = 128 ounces = 3.785 liters = 8.34 lbs.
1 cubic foot water = 62.41 lbs. at 10°C

1 gallon = 4 quarts
1 quart = 4 cups = 32 fluid ounces
2 cups = .47 liters

1 cubic inch = 16.3872 cubic centimeters
1 cubic foot = .0283 cubic meters
1 cubic yard = .7646 cubic meters

1 cubic yard = 324 square feet 1 inch deep
             81 square feet 4 inches deep
             27 square feet 12 inches deep

1 cubic centimeter = .0610 cubic inches
1 cubic meter = 35.3145 cubic feet
1 cubic meter = 1.3079 cubic yards

1 pound = 16 ounces = .4536 kilograms = 454 grams
1 ounce = 28.35 grams

1 net ton = 2,000 pounds = .907 metric tons
1 metric ton = 1,000 kilograms
1 gross ton = 2,240 pounds = 1.016 tons

Temperature

$$0°K = -273°C \qquad C = 5/9 \, (F - 32)$$
$$F = 9/5 \, (C + 32)$$

0°C =	32°F	0°F = -17.8 C
5°C =	41°F	5°F = -15.5°C
10°C =	50°F	10°F = -12.2°C
15°C =	59°F	15°F = -9.4°C
20°C =	68°F	20°F = -6.7°C
25°C =	77°F	25°F = -3.9°C
30°C =	86°F	30°F = -1.1°C
35°C =	95°F	35°F = 1.6°C
40°C =	104°F	40°F = 4.4°C
45°C =	113°F	45°F = 7.2 C
50°C =	122°F	50°F = 10.0°C
100°C =	212°F	60°F = 18.3°C
200°C =	392°F	70°F = 21.2°C
80°F =	26.7°C	
90°F =	32.2°C	
95°F =	35.0°C	
100°F =	37.8°C	
120°F =	48.9°C	
140°F =	60.0°C	
160°F =	71.1°C	
180°F =	82.2°C	
200°F =	93.4°C	

Heat Energy

1 calorie = the heat energy to raise to 1 gram (1 ml) H_2O 1°C
1 BTU = 252 calories = the heat energy to raise 1 pound H_2O 1°F

Light

1 foot-candle = 10.7639 lux = the amount of light from 1 lumen from a distance of 1 foot over the
surface area of 1 square foot

1 lux = 1 lumen at a distance of 1 meter over a surface area of 1 square meter

Pressure and Power

1 horsepower = 746 watts
1 kilogram/square centimeter = 14.233 pounds per square inch
1 pound/square inch = .0703 kilograms per square centimeter
1 kilogram per square meter = .2048 pounds per square foot

1 atmosphere pressure = 1.0332 kilograms per square centimeter
= 4.696 pounds per square inch
= 1.0133 bars

Miscellaneous Data

1 level tablespoon of malt/agar (50:50) media = approximately 7 grams

1 US 5-cent piece (nickel) = approximately 5 grams

50 pounds of dry rye grain = approximately 125 cups
100 grams of dry rye grain = approximately 125 milliliters
1 standard US glass gallon "mayonnaise" jar = 3,800 milliliters when filled to brim

1 dry ton finely chopped wheat straw, when wetted and compressed,
    occupies approximately 200–250 cubic feet

1 ton compost (straw/manure) = 2 cubic yards = 70 square feet of beds (10–12 inches deep)

1 yard fresh alder sawdust = approximately 700 lbs.

household bleach = 5% sodium hypochlorite
1 tablespoon bleach/gallon water = 200 ppm chlorine
1 cup household bleach/gallon = 3,200 ppm chlorine

Glossary

A

agar: a product derived from seaweed, valued for its gelatinizing properties, and commonly used to solidify nutrified media for sterile tissue culture.

agarics: mushrooms with gills.

anamorph: The asexual form or morph of a fungus characterized by the presence of conidia.

anastomosis: the fusion of hyphal cells followed by an exchange of cellular contents between two mycelial networks.

annulus: a ring, a collar, or cellular skirt forming on the stem, typically originating from a portion of the partial veil.

appressed: flattened.

ascomyces: the larger group of fungi hosting those which reproduce using an ascus for spore generation.

ascus, asci: a sac-like cell typical of the class Ascomycetes, usually containing 6 or 8 spores. Most cup fungi and morels (*Morchella*) belong to this group.

autoclave: a steam-pressurized vessel used for heat-treating.

B

basidia: the club-like cells that give rise to 4 (more rarely 2 or 6) spores.

Basidiomycetes: the class of fungi producing spores on basidia. The gilled, pored, teethed, and some cup mushrooms are basidiomycetes.

biological efficiency: the percentage measurement of the yield of fresh mushrooms from the dry weight of the substrate. (See page 55.) Biological efficiency of 100% is equivalent to saying that from a substrate with a moisture content of 75%, 25% of its mass will yield fresh mushrooms having a moisture content of 90%.

bleach bombing: an industry-used phrase to describe the use of bleach sprayed on the walls and floors. The rooms so treated are usually sealed tight for 24 hours, allowing the chlorine gas to thoroughly disinfect the environment.

block: a term used in mushroom culture, referring to the cube-shaped mass of sawdust substrate contained within plastic bags. Once the mycelium has grown through the substrate, the plastic can be stripped off, and the mycelium holds the mass together. Blocks can be used individually or collectively to build "walls" of mushroom mycelium.

brown rot: a condition caused by the degradation of cellulose by fungi, which leaves the substrate brown in color. The brown color is largely due to undecomposed lignin. Solid blocks of wood are used for testing whether or not a fungus causes "brown rot" or "white rot."

C

capitate: having a swollen head.

carpophore: the fruiting body of higher fungi.

casing: a layer of water-retentive materials applied to a substrate to encourage and enhance fruit-body production.

cheilocystidia: variously shaped, sterile cells on the gill edge of mushrooms.

chlamydospores: thick walled, secondary spores developing from hyphae but not from basidia, nor from chlamydospores.

clamp connection: a small, semicircular, hollow bridge that is laterally attached to the walls of two adjoining cells and spanning the septum between them. See page 134.

collyboid: resembling mushrooms typical of the genus *Collybia*-groups of mushrooms clustered together at the base and having convex to plane caps.

conidia: a uninucleate, exteriorly borne cell formed by constriction of the conidiophore.

conidiophore: a specialized stalk arising from mycelium upon which conidia are borne.

conspecific: equal to, e.g., two conspecific taxa are in fact the same species.

contamination: any organism other than the one desired to cultivated.

context: the internal flesh of mushroom, existing between the differentiated outer layers of the mushroom.

coprophilic: dwelling on and having an affection for manure.

coremia: a bundle of reproductive structures (conidiophores). Some *Pleurotus* species (*P. cystidiosus, P. abalonus,* and *P. smithii*) produce coremic structures in culture-often black droplets of spore bundles on relatively long stalks. Once the droplets dry, the spores become airborne. See page 70.

cystidia: microscopic, sterile cells arising from the gill, cap, or stem.

D

deciduous: used to describe trees that seasonally shed their leaves.

decurrent: the attachment of the gill plates to the stem of a mushroom where the gills partially run down the stem.

deliquescing: the process of auto-digestion by which the gills and cap of a mushroom melt into a liquid. Typical of some members in the genus *Coprinus.*

dikaryotic: the state wherein two individual nuclei are present in each fungal cell.

dimitic hyphae: fungal flesh typified by two kinds of hyphae.

dimorphic: having two forms.

diploid: a genetic condition wherein each cell has a full complement of chromosomes necessary for sexual reproduction, denoted as 2N.

disk: the central portion of the mushroom cap.

E

eccentric: off-centered.

ellipsoid: oblong-shaped.

endospores: spores formed internally.

entheogen: any naturally occurring substance that, when ingested, produces a profound religious state of mind, often described as God-like.

evanescent: fragile and soon disappearing.

F

farinaceous: grain-like, usually in reference to the scent of mycelium or mushrooms.

fermentation: the state of actively growing micro-organisms, usually in a liquid environment.

fibrillose: fine, thin, hair-like filaments.

filamentous: composed of hyphae or thread-like cells.

flexuose, flexuous: bent alternatively in opposite directions.

floccose: having cottony patches of tissue.

flush: a crop of mushrooms, collectively forming within a defined time period, often repeating in a rhythmic fashion.

foot-candle: a measurement of the intensity of light, equivalent to 10.7639 lux. A foot-candle is the amount of light from 1 lumen at a distance of 1 foot over a surface area of 1 square foot.

fruitbody: the mushroom structure.

fruiting: the event of mushroom formation and development.

G

generative hyphae: the thin-walled, branched, and narrow cells that give rise to the spore-producing layers and surface tissues. Species typified by clamp connections will have clamps at the septa of the generative hyphae.

genotype: the total genetic heritage of constitution of an organism, from which individual phenotypes are expressed.

Gram (Gram's Stain): A method for separating bacteria whereby bacteria are stained first with crystal violet (a red dye) and then washed with an iodine solution. Gram positive bacteria retain the dye. Gram negative bacteria lose the dye.

gypsum: calcium sulfate; $CaSO_4 \times 2\ H_2O$. A buffer used in spawn making to keep grain kernels separated. Calcium sulfate slightly acidifies a substrate as sulfuric acids evolve.

H

heterothallic: having two or more morphologically similar pairs of strains within the same species. The combination of compatible spore types is essential for producing fertile offspring.

homothallic: having one strain type that is dikaryotic and self-fertile, typically of mushrooms that produce two spores on a basidium.

hygrophanous: fading markedly in color upon drying.

hymenium: the fertile outer layer of cells from which basidia, cystidia, and other cells are produced.

hymenophore: the fertile portion of the mushroom bearing the hymenium.

hypha, hyphae (pl.): the individual fungal cell or cells.

hyphal aggregates: visible clusters of hyphae, resembling cottony tufts of mycelium, often preceding but not necessarily leading to primordia formation.

hyphosphere: the microscopic environment in direct proximity to the hyphae.

K

karyogamy: the fusion of two sexually opposite nuclei within a single cell.

L

lageniform: thin and sinuous.

lamellae: the gills of a mushroom, located on the underside of the cap.

lamellulae: the short gills, originating from the edge of the outer peripheral edge of cap but fully extending to the stem.

lignicolous: growing on wood or a substrate composed of woody tissue.

lignin: the organic substance that, with cellulose, forms the structural basis of most woody tissue.

lumen: the amount of the flow of light emitted from a single international foot-candle.

lux: a measurement of light received by a surface equal to 1 lumen at a distance of 1 meter over a surface area of 1 square meter.

M

macroscopic: visible to the naked eye.

membranous: being sheath-like in form.

meiosis: the process of reduction division by which a single cell with a diploid nucleus subdivides into four cells with one haploid nucleus each.

mesophile: an organism thriving in moderate temperature zone, usually between 40–90°F (4–32°C).

micron: 1,000,000th of a meter.

mitosis: the nonsexual process of nuclear division in a cell by which the chromosomes of one nucleus are replicated and divided equally into two offspring nuclei.

monokaryon: the haploid state of the mushroom mycelium, typically containing one nucleus.

monomitic: fungal flesh consisting only of thin-walled, branched, and narrow (generative) hyphae.

mycelium: a fungal network of thread-like cells.

mycology: the study of fungi.

mycophagist: a person or animal that eats fungi.

mycophile: a person who likes mushrooms.

mycophobe: a person who fears mushrooms.

mycorrhizal: a symbiotic state wherein mushroom mycelium forms on or in the roots of trees and other plants.

mycosphere: the environment in which the mycelium operates.

mycotopia: a term coined by Paul Stamets to describe an environment in which fungi are actively used to enhance and/or preserve ecological equilibrium.

N

natural culture: the cultivation of mushrooms outdoors, benefiting from natural weather conditions.

nucleus: a concentrated mass of differentiated protoplasm in cells containing chromosomes and playing an integral role in the reproduction and continuation of genetic material.

O

oidia: conidia (spores) borne in chains.

P

pan/panning: the dieback of mycelium caused by a variety a reasons, primarily sudden drying after wetting.

parasite: an organism living on another living species and deriving its sustenance to the detriment of the host.

partial veil: the inner veil of tissue extending from the cap margin to the stem and at first covering the gills of mushrooms.

pasteurization: the rendering of a substrate to a state where competitor organisms are at a disadvantage, allowing mushroom mycelium to flourish. Steam or hot water is usually used; biological and chemical pasteurization are alternative methods.

phenotype: the observable physical characteristics expressed from the genotype.

photosensitive: sensitive to light.

phototropic: growing towards light.

pileocystidia: sterile cells on the surface of the cap.

pileus: the mushroom cap.

pinhead: a dot-like form that develops into a mushroom. The pinhead is the earliest visible indication of mushroom formation.

pleurocystidium, pleurocystidia (pl.): the sterile cell(s) on the surface of mushroom gills, distinguished from those sterile cells occurring on their outer edges.

primordium, primordia(pl.): the mushroom at the earliest stage of growth, synonymous with "pinhead."

psilocybian: containing psilocybin and/or psilocin.

R

radicate: tapering downwards. Having a long root-like extension of the stem.

rhizomorph: a thick string-like strand of mycelium. A rhizomorph can consist of one enlarged cell or many, usually braided.

rhizosphere: the space encompassing the rhizomorph or the zone around the roots of plants.

S

saprophyte: a fungus that lives on dead organic matter.

scabrous: roughened with tufted, short ridges.

sclerotium, sclerotia (pl.): a resting stage of mycelium typified by a mass of hardened mycelium resembling a tuber and from which mushrooms, mycelia, or conidia can arise. Sclerotia are produced by both ascomycetes and basidiomycetes.

sector: usually used to describe fans of mycelium morphologically distinct from the type of mycelium preceding and bordering it.

senescence: the state whereby a living organism declines in vigor due to age, a consequence of limitation of cell divisions due to errors in replicating DNA.

septate: cells with distinct walls.

septum, septa (pl.): structural divisions between cells, i.e., cell walls.

skeletal hyphae: coarse, inflated cellular network consisting of thick-walled, unbranched cells lacking cross-walls. Skeletal hyphae give mushrooms a tough, fibrous texture, especially at the stem base. Except for the basal cell, they are typically clampless.

spawn: any material impregnated with mycelium, the aggregation of which is used to inoculate more massive substrates.

species: a biologically discrete group of individuals that are cross-fertile, and give rise to fertile progeny.

sporeless strains: Strains that do not produce spores. Sporeless Oyster strains are highly sought after given the health problems associated with growing these mushrooms indoors.

spores: discrete cells that are used to spread fungi to new ecological niches, and are essential in the recombination of genetic material.

sporocarps: any fruitbody that produces spores.

sterilization: the rendering of a substrate to a state where all life-forms have been made inviable. Sterilization by heat (steam) is more commonly employed in mushroom cultivation than chemical, gas, UV, or radioactive means. Sterilization usually implies prolonged exposure to temperatures at or above the boiling point of water (212°F/100°C) at or above atmospheric pressure.

stipe: the stem of a mushroom.

strain: a race of individuals within a species sharing a common genetic heritage but differing in some observable set of features, which may or may not be taxonomically significant.

stroma: a dense, cushion-like aggregation of mycelium forming on the surface of substrate, which generally does not lead to fruitbody formation.

subhymenium: the layer of cells directly below the hymenium.

substrate: straw, sawdust, compost, soil, or any organic material on which mushroom mycelium will grow.

super-pasteurization: prolonged pasteurization utilizing steam. Super-pasteurization typically is for 12–48 hours at or near 100°C (212°F) at or near atmospheric pressure. Super-pasteurization is a method commonly used to render sawdust substrates, in bulk, into a form usable for the cultivation of Shiitake, Oyster, and/or similar mushrooms.

T

taxon, taxa (pl.): a taxonomic unit, usually in reference to a species.

thermogenesis: the natural and spontaneous escalation of temperature occurring in substrates as fungi, bacteria, and other microorganisms flourish.

through-spawning: mixing spawn evenly throughout the substrate.

top-spawning: placing spawn as a layer on the top of a substrate.

trama: the internal layers of cells between the gills of mushrooms.

U

universal veil: an outer layer of tissue enveloping the cap and stem of some mushrooms, best seen in the youngest stages of fruitbody development.

V

variety: a subspecies epithet used to describe a consistently appearing variation of a particular mushroom species.

vector: the pathway through or carrier on which an organism travels.

veil: a tissue covering mushrooms as they develop.

W

wedge-transfer: the cutting of triangular-shaped sections of mycelium from agar and their transfer into other vessels or substrates.

white rot: a condition whereby a substrate is rendered light in color from the decomposition of lignin ("delignification") and/or cellulose and hemicellulose from fungi. Solid blocks of wood are used for testing.

Bibliography

Adachi, K., H. Nanba & H. Kuroda, 1987. Potential of host-mediated antitumor activity in mice by beta-glucan obtained from *Grifola frondosa* (Maitake). *Chem. Pharm. Bull.* 35: 262.

Adachi, K., H. Nanba, M. Otsuka & H. Kuroda, 1988. Blood pressure-lowering activity present in the fruit body of *Grifola frondosa* (Maitake). *Chem. Pharm. Bull.* 36(3):1000–1006.

Adaskaveg, J. E. & R. L. Gilbertson, 1986. *Mycologia* 78: 694–705.

_____, 1987. Vegetative incompatibility between intraspecific pairings of *Ganoderma lucidum* and *G. tsugae*. *Mycologia* 79: 603–613.

Ainsworth, G. C., 1971. *Dictionary of Fungi*, 6th ed. Commonwealth Mycological Institute, Kew, Surrey, England.

Andersson, H. C., J. Hajslova, V. Schulzova, Z. Panovksa, L. Hajkova, & J. Gry, 1999. Agaritine content in processed foods containing the cultivated mushroom *(Agaricus bisporus)* on the Nordic and Czech market. Food Additive Contamination Oct: 16(10):439–446.

Ando, M., 1974. Fruitbody formation of *Lentinus edodes* (Berk.) Sing. on the artificial media. *Mushroom Science IX* (Part I). Proceedings of the Ninth International Scientific Congress on the Cultivation of Edible Fungi, Tokyo.

Anselmi, N. & G. Deandrea, 1979. Culture de *Pleurotus ostreatus* (Jacq.) Quel. sur du bois de salicicacees: applications pratiques et risques eventuels de diffusion hemiparasitaire. *Mushroom Science X* Part 2: 451–461.

Arita, I., 1979. The mechanism of spontaneous dedikaryotization in hyphae of *Pholiota nameko*. *Mycologia* 71: 603–611.

Arnolds, E., 1992. Mapping and monitoring of macromycetes in relation to nature conservation. *McIlvainea* 10: 2, 4–27.

Arora, D., 1979 (2nd ed., 1986). *Mushrooms Demystified*. Ten Speed Press, Berkeley, California.

_____, 1992. *All that the Rain Promises*. Ten Speed Press, Berkeley, California.

Atkins, F., 1966. *Mushroom Growing Today*. Macmillan Publishing Co., New York, New York.

Azizi, K. A. & T. R. Shamala, 1990. Cultivation of *Pleurotus sajor-caju* on certain agro-industrial wastes and utilization of the residues for cellulase and D-xylanase production. *Mushroom Journal for the Tropics* 10(1): 21–26.

Badham, E. R., 1985. The influence of humidity upon transpiration and growth in *Psilocybe cubensis*. *Mycologia* 77: 932–939.

_____, 1988. Is autoclaving Shiitake substrate necessary? *Mushroom Journal for the Tropics* (8): 129–136.

Bankhead, C., 1999. Mushrooms may play role in breast cancer prevention and treatment. *Medicine Science News* Dec. 10.

Bano, Z., S. Rajarathnam, 1981. Studies on the cultivation of *Pleurotus sajor-caju*. *Mushroom Journal* 101: 243–245.

Bano, Z., S. Rajarathnam & N. Nagaraja, 1978. Some aspects on the cultivation of *Pleurotus flabellatus* in India. *Mushroom Science X*, Part 2: 597–607.

Benjamin, D., 1995. *Mushrooms: Poisons & Panaceas*. W.H. Freeman & Co., New York.

Bessette, A., 1988. *Mushrooms of the Adirondacks: A Field Guide*. North Country Books, Utica, New York.

Blanchart, R. A., B. Compton, N. Turner & R. Gilbertson, 1992. Nineteenth-century shaman grave guardians are carved *Fomitopsis officinalis* sporophores. *Mycologia* 84(1): 119–124.

Bo, L. & B. Yun-sun, 1980. *Fungi pharmacopoeia (Sinica)*. The Kinoko Company, Oakland, California.

Bobek, P., L. Ozdin & S. Galbavy, 1998. Dose and time-dependent hypocholesterolemic effect of oyster mushroom (*Pleurotus ostreatus*) in rats. *Nutrition* March 14(3): 282–286.

Bobek, P., O. Ozdin & M. Mikus, 1995. Dietary oyster mushroom (*Pleurotus ostreatus*) accelerates plasma cholesterol turnover in hypercholesterolaemic rat. *Physiological Research* 44(5): 287–291.

Bononi et al. 1991. *Pleurotus ostreatoroseus* cultivation in Brazil. *Mushroom Science XI*. A. A. Balkema, Netherlands.

Borchers, A. T., J. S. Stern, R. M. Hackman, C. L. Keen & M. E. Gershwin, 1999. Mushrooms, tumors and immunity. *Proc. Soc. Exp. Biol. Med*. Sept. 221(4):281–293.

Bresinski, A., O. Hilber & H. P. Molitoris, 1977. The genus *Pleurotus* as an aid for understanding the concept of species in basidiomycetes. *The Species Concept in Hymenomycetes* 229–259. Cramer, Valduz.

Bresinski, A., M. Fischer, B. Meixner & W. Paulus, 1987. Speciation in *Pleurotus*. *Mycologia* 79: 234–245.

Bressa, G., L. Cima & P. Costa, 1988. Bioaccumulation of Hg in the mushroom *Pleurotus ostreatus*. *Ecotoxicology and Environmental Safety* Oct. 16(2): 85–89.

Brooke-Webster, D., 1987. The use of polyethylene film to control fructification of *Pleurotus* species on horizontal trays. *Developments in Crop Science X: Cultivating Edible Fungi*. Elsevier, Oxford.

Brough, J., 1971. Soma and *Amanita muscaria*. *Bulletin of the School of Oriental and African Studies (BSOAS)* 34(2): 331–362.

Buller, A. K., 1934. *Researches on Fungi 6*. Longmans, Green & Co., London, 310–324.

Burdsall, H., T. J. Volk & J. Ammirati, 1996. Bridgeoporus, a new genus to accommodate *Oxyporus nobilissimus* (Basidiomycotina, Polyporaceae) *Mycotaxon* LX., Oct.-Dec.: 387–395.

Callac, P., C. Billette, M. Imbernon & R. W. Kerrigan, 1993. Morphological, genetic, and interfertility analyses reveal a novel, tetrasporic variety of *Agaricus bisporus* from the Sonoran desert of California. *Mycologia* 85(5): 835–851.

Calzada, J. F., E. de Porres, R. de Leon, C. Rolz & L. F. Franco, 1987. Production of food and feed from wheat straw by *Pleurotus sajor-caju*. *Mushroom Journal of the Tropics* (7): 45–46.

Chang, S. T., 1972. *The Chinese Mushroom (Volvariella volvacea): Morphology, Cytology, Genetics, Nutrition and Cultivation*. The Chinese University of Hong Kong, Hong Kong.

Chang, S. T. & W. A. Hayes, 1978. *The Biology and Cultivation of Edible Mushrooms*. Academic Press, New York.

Chang, S. T. & P. G. Miles, 1987. Historical record of the early cultivation of *Lentinus* in China. *Mushroom Journal of the Tropics* 7: 47.

_____, 1989. *Edible Mushrooms and Their Cultivation*. CRC Press, Boca Raton, Florida.

Chang, S. T., J. A. Buswell & P. G. Miles (eds.), 1992. *Genetics and Breeding of Edible Mushrooms*. Gordon & Breach Science Publishers, New York.

Chantarasnit, A., 1989. Factors affecting contamination in plastic bag cultivation of the black mushroom (*Lentinus edodes*). *Mushroom Journal of the Tropics* (9): 15–20.

Chen, A. W., 2000. Mixed culture cultivation by synthetic logs of *Tremella fuciformis* Berk. In press.

Chen, A. W., P. Stamets & N. L. Huang, 1999. Compost-substrate fermentation and crop management for successful production of *Agaricus blazei*. In press.

Chen, Guo-Liang, 1992. Studies on the cultivation and medicinal efficacy of *Hericium erinaceus*. Translation. The Edible Fungi Research Institute of The Shanghai Academy of Agricultural Science, China.

Chihara, G., 1978. Antitumor and immunological properties of polysaccharides from fungal origin. National Cancer Institute, Tokyo. *Proceedings of the Tenth International Congress on the Science and Cultivation of Edible Fungi,* France.

Chilton, J., 1986. Grow the garden giant. *Mushroom* 14:5(1): 17–19.

Chu-Chous, M., 1983. Cultivating edible forest mushrooms. Vol. 119. *What's New in Forest Research,* Forest Research Institute, Rotorua, New Zealand.

Clemencon, H. & J. M. Moncalvo, 1990. Taxonomic analysis of cultural characters in the group *Lyophyllum shimeji* (Agaricales, Basidiomycetes) from Japan. *Transactions of the Mycological Society of Japan* 31:479–488.

Cochran, K. W., 1978. Medical effects. *The Biology & Cultivation of Edible Mushrooms*. Academic Press, New York.

Cochran, K., 1989. Personal communication. Mushrooms and cancer. Unpublished bibliography on the anticancer properties of mushrooms.

Collins, R. A. & T. B. Ng, 1997. Polysaccharopeptide from *Coriolus versicolor* has potential for use against human immunodeficiency virus type 1 infection. *Life Sciences* 60(25): PL383–7.

Corner, E. J. H., 1981. The Agaric Genera *Lentinus, Panus,* and *Pleurotus*. Nova Hedwigia, J. Cramer.

Costanin, J. 1936. La culture de la morille et sal forme conidienne. *Ann. Sci. Nat. Bot.* (Ser.10)18: 111–129.

Crisan, E. V. & A. Sands, 1978. Nutritional value. *The Biology & Cultivation of Edible Mushrooms*. Academic Press, New York.

Czarnecki, J., 1988. *Joe's Book of Mushroom Cookery*. Atheneum/Macmillan Publishing Co., New York.

Dahlstrom, J. L., J. E. Smith & N. S. Weber, 2000. Mycorrhiza-like interaction by *Morchella* with species of the *Pinaceae* in pure culture synthesis. *Mycorrhiza* 9: 279–285.

Danell E., 1994. *Cantharellus cibarius:* mycorrhiza formation and ecology. Acta Universitatis Upsaliensis. Comprehensive Summaries of Uppsala Dissertations from the Faculty of Science and Technology 35. 75 pp.

Danell E. & F. Camacho, 1997. Successful cultivation of the golden chanterelle. *Nature,* 385: 303.

Dong, Y., C. Y. Kwan, S. N. Chen & M. Yang, 1996. Antitumor effects of a refined polysaccharide peptide fraction isolated from *Coriolous versicolor:* In Vitro and In Vivo Studies. *Research Communications in Molecular Pathology and Pharmacology,* May, 92(2): 140–147.

Dong, Y., M. M. Yang & C. Y. Kwan, 1997. In vitro inhibition of proliferation of HL-60 cells by tetrandrine and *Coriolus versicolor* peptide derived from Chinese medicinal herbs. *Life Sciences* 60(8): PL135–40.

Donoghue, D. C., 1962. New light on fruitbody formation. *Mushroom Science V* : 247–249.

Eger, G., 1974. The action of light and other factors on sporophore initiation in *Pleurotus ostreatus. Mushroom Science IX* (I): 575–583.

_____, 1980. Blue light photomorphogenesis in mushrooms. *The Blue Light Syndrome:* 556-562. Springer-Verlag, Berlin.

Eger, G., G. Eden & E. Wissig, 1976. *Pleurotus ostreatus*-breeding potential of a new cultivated mushroom. *Theor. Appl. Genet.* 47: 155–163.

Eger, G., S. F. Li, and H. Leal-lara, 1979. Contribution to the discussion on the species concept in the *Pleurotus ostreatus* complex. *Mycologia,* 71: 577–588.

El-Kattan, M. H., Y. Gali, E. A. Abdel-Rahim & A. Z. M. Aly, 1990. Submerged production of *Pleurotus sajor-caju* on bagasse hydrolyzate medium. *Mushroom Journal of Tropics* 10: 105–114.

El-Kattan, M. H., Z. A. Helmy, M. A. E. El-Leithy & K. A. Abdelkawi, 1991. Studies on cultivation techniques and chemical composition of Oyster mushrooms. *Mushroom Journal of the Tropics* (11): 59–66.

Eugenio, C. P. & N. A. Anderson, 1968. The genetics and cultivation of *Pleurotus ostreatus. Mycologia* 60: 627–634.

Farr, D., 1983. Mushroom industry: diversity with additional species in the United States. *Mycologia* 75(2): 351–360.

Field, J. A., E. de Jong, G. Feijoo-Costa & J. A. de Bont, 1992. Biodegradation of polycyclic aromatic hydro-carbons by new isolates of white rot fungi. *Applied Environmental Microbiology* July 58(7): 2219–2226.

Finchman, J. R. S., P. R. Day & A. Radford, 1979 (4th ed.). *Fungal Genetics.* Blackwell Scientific Publications, Oxford, England.

Finkenstein, David B. & C. Ball, 1991. *Biotechnology of Filamentous Fungi.* Butterworth-Heinemann, Boston, Massachusetts.

Fisher, D.W. & Alan E. Bessette, 1992. *Edible Mushrooms of North America: A Field-to-Kitchen Guide.* University of Texas Press, Austin, Texas.

Flegg, P. B., D. M. Spencer & D. A. Wood, 1985. *The Biology and Technology of the Cultivated Mushroom.* John Wiley & Sons, Chichester, U.K.

Fletcher, J. T., R. F. White & R. H. Gaze, 1986. Mushrooms: pest and disease control. *Intercept,* Andover, Hants, U.K.

Fox, F. M., 1983. Role of basidiospores as inocula for mycorrhizal fungi of birch. *Tree Root Systems and Their Mycorrhizas.* Nijhoff, The Hague.

Fujii, T., H. Maeda, F. Suzuki & N. Ishida, 1978. Isolation and characterization of a new antitumor polysaccharide, KS-2, extracted from cultured mycelia of *Lentinus edodes. Journal of Antibiotics* 31(1): 1079–1090.

Fujimiya, Y., Y. Suzuki, K. Oshiman, H. Kobori, K. Moriguchi, H. Nakashima, Y. Matumoto, S. Takahara, T. Ebina & R. Katakura, 1998. Selective tumoricidal effect of soluble proteoglucan extracted from basidiomycete, *Agaricus blazei* Murrill, mediated via natural killer cell activation and apoptosis. *Cancer Immunololgy & Immunotherapy* 46: 147–159.

Fujimiya, Y., Y. Suzuki, R. Katahura & T. Ebina, 1999. Tumor-specific cytocidal and immunopotentiating effects of relatively low molecular weight products derived from the basidiomycete, *Agaricus blazei* Murrill. *AntiCancer Research* Jan-Feb 19(1A):113–118.

Fujimoto, T., 1989. High-speed year-round shiitake cultivation. *Shiitake News 5 & 6.*

Fukuoka, M. 1978. *The One-Straw Revolution.* Rodale Press, Emmaus, Pennsylvania.

Ghoneum, M., 1994. NK-Immunomodulation by active hemicellulose compound (AHCC) in 17 cancer patients. Abstract of 2nd Meeting Society of Natural Immunity, Taormina, Italy, May 25–28.

_____, 1995. Immunomodulatory and anti-cancer properties of (MGN-3), a modified xylose from rice bran, in 5 patients with breast cancer. Abstract, 87th Meeting Am. Ass. Cancer Res. (AACR) special conference: The Interference between Basic and Applied Research November 5–8. Baltimore, Maryland.

_____, 1998. Enhancement of human natural killer cell activity by modified arabinoxylane from rice bran (MGM-3). *International Journal of Immunotherapy* XIV (2): 89–99.

_____, 1998. Anti-HIV activity in vitro of MGN-3, an activated arabinoxylane from rice bran. *Biochemical and Biophysical Research Communications* 243: 25–29.

Ghoneum, M. & G. Namatalla, 1996. Natural killer immunomodulatory function in 27 cancer patients by MGN-3, a modified arabinoxylane from rice bran. Abstract, Ann. Assoc. Cancer Res., April 2–4. Washington, D.C.

Ghoneum, M., M. Wimbley, F. Salem, A. Mcklain, N. Attallan, G. Gill., 1995. Immunomodulatory and anti-cancer effects of active hemicellulose compound (AHCC). *Int. Journal of Immunotherapy* XI (1): 23–28.

Gilbert, F.A., 1960. The submerged culture of *Morchella. Mycologia* 52: 201–209.

Gilbertson R. & Ryvarden L., 1986. *North American Polypores: Vol. I & II.* FungiFlora, Oslo, Norway.

Gordon, M., B. Bihari, E. Goosby, R. Gorter, M. Greco, M. Guralnik, T. Mimura, V. Rudinicki, R. Wong & Y. Kaneko, 1998. A placebo-controlled trial of the immune modulator, lentinan, in HIV-positive patients: a phase I/II trial. *Journal of Medicine* 29(5–6): 305–330.

Gormanson, D. & M. Baughman, 1987. *Financial Analysis of Three Hypothetical, Small-Scale Shiitake Mushroom Production Enterprises.* University of Minnesota Department of Forest Resources.

Graham, K. M., R. M. Herbagiandono & M. E. Marvel, 1980. Cultivation of *Pleurotus flabellatus* on agricultural wastes in Indonesia. *Mushroom Newsletter for the Tropics* 1(1): 17–18.

Guiochon, P. F. H. G. F., 1958. U.S. #2,851,821, U.S. Patent Office, Washington, D.C.

Gunde-Cimerman, N. G. & A. Cimerman, 1995. *Pleurotus* fruiting bodies contain the inhibitor of 3-hydroxy-3-methylglutaryl-Coenzyme A Reductase-Lovastatin. *Experimental Mycology* 19: 1–6.

Gunde-Cimerman, N., 1999. Medicinal value of the genus *Pleurotus* (Fr.) P. Kast. (Agaricales s.l., Basidiomycetes). *International Journal of Medicinal Mushrooms* 1: 69–80.

Guzman, G. 1983. *The Genus Psilocybe.* J. Cramer, Lichtenstein.

Guzman, G. & J. Ott, 1976. Description and chemical analysis of a new species of hallucinogenic *Psilocybe* from the Pacific Northwest. *Mycologia* 68: 1261–1267.

Guzman, G., V. M. Bandala & L. Montoya, 1991. A comparative study of telemorphs and anamorphs of *Pleurotus cystidiosus* and *Pleurotus smithii. Mycological Research* 95: 1264–1269.

Guzman, G., J. Ott, J. Boydston & S. H. Pollock, 1976. Psychotropic mycoflora of Washington, Idaho, Oregon, California, and British Columbia. *Mycologia* 68: 1267–1272.

Guzman, G., L. Montoya, D. Salmones & V. M. Bandala, 1993. Studies of the genus *Pleurotus* (Basidiomycotina) II. *P. djamour* in Mexico and in other Latin-American countries, taxonomic confusions, distribution, and semi-industrial culture. *Journal of Cryptogamic Botany* 3: 213–220.

Hammerschmidt, D.E., 1980. Szechwan purpura. *New England Journal of Medicine* 302: 1191–1193.

Hanada, K. & I. Hashimoto, 1998. Flagellate mushroom (Shiitake) dermatitis and photosensitivity *Dermatology* 197(3): 255–257.

Hardenburg, R. E., A. E. Watada & C. Y. Wang, 1993. The commercial storage of fruits, vegetables, and florist and nursery stocks. U.S.D.A., Government Printing Office, Washington, D.C.

Harrington, P., 1999. Fungus does the dirty work. *Breakthroughs* Winter 1999–2000: 15.

Harris, B., 1976. *Growing Wild Mushrooms.* Wingbow Press, Berkeley, California.

_____, 1986. *Growing Shiitake Commercially.* Science Tech Publishers, Madison, Wisconsin.

Hashida, C., K. Hayashi, L. Jie, S. Haga, M. Sakurai & H. Shimizu, 1990. Quantities of agaritine in mushrooms (Agaricus bisporus) and the carcinogenicity of mushroom methanol extracts on the mouse bladder epithelium *Nippon Kosh Eisei Zasshi June:* 37(6):400–405.

Hawksworth, D. L., P. M. Kirk, B. C. Sutton & D. N. Pegler, 1995. *Ainsworth and Bisby's Dictionary of the Fungi* (8th ed.) CAB International, Wallingford, United Kingdom.

Hayes, W. A., P. E. Randle & F. T. Last, 1969. The nature of the microbial stimulus affecting sporophore formation in *Agaricus bisporus* (Lange) Sing. *Annals of Applied Biology* 64: 177–181.

Heim, R. & R. G. Wasson, 1958. *Les Champignons Hallucinogene du Mexique.* Editions du Museum National D'Histoire Naturelle, Extrait des Archives. Paris.

Heim, R., R. Cailleux, R. G. Wasson & P. Thevenard, 1967. *Nouvelles Investigations sur les Champignons Hallucinogenes.* Editions du Museum National D'Histoire Naturelle, Extrait des Archives. Paris.

Hengshan, C. et al. 1991. Log cultivation of *Ganoderma lucidum. Edible Fungi of China,* 10(2): 29–32.

Hibbett, D. S. & R. Vilgalys, 1991. Evolutionary relationships of *Lentinus* to the Polyporaceae: Evidence from restriction analyses of enzymatically amplified ribosomal DNA. *Mycologia* 83(4): 425–439.

Hibbett, D. S. & R. G. Thorn, 1994. Nematode trapping in *Pleurotus tuberregium. Mycologia* 86: 696–699.

Hilber, O., 1982. Die Gattung *Pleurotus* (Fr.) Kummer. *Bibliotheca Mycologica* 87. J. Cramer, Vaduz.

_____, 1989. Valid, invalid, and confusion taxa of the genus *Pleurotus.* Mushroom Science XII (Part II). Proceedings of the Twelfth International Congress on the Science and Cultivation of Edible Fungi, Braunschweig, Germany.

_____, 1997. *The Genus Pleurotus* (Fr.) Kummer. Erschienen im Selbstverlag, Kelheim, Germany.

Hirasawa, M., N. Shouji, T. Neta, K. Fukushima & K. Takada, 1999. Three kinds of antibacterial substances from *Lentinus edodes* (Berk.) Sing. (shiitake, an edible mushroom). *International Journal of Antimicrobial Agents* Feb. 11(2): 151–157.

Hirotani, M., T. Furuya & M. Shiro, 1985. A ganoderic acid derivative, a highly oxygenated lanostne-type triterpenoid from *Ganoderma lucidum. Phytochemistry,* 24(9): 2055–2061.

Hishida, I., H. Nanba & H. Kuroda, 1988. Antitumor activity exhibited by orally administered extract from fruitbodies of *Grifola frondosa* (maitake). *Chem. Pharm. Bull.* 36(5): 819–1827.

Ho, M. S., 1971. Straw mushroom cultivation in plastic house. Proceedings from the VIIIth International Congress on Mushroom Science, London, pp. 257–263.

Hobbs, C., 1986. *Medicinal Mushrooms.* Botanica Press, Santa Cruz, California.

Hobbs, C., 1995. *Medicinal Mushrooms: An Exploration of Tradition, Healing & Culture.* Botanic Press, Santa Cruz, California.

Hodge, K. T., R. A. Humber & C. A. Wozniak, 1998. *Cordyceps variabilis* and the genus *Synglocladium. Mycologia* 90(5): 743–753.

Hodge, K. T., S. B. Krasnoff & R. A. Humber, 1996. *Tolypocladium inflatum* is the anamorph of *Cordyceps subsessilis. Mycologia* 88(5): 715–719.

Horner, W. E., E. Levetin & S. B. Lehrer, 1993. Basidiospore allergen release: elution from intact spores. *Journal of Allergy and Clinical Immunology* 92(2): 306–312.

Houdeau, G. & J. M. Olivier, 1992. *Pathology of Cultivated Mushrooms.* OK Press, Bologna, Italy.

Hseu, R. S. & H. H. Wang, 1991. A new system for identifying cultures of *Ganoderma* species. *Science and Cultivation of Edible Fungi,* ed. Maher. Balkema, Rotterdam.

Huguang, D. 1992. High yield and quality cultivation of *Hericium erinaceus* under new technology (sic). *Edible Fungi of China,* II(4): 40–43; (5): 29–30.

Ikekawa, J., M. Nakamishi, N. Uehara, G. Charara & F. Fukuoka, 1968. Antitumor action of some basidiomycetes, especially *Phellinus linteus. Gann.* 59: 155–157.

Ikekawa, J., N. Vehara, Y. Maeda, M. Nakanishi, 1969. Twenty years of studies on antitumor activities of mushrooms. *Cancer Research* 29: 734–735.

Imazeki, R., 1937. Reishi and *Ganoderma lucidum* that grow in Europe and America: their differences. *Botany & Zoology* May, 5(5).

_____, 1943. Genera of polyporaceae of Nippon. *Bull. Tokyo Sci. Mus.* 6: 1–111.

_____, 1973. Japanese mushroom names. *The Transactions of the Asiatic Society of Japan.* Third Series, Vol. XI, Tokyo.

Imazeki, R., Y. Otani & N. Mizuno, 1988. *Fungi of Japan.* Yama-Kei Publishers, Tokyo.

Imbernon, M. & J. Labarere, 1989. Selection of sporeless or poorly spored induced mutants from *Pleurotus ostreatus* and *Pleurotus pulmonarius* and selective breeding. *Mushroom Science XII.* 1: 109–123.

Imbernon, M. & G. Houdeau, 1991. *Pleurotus pulmonarius* 3300 INRA-Somycel: A new poorly spore-producing strain. *Science and Cultivation of Edible Fungi,* ed. Maher, pp. 555–559. Balkema, Rotterdam.

Ingle, S., 1988. Mycotopia: Paul Stamets, visionary, is perfecting the art of mushroom cultivation. *Harrowsmith Magazine* May/June, 3(15): 68–73.

Ishikawa, H. 1967. Physiological and ecological studies on the *Lentinus edodes* (Berk.) Sing. *J. Agri. Lab. (Abiko)* Japan (8): 1–57.

Isikhuemhen, O. S. & J. A. Okhuoya, 1995. A low-cost technique for the cultivation of *Pleurotus tuberregium* in developing tropical countries. *Mushroom Growers Newsletter* 4: 2–4.

_____, 1996. Cultivation of edible sclerotia of *Pleurotus tuberregium* (Fr.) Singer on agricultural wastes. In *Proceedings of the 2nd International Conference on Mushroom Biology and Mushroom Products,* June 9–12, 1996., ed. D. J. Royse, Pennsylvania State University, University Park, Pennsylvania, 429–436.

Isikhuemhen, Omoanghe S., 1996. Studies on an edible tropical white rot fungus *Pleurotus tuberregium* (Fr.) Sing. Ph.D. thesis, Institute of Microbiology, Academy of the Czeck Republic, Prague, Czech Republic.

Isikhuemhen, Omoanghe S. & F. Nerud, 1999. Preliminary studies on the ligninolytic enzymes produced by the tropical fungus *Pleurotus tuberregium* (Fr.) Sing. *Antonie van Leeuwenhoek.* In press.

Isikhuemhen, O. S., Monclavo, J. M., F. Frantisek & R. Vilgalys, 2000. Mating compatibility and phylo-geography in *Pleurotus tuberregium. Mycological Research:*104 (1). In press.

Ito, T., 1978. Cultivation of *Lentinus edodes. The Biology and Cultivation of Edible Mushrooms,* eds. S. T. Hayes & W. A. Hayes. Academic Press, New York, pp. 461–473.

Ito, H., K. Shimura, H. Itoh & M. Kawade, 1997. Antitumor effects of a new polysaccharide-protein complex (ATOM) prepared from *Agaricus blazei* (Iwade Strain 101) "Himematsutake" and its mechanisms in tumor-bearing mice. *Anticancer Research* 17: 277–284.

Itoh, H., H. Ito, H. Amano & H. Noda, 1994. Inhibitory action of a (1>6)-ß-D-Glucan-Protein Complex (FIII-2-b) isolated from *Agaricus blazei* Murrill and its antitumor mechanism. *Japanese Journal of Pharmacology* 66:265–271.

Iwade, I. & T. Mizuno, 1997. Cultivation of Kawariharatake *(Agaricus blazei). Food Reviews International* 13(3): 383–390.

Jalc, D., F. Nerud, R. Zitnan & P. Siroka, 1996. The effect of white-rot basidiomycetes on chemical composition and in vitro digestibility of wheat straw. *Folia Microbiologica* 41(1): 73–75.

Jianjung, Z., 1991. Bottle cultivation techniques of *Auricularia auricula* with straw won the Golden Medal in Tailand (sic). *Edible Fungi of China,* 10(2): 48.

Jifeng, W., Z. Jiajun & C. Wenwei, 1985. Study of the action of *Ganoderma lucidum* on scavenging hydroxyl radical from plasma. *Journal of Traditional Chinese Medicine* 5(1): 55–60.

Jinxia, Z. & S. T. Chang, 1992. Study on productivity and quality of *Volvariella volvacea* (Bull. ex. Fr.) Sing. stock culture after storage. *Edible Fungi of China* 11(4): 3–9.

Jones, K., 1992. *Reishi, Ancient Herb for Modern Times.* Sylvan Press, Issaquah, Washington.

_____, 1995. *Shiitake: The Healing Mushroom.* Healing Arts, Rochester, NY

Jong, S. C. & J. T. Peng, 1975. Identity and cultivation of a new commercial mushroom in Taiwan. *Mycologia* 67: 1235–1238.

Jong, S. C., 1989. Commercial cultivation of the Shiitake mushroom on supplemented sawdust. *Mushroom Journal of the Tropics* (9): 89–90.

_____, 1991. Immunomodulatory substances of fungal origin. *Journal of Immunol. Immunopharmacol.* 11(3).

Jong, S. C. & J. M. Birmingham, 1992. Medicinal benefits of the mushroom Ganoderma. *Advances in Applied Microbiology* 37: 101–133.

Jung-lieh, Chang, 1983. Experimental study of antitumor effect of an extract derived from Zhu-Ling *(Polyprous umbellatus).* Institute of Material Medica, Academy of Traditional Chinese Medicine, Peking, China.

Jwanny, E. W., M. M. Rashad & H. M. Abdu, 1995. Solid-state fermentation of agricultural wastes into food through Pleurotus cultivation. *Applied Biochemistry & Biotechnology* Jan 50(1): 71–78.

Kabir, Y. M. & S. Kimura, 1987. Effect of shiitake *(Lentinus edodes)* and maitake *(Grifola frondosa)* mushrooms on blood pressure and plasma lipids of spontaneously hypertensive rats. *J. Nutr. Sci. Vitaminol.* 33: 341–346.

Kalberer, P. P., 1974. The cultivation of *Pleurotus ostreatus:* Experiments to elucidate the influence of different culture conditions on the crop yield. *Mushroom Science* IX: 653–662.

Kamm, Y. J., H. T. Folgering & H. G. van den Bogart, 1991. Provocation tests in extrinsic allergic alveolitis in mushroom workers *Nethlerlands Journal of Medicine* Feb. 38(1-2): 59–64.

Kariya, K., K. Nakamura, K. Nomoto, S. Matama & K. Saigneji, 1992. Mimicking of superoxide dismutase activity by protein-bound polysaccharide of *Coriolus versicolor* QUEL., and oxidative stress relief for cancer patients. *Molecular Biotherapy* March 4(1): 40–6.

Katagiri, N., Y. Tsutsumi & T. Nishida, 1995. Correlation of brightening with cumulative enzyme activity related to lignin biodegradation during biobleaching of kraft pulp by white rot fungi in the solid-state fermentation system. *Applied Environmental Microbiology* Feb. 61(2): 617–622.

Kawaano et al. 1990. Novel strains of *Lyophyllum ulmarium* (sic). United States Patent #4,940,837. U.S. Patent Office, Washington, D.C.

Kawagishi, H. et al. 1991. Hericenones C, D, and E, Stimulators of nerve growth factor (NGF)-synthesis from the mushroom *Hericium erinaceum*. *Tetrahedron Letters,* 32(35): 4561–4564.

Kawagishi, H. et al. 1994. Erinacines A, B, C, strong stimulators of nerve growth factor synthesis, from the mycelia of *Hericium erinaceum*. *Tetrahedron Letters* 35(10): 1569–1572.

Kawai, G., H. Koboyashi, Y. Fukumisha & K. Ohsaki, 1996. Effect of liquid mycelial culture used as spawn on sawdust cultivation of shiitake *(Lentinula edodes)*. *Mycoscience* 37: 201–207.

Kay, E. & R. Vilgalys, 1992. Spatial distribution and genetic relationships among individuals in a natural population of the oyster mushroom, *Pleurotus ostreatus*. *Mycologia* 84(2): 173–182.

Kerrigan, R., 1982. Is Shiitake Farming for You? Far West Fungi, South San Francisco, California.

Kerrigan, R. W. & I. K. Ross, 1989. Allozymes of a wild *Agaricus bisporus* population: new alleles, new genotypes. *Mycologia* 81(3): 433–443.

Kerrigan, R. W., 1993. New prospects for *Agaricus bisporus* strain improvement. *Rept. Tottori Myco. Inst.* 31: 188–200.

Kerrigan, R. W., D. B. Carvalho, P. A. Horgen & J. B. Anderson, 1998. The indigenous coast California population of the mushroom *Agaricus bisporus,* a cultivated species, may be at risk of extinction. *Molecular Ecology* 7: 35–45.

Khan, S. M., J. H. Mirza & M. A. Khan, 1991. Physiology and cultivation of wood's ear mushroom *[Auricularia polytricha (Mont.)] Sacc. Mushroom Science* XII: 503–508.

Kim, B. K., H. W. Kim & E. C. Choi, 1994. Anti-HIV effects of *Ganoderma lucidum*. Ganoderma: Systematics, Phytopathology & Pharmacology. Proceedings of Contributed Symposium 59 A, B at the 5th International Mycological Congress, Vancouver, Canada, August 14–21, pg. 115. (Short Communication.)

Kim, B. K., H. W. Kim and E. C. Choi, 1994. Anti-HIV effects of *Ganoderma lucidum*. in Ganoderma: Systematics, Phytopathology & Pharmacology: Proceedings of Contributed Symposium 59 A, B. 5th International Mycological Congress, Vancouver, Canada.

Kim, B. K., R. S. Kim & H. W. Kim, 1997. Effects of *Ganoderma lucidum* on human leukocytes. In Proceedings of the 1st International Symposium on *Ganoderma lucidum* in Japan, Nov. 17–18, Tokyo, 87–89.

Kim, H. S., S. Kacew & B. M. Lee, 1999. In vitro chemopreventive effects of plant polysaccharides (*Aloe barbadensis* Miller, *Lentinus edodes, Ganoderma lucidum* and *Coriolus versicolor*). *Carcinogenesis* Aug. 20(8): 1637–1640.

Kim, K. C. & I. G. Kim, 1999. *Ganoderma lucidum* extract protects DNA from strand breakage caused by hydroxyl radical and UV irradiation. *International Journal of Molecular Medicine* Sept; 4(3): 273–277.

Kim, M. J. & Y. S. Kim, 1986. Studies on safety of *Ganoderma lucidum*. *Korean Journal of Mycology* 14(1): 49–60.

Kio, T., M. Sakushima, S. R. Wang, K. Nagai & S. Ukai, 1991. Polysaccharides in fungi XXVI. Two branched 1-3-beta-D-glucans from hot water extract of Yuer. *Chem. Pharm. Bull.* (Tokyo) March 39(3): 798–800.

Kobayashi, H., K. Matsunga & M. Fujii, 1993. PSK as a chemopreventive agent. *Cancer Epidemiology, Biomarkers and Prevention* May-June 2(3): 271–276.

Kobayashi, H., K. Matsunaga & Y. Oguchi, 1995. Antimetastatic effects of PSK (Krestin), a protein-bound polysaccharide obtained from basidiomycetes: an overview. *Cancer Epidemiology, Biomarkers and Prevention* 4(3): 275–81.

Komatsu, M., Y. Nozaki, A. Inoue & M. Miyauchi, 1980. Correlation between temporal changes in moisture contents of the wood after felling and mycelial growth of *Lentinus edodes* (Berk.) Sing. *Report of the Tottori Mycological Institute* 18: 169–187.

Komatusu, M. & K. Tokimoto, 1982. Effects of incubation temperature and moisture content of bed logs on primordium formation of *Lentinus edodes* (Berk.) Sing. *Report of the Tottori Mycological Institute* 20: 104–112.

Kozak, M & J. Krawcyk, 1989. *Growing Shiitake Mushrooms in a Continental Climate*. Field & Forest Products, Peshtigo, Wisconsin.

Kruger, E. 1992. A new status for the lowly ironwood. *Shiitake News* (9): 2.

Kuck, U., H. D. Osiewacz, U. Schmidt, B. Kappelhoff, E. Schulte, U. Stahl & K. Esser, 1985. The onset of senescence is affected by DNA rearrangements of a discontinuous mitochondrial gene in *Podospora anserina*. *Current Genetics* 9: 373–382.

Kunno, S., S. A. Fullerton, A. A. Samadi, D. G. Tortorelis, C. Mallouh & H. Tazaki, 2000. Apoptosis in prostatic cancer cells with maitake mushroom extract: potential alternative therapy. In press.

Kuo, D. D. & M. H. Kuo, 1983. *How to Grow Forest Mushroom (Shiitake)*. Mushroom Technology Corp., Naperville, Illinois.

Kweon, M. H., 1998. Studies on bioactive polysaccharide isolated from *Agaricus bisporus* (Kor, Eng abst). *Agr. Chem. Biochem.* 41: 60–6.

Ladanyi, A., J. Timar & K. Lapis, 1993. Effect of lentinan on macrophage cytotoxicity against metastatic tumor cells. *Journal of Cancer Immunology* 36: 123–126.

Leatham, G. 1982. Cultivation of Shiitake, the Japanese forest mushroom, on logs: a potential industry for the United States. *Forest Products Journal* 332: 29–35.

Leatham, G. & T. J. Griffin, 1984. Adapting liquid spawn of *Lentinus edodes* to oak wood. *Applied Microbiology and Biotechnology* 20: 360–364.

Leatham, G. & M. A. Stahlman, 1987. Effect of light and aeration on fruiting of *Lentinus edodes*. Transactions of the British Mycological Society, 88: 9–20.

_____, 1989. The effects of common nutritionally-important cations on the growth and development of the cultivated mushroom *Lentinula edodes*. *Mushroom Science XII*: 253–265.

Leck, C., 1991. Mass extinction of European fungi. *Trends in Ecology and Evolution*. June 6(6).

Leonard, T. J. & T. J. Volk, 1992. Production of Specialty Mushrooms in North America: Shiitake & Morels. *Frontiers in Industrial Mycology*, Chapman Hall, New York.

Leon-Chocoo, R. de, G. Guzman & D. Martinez-Carrera, 1988. Planta productora de hongos comestibles (*Pleurotus ostreatus*) en Guatemala. *Rev. Mex. Mic.* 4: 297–301.

Lhote, H., 1987. Oasis of art in the Sahara. *National Geographic* August, pp. 180–188.

Li, S. & G. Eger, 1978. Characteristics of some *Pleurotus* strains from Florida, their practical and taxonomic importance. *Mushroom Science X*: 155–169.

Li, Y. Y., L. Wang, Q. Y. Jia, J. Li & Y. Wu, 1988. A study on the cultivation of straw mushrooms with wheat straw. *Mushroom Journal of the Tropics* (8): 67–72.

Lincoff, G. & D. H. Mitchel, 1977. *Toxic and Hallucinogenic Mushroom Poisoning: A Handbook for Physicians and Mushroom Hunters*. Van Nostrand Reinhold, New York.

Lincoff, G., 1981. *The Audubon Society Field Guide to North American Mushrooms*. Alfred A. Knopf, New York.

Lizon, P., 1993. Decline of Macrofungi in Europe: An Overview. *Transactions of Mycological Society of the Republic of China* 8(3–4): 21–48.

_____, 1995. Preserving the biodiversity of fungi. *Inoculum* 46(6): 1–4.

Lomax, K. M., 1990. Heat transfer for fresh-market mushrooms. *Shiitake News* 7: 2.

Lonik, L., 1984. *The Curious Morel: Mushroom Hunter's Recipes, Lore and Advice*. RKT Publishers, Royal Oak, Michigan.

Lovy, A., B. Knowles, R. Labbe & L. Nolan, 1999. Activity of edible mushrooms against the growth of human T4 leukemia cancer cells, and *Plasmodium falciparum*. *Journal of Herbs, Spices and Medicinal Plants*. 6(4): 49–57.

Ma, L. & Z. B. Lin, 1992. Effect of *Tremella* polysaccharide on IL-2 production by mouse splenocytes. *Acta Pharmacologica Sinica* 27(1).

Maher, M. J., 1991. *Mushroom Science XIII: Science and Cultivation of Edible Fungi*. vol. I & II. A.A. Balkema, Rotterdam, Netherlands.

Martinez, D., G. Guzman & C. Soto, 1985. The effect of fermentation of coffee pulp in the cultivation of *Pleurotus ostreatus* in Mexico. *Mushroom Newsletter for the Tropics*, 1(6): 21–28.

Martinez-Carrera, D., 1987. Design of a mushroom farm for growing *Pleurotus* on coffee pulp. *Mushroom Journal of the Tropics* 7: 13–23.

Martinez-Carrera, D., P. Morales & M. Sobal, 1988. Cultivo de diversas cepas mexicanas de *Pleurotus ostreatus* sobre pulpa de café de cebada. *Rev. Mex. Mic.* 4: 153–160.

Matsumoto II, K., 1979. *The Mysterious Reishi Mushroom*. Woodbridge Press Publishing Co., Santa Barbara, California.

Matsumoto, K., M. Ito, S. Tagyu, H. Ogino & I. Hirono, 1991. Carcinogenicity examination of *Agaricus bisporus*, an edible mushroom, in rats. *Cancer Letters* 58(1-2): 87–90.

Mayer J. & J. Drews, 1980. The effect of protein-bound polysaccharide from *Coriolus versicolor* on immunological parameters and experimental infections in mice. *Infection* 8: 13–21.

McKenny, M., D. Stuntz & J. Ammirati, 1987. *The New Savory Wild Mushroom*. University of Washington Press, Seattle, Washington.

McKnight, K. B., 1985. The adaptive morphology of *Flammulina velutipes* with respect to water stress. Ph.D. Thesis. University Microfilms, Ann Arbor, Michigan.

_____, 1990. Effect of low humidity on spore production and basidiocarp longevity among selected isolates of *Flammulina velutipes*. *Mycologia* 82: 379–384.

_____, 1992. Evolution of the *Flammulina velutipes* basidiocarp size with respect to relative humidity. *Mycologia* 84(2): 219–228.

McKnight, K. B. & G. F. Estabrook, 1990. Adaptations of sporocarps of the basidiomycete *Flammulina velutipes* (Agaricales) to lower humidity. *Bot. Gas. (Crawfordsville)* 151: 528–537.

Miles, P. G. & S. T. Chang, 1986. Application of biotechnology in strain development of edible mushrooms. *Asian Food Journal* 2(1): 3–10.

Miller, O. K., 1969. A new species of *Pleurotus* with a coremoid imperfect stage. *Mycologia* 61: 887–893.

_____, 1977. *Mushrooms of North America*. E. P. Dutton, New York.

Miller, M. W. & S. C. Jong, 1986. Commercial cultivation of shiitake in sawdust-filled plastic bags. Proceedings of the International Symposium on Scientific and Technical Aspects of Cultivating Edible Fungi. July, University Park, Pennsylvania. Elsevier Science Publishers. pp. 421–426.

Mimura, S., 1904. Notes on Shiitake culture. *Journal of the Forestry Society of Japan*, Vol. 4. Tokyo.

_____, 1915. Notes on Shiitake (*Cortinellus Shiitake* Schrot) culture. *Bulletin of the Forest Experiment Station*, Meguro, Tokyo.

Misaki, A. & M. Kakuta, 1995. Kikurage (Tree Ear) and Shirokikurage (White Jelly Leaf): *Auricularia auricula* and *Tremella fuciformis*. *Food Review International* 11(1): 211–218.

Miyasaki, T., 1983. Antitumor activity of fungal crude drugs. *Journal of Traditional Sino-Japanese Medicine* 4(1): 61–65.

Mizuno, M., M. Morimoto, K. Minato & H. Tsuchida, 1998. Polysaccharides from *Agaricus blazei* stimulate lymphocyte T-cell subsets in mice. *Bioscience Biotechnology & Biochemistry* March 62(3): 434–7.

Mizuno. M., K. Minato, K. Ito, M. Kawade, H. Terai & H. Tsuchida, 1999. Antitumor polysaccharide from the mycelium of liquid-cultured *Agaricus blazei* Murrill. *International Journal of Biochemistry & Molecular Biology* April 47(4): 707–714.

Mizuno, T., 1988. Development and utilization of bioactive substances from medicinal and edible mushroom fungi II. *Ganoderma lucidum. Chemical Times* 1989(3): 50–60.

_____, 1989. YamabuShiitake, *Hericium erinaceum*: bioactive substances and medicinal utilization. *Food Reviews International* 11(1): 173–178.

_____, 1989. Pharmacological and gastronomic effects of fungi and its applications. *Chemical Times* 1: 12–21.

_____, ed., 1995. Mushrooms: The versatile fungus-food and medicinal properties. In *Food Reviews International*, 11(1). Marcel Dekker, Inc., New York.

_____, 1996. *A Medicinal Mushroom, Ganoderma lucidum*. Il-Yang Pharm. Co., Seoul, Korea.

Mizuno, T, H. Ito, K. Shimura, T. Hagiwara & T. Nakamura, 1989. Preparation method of antitumor neutral polysaccharides from Himematsutake. *Japan Kokai Tokkyo Koho* (A), Sho64-67195, March 13, 565–568.

_____, 1989. Preparation method of antitumor acidic glycans from Himematsutake. *Japan Kokai Tokkyo Koho* (A), Sho64-67194, March 13, 561–564.

Mizuno, T, T. Hagiwara, T. Nakamura, H. Ito, K. Shimura, T. Sumiya & A. Asakura, 1990. Antitumor activity and some properties of water-soluble polysaccharides from "Himematsutake," the fruiting body of *Agaricus blazei* Murrill. *Agricultural & Biological Chemistry* 54. Tokyo. 2889–2896.

_____, 1990. Antitumor activity and some properties of water-insoluble hetero-glycans from "Himematsutake," the fruiting body of *Agaricus blazei* Murrill. *Agricultural & Biological Chemistry* 54. Tokyo. 2897–2905.

Mizuno, T, T. Waa, H. Ito, C. Suzuki & N. Ukai, 1992. Antitumor-active polysaccharides isolated from the fruitbody of *Hericium erinaceum*, an edible and medicinal mushroom called yamabushiitake or houtou. *Biosci. Biotech. Biochem.* Feb. 56(2): 347–348.

Mizuno, T, H. Saito, T. Nishitoba & H. Kawagishi, 1995. Antitumor active substances from mushrooms. *Food Reviews International* 111: 23–61.

Mizuno, T. & C. Zhuang, 1995. Maitake, *Grifola frondosa*, pharmacological effects. *Food Reviews International* 111: 135–149.

Mizushina, Y., L. Hanashima, T. Yamaguchi, M. Takemura, F. Sugawara, M. Saneyoshi, A. Matsukage, S. Yoshida & K. Sakaguchi, 1998. A mushroom fruiting body-inducing substance inhibits activities of replicative DNA polymerases. *Biochemical & Biophyscial Research Communications* Aug. 10; 249(1): 17–22.

Mollison, B., 1990. *Permaculture: A Practical Guide for a Sustainable Future.* Island Press, Washington D.C.

_____, 1993. *The Permaculture Book of Ferment and Human Nutrition.* Tagari Publications, Tyalgum, Australia.

Money, N., 1998. More g's than the space shuttle: ballistospore discharge. *Mycologia* 90(4): 547–558.

Morales, P., D. Martinez-Carrera & W. Martinez-Sanchez, 1991. Cultivo de Shiitake sobre diversos substratos en Mexico. *Micologia Neotropical Aplicada* 4: 75–81.

Mori, K., 1974. *Mushrooms as Health Foods.* Japan Publications, Tokyo.

Mori, K., T. Toyomasu, H. Nanba & H. Kuroda, 1987. Antitumor activity of fruitbodies of edible mushrooms orally administered to mice. *Mushroom Journal of the Tropics* 7: 121–126.

Mori, S., K. Nakagawa, Yoshida, H. Tschihasi et al. 1998. Mushroom worker's lung resulting from indoor cultivation of *Pleurotus ostreatus*. *Occupational Medicine* (Oxford) 48(7).

Morigawa, A., K. Kitabatake, Y. Fujimoto & N. Ikekawa, 1986. Angiotensin converting enzyme-inhibitory triterpenes from *Ganoderma lucidum*. *Chemical and Pharmaceutical Bulletin* 34 (7): 3025–3028.

Moser, M., 1978. *Keys to Agarics and Boleti.* Roger Phillips, London.

Motohashi, T., 1993. Personal communication.

Mueller, J. C. & J. R. Gawley, 1983. Cultivation of phoenix mushrooms on pulp mill sludges. *Mushroom Newsletter for the Tropics* 4(1): 3–12.

Mueller, J. C., J. R. Gawley & W. A. Hayes, 1985. Cultivation of the Shaggy Mane mushroom (*Coprinus comatus*) on cellulosic residues from pulp mills. *Mushroom Newsletter for the Tropics* 1(6): 15–20.

Murr, D. P. & L. Morris, 1975. Effects of storage temperature on post-harvest changes in mushrooms. *Journal of American Society of Horticultural Science* 100: 16–19.

_____, 1975. Effects of storage atmosphere on post-harvest growth of mushrooms. *Journal of American Society of Horticultural Science* 100: 298–301.

Murrill, W. A., 1945. New Florida Fungi. *Journal of Florida Academy of Sciences* 8(2): 175–198.

Nagasawa, E. & I. Arita, 1988. A note on *Hypsizygus ulmarius* and *H. marmoreus*. *Rept. Tattori Myc. Inst.* 26: 71–78.

Nair, L. N. & V. P. Kaul, 1980. The anamorphs of *Pleurotus sajor-caju* (Fr.) Singer and *Pleurotus gemmellarii* (Inzeng.) Sacc. *Sydowia* 33: 221–224.

Nakazato, H., A. Koike, S. Saji, N. Ogawa & J. Sakamoto, 1994. Efficacy of immunotherapy as adjuvant treatment after curative resection of gastric cancer. *The Lancet* May 7, 343: 1122–1126.

Nanba, H. 1992. Maitake: The king of mushrooms. *Explore* 3(5): 44–46.

Ng, T. B., 1998. A review of research on the protein-bound polysaccharide (polysaccharopeptide, PSP) from the mushroom *Coriolus versicolor*. *General Pharmacology* Jan. 30(1): 1–4.

Nianlai, H., M. Wang, M. Guo & L. Huang, 1998. A preliminary study on *Pleurotus tuberregium* (Fr.) Sing. cultivation in Fujan Province. Proceedings of the '98 Nanjing International Symposium on the Science and Cultivation of Mushrooms, Nanjing, China, pp. 40–41.

Nishitoba, T., H. Sato, T. Kasai, H. Kawagishi & S. Sakamura, 1984. New bitter C27 and C30 Terpenoids from the fungus, *Ganoderma lucidum* (Reishi). *Agric. Biol. Chem.* 48(11): 2905–2907.

Nishitoba, T., H. Saot, S. Shirasu & S. Sakamura, 1987. Novel triterpenoids from the mycelial mat at the previous stage of fruiting of *Ganoderma lucidum*. *Agricultural & Biological Chemistry* 51(2): 619–622.

Odani, S., K. Tominaga, S. Kondou, H. Hori, T. Koide, S. Hara, M. Isemura & S. Tsunasawa, 1999. The inhibitory properties and primary structure of a novel serine proteinase inhibitor from the fruiting body of the basidiomycete, *Lentinus edodes*. *European Journal of Biochemistry*, June 262(3): 915–923.

Oei, P. 1991. *La Culture des Champignons*. Tool, Amsterdam.

_____, 1991. *Manual on Mushroom Cultivation*. Tool, Amsterdam.

Ohno, N., K. Iino, T. Takeyama, I. Suzuki, K. Sato, S. Oikawa, T. Miyazaki & T. Yadomae, 1985. Structural characterization and antitumor activity of the extracts from matted mycelium of cultured *Grifola frondosa*. *Chem. Pharm. Bull.* 33(8): 3395–3401.

Okhuoya, J. A. & C. Ajerio, 1988. Analysis of sclerotia and sporophores of *Pleurotus tuberregium* Fr. an edible mushroom in Nigeria. *Kor. J. Mycol.* 6(4): 204–206.

_____, 1988. Sporocarp development of *Pleurotus tuberregium* Fr. under different watering systems. *Kor. J. Mycol.* 16(4): 207–209.

Okhuoya, J. A. & F. O. Okogbo, 1991. Cultivation of *Pleurotus tuberregium* (Fr.) Sing. on various farm wastes. *Proc. Okla. Sci.* 71: 1–3.

Okwujiako, I.A., 1990. The effect of vitamins on the vegetative growth and fruitbody formation of *Pleurotus sajor-caju* (Fr.) Singer. *Mushroom Journal of the Tropics* (10): 35–39.

Omoanghe, S., 1992. Studies on the cultivation of edible sclerotia of *Pleurotus tuberregium* (Fr.) Sing. on various farm wastes. Master's Thesis, University of Benin, Benin City, Nigeria.

Oso, B. A., 1977. *Pleurotus tuberregium* from Nigeria. *Mycologia* 69: 271–279.

Oss, O. & E. Oeric, 1976. *Psilocybin Magic Mushroom Grower's Guide*. And/Or Press, Berkeley, California.

Ott, J., 1976. *Hallucinogenic Plants of North America*. Wingbow Press, Berkeley, California.

_____, 1978. Mr. Jonathan Ott's rejoinder to Dr. Alexander Smith. *Ethnomycological Studies No. 6*. Botanical Museum of Harvard University, Cambridge, Massachusetts.

_____, 1993. *Pharmacotheon: Entheogenic Drugs, Their Plants and History*. Natural Products Co., Kennewick, Washington.

Ott, J. & J. Bigwood, ed. 1978. *Teonanacatl: Hallucinogenic Mushrooms of North America: extracts from the second international conference on hallucinogenic mushrooms*. Oct. 27–30, Madrona Publishers, Seattle, Washington.

_____, 1982. Notes on the development of the Morel ascocarp: *Morchella esculenta. Mycologia* 74: 142–144.

Ower, R., G. Mills & J. Malachowski, 1986. Cultivation of *Morchella*. U.S. Patent No. 4,594,809.

_____, 1988. Cultivation of *Morchella*. U.S. Patent No. 4,757,640.

Papaparaskeva, C., C. Ioannides & R. Walker, 1991. Agaritine does not mediate the mutagenicity of the edible mushroom *Agaricus bisporus. Mutagenesis* 6(3): 213–218.

Pegler, D., 1975. The classification of the genus *Lentinus* Fr. (Basidiomycota). *Kavaka* 3: 11–20.

_____, 1983. *The Genus Lentinus: A World Monograph*. Kew Bulletin Additional Series X, London.

_____, 1983. The genus *Lentinula* (Tricholomataceae tribe Collybiaeae). *Sydowia* 36: 227–239.

Peintner, U., R. Poder & T. Pumpel, 1998. The Ice Man's fungi. *Mycological Research* 102: 1153–1162.

Perrin, P.S. & B.J. Macauley, 1995. Positive aeration of conventional (Phase I) mushroom compost stacks for odor abatement and process control. In the *Science and Cultivation of Edible Fungi*, (ed.) Elliott. Balkema, Rotterdam.

Petersen, R., 1992. Neohaploidization and neohaplont mating as a means of identification of *Pleurotus* cultures. *Mycosystema* 5: 165–170.

Pilegaard, K., E. Kristianson, O. A. Meyer and J. Gry, 1997. Failure of the cultivated mushroom (Agaricus bisporus) to induce tumors in the A/J mouse lung tumor model. *Cancer Letters* Nov. 25: 120(1):79–85.

Piraino, F. & C. R. Brandt, 1999. Isolation and partial characterization of an antiviral, RC-183, from the edible mushroom, *Rozites caperata. Antiviral Research* 43: 67–78.

Pischl, C., 1999. Die Auswirkungen von Pflanzen-Pilzmischkulturen auf den Bodennaehrstoffgehalt und die Ernteertraege. Master's Thesis, Leopold-Franzens-Universitat Innsbruck.

Platt, M., I. Chet & Y. Henis, 1982. Growth of *Pleurotus ostreatus* on cotton straw. *Mushroom Journal* 20: 425–426.

Pollock, S. H., 1977. *Magic Mushroom Cultivation*. Herbal Medicine Research Foundation, San Antonio, Texas.

Pool-Kobel, B. L., 1990. Mutagenic and genotoxic activities of extracts derived from the cooker and raw edible mushroom *Agaricus bisporus*. *Journal of Cancer Research & Clinical Oncology* 116(5): 475–479.

Prance, G. T., 1984. The use of edible fungi by Amazonian Indians. *Advances in Economic Botany* 1: 127–139.

Price, R. J., D. G. Walters, C. Hoff, H. Misrty, A. B. Renwick, P. T. Wield, J. A. Beamand & B. G. Lake, 1996. Metabolism of (ring-U-14C) agaritine by precision-cut rate, mouse and human liver and lung slices. *Food Chem Toxicology* July: 34(7):603–609.

Przybylowicz, P. & J. Donoghue, 1989. *Shiitake Grower's Handbook: The Art and Science of Mushroom Cultivation*. Kendall Hunt, Dubuque, Iowa.

Qingtian, Z. et al. 1991. Antitumor activity of *Flammulina velutipes* polysaccharide (FVP). *Edible Fungi of China* 10(2): 11–15.

Raaska, L., 1990. Production of *Lentinula edodes* mycelia in liquid media: Improvement of mycelial growth by medium modification. *Mushroom Journal of the Tropics* (10): 79–92.

Rai, R. D., S. Saxena, R. C. Upadhyay & H. S. Sohi, 1988. Comparative nutritional value of various *Pleurotus* species grown under identical conditions. *Mushroom Journal of the Tropics* (8): 93–98.

Rathke, D. M. & M. J. Baughman, 1993. Can shiitake production be profitable? *Shiitake News* 10(1): 1–10.

Rayner, A., 1988. Life in a collective. *New Scientist*, Nov. 19, pp. 49–53.

Redhead, S., 1984. Mycological observations 13-14: on *Hypsizygus* and *Tricholoma*. *Trans. Myco. Soc. Japan* 25: 1–9.

_____, & J.H. Ginns, 1985. A reappraisal of agaric genera associated with brown rots of wood. *Trans. Myco. Soc. Japan* 26: 349–381.

_____, 1986. Mycological observations 15–16: on *Omphalina* and *Pleurotus*. *Mycologia* 78(4): 522–528.

Reshef, A., I. Moulalem & P. Weiner. Acute and long-term effect of exposure to basidiomycetes spores to mushroom growers. *Journal of Aller. Clin. Immunol.* 81(1): 275.

Rice, M. C., 1980. *Mushrooms for Color*. Mad River Press, Eureka, California.

Roberts, T., 1988. Review of "Financial analysis of three hypothetical, small-scale Shiitake mushroom production enterprises." L. F. Lambert Spawn Co., Coatesville, Pennsylvania.

Robinson, W., 1885. *Mushroom Culture: Its Extension & Improvement*. David McKay, Philadelphia, Pennsylvania.

Romaine, C. P. & B. Schlagnhaufer, 1992. Characteristics of a hydrated, alginate-based delivery system for the cultivation of the button mushroom. *Applied and Environmental Microbiology*, pp. 3060–3066.

Roxon, J. E. & S. C. Jong, 1977. Sexuality of an edible mushroom, *Pleurotus sajor-caju*. *Mycologia* 69 (1): 203–205.

Royse, D. J., 1985. Effect of spawn run time and substrate nutrition on yield and size of the Shiitake mushroom. *Mycologia* 77: 756–762.

Royse, D. J., L. C. Schisler & D. A. Diehle, 1985. Shiitake mushrooms: consumption, production, and cultivation. *Interdisciplinary Science Reviews* 10(4): 329–335.

Royse, D. J. & C. C. Bahler, 1986. Effects of genotype, spawn run time, and substrate formulation on biological efficiency of Shiitake. *Applied and Environmental Microbiology* 52: 1425–1427.

Royse, D. J., & J. E. Sanchez-Vasquez, 2000. Influence of wood chip particle size used in substrate on biological efficiency and post-soak log weights of shiitake. *Science & Cultivation of Edible Fungi*, Van Griensven (ed.) 367–373.

Royse D. J. & L. C. Schisler, 1987a. Yield and size of *Pleurotus ostreatus* and *Pleurotus sajor-caju* as affected by delayed-release nutrient. *Applied Microbiology and Biotechnology* 26: 191–194.

_____, 1987b. Influence of benomyl on yield response of *Pleurotus sajor-caju* to delayed release nutrient supplementation. *HortScience* (22): 60–62.

Royse, D. J. & B. D. Bahler, 1988. The effect of alfalfa hay and delayed release nutrient on biological efficiency of *Pleurotus sajor-caju*. *Mushroom Journal of the Tropics* (8): 59–65.

_____, 1989. Yield and size of Shiitake as influenced by synthetic log diameter and genotype. *Mushroom Journal for the Tropics* (9): 109–113.

Royse, D. J., B. D Bahler & C. C. Bahler, 1990. Enhanced yield of Shiitake by saccaride amendment of the synthetic substrate. *Journal of Applied Environmental Microbiology*. Feb.: 479–482.

Royse, D. J., 1991. Yield stimulation of *Pleurotus flabellatus* by dual nutrient supplementation of pasteurized wheat straw. *Science and Cultivation of Edible Fungi*, ed. Maher, Blakema, Rotterdam.

Royse, D. J. & S. A. Zaki, 1991. Yield stimulation of *Pleurotus flabellatus* by dual nutrient supplementation of pasteurized wheat straw. *Science and Cultivation of Edible Fungi*, ed. Maher, pp. 545–547.

Royse, D. J. & B. May, 1982. Use of Isozyme variation to identify genotypic classes of *Agaricus brunnescens*. *Mycologia* 74: 93–102.

Sakagami, H. & M. Takeda, 1993. Diverse biological activity of PSK (Krestin): a protein bound poly-saccharide from *Coriolous versicolor* (Fr.) Quel. In *Mushroom Biology & Mushroom Products*, eds. S.T. Chang, J.A. Buswell & Siu-wai Chiu, The Chinese University Press, Hong Kong. pp. 237–245.

Sakagami, H., K. Sugaya, A. Utsumi, S. Fujinaga, T. Sato & M. Takeda, 1993. Stimulation by PSK of interleukin-1 production by human peripheral blood mononuclear cells. *Anticancer Research*, May–June 13(3): 671–5.

Samajpati, N., 1979. Nutritive value of some Indian edible mushrooms. *Mushroom Science X* (Part II): 695–703.

San Antonio, J. P., 1971. A laboratory method to obtain fruit from cased grain spawn of the cultivated mushroom *Agaricus bisporus*. *Mycologia* 63: 16–21.

_____, 1981. Cultivation of the Shiitake mushroom. *HortScience* 16(2): 151–156.

_____, 1984. Using basidiospores of the Oyster mushroom to prepare grain spawn for mushroom cultivation. *HortScience* 19(5): 684–686.

San Antonio, J. P. & P. K. Hanners, 1983. Spawn disk inoculation of logs to produce mushrooms. *HortScience* 18(5): 708–710.

San Antonio, J. P., F. B. Abeles & P. K. Hanners, 1989. Shiitake mushroom production on oak logs inoculated with grain-spawn disks by 23 different Shiitake strains. *Mushroom Journal of the Tropics* 9: 161–164.

Sanderson, R., 1969. Some field and laboratory observations on *Morchella*. Master's thesis. South Dakota State University.

Sarkar, S., J. Koga, R. J. Whitley & S. Chatterjee, 1993. Antiviral effect of the extract of culture medium of *Lentinus edodes* mycelia on the replication of herpes simplex virus type 1. *Antiviral Research* April 20(4): 293–303.

Sato, H., T. Nishitoba, S. Shirasu, K. Oda & S. Sakamura, 1986. Ganoderiol A & B, new triterpenoids from the fungus *Ganoderma lucidum* (Reishi). *Agric. Biol. Chem.* 50(11): 2887–2890.

Schenck, N.C., 1982. *Methods and Principles of Mycorrhizal Research*. The American Phytopathological Society, St. Paul, Minnesota.

Sharma, A. D. & C. L. Jandalk., 1985. Studies on recycling Pleurotus waste. *Mushroom Journal of the Tropics*, (6)2: 13–15.

Sharman, M., A. L. Patsy, & J. Gilbert, 1990. A survey of the occurrence of agaritine in U.K. cultivated mushrooms and processed mushroom products. *Food Additives and Contaminants.* 7: 649–656.

Shepard, S. E., D. Gunz, & C. Schlatter, 1995. Genotoxicity of agaritine in the lacI transgenic mouse mutation assay: evaluation of health risk of mushroom consumption. *Food Chem Toxicology.* Apr: 33(4):257–264.

Sia, G.M. & J. K. Candlish, 1999. Effect of shiitake (*Lentinus edodes*) extract on human neutrophils and the U937 monocytic cell line. *Phyotherapy Research* March 13(2): 133–137.

Singer, R. & A. H. Smith, 1958. New species of *Psilocybe. Mycologia* 50: 141–142.

_____, 1958. Mycological investigations on *Teonanacatl*, the Mexican halucinogenic mushroom. Part I: The history of *Teonanacatl*, field work and culture work. Part II: A taxonomic monograph on *Psilocybe*, section Caerulescentes. *Mycologia* 50: 239–303.

Singer, R. 1961. *Mushrooms and Truffles: Botany, Cultivation and Utilization*. Interscience Publishers, New York.

_____, 1986. *The Agaricales in Modern Taxonomy* (4th ed). Koeltz Scientific Books, Germany.

Smith, A. H., 1959. *Mushrooms in Their Natural Habitats*. Hafner Press, New York.

_____, 1977. Comments on hallucinogenic agarics and the hallucinations of those who study them. *Mycologia* 69: 1196–1200.

_____, 1979. Notes on the Strophariaceae. *Taxon* 28: 19–21.

Snell, W. H. & Dick, E. A., 1971. *A Glossary of Mycology*. Harvard University Press, Cambridge, Massachusetts.

Stamets, P., 1978. *Psilocybe Mushrooms and Their Allies*. Homestead Book Co., Seattle, Washington.

Stamets, P., M. W. Beug & G. Guzman, 1980. A new species and a new variety of *Psilocybe* from North America. *Mycotaxon* 11: 476–484.

Stamets, P. & J. S. Chilton, 1983. *The Mushroom Cultivator*. Agarikon Press, Olympia, Washington.

Stamets, P. & J. Gartz, 1995. A new caerulescent Psilocybe from the Pacific Coast of northwestern North America. *Mycotaxon* 11: 476–484.

Stamets P., A. Weil, et al. 1983. Personal communication. Mushroom study tour of China, Fungophile, Denver, Colorado.

Stamets, P., 1989. How to grow mushrooms: a simplified overview of mushroom cultivation strategies. *Shiitake News* 6(1): 11–13.

_____, 1990. A discussion on the cultivation of *Ganoderma lucidum* (Curtis:Fr.) Kar., the Reishi or Ling Zhi, mushroom of immortality. *McIlvainea* 9(2): 40–50.

_____, 1990. Potential uses of saprophytic fungi in the recycling of wood wastes from the forest environment. Special Forest Products Conference. Portland, Oregon.

_____, 1993. Mycofiltration of gray water run-off utilizing *Stropharia rugosoannulata*, a white rot fungus. (Unpublished Research Proposal awarded a grant by the Mason County Water Conservation District, Shelton, Washington.)

_____, 1993. Permaculture with a mycological twist. *Mushroom, the Journal* 40, 11(3): 5–7.

_____, 1999. Earth's natural Internet. *Whole Earth Review*, Fall, pp. 74–77.

_____, 1999. *Psilocybin Mushrooms of the World*. Ten Speed Press, Berkeley, California.

Stamets, P. & D. Yao, 1999. *Mycomedicinals: an informational booklet on the medicinal properties of mushrooms*. Mycomedia Productions, Fungi Perfecti, Olympia, Washington.

Stavinoha, W. B., S. Weintraub, T. Opham, A. Colorado, R. Opieda & J. Slama, 1990. Study of the anti-inflammatory activity of *Ganoderma lucidum*. Proceedings from the Academic/Industry Conference (AIJC), August 18–20, Sapporo, Japan.

Stijve, T., 1992. Certain mushrooms do accumulate heavy metals. *Mushroom, the Journal of Wild Mushrooming* 38(1): 9–14.

Stijve, T., R. Fumbaux, & G. Philippossian, 1986. Agaritine, a p-hydroxymethylphenylhydrazine derivative in cultivated mushrooms *(Agaricus bisporus)*, and in some of its wild-growing relatives. *Deutche Lebensmittel-Rundschau* 82:243–248

Stijve, T. & R. Roschnik, 1974. Mercury and methyl mercury content of different species of Fungi. *Trav. chim. alimen. Hyg.* 65: 209–220.

Stijve, T. & R. Besson, 1976. Mercury, cadmium, lead and selenium content of mushroom species belonging to the genus *Agaricus*. *Chemosphere* 2: 151–158.

Stoller, B. B., 1962. Some practical aspects of making mushroom spawn. *Mushroom Science* V: 170–184.

Stone, R., 1993. Surprise! A fungus factory for taxol? *Science* April, 260: 9.

Streeter, C. L., K. E. Conway & G. W. Horn, 1981. Effect of *Pleurotus ostreatus* and *Erwinia caratovora* on wheat straw digestibility. *Mycologia*, 73(6): 1040–1048.

Styler, J. F., 1933. *Modern Mushroom Culture*. Edw. H. Jacob, West Chester, Pennslyvania.

Sugimachi, K., Y. Maehara, M. Ogawa, T. Kakegawa & M. Tomita, 1997. Dose intensity of uracil and tegafur in postoperative chemotherapy for patients with poorly differentiated gastric cancer. *Cancer Chemotherapy and Pharmacology* 40(3): 233–238.

Tan, T. & S. T. Chang, 1989. Yield and mycelial growth response of the Shiitake mushroom, *Lentinus edodes* (Berk.) Sing to supplementation on sawdust media. *Mushroom Journal of the Tropics* (9): 1–14.

Thaithatgoon, S. T. & S. Triratana, 1993. Wild, naturally grown Reishi: cultivation, breeding, identification and their development in Thailand. In *A Medicinal Mushroom, Ganoderma lucidum, Polyporaceae, and others. Oriental Tradition, Cultivation, Breeding, Chemistry, Biochemistry and Utilization*, Takashi Mizuno, B. K. Kim & Il-Yang, eds. Pharm.Co., Ltd.

Thielke, C., 1989. Cultivation of edible fungi on coffee grounds. *Mushroom Science* XII (Part II): 337–343.

Thomas, S., M. Pinza, P. Becker, A. Drum & J. Word, 1998. Bioremediation: mycofiltration study for the cleanup of oil-contaminated soil. Internal document. Battelle Marine Science Laboratory, Sequim, Washington.

Thorn, R. G. & G. L. Barron, 1984. Carnivorous mushrooms. *Science,* 224: 76–78.

Thorn, R. G., J. M. Mocalvo, C. A. Reddy & R. Vilgalys, 2000. Phylogenetic analyses and the distribution of nematophagy support a monophyletic Pleurotaceae within the polyphyletic pleurotoid-lentinoid fungi. *Mycologia* 92(2): 241–252.

Tomasi, A., E. Albano, B. Botti, & V. Vannini, 1987. Detection of free radical intermediates in the oxidative metabolism of carcinogenic hydrazine derivatives. *Toxicolog. Pathology* 15(2):178–83.

Toth, B., 2000. A review of the natural occurrence synthetic production and use of carcinogenic hydrazines and related chemicals. *In-Vivo.* 14: 299–320.

Toth, B., P. Gannet, W. Visek & K. Patil, 1998. Carcinogenesis studies with the lyophilized mushroom *Agaricus bisporus* in mice. *In-Vivo* March-April 12(2): 239–244.

Triratana, S., S. Thaithatgoon & M. Gawgla. 1991. Cultivation of *Ganoderma lucidum* in sawdust bags. *Mushroom Science XII; Science and Cultivation of Edible Fungi*. A. A. Balkema, Rotterdam.

Ukai, S., K. Hirose, T. Kiho, C. Hara & T. Irikura, 1972. Antitumor activity on sarcoma 180 of the polysaccharides from *Tremella fuciformis* Berk. *Chemical and Pharmaceutical Bulletin* Oct. 20(10): 2293–2294.

Ukai, S., H. Kiriki, K. Nagai & T. Kiho, 1992. Synthesis and antitumor activities of conjugates of mitomycin C-polysaccharide from *Tremella fuciformis*. *Journal of the Pharmaceutical Society of Japan* Sept. 112(9): 663–668.

Upadhyay, R.C. & B. Vijay, 1991. Cultivation of *Pleurotus* species during winter in India. *Mushroom Science XI,* Proceedings of the 13th International Congress on the Science and Cultivation of Edible Fungi. A.A. Balkema, Netherlands.

Van de Bogart, F., 1976–1979. The genus *Coprinus* in Western North America I: *Mycotaxon* 4: 233–275; II: *Mycotaxon* 8: 243–291. *Mycotaxon* 10: 154–174.

Vasilov, P. B. 1955. Abriss der geographischen verbreitung der hutpilze in der Sowjetunion. Moscow-Leningrad.

Vedder, P. J. C., 1978. *Modern Mushroom Growing*. Educaboek, Culemborg, Netherlands.

Vetter, J., 1995. Mineral and amino acid contents of edible, cultivated shiitake mushrooms (*Lentinus edodes*). *Zeitschrift fur Lebensmittel Untersuchung und Forschung*. July 201(1): 17–19.

Vela, R. M. & D. Martinez-Carrera, 1989. Cultivation of *Volvariella bakeri* and *V. volvacea* in Mexico: a comparative study. *Mushroom Journal for the Tropics* (9): 99–108.

Venter, A. J., 1999. Fungus could combat chemical weapons. *Jane's Defense Weekly*. August 18.

Vilgalys, R. & M. Hester, 1990. Rapid genetic identification and mapping of enzymatically amplified ribosomal DNA from several *Cryptococcus species*. *J. Bacteriol*. 172: 4238–4246.

Vilgalys, R., A. Smith, B. L. Sun & O. K. Miller, 1993. Intersterility groups in the *Pleurotus ostreatus* complex from the continental United States and adjacent Canada. *Canadian Journal of Botany* 71:113–128.

Visccher, H. R., 1978. Fructification of Agaricus bisporus (Lge.) Imb. In relation to the relevant microflora in the casing soil *Mushroom Science X* (Part I). Proceedings of the Tenth International Congress on the Science and Cultivation of Edible Fungi, France.

Volk, T. J. & T. J. Leonard, 1989. Experimental studies on the morel. I. Heterokaryon formation between monoascosporous strains of *Morchella*. *Mycologia* 81(4): 523–531.

_____, 1989. Physiological and environmental studies of sclerotium formation and maturation in isolates of *Morchella crassipes*. *Applied and Environmental Microbiology*, Dec. 55: 3095–3100.

_____, 1990. Cytology of the life-cycle of *Morchella*. *Mycological Research* 94(3): 399–406.

Volk, T. J., 1990. The current state of Morel cultivation. *Mushroom News* 38: 24–27.

Volz, P. A., 1972. Nutritional studies on species and mutants of *Lepista, Cantharellus, Pleurotus* and *Volvariella. Mycopathol. Mycol. Appl*. 48: 175–185.

Wang, S. Y., M. L. Hsu, H. C. Hsu, C. H. Tzeng, S. S. Le, M. S. Shiao & C. K. Ho, 1997. The anti-tumor effect of *Ganoderma lucidum* is mediated by cytokines released from activated macrophages and T lymphocytes. *International Journal of Cancer* 70(6): 669–705.

Walton, K., M. M. Coombs, R. Walker & C. Ioannides, 1997. Bioactivation of mushroom hydrazines to mutagenic products by mammalian and fungal enzymes. *Mutation Research* Nov. 19; 381(1): 131–139.

Walton, K., M. M. Coombs, F. S. Catterall, R. Walker & C. Ioannides, 1997. Bioactivation of the mushroom hydrazine, agaritine, to intermediates that bind covalently to proteins and induce mutations in the Ames test. *Carcinogenesis* Aug; 18(8): 1603–1608.

Walton, K., R. Walker & C. Ioannides, 1998. Effect of baking and freeze-drying on the direct and indirect mutagenicity of extracts from the edible mushroom *Agaricus bisporus. Food. Chem. Toxicology* April 36(4): 315–20.

Wang, B. C., C. T. Lin & S. C. Jong, 1990. Investigation on preservation of *Agaricus bisporus* and some other mushrooms. *Mushroom Journal of the Tropics*. 10(1): 1–8.

Wasson, R. G. & R. Heim, 1958. *Les champignons hallucinogenes du Mexique.* Editions du Museum National D'Histoire Naturelle, Extrait des Archives series 7, no., 6. Paris.

Wasson, R. 1972. *Soma and the Fly-Agaric: Mr. Wasson's Rejoinder to Professor Brough.* Botanical Museum of Harvard University, Cambridge, Massachusetts.

_____, 1973. Mushrooms and Japanese culture. *The Transactions of the Asiatic Society of Japan.* Third Series., Vol. II, Dec. (April 1975), pp. 305–324.

_____, 1976. *Soma: Divine Mushroom of Immortality.* Harcourt, Brace & Jovanovich, New York.

_____, 1978. *The Wondrous Mushroom: Mycolatry in Mesoamerica.* McGraw-Hill, New York.

Wasson, R. G., A. Hofmann & C. Ruck, 1978. *The Road to Eleusis: Unveiling the Secret of the Mysteries.* Harcourt Brace Jovanovich, New York and London.

Watling, R., 1982. Bolbitaceae: Agrocybe, Bolbitious & Conocybe. *British Fungus Flora 3.* Royal Botanic Garden, Edinburgh, Scotland.

_____, 1984. *How to Identify Mushrooms to Genus V: Cultural and Developmental Features.* Mad River Press, Eureka, California.

Watling, R. & N. M. Gregory, 1989. Crepidotaceae: Pleurotaceae and other pleurotoid agarics. *British Fungus Flora.* Agarics and Boleti 6. Royal Botanic Garden, Edinburgh, Scotland.

Wayne, R., 1999. *Growing Mushrooms the Easy Way: Home Mushroom Cultivation with Hydrogen Peroxide.* Rush Wayne Enterprises, Eugene, Oregon.

Weil, A., 1987. Recipes: a mushroom a day. *American Health Magazine* 6:4, May.

_____, 1980. *The Marriage of the Sun and the Moon.* Houghton Mifflin, Boston, Massachusetts.

_____, 1991. *Natural Health, Natural Medicine.* Houghton Mifflin, New York.

_____, 1993. Boost immunity with medicinal mushrooms. *Natural Health Magazine,* May–June, pp. 12–17.

_____, 1995. *Spontaneous Healing.* Knopf, New York.

Weir, J. R., 1917. Montana forest tree fungi-I. Polyporaceae. *Mycologia* 9: 129–137.

Willard, T., 1990. *The Reishi Mushroom: Herb of Spiritual Potency and Medical Wonder.* Sylvan Press, Vancouver, British Columbia.

Word, J. Q., S. A. Thomas, P. Becker, M. Huesemann, T. E. Devine, F. Roberto & P. Stamets, 1997. Adaptation of mycofiltration phenomena for wide-area and point-source decontamination of CW/BW agents: proof of concept. Internal Report. Battelle MSL, Sequim, Washington.

Xia, D. & Z. B. Lin, 1989. Effect of Tremella polysaccharides on immune function in mice. *Acta Pharmacologica Sinica* 10: 453.

Xiang, Y., 1991. A new granular structure medium for spawn manufacture and the preservation of strains. *Mushroom Science* XIII, (1): 123–124.

Yamada, Yuko, Hioraki, Nanba, Kuroda, 1990. Antitumor effect of orally administered extracts from fruitbody of *Grifola frondosa* (Maitake). *Chemotherapy* 38: August.

Yamamoto, Y., H. Shirono, K. Kono & Y. Ohashi, 1997. Immunopotentiating activity of water-soluble lignin rich fraction prepared from LEM-the extract of the solid culture medium of *Lentinus edodes* mycelia. *Bioscience, Biotechnology and Biochemistry* Nov. 61(11): 1909–1912.

Yang, Q. Y. & S. C. Jong, 1987. A quick and efficient method of making mushroom spawn. *Mushroom Science* XII, pp. 317–324.

Yang, M. M., Z. Chen & J. S. Kwok, 1992. The antitumor effect of a small polypeptide from *Coriolus versicolor* (SPCV). *American Journal of Chinese Medicine* 20(3-4): 221–32.

Yang, Q. Y. & M. M. Wang, 1994. The effect of *Ganoderma lucidum* extract against fatigue and endurance in the absence of oxygen. Proceedings of Contributed Symposium 59 A, B at the 5th International Mycological Congress, Vancouver, British Columbia, August 14–21.

Yang, Q. Y., J. N. Fang & X. T. Yang, 1994. The isolation and identification of two polysaccharides of *Ganoderma lucidum* (GL-A, GL-B). Proceedings of Contributed Symposium 59 A, B at the 5th International Mycological Congress, Vancouver, British Columbia, August 14–21.

Ying, J., 1987. *Icons of Medicinal Fungi*. Science Press, Beijing.

Yuan, Z., P. He, J. Cui & H. Takeuchi, 1998. Hypoglycemic effect of water-soluble polysaccharide from *Auricularia auricula-judae* Quel. on genetically diabetic KK-ay mice. *Bioscience, Biotechnology, Biochemistry* Oct. 62(10): 1898–1903.

Yun, T. K., S. H. Kim & Y. S. Lee, 1995. Trial of a new medium-term model using benzo(a)pyrene induced lung tumor in newborn mice. *Anticancer Research* 15: 839–846.

Zadrazil, F. 1976. The ecology and industrial production of *Pleurotus ostreatus, P. florida, P. cornucopiae*, and *P. eryngii. Mushroom Science* IX (Part I): 621–652.

_____, 1977. The conversion of starch into feed by basidiomycetes. *European Journal of Applied Microbiology*, 4: 273.

_____, 1980. Conversion of different plant waste into feed by Basidiomycetes. *European Journal of Applied Microbiology* 9: 243–248.

_____, 1980. Influence of ammonium nitrate and organic supplements on the yield of *Pleurotus sajor-caju* (Fr.) Singer. *European Journal of Applied Microbiology Biotechnology* 9: 243–248.

Zhao, T. F., C. X. Xu, Z. W. Li, F. Xie, Y. T. Zhao, S. Q. Wang, C. H. Luo, R. S. Lu, G. L. Ni, Z. Q. Ku, Y. F. Ni, Q. Qian & X. Q. Chen, 1982. Effect of *Tremella fuciformis* Berk on acute radiation sickness in dogs. *Chinese Medical Journal* Feb. 4(1): 20–23.

Zeng, Q., J. Zhao & Z. Deng, 1990. The antitumor activity of *Flammulina velutipes* Polysaccharide (FVP). *Edible Fungi of China*, vol. (10): 2. Szechwan Institute of Materia Medica.

Zhao, J. D., 1989. The Ganodermataceae in China. *Bibliotheca Mycologica* 132.

Zhuliang, Y. & Y. Chonglin, 1992. Recognization of *Hypsizgus marmoreus* (Peck) Bigelow and its cultivation. *Edible Fungi of China*, 2(55): 19–20.

Zusman, I., R. Reifen, O. Livini, P. Smirnoff, P. Gurevich, B. Sandler, A. Nyska, R. Gal, Y. Tendler & Z. Madar, 1997. Role of apoptosis, proliferating cell nuclear antigen and p53 protein in chemically induced colon cancer in rats fed corncob fiber treated with the fungus *Pleurotus ostreatus. Anticancer Research* May-June 17(3C): 2105–13.

Photo Credits

Black and white photographs
Rodney Barrett, page 194
Kit Scates Barnhart, pages 9, 30, 135 (top), and 192 (top left)
Alan Bessette, pages 368–370
Michael Beug, pages 247 and 336
David Bill, pages 36 and 378
Davel Brooke-Webster, page 182
Jeff Chilton, page 352 and spine image
Eric Danell, page 8
Kevin Delahay, page 350 (bottom)
Bill Freedman, page 353 (top)
Peter Furst, page 3, 466
Guy Gardiner, page 356
Bob Harris, page 235
Jim Haseltine, pages 191, 251 (bottom and left), and 253 (bottom)
Omon Isikhuemhen, page 275
Daniel Job, page 412 (bottom)
Bob Johns, page 184
S. C. Jong, page 293 (bottom)
Paul Lewis, pages 357 and 485 (left)
Lim Chea Lim, page 360
Gary Lincoff, pages 237, 286, and 288
Hughes Massicote, page 7
Gary Menser, page 6
Orson Miller, page 293 (top)
Minnesota Forest Resource Center, page 263 (right)
T. Motohashi, page 246
Perry Muleazy, page 420 (right and bottom)
The Mushroom Council, page 217, 219, and 220
Takashi Nakazawa, page 236 (bottom), 251 (bottom right), 318, 355 (bottom), and 374
S. Omoanghe, 64 (left)
Paul Pryzbylowicz, page 263 (top left)
Ed Sliffe, page 42 (lower left)
Glenn Tamai, pages 185, 269 (right)
Ralph Tew, page 485
Satit Thaithatgoon, pages 121 (bottom), 183, 192 (top right and lower right), 234 (right), 297, 343, 345 (top left), 346, 361, 362, 398, 403, and 436
Spencer Throckmorton, page 4
Susan Thomas, page 16
Steve Vento, page 147
Thomas Volk, page 418

Dusty Wu Yao, pages xviii and 348
F. Zadrazil, page 40
(All other black and white images are by Paul Stamets.)

Color insert photographs
Alan Bessette, page 17 (top right and lower left)
Michael Beug, page 4 (lower left)
Lance Howell, page 7 (bottom)
Omon Isikhuemhen, pages 10 (bottom right) and 22 (bottom)
Paul Lewis, page 15 (lower right)
Gary Lincoff, page 8 (top left and middle)
Rich Lucas, page 12 (bottom)
The Mushroom Council, page 2 (left)
Takeshi Nakazawa, pages 3 (bottom), 8 (bottom), 15 (lower left), and 16 (bottom)
Luiz Amaro Pachoa de Silva, page 6 (lower left)
Paul Stamets, pages 1, 2 (upper right and bottom), 3 (top left and top right), 4 (top left, top right, and lower right), 5, 6 (top left, top right, and lower right), 7 (top left and top right), 8 (top right), 9, 10 (top left and top right), 11, 12 (top left and top right), 13, 15 (top left), 16 (top left and top right), 17 (top left and lower right), 18, 19, 21 (top), 23, and 24
Satit Thaithatgoon, pages 10 (bottom left), 14, 15 (top right), 20, and 22

Illustration Credits

Kristine M. Adams, page 423
Kandis Elliot, page 419
Linda Greer, pages 20, 31, 32, 35, 37, 38, 43, 381, and 440
Ann Gunter, pages 62, 80, 89, and 136
Kathleen Harrison, page vi
Andrew Lenzer, pages 56 and 118
Mark Sanford, pages 81, 83, 90, 139, 140, and 470
Trace Salmon, pages 57, 156, 157, 159, 172, 174, 204, 424, and 468–470 (top)

Index